Polar Cap Boundary Phenomena

NATO ASI Series

Advanced Science Institutes Series

A Series presenting the results of activities sponsored by the NATO Science Committee, which aims at the dissemination of advanced scientific and technological knowledge, with a view to strengthening links between scientific communities.

The Series is published by an international board of publishers in conjunction with the NATO Scientific Affairs Division

A Life Sciences	Plenum Publishing Corporation
B Physics	London and New York
C Mathematical and Physical Sciences	Kluwer Academic Publishers
D Behavioural and Social Sciences	Dordrecht, Boston and London
E Applied Sciences	
F Computer and Systems Sciences	Springer-Verlag
G Ecological Sciences	Berlin, Heidelberg, New York, London,
H Cell Biology	Paris and Tokyo
I Global Environmental Change	

PARTNERSHIP SUB-SERIES

1. Disarmament Technologies	Kluwer Academic Publishers
2. Environment	Springer-Verlag / Kluwer Academic Publishers
3. High Technology	Kluwer Academic Publishers
4. Science and Technology Policy	Kluwer Academic Publishers
5. Computer Networking	Kluwer Academic Publishers

The Partnership Sub-Series incorporates activities undertaken in collaboration with NATO's Cooperation Partners, the countries of the CIS and Central and Eastern Europe, in Priority Areas of concern to those countries.

NATO-PCO-DATA BASE

The electronic index to the NATO ASI Series provides full bibliographical references (with keywords and/or abstracts) to more than 50000 contributions from international scientists published in all sections of the NATO ASI Series.
Access to the NATO-PCO-DATA BASE is possible in two ways:

– via online FILE 128 (NATO-PCO-DATA BASE) hosted by ESRIN,
Via Galileo Galilei, I-00044 Frascati, Italy.

– via CD-ROM "NATO-PCO-DATA BASE" with user-friendly retrieval software in English, French and German (© WTV GmbH and DATAWARE Technologies Inc. 1989).

The CD-ROM can be ordered through any member of the Board of Publishers or through NATO-PCO, Overijse, Belgium.

Series C: Mathematical and Physical Sciences – Vol. 509

Polar Cap Boundary Phenomena

edited by

J. Moen

University Courses on Svalbard,
Longyearbyen, Norway

A. Egeland

Department of Physics,
University of Oslo,
Oslo, Norway

and

M. Lockwood

Rutherford Appleton Laboratory,
Chilton, Oxfordshire, U.K.

Springer-Science+Business Media, B.V.

Proceedings of the NATO Advanced Study Institute on
Polar Cap Boundary Phenomena
Longyearbyen, Svalbard, Norway
4–13 June 1997

A C.I.P. Catalogue record for this book is available from the Library of Congress.

ISBN 978-94-010-6195-7   ISBN 978-94-011-5214-3 (eBook)
DOI 10.1007/978-94-011-5214-3

Printed on acid-free paper

TABLE OF CONTENTS

PREFACE

These proceedings are based on the invited talks and selected research reports presented at the NATO Advanced Study Institute (ASI) on "POLAR CAP BOUNDARY PHENOMENA" held at Longyearbyen, Svalbard, June 4 - 13, 1997.

The role of the polar cap and its boundary is very substantial in solar-terrestrial physics. At this NATO ASI a major change in thinking on the "cusp" precipitation region in the high-latitude dayside upper atmosphere was reflected, at least for intervals when the interplanetary magnetic field (IMF) is directed southward. It is likely that this has implications for northward IMF as well. The change comes from the now almost complete acceptance of the concept of magnetosheath particle entry along open magnetic field lines and the evolution of the precipitation into the upper atmosphere with time elapsed since magnetic reconnection which opened the field line. A key prediction of this view is that the low-latitude boundary layer (LLBL) is on open field lines.

The meeting discussed an "intermediate" state of the dayside boundary when IMF is weakly northward. Presented data sets revealed a combination of effects seen during northward and southward IMF. Many other phenomena, such as travelling convection vortices, flow channels, optical transients were widely discussed. Other important topics addressed were: magnetic topology and the occurrence and meaning of conjugate phenomena; the scale sizes involved in the magnetospheric boundary regions and their relation to energy transfer; observation and identification of boundaries including nightside polar cap boundary structure and dynamics, particularly in relation to substorms. In all these areas the need for improved, more co-ordinated ground and satellite/rocket in-situ observations was stressed.

By studying and comparing the books from the two foregoing NATO Workshops (Physical Signatures of Magnetospheric Boundary Layer Procesesses, Eds. J. A. Holtet and A. Egeland, and Electromagnetic Coupling in the Polar Cleft and Caps, Eds. P. E. Sandholt and A. Egeland) with the present one, one will fortunately discover that research related to the Earth's plasma environments at polar latitudes are rapidly progressing. These proceedings should be valuable textbooks for both graduate students and researchers. They contain basic information in the general study of solar-terrestrial physics, but viewed from a variety of disciplines.

We gratefully acknowledge the funds provided by the NATO Scientific Affairs Division to this ASI. This was the first NATO meeting arranged at Svalbard, and we would like to thank the director of The University Courses on Svalbard (UNIS), Jarle Nygard, his staff and students for providing excellent working conditions for the ASI. We also thank the local travel agency, Spitsbergen Travel A/S, who organised all domestic travel arrangements including full board accomodation at Longyearbyen, and in their professional manner making a reliable framework in which to carry out the scientific programme.

Finally, we thank all the participants for their active participation, and thanks to all the contributors who made this NATO ASI book possible.

Longyearbyen, November 1997

Jøran Moen   Alv Egeland

LIST OF PARTICIPANTS

Anderson, David
danderson@plh.af.mil

PL/GPSM
Hanscom AFB
Bedford, Mass 01731, USA

Andersson, Laila
laila@irf.se

Swedish Institute of Space Physics
Box 812, S-981 28 Kiruna, Sweden

Asai, Keiko
asai@iksun1.phys.nagoya-u.ac.jp

Dept. of Physics, Faculty of Science
Nagoya University
Chikusa, Nagoya 464-01, Japan

Berry, Simon
stb93@aber.ac.uk

Dept. of Physics
University of Wales, Aberystwyth
Ceredigion, Wales SY23 3BZ, UK

Besser, Veronika
veronika@hoffa.gi.alaska.edu

Geophysical Institute
University of Alaska, Fairbanks
903 Koyukuk Drive
Fairbanks, Alaska 99775-7320, USA

Biernat, Helfried K.
biernat@bkfng.kfunigraz.ac.at

Space Research Institute
Austrian Academy of Sciences
Lustbühelstr. 46, A-8042, Graz, Austria

Brekke, A.
asgeir.brekke@phys.uit.no

The Auoral Observatory
University of Tromsø
N-9037 Tromsø, Norway

Brittnacher, Mitchell
britt@geophys.washington.edu

Geophysics Program
Box 351650, University of Washington
Seattle, WA 98195, USA

Bösinger, Tilmann
Tilmann.Bosinger@oulu.fi

University of Oulu
Dept. of Physical Sciences
Division of Physics, P.O. Box 333
FIN-90571 Oulu, Finland

Bøen, Kjell

Andøya Rocket Range
P.O.Box 54
N-8480 Andenes, Norway

Carlson, H.C. Jr.
carlsonh@plh.af.mil

AFGL/LIS
Hanscom AFB
Bedford, MA 01731, USA

Cowley, S.W.H.
e-mail, swhc1@ion.le.ac.uk

Dept. of Physics and Astronomy
University of Leicester
Leicester, LE1 7RH, UK

Dahl, Astrid

The University Courses on Svalbard
N-9170 Longyearbyen, Norway

Dyrud, Lars
dyrud@auggie.dnet.nasa.gov

Augsburg College
2211 Riverside Avenue
Minneapolis, MN 55454, USA

Egeland, Alv
alv.egeland@fys.uio.no

Dept. of Physics, University of Oslo
P.O.Box 1048, Blindern
N-0316 Oslo, Norway

Engebretson, M.J.
engebret@geophys.nat.tu-bs.de

Augsburg College
2211 Riverside Avenue
Minneapolis, MN 55454, USA

Erkaev, Nikolai V.
erkaev@cckr.krasnoyarsk.su

Computer Center
Siberian Division, 660036 Krasnoyarsk 36
Akademgorodok, Russia

Ernström, Per
ernstrom@dmi.dk

Danish Meteorological Institute
Lyngbyvej 100
DK-2100 København, Denmark

Farrugia, C.J.
ferrugia@unhedi2.sr.unh.edu

University of New Hampshire
39 College Road, SSC/Morse Hall
Durham, NH 03824, USA

Friis-Christensen, E.
eigil.friis@dmi.dk

Danish Meteorological Institute
Lyngbyvej 100
DK-2100 København, Denmark

Fritz, Theodore A.
fritz@bu.edu

Center for Space Physics
Boston University
Boston, MA 02215, USA

Fuselier, Stephen
fuselier@space.lockheed.com

Lockheed Martin
Dept. H1-11, Bldg. 252
3251 Hannover St.,Palo Alto
CA 94304, USA

Fälthammar, Carl-Gunne
falthammar@plasma.kth.se

The Royal Institute of Technology
Division of Plasma Physics
Alfvèn Laboratory
S-100 44 Stockholm, Sweden

Hall, Adrian M.
amhl@ion.le.ac.uk

Dept. of Physics and Astronomy
University of Leicester
Leicester, LE1 7RH, UK

Heikkila Walter
heikkila@utdssa.utdallas.edu

Center for Space Science
The University of Texas at Dallas
MS FO22, Richardson
Texas 75083 - 0688, USA

Holter, Øivin
oivin.holter@fys.uio.no

Dept. of Physics, University of Oslo
P.O.Box 1048 Blindern
N-0316 Oslo, Norway

Holtet, Jan A.
j.a.holtet@fys.uio.no

Dept. of Physics, University of Oslo
P.O.Box 1048 Blindern
N-0316 Oslo, Norway

Huba, D.
huba@ppd.nrl.navy.mil

Code 6790
Naval Research Laboratory
Washington DC 20375-5320, USA

Håland, Stein
stein.haaland@fi.uib.no

Department of Physics
University of Bergen
Allégt. 55 - N-5007 Bergen, Norway

Jacobsen, Bjørn
bjacob@estec.esa.nl

Solar System Division
Space Science Department
ESTEC, Postbus 299
2200 AG Noordwijk, The Netherlands

Jørgensen, Peter Siegbjørn
psj@gfy.ku.dk

Brydes Allé 23
Vær 429, DK-2300 Copenhagen, Denmark

Kandemir, Gulcin
kandemir@sariyer.cc.itu.edu.tr

Istanbul Teknik Üniversitesi
Fen-Ed. F., Fizik BI.
Maslak 80626 Istanbul,Turkey

Kersley, Len
lek@aber.ac.uk

Dept. of Physics
University of Wales, Aberystwyth
Ceredigion, Wales SY23 3BZ, UK

Kjus, Solveig Helene
s.h.kjus@fys.uio.no

Dept. of Physics, University of Oslo
P.O.Box 1048 Blindern
N-0316 Oslo, Norway

Kozlovsky, Alexander
skoz@pgi-ksc.murmansk.su

Polar Geophysical Institute
26A, Ferman st.
Apatity - 184200, Russia

Kuznetsov, Sergey
kuznets@srdlan.npi.msu.su

Institute of Nuclear Physics
Moscow State University
119899 Moscow, Russia

Lahtinen, Tero Tapio

University of Oulu
Dept. of Physical Sciences
Division of Physics, P.O. Box 333
FIN-90571 Oulu, Finland

Lanchester, Betty
bsl@phys.soton.ac.uk

Dept. of Physics
University of Southampton
Southampton, SO17 1BJ, UK

Lemaire, J.
jl@plasma.oma.be

I.A.S.B..
3 Avenue Circulaire
B-1180 Bruxelles, Belgium

Lester, M.
mle@ion.le.ac.uk

Dept. of Physics and Astronomy
University of Leicester
Leicester, LE1 7RH, UK

Lockwood, M.
m.lockwood@rl.ac.uk

Rutherford Appleton Lab.
Chilton, Didcot, Oxon
OX11 OQX, UK

Lorentzen, Dag
dagl@unis.no

The University Courses on Svalbard
N-9170 Longyearbyen, Norway

Maclennan, Carol
cgm@bell-labs.com

Room 1E-436, Bell Labs
Lucent Technologies
600 Mountain Avenue, Murray Hill
NJ 07974, USA

Maezawa, Kiyoshi
maezawa@iksun1.phys.nagoay-u.ac.jp

Department of Physics
Faculty of Science, Nagoya University
Chigusa-ku, Nagoya 464-01, Japan

Makita, Kazuo
kmakita@nipr.ac.jp

815-1 Taiemachi
Hachiosi, Tokyo 173, Japan

Maynard, N.C.
nmaynard@mrcnh.com

Mission Research Corporation
One Tara Blvd. Suite 302
Nashua, NH 03062-2801, USA

McEwen, Donald
mcewen@dansas.usask.ca

Univ. of Saskatchewan, Saskatoon
Dept. of Physics + Engineering Physics
S7N 5E2, Canada

Meng, C.I.
Ching_Meng@jhuapl.edu

The Johns Hopkins University
Applied Physics Laboratory
Johns Hopkins Road
Laurel, MD 20723, USA

Moretto, Therese
moretto@dmi.dk

Danish Meteorological Institute
Lyngbyvej 100
DK-2100 København, Denmark

Moen, J.
jmoen@unis.no

The University Courses on Svalbard
N-9170 Longyearbyen, Norway

Mukai, Toshifumi
mukai@fujitubo.gtl.isas.ac.jp

Inst. of Space andAstronautical Science
3-1-1 Yoshinodai Sagamihara
Kanagawa 229, Japan

Nagatsuma Tsutomu
tnagatsu@crl.go.jp

Hiraiso Solar Terr. Res. Cen.
3601 Isozaki, Hitachinaka
Ibaraki 311-12, Japan

Newell, P.T.
patrick.newell@jhuapl.edu

The Johns Hopkins University
Applied Physics Laboratory
Johns Hopkins Road
Laurel, Maryland 20723-6099, USA

Nilsson, Hans
Hans.Nilson@irf.se

IRF
Box 812
S-981 28 Kiruna, Sweden

Oksavik, Kjellmar
kjellmar@fi.uib.no

Department of Physics
University of Bergen
Allégt. 55, N-5007 Bergen, Norway

Eivind Osnes
eivind.osnes@fys.uio.no

Dept. of Physics, University of Oslo
P.O.Box 1048 Blindern
N-0316 Oslo, Norway

Pedersen, Arne
arne.pedersen@fys.uio.no

Dept. of Physics, University of Oslo
P.O.Box 1048 Blindern
N-0316 Oslo, Norway

Partamies, Noora

The University Courses on Svalbard
N-9170 Longyearbyen, Norway

Ine-Therese Pedersen

The University Courses on Svalbard
N-9170 Longyearbyen, Norway

Pfaff Rob
rob.pfaff@gsfc.nasa.gov

Lab. For Extraterr. Physics
Code 690 (Bld. 2, Rm 115)
NASA/GSFC - Greenbelt, MD20071, USA

Phan, Thai
phan@ssl.berkeley.edu

University of California
Space Sciences Lab.
Berkeley, CA 94720, USA

Potemra, T.A.
tom_potemra@jhuapl.edu

The Johns Hopkins University
Applied Physics Laboratory
Johns Hopkins Road
Laurel, Maryland 20723, USA

Pudovkin, Mikhail
pudovkin@snoopy.phys.spbu.ru

Inst. of Physics, St. Petersburg University.
Petrodorets, Stary Petergof
St. Petersburg 198904, Russia

Roger, Alan S.
a.rodger@bas.ac.uk

British Antartic Survey
High Cross, Madingley Road
Cambridge, CB3 OET, UK

Saito, Yoshifumi
saito@fujitubo.gtl.isas.ac.jp

Inst. of Space and Astron. Sci.
3-1-1 Yoshinodai, Sagamihara
Kanagawa 229, Japan

Samson, John
samson@space.ualberta.ca

University of Alberta
Physics Dept., Edmonton
Alberta T6G 2J1, Canada

Sandholt, P. E.
p.e.sandholt@fys.uio.no

Dept of Physics, University of Oslo
P.O.Box 1048 Blindern
N-0316 Oslo, Norway

Saviaro, Erkko
erkko@skynet.oulu.fi

University of Oulu
Dept. of Physical Sciences
Division of Physics, P.O. Box 333
FIN-90571 Oulu, Finland

Shumilov, Nikita

The University Courses on Svalbard
N-9170 Longyearbyen, Norway

Smith, R. W.
bblw@geewiz.gi.alaska.edu

Geophysical Institute
University of Alaska
Fairbanks, Alaska 99775 - 7320, USA

Southwood, D.
d.southwood@ic.oc.uk

Blackett Laboratory, Imperial College
SW7 2BZ London, England

Stadsnes, Johan
johan.stadsnes@fi.uib.no

Dept. of Physics, University of Bergen
Allégt. 55, N-5007 Bergen, Norway

Stauning, Peter
pst@dmi.min.dk

Danish Meteorological Institute
Lyngbyvej 100
DK-2100 Copenhagen, Denmark

Svenes, Knut
knut.svenes@ffi.no

NDRE
Division for electronics
Pb. 25 - 20076 Kjeller, Norway

Søraas, Finn
finn.soraas@fi.uib.no

Dept. of Physics, University of Bergen
Allégt. 55, N-5007 Bergen, Norway

van Eyken, Tony
tony@eiscat.no

EISCAT, Ramfjordmoen
N-9027 Ramjordbotn, Norway

Vasyliunas, V.M.
vasyliunas@linax1.mpae.gwdg.de

Max-Planck Institut für Aeronomie
Postfach 20
D-37189 Katlenburg-Lindau, Germany

Villain, Jean-Paul
jvillain@cnrs-orleans.fr

3A Av. Recherche Scientifique
45071 Orleans Cedex, France

Vondrak, R. R.
vondrak@lepvax.gsfc.nasa.gov

Lab. for Extraterrestrial Physics
Code 690 (Bldg. 2, Rm 115)
NASA/GSFC - Greenbelt, MD20071, USA

Walker, A.D.M.
walker@ph.und.ac.za

Space Physics Research Inst.
Dept. of Physics, University of Natal
Durban, 4041 South Africa

Walker, Ian
ixw@aber.ac.uk

Dept. of Physics
University of Wales, Aberystwyth
Ceredigion, Wales SY23 3BZ, UK

Woch, Joachim
woch@linax1.mpae.gwdg.

MPI für Aeronomie
Postfach 20
D-37189 Katlenburg-Lindau, Germany

Yahnin, Alexander
yahnin@pgi-ksc.murmansk.su

Polar Geophysical Institute
Apatity, Murmansk region
184200 Apatity, Russia

Yamagishi, Hisao
yamagisi@nipr.ac.jp

National Inst. of Polar Res.
9-10 Kaga 1-Chome, Itabashi-ku
Tokyo 173 - Japan

Yamauchi, M.
yamau@irf.se

Swedish Inst. of Space Physics
Box 812 , S-981 28 Kiruna, Sweden

Øieroset, Marit
oieroset@issi.unibe.ch

International Space Science Institute
Hallerstrasse 6
CH-3012 Bern, Switzerland

RECENT WORK ON THE KELVIN-HELMHOLTZ INSTABILITY AT THE DAYSIDE MAGNETOPAUSE AND BOUNDARY LAYER

C. J. FARRUGIA
Institute for the Study of Earth, Oceans, and Space, University of New Hampshire, Durham, NH 03824, USA

F. T. GRATTON, L. BENDER
Instituto de Fisica del Plasma, (CONICET-FCEyN/UBA), Ciudad Universitaria, Pab.1, 1428 Buenos Aires, Argentina

J. M. QUINN, R. B. TORBERT
Institute for the Study of Earth, Oceans, and Space, University of New Hampshire, Durham, NH 03824, USA

N. V. ERKAEV
Computing Center of the Russian Academy of Sciences, 660036, Krasnoyarsk, Russia

AND

H. K. BIERNAT
Space Research Institute, Austrian Academy of Sciences, A-8042, Graz, Austria

1. Introduction

After Dungey [1] first pointed out that velocity shears may render the magnetopause Kelvin-Helmholtz (KH) unstable and provide an energy source for geomagnetic pulsations, the possibility of transferring solar wind momentum and energy to the magnetosphere by this means has been pursued in many studies (see, e.g., [2, 3], and reviews in [4-7]). Few would doubt that the KH instability does occur since, to mention but one example, a relation between the amplitude and polarization features of certain types of pulsations of the geomagnetic field (Pc 5's) and a putative KH cause seems well established ([8 - 10]; see reviews in [11, 12]) A correlation between Pc 5's and solar wind speed, attributed to KH enhancements, has also been

1

J. Moen et al. (eds.), Polar Cap Boundary Phenomena, 1–14.

established [13]. Nonetheless, the role of the KH instability in providing anything but marginal momentum transfer has also been maintained.

The field and plasma environment of the magnetopause is complex. The common presence of a boundary layer attached to its earthward side, across which the velocity but not the magnetic field is sheared, implies that the inner edge of the boundary layer (IEBL), too, may go unstable, either independently of, or together with, the magnetopause. Another consideration is accelerated motions of the magnetopause, which, when directed sunward, enhance KH growth rates and render unstable previously KH-stable regions [14, 15]. Indeed, Gratton *et al.* [15] showed that under well-specified conditions, a rapid and large drop in solar wind dynamic pressure may render the magnetopause prone to the Rayleigh-Taylor (RT) instability. Interestingly, therefore, coherent, radial motions of the magnetopause, which data analyses are careful to distinguish from surface waves, may yet be important in considerations of a possible KH source of surface waves.

In this article we report on some recent work on the KH instability at the magnetopause and its boundary layers. We first discuss some recent observational work, highlighting a number of difficulties encountered in interpretative efforts. In a brief review of recent theoretical and simulation developments, which for reasons of space is necessarily very selective, we shall focus on topics which are still open to discussion and which are connected with observable effects. In a subsequent section we describe aspects of a recent study applicable to strongly northward-pointing interplanetary magnetic field (IMF) and which considers (a) the magnetopause as being embedded in a magnetohydrodynamic (MHD) flow, which flow is strongly influenced by the IMF; and (b) accelerated magnetopause motions.

2. Observations

An observational datum of long standing is that of surface waves on the magnetopause, as inferred, e.g., from multiple crossings during a single spacecraft traversal of the magnetopause. Three main causative mechanisms for these oscillations have been proposed. (a) Dynamic pressure changes of solar wind or bow shock origin where, after the initial impact of the pressure front on the magnetopause, the latter is set "ringing" [16, 17]. (Using *in situ* ISEE 1 and 2 data, Farrugia *et al.* [18] gave one such example.) Flux transfer events (FTE's, [19]; where it is thought that the reconnected magnetic flux tubes produced by sporadic enhancements in the reconnection rate perturb the boundary while convecting tailward; and, finally, (c) The KH instability.

Song *et al.* [16] carried out a large-scale survey of magnetopause oscillations using multiple magnetopause crossings made by ISEE 1 and 2 over their 10-year operational lifetime. The analysis is sensitive to waves of periods between 2 and 30 min. Song *et al.* find that the magnetopause is more oscillatory when the IMF points southward than when it points northwards. For IMF south, Song *et al.* conclude that dynamic pressure pulses alone cannot account fully for the oscillatory magnetopause, and they thus propose an additional cause, namely, FTE's, a conclusion which is further supported by the fact that, since FTE's grow as they travel away from the reconnection site [20, 21] they should render the magnetopause increasingly more undulatory as the flanks are approached, in agreement with the statistics of the ISEE observations [16]. For IMF north, Song *et al.* find that dynamic pressure variations are a sufficient cause of the surface waves. They thus conclude that the KH instability is at best a very minor contributor to the generation of surface waves at the magnetopause.

There are, however, a number of individual studies of surface perturbations where, through a fairly convincing exclusion of other causative mechanisms, the case for a KH source is quite solid. A selection is discussed briefly below. Our choice has also been influenced by a wish to highlight dilemmas in the interpretation of data. The reader is referred to the reviews in [6, 7] for other examples.

Chen and Kivelson [22] and Chen *et al.* [23] studied a number of cases of multiple magnetopause/boundary layer crossings on the flanks of the magnetosphere using ISEE 1 and 2 magnetic field and plasma data. Some of these encounters lasted for several hours. The majority of these examples were for strongly northward IMF, an orientation which does not favour a reconnection interpretation. The multiple crossings were shown to be due to tailward propagating surface perturbations. A careful exclusion of a possible solar wind dynamic pressure cause was made. Thus, by elimination, the KH instability emerged as the likely source. In these studies, the magnetopause was generally found to be the unstable boundary. This body of work constitutes a strong contribution to the study of "viscous" coupling of magnetosheath energy and momentum at the magnetosphere flanks. Among several noteworthy results, one can mention the strongly non-sinusoidal wave forms inferred from the analysis of the observations, where the sunward-facing edge is steepest. Such a wave form allows for the extraction of momentum from the magnetosheath flow in a natural way.

In an elegant analysis, Ogilvie and Fitzenreiter [24] used measurements made by the vector electron spectrometer on ISEE 1 at good temporal resolution (few s) during traversals of the magnetopause and the low latitude boundary layer. Using the KH instability criterion in the incompressible limit, Ogilvie and Fitzenreiter [24] inquired about the stability of magnetopause and IEBL as well as of structures within the boundary layer. The authors subdivided boundary layer/magnetopause crossings into two classes, smooth and pulsed, according to whether the electron temperature and density in the boundary layer changed smoothly or in discrete pulses during a spacecraft traversal of the magnetopause. Among other results, they find that for pulsed passes the IEBL is likely to be unstable, whereas for smooth passes both IEBL and magnetopause are generally stable.

One focus of recent observational work on the KH instability has been to decide which boundary has gone unstable, the magnetopause or the IEBL. As evidenced by the number of contrasting interpretations given to the same event by different groups of workers, these issues are not easy to settle observationally. We concur with the view of Fitzenreiter and Ogilvie [6] that this ambiguity is partly due to the relative lack of observations and partly also to the difficulty in determining the sources of the low latitude boundary layer. To this we would add the following. In a much-studied example, Sckopke *et al.* [25] broached the possibility of coupled motions of the magnetopause and IEBL (see their Figure 7). As is implicit in the oft-posed question as to which boundary is unstable, with its either-or implication, this possibility has not been given much attention in recent data work. Nonetheless, keeping it in mind might throw fresh light on some dilemmas plaguing interpretational efforts. We return to this point below.

3. Theory

Corrections to the ordinary KH modes due to finite compressibility are ordinarily minor effects. However, at large velocity differences (supersonic), compressibility may restore stability. For this reason, instability may be excited only at the dayside magnetopause, since, more tailward, the Alfvén Mach number becomes too large for the instability to arise. For the magnetopause, the physical picture that emerges is one where the KH instability develops on the frontside and from there large amplitude waves are convected toward the tail [26]. There exists also another set of KH modes, which depends critically on compressibility and disappears when the speed of sound is set to infinity (see e.g., [27, 28]). Although these compressible KH modes have much smaller growth rates than ordinary KH modes, they are interesting because they are associated with compressional Alfvénic perturbations. Theory has been developed to understand how magnetopause perturbations can couple with resonances in the magnetospheric magnetic field. These perturbations are transmitted to magnetic shells (L shells) where conditions for shear Alfvén resonances are matched [9, 29]. The wave that couples magnetopause perturbations to resonant field lines must be a compressional mode. This is the concept of field-line resonance for ULF magnetospheric waves that received confirmation from ground and spacecraft observations. We noted that the motion of the magnetopause under sudden changes of the solar wind dynamic pressure is also a source of surface perturbations. The compressional MHD waves generated in the magnetosphere in this way are partially confined between the near equatorial magnetopause and the ionosphere. These modes can again resonate with shear Alfvén modes at special field lines.

Associated with resonant couplings there are other effects due to the existence of a continuum spectrum. The ubiquitous appearance of non dissipative damping in inhomogeneous plasmas is connected with the presence of a continuum wave spectrum. For Alfvén waves, this is related to the possibility of resonant absorption, and, when velocity gradients are present, to a resonant instability. These effects are relevant for solar physics, magnetopheric physics, and laboratory plasma physics (see, for instance, [30-32] and references therein). A surface wave may resonate with a shear Alfven wave in an inner layer of the thin gradient stratum at the magnetopause. At present, the subject has been developed mainly for resonant absorption and heating by Alfvén waves in inhomogeneous systems, and for Alfvén resonant coupling under magnetospheric conditions. For the KH problem, conditions are possible in which a resonant instability (different from the ordinary KH surface modes [30]) occurs. The subject is still in progress for the KH theory at the magnetopause.

There are many results on 2D and 3D simulations with MHD, and more recently with kinetic models. These studies are very important to establish (i) the level and conditions at which the instability saturates; and (ii) the amount of momentum and energy transport. Considerable advances in the non linear dynamics of the KH instability for magnetopause conditions and its consequences for transport were achieved in numerical simulations by Miura. (See [5] for a review). The transport properties depend importantly on the value of a special Alfvenic Mach number defined with the component of the magnetosheath magnetic field parallel to the magnetosheath velocity. The momentum transfer at the magnetopause obtained in the simulations was then embodied into an effective viscosity, generated by the KH instability. The level of viscosity is adequate to explain the drag of the magnetosphere by the magnetosheath flow. Regarding the spectrum formation and non linear saturation, objections to Miura's simulations were formulated. These refer to the use of periodic boundary conditions: the instability, being of convective nature, should be studied with a long simulation box [4, 26]. In the non linear stage the convective KH can grow to a much larger amplitude than in periodic systems, and new phenomena were noted in the simulations. The non linear mode couplings in 2D, MHD simulations, with non periodic boundary conditions (convective instability) by [4], lead to the concept of inverse-λ cascade, i.e., a trend towards long λ's. This effect is assumed to be responsible for the very long wavelength surface waves often observed at the flanks, which apparently could not be explained by linear theory. According to these authors, there would be a factor of ten difference from the shorter wavelengths predicted by linear KH theory. We may note, however, that even in the linear phase there is a stretching of λ's effect, due to the increase of flow speed from frontside to the flanks (not included in the numerical simulations). For instance, starting near the subsolar point with an eastward pointing $\mathbf{k}$, the leading front of the perturbation is convected faster than the rear front [23]. Thus an increase of λ takes place.

Thomas and Winske [33] carried out sophisticated 2D hybrid simulations with particle ions and fluid electrons. They were able to study the kinetic non linear evolution of the KH instability. Essentially, they confirmed the main results of MHD codes, including the inverse cascade effect to longer wavelengths. But, of course, they were also able to study other high frequency effects that MHD simulations cannot provide. *Thomas* [34] has recently reported some kinetic simulations in 3D, where the effect of magnetic shear is taken into account (the relative angle of the magnetic field and the velocity vector is varied). The instability growth is strongly reduced during the non linear stage due to the presence of even modest amounts of rotation of the magnetic field across the velocity shear layer. As a consequence, Thomas [34] believes that the occurrence of the KH instability at the dayside magnetopause is greatly inhibited. We address the question of magnetic shear at the frontside magnetopause in the next section.

4. Charts of Kelvin-Helmholtz activity

In our study of the distribution of KH growth rates over the frontside magnetopause [35] we considered two orientations of the IMF: due north and 30° west of north (dawnward). Under these conditions, a plasma depletion layer (PDL; see, e.g., [36, 37] forms adjacent to the sunward side of the magnetopause where the flow is of the stagnation line type [38, 39]. We study the instability using PDL parameters as obtained by a code describing the MHD flow past a paraboloidal magnetosphere with a magnetopause stand-off distance equal to the radius of curvature at the subsolar point (see references in [35]). In practice, the instability theory for a statified model was applied in the local tangent plane at each point of a calculation grid covering the whole dayside, and maps of maximum growth rates were produced. A further novel feature of the model is that magnetopause accelerations are incorporated in a generalized KH+RT dispersion relation. A boundary layer is included.

For the dispersion relation of the instability, we used a model with piecewise constant functions. Let axis $\bar{y}$ be along the magnetopause normal, positive sunwards, and denote quantities in the PDL ($\bar{y} > 0$), boundary layer interior (-h $< \bar{y} <$ 0, h is the thickness of the boundary layer) and magnetosphere ($\bar{y} <$ -h) by suffices 1, 2, 3, respectively. (The vector fields are in ($\bar{x}$, $\bar{z}$) planes.) We used the following relations for the mass density, ρ, velocity, $\mathbf{v}$, and magnetic field, $\mathbf{B}$: $\rho_2 = \rho_1/2$, $\mathbf{v_2} = \mathbf{v_1}/2$; $\rho_3 = \rho_2/5$, $\mathbf{v_3} = 0$, $\mathbf{B_3} = \mathbf{B_2}$. The magnetic shear is thus concentrated at the magnetopause, and the boundary layer flow is aligned with that in the PDL.

It can be shown [35] that the dispersion relation for incompressible surface modes is

$$D(\omega, \mathbf{k}) = H_2(H_1 + H_3 + G_{13}) + F(H_1 + H_2 + G_{12})(H_2 + H_3 + G_{23}) = 0, \quad (1)$$

where $H_i = \rho_i[\bar{\omega}_i{}^2 - (kV_{Ai}\cos(\angle kB_i))^2]$, $\bar{\omega}_i = \omega - kv_i\cos(\angle kv_i)$, $G_{ij} = gk(\rho_i - \rho_j); i, j = 1, 2, 3; i \neq j$; $\mathbf{g} = (0, \text{-g}, 0)$ and $F = \frac{1}{2}[\exp(2kh) - 1]$. Dispersion relation (1) thus includes both KH and RT effects. For a stationary magnetopause ($G_{ij} = 0$) this model was first studied in [40].

Relation (1) allows for coupled magnetopause and IEBL modes, and is referred to as the "thick" model. Two limiting cases may be noted: very short λ ($kh >> 1$) and very long λ's ("thin" approximation, $kh << 1$). For wavelengths smaller than, or even $\sim$ h, the dispersion relation (1) may be expressed as $H_1 + H_2 + G_{12} = 0$, and $H_2 + H_3 + G_{23} = 0$, i.e., as two dispersion relations for surface waves on the magnetopause and the inner edge of the boundary layer, respectively, with no coupling between the two interfaces. In the thin model, the dispersion relation (1) reduces to $H_1 + H_3 + G_{13} = 0$ (here magnetopause and IEBL oscillations are coupled), and $H_2 = 0$. The former equation connects directly the magnetosheath with the magnetosphere and would be appropriate also to a magnetopause without boundary layer. The latter equation $H_2 = 0$, is the dispersion relation for shear Alfvén waves propagating inside the boundary layer.

The procedure may be briefly described as follows. We define a non-dimensional growth rate, p as p = I(ω/kV_∞), where V_∞ is the solar wind speed, and normalize quantities to their solar wind values [35]. For the thin model, where p can be obtained analytically, and for a given λ (= mR_e, in earth radii), it is found that the RT effect may be expressed by a G-term where $G = (g/2\pi(V_\infty)^2)mR_e$, i.e., the RT effect increases with wavelength of the perturbative mode. For a given m, the maximum growth rate is found by rotating the wave vector $\mathbf{k}$ by 360 deg in the local magnetopause plane. Results are then shown as charts. A similar procedure is employed in the thick model, except that this time a value of kh is specified. The appropriate dispersion relation (see above) is then solved numerically for the non-dimensional growth rates, and two growth rate charts are compiled, one each for the magnetopause and IEBL modes.

We now illustrate some of the results obtained. Figure 1 shows the distribution of magnetic shear across the dayside magnetopause for an IMF pointing due north. The whole equatorial region up to 30-40° latitude and higher latitudes nearer the noon meridian are low shear, indeed, very low shear ($\leq 10°$). Perturbations in near-equatorial regions travelling towards the flanks with a **k** vector aligned generally with the flow (which in stagnation line flow is mostly east-west) just flute the fields. Specific examples show that short wavelength perturbations reach a non-linear stage by the time they arrive at the flanks. Longer wavelengths grow less. At higher latitudes, perturbations will grow until they leave the wave amplifying region, after which they propagate further downtail as surface waves.

By contrast, Figure 2 shows the dramatic change in the map of magnetic shear when the IMF is tilted west of north by 30°. The subsolar region has now $\sim 30°$ shear. Two "strips" of very low shear are evident: on the duskside in the northern, and on the dawnside in the southern, hemispheres. This results in a dawn-dusk asymmetry in KH activity. $\mathbf{B_1}$ is roughly aligned along each of these strips, and, in stagnation line flow, the magnetosheath velocity $\mathbf{v_1}$ is almost orthogonal to it. Therefore, a **k** vector perpendicular to the strip and pointing tailward, will practically switch off the magnetic tension and, in addition, it is more or less aligned with the flow field. KH instability will thus develop at mid-latitudes. Since wave growth occurs only in the unstable strips, only relatively short-λ perturbations will finally emerge from the unstable regions. These will travel as stable waves rippling the mid- to high-latitude magnetopause.

The growth rate map in the thin model approximation for this case indicates positive growth confined to the very low shear strips. Peak growth rates on the dayside are lower than peak growth rates for a due-northward IMF by an overall factor of $\sim$2. Clearly, then, in this approximation appreciable KH activity accurs only for strongly northward-pointing IMF.

We now discuss some results in the thick model. We present first results for a due northward IMF orientation and for a sunward accelerating magnetopause. We then discuss results for the inclined IMF orientation and a stationary magnetopause. In the thick model we consider long and short λ's separately. To fix ideas, we take kh = 0.4 for the former, and hk = 5 for the latter. (With a boundary layer width of 0.3 R_e, the average value obtained by AMPTE-IRM experimenters, these values translate to $\sim$ 5 and 0.4 R_e, respectively [41]). In calculating the RT effect, we shall take g = 1 km s^{-2} (see [15] and references therein) and $V_\infty = 500$ km s^{-1}. From these values the respective G-parameters may be obtained. We concentrate first on long-λ perturbations. Calculations of the ratio, r, of the amplitude of the IEBL perturbations to those of the magnetopause along $\bar{y}$ show that $r \approx 1$, signifying that both interfaces move together with similar amplitudes: the

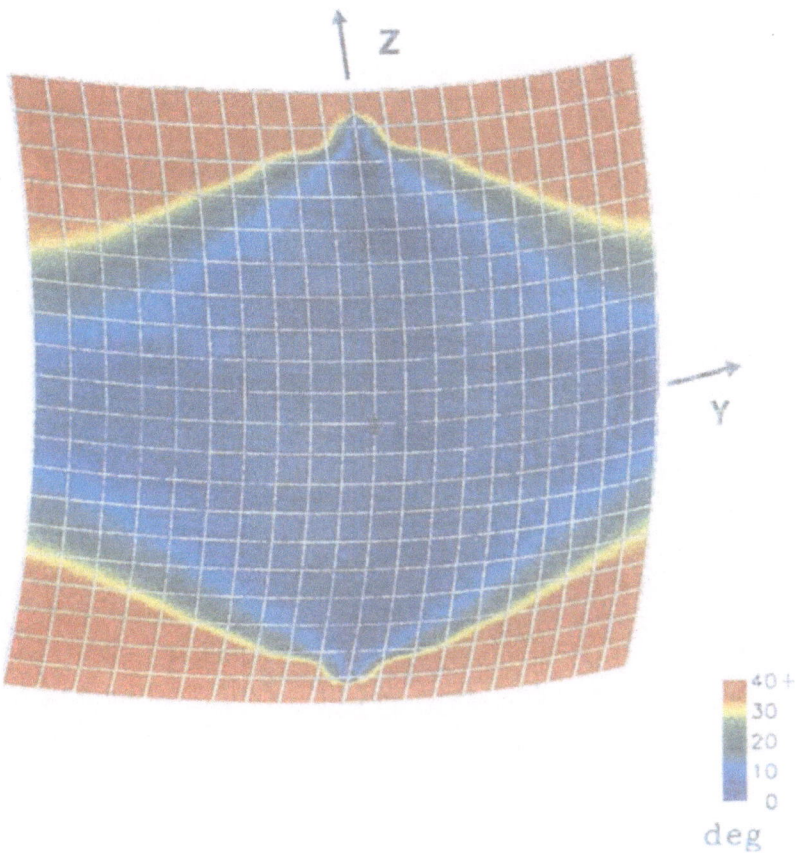

Figure 1. Distribution of magnetic shear across the dayside magnetopause for an IMF pointing due north.

surfaces are thus coupled, with the dominant perturbation growing in time with the fastest of the two growth rates. Both magnetopause and the IEBL are unstable over the entire dayside. The growth rate charts are shown in Figure 3 bottom panels. When the magnetopause is stationary and only the KH instability is acting (Figure 3, upper two panels), we have that the magnetopause is active at low latitudes and away from noon, while the IEBL is KH-active up to higher latitudes but with the activity concentrated closer to the flanks.

For short λ's, however, the two surfaces can oscillate independently of each other to different unstable modes. The low-latitude magnetopause ($\leq$

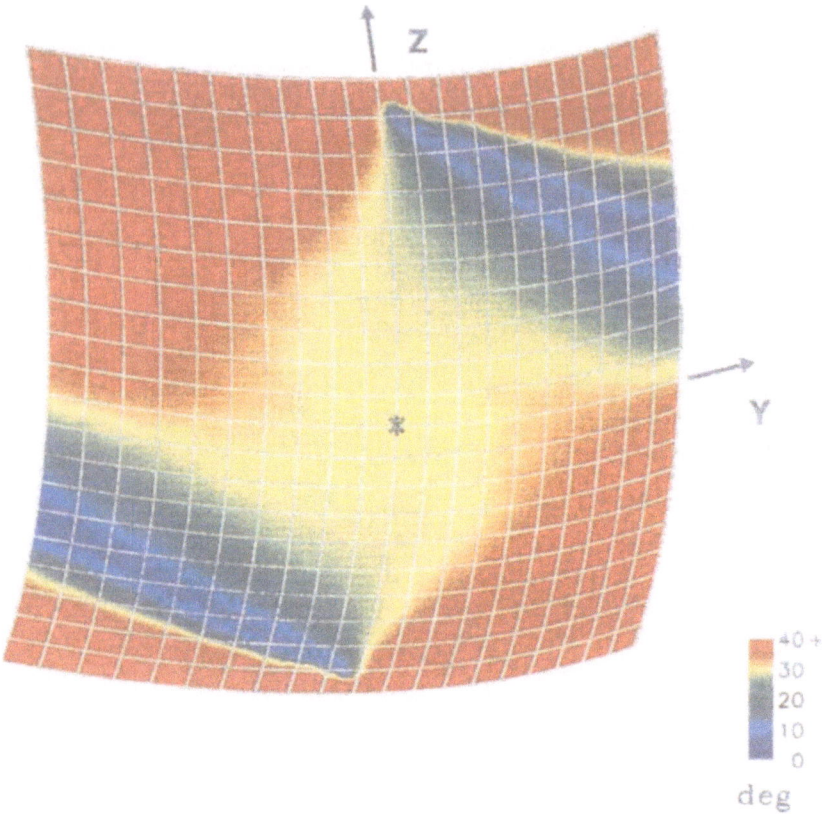

Figure 2. Distribution of magnetic shear across the dayside magnetopause for an IMF inclined at 30° west of north.

30°) is unstable for all local times, and further instability with decreased growth rates extends to higher latitudes as the noon meridian is approached (Figure 4, bottom left panel). Maximal growth rates are enhanced with respect to stationary conditions (Figure 4, upper left panel). By contrast, the entire IEBL is unstable to the combined KH+RT instabilities (Figure 4, lower right panel). As shown by the right-hand, top panel in Figure 4, in the absence of acceleration, the IEBL is unstable only well away from noon.

Our final comment refers to the KH instability (i.e., stationary magnetopause) for an IMF orientation tilted westward (dawnward) from north by 30°. Growth rate maps are shown in [35] (their Figures 12 and 13). For the magnetopause and for short wavelengths, activity is confined to two

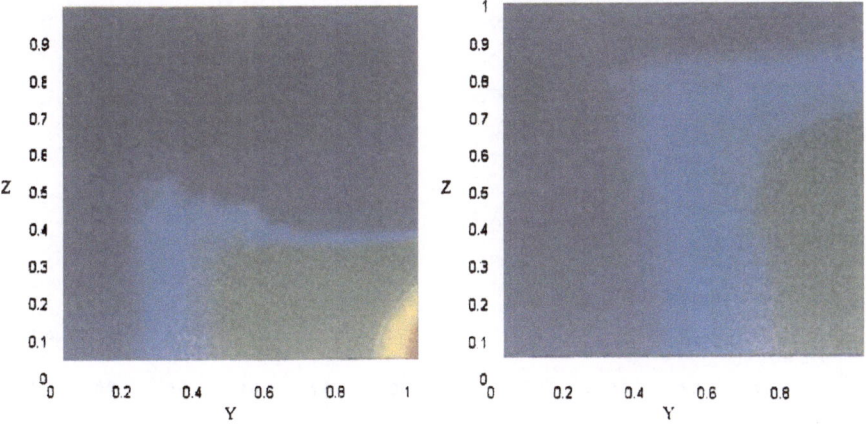

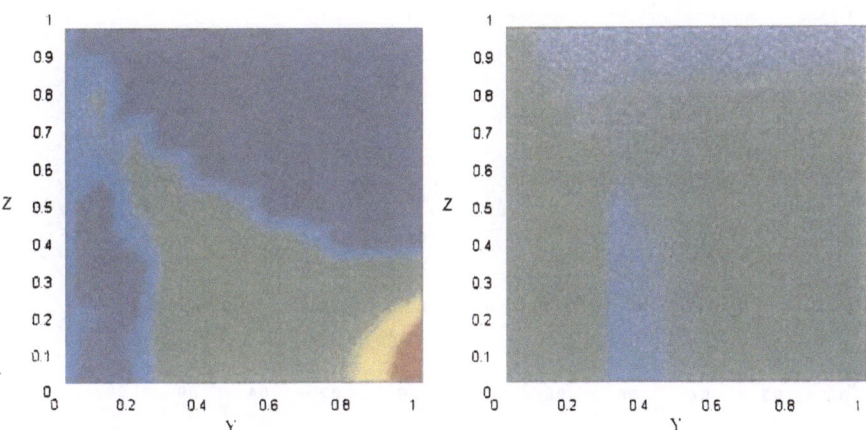

Figure 3. Top two panels: Maximum growth rates on a static magnetopause (top left) and the IEBL (top right) for long-λ (kh = 0.4) perturbations. The lower two panels give the corresponding results for a sunward accelerating magnetopause. The IMF points due north.

strips as in the thin model, giving rise to ripples at high latitudes. The magnetopause is stable with respect to long λ perturbations, which do not fit into the strips. For the IEBL we have the following growth rate picture. For long λ's, the instability in the post-noon sector resides in a "fan" which broadens at high northern latitudes. The KH instability in the pre-noon sector is obtained from post-noon by a reflection about the subsolar point. For short λ's, the unstable region widens and growth rates are higher than

12

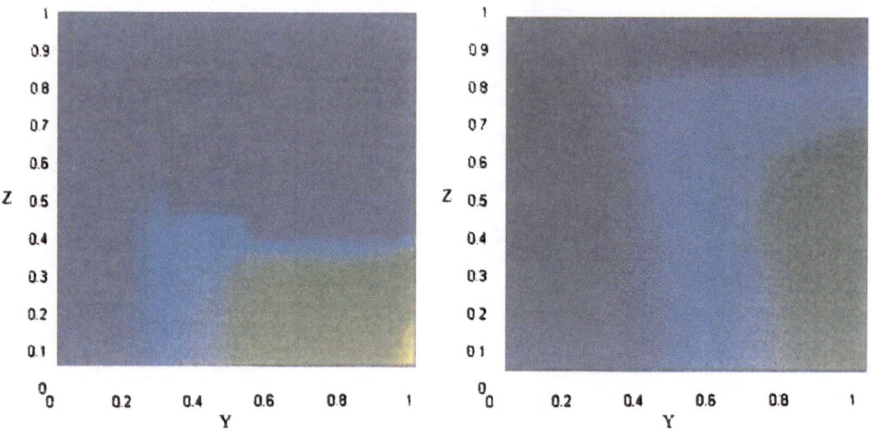

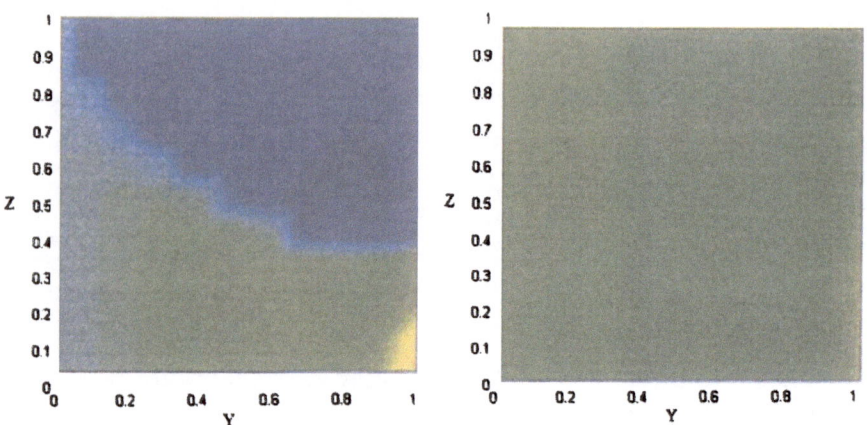

Figure 4. Similar to Figure 3, but for short λ modes (kh = 5). The IMF points due north.

for long λ's. The IMF B_y component thus unfluences the KH stability of the IEBL, even though this surface is not in contact with the PDL. PDL flows ($\mathbf{v_1}$) *are* influenced by the IMF orientation and the IEBL "inherits" this dependence through the further, model assumption $\mathbf{v_2} \sim \mathbf{v_1}$. The major difference for both λ's when the magnetopause accelerates sunward is that, at the magnetopause, the active strips are broader and growth rates are higher, while the IEBL is completely destabilized.

Acknowledgments.
This work was partially supported by NASA grant NAG5-2834, by a BID-CONICET grant PIP046, by UBA grant EX092, and by grant 95-05-14192 from the Russian Foundation of Basic Research. LB was supported by a CONICET fellowship. CJF would like to thank the organizing committee for an excellent conference and for their generous support of his participation.

References

1. Dungey, J. W. (1954) Electrodynamics of the outer atmospheres, *Penn. State Ionos. Res. Lab. Sci. Rep* **69**.
2. Axford, W.I., and Hines, C. O. (1961) A unifying theory of high-latitude geophysical phenomena and geomagnetic storms, *Can. J. Phys.*, **39**, 1433.
3. Southwood, D. J. (1968) The hydrodynamic stability of the magnetospheric bound ary, *Planet. Space Sci.*, **16**, 587.
4. Belmont, G, and Chanteur, G. (1989) Advances in Magnetopause Kelvin-Helmholtz instability studies, *Physica Scripta*, **124**, 124.
5. Miura, A. (1995) Kelvin-Helmholtz instability at the magnetopause: Computer Simulations, in *Physics of the Magnetopause* edited by Song, P., Sonnerup, B. U. O., and Thomsen, M. F., Geophysical Monograph **90**, American Geophysical Union, Washington, D. C., 285.
6. Fitzenreiter, R. J., and Ogilvie, K. W. (1995) Kelvin-Helmholtz instability at the magnetopause: Observations, in *Physics of the Magnetopause* edited by Song, P., Sonnerup, B. U. O., and Thomsen, M. F., Geophysical Monograph **90**, American Geophysical Union, Washington, D. C., 277.
7. Kivelson, M. G., and Chen, S.-H. (1995) The magnetopause: Surface waves and instabilities and their possible dynamical consequences, in *Physics of the Magnetopause,* edited by Song, P., Sonnerup, B. U. O., and Thomsen, M. F., Geophysical Monograph **90**, American Geophysical Union, Washington, D. C., 257.
8. Samson, J. C., Jacobs, J. A., and Rostoker, G. (1971) Latitude-dependent character istics of long-period micropulsations, *J. Geophys. Res.*, **76**, 3675.
9. Southwood, D.J. (1974) Some features of field resonances in the magnetosphere, *Planet Space.Sci.*, **22**, 483.
10. Chen, L., and Hasegawa, A. (1974) A theory of long-period magnetic pulsations, 1, Steady state excitation of field line resonances, *J. Geophys. Res.*, **79**, 1024.
11. Southwood, D. J. (1979) Magnetopause Kelvin-Helmholtz instability, in *Magnetospheric Boundary Layers*, ESA publications SP-148, 357.
12. Samson, J. C. (1991) Geomagnetic pulsations and plasma waves in the Earth's magnetosphere, in *Gemagnetism*, **vol. 4**, edited by Jacobs, J. A., Academic Press, New York, 481.
13. Engebretson, M., Glassmeier, K.-H., Stellmacher, M., Hughes, W. J., and Luehr, H. (1997) The dependence of high latitude Pc 5 wave power on solar wind velocity and on the phase of high speed solar wind streams, *J. Geophys. Res.*, in press.
14. Mishin, V. V. (1993) Accelerated motions of the magnetopause as a trigger of the Kelvin-Helmholtz instability, *J. Geophys. Res.*, **98**, 21,365.
15. Gratton, F. T., Farrugia, C. J., and Cowley, S. W. H. (1996) Is the magnetopause Rayleigh-Taylor unstable sometimes ?, *J. Geophys. Res.*, **101**, 4929.
16. Song, P., Elphic, R. C., and Russell, C. T. (1988) ISEE 1 and 2 observations of the oscillating magnetopause, *Geophys. Res. Lett.*, **15**, 744-747.
17. Freeman, M. P., and Farrugia, C. J. (1997) Magnetopause motions in a Newton-Busemann approach, *These Proceedings*.

14

18. Farrugia, C.J., Freeman, M. P., Cowley, S.W.H., Southwood, D. J., Lockwood, M., and Etemadi, A. (1989) Pressure-driven magnetopause motions and attendant response on the ground *Planet. Space Sci.*, **37**, 589.

19. Russell, C. T., and Elphic, R. C. (1978) Initial ISEE magnetometer results: Magnetopause observations, *Space Sci. Rev.*, **22**, 681.

20. Rijnbeek, R. P., Cowley, S.W.H., Southwood, D.J., and Russell, C.T. (1984) A survey of dayside flux transfer events by the ISEE-1 and -2 magnetometers, *J. Geophys. Res.*, **89**, 786.

21. Berchem, J., and C.T. Russell (1984) Flux transfer events on the magnetopause: Spatial distribution and controlling factors, *J. Geophys. Res.*, **89**, 6689.

22. Chen, S.-H., and Kivelson, M. G. (1993) On nonsinusoidal waves at the Earth's magnetopause, *Geophys. Res. Lett.*, **20**, 2699.

23. Chen, S.-H., Kivelson, M. G., Gosling, J. T., Walker, R. J., and Lazarus, A. J.(1993) Anomalous aspects of magnetosheath flow and of the shape and oscillations of the magnetopause during an interval of strongly northward interplanetary magnetic field, *J. Geophys. Res.*, **98**, 5727.

24. Ogilvie, K. W., and Fitzenreiter, R. J. (1989) The Kelvin-Helmholtz instability at the magnetopause and inner boundary layer surface, *J. Geophys. Res.*, **94**, 15,113.

25. Sckopke, N., Paschmann, G., Haerendel, G., Sonnerup, B. U. O., Bame, S. J., Forbes, T. G., Hones, E. W., Jr., and Russell, C. T. (1981) Structure of the low-latitude boundary layer, *J. Geophys. Res.*, **86**, 2099.

26. Wu, C. C. (1986) Kelvin-Helmholtz instability at the magnetopause boundary, *J. Geophys. Res.*, **91** 3042.

27. Kivelson, M. G., and Pu, Z. Y. (1984) The Kelvin Helmholtz instability on the magnetopause, *Planet. Space Sci.*, **32**, 1335.

28. González, A.G., and Gratton, J.(1994) The role of a density jump in the Kelvin-Helmholtz instability of a compressible plasma *J.Plasma Phys.*, **52**, 223.

29. Kivelson, M. G., and Southwood, D. J. (1986) Coupling of global magnetospheric MHD eigenmodes to field line resonances *J. Geophys. Res.*, **91**, 4345.

30. Hollweg, J. V., Yang, G., Cadez, V., and Gakovic, B. (1989) Surface waves in an incompressible fluid: resonant instability due to velocity shear, *Astrophys.J.*, **336**, 430.

31. Sedlacek, Z. (1995) Continuum damping in plasma physics, *AIP Conf.Proc. 345, (ICPP 1994)*, AIP Press, 119.

32. Uberoi, C. (1995) Alfvén waves in space plasmas, *AIP Conf.Proc. 345, (ICPP 1994)*, AIP Press, 383.

33. Thomas, V. A., and Winske, D. (1993) Kinetic simulations of the Kelvin-Helmholtz instability at the magnetopause, *J. Geophys. Res.*, **98**, 11,425.

34. Thomas, V.A. (1995) Three-dimensional kinetic simulation of the Kelvin-Helmholtz instability, *J. Geophys. Res.*, **100**, 19,429.

35. Farrugia, C. J., Gratton, F. T., Bender, L., Biernat, H. K., Erkaev, N. V., Denisenko, V., Torbert, R. B., and Quinn, J. M. (1997) Charts of joint Kelvin-Helmholtz and Rayleigh-Taylor instability growth rates at the dayside magnetopause for strongly northward IMF, *J. Geopys. Res.*, (in press).

36. Zwan, B. J., and Wolf, R. A. (1976) Depletion of solar wind near a planetary boundary, *J. Geophys. Res.*, **81**, 1636.

37. Erkaev, N. V., Farrugia, C. J., and Biernat, H. K. (1997) Comparison of gas dynamics and MHD predictions for magnetosheath flow, *These Proceedings*.

38. Pudovkin, M.I. and Semenov, V.S. (1977) Stationary frozen-in coordinate system, *Ann. Geophys.*, **33**, 429.

39. Sonnerup, B. U. O. (1974) The reconnecting magnetopause, in *Magnetospheric Physics*, edited by B. M. McCormac, D. Reidel, Norwell, Mass., 23.

40. Lee, L. C., Albano, R. K., and Kan, J. R. (1981) Kelvin-Helmholtz instability in the magnetopause-boundary layer region, *J. Geophys. Res.*, **86**, 54.

41. Paschmann, G., W. Baumjohann, N. Sckopke, T.-D. Phan, and L. Luehr (1993) Structure of the dayside magnetopause for low magnetic shear, *J. Geophys. Res.*, **98**, 13409.

MAGNETOPAUSE MOTIONS IN A NEWTON-BUSEMANN APPROACH

M. P. FREEMAN
British Antarctic Survey
Madingley Road, Cambridge, CB3 0ET, U. K.

C. J. FARRUGIA
University of New Hampshire
Morse Hall, Durham, NH 03824-3525, U. S. A.

1. Introduction

The buffetting of the magnetosphere by the solar wind is the cause of many phenomena in the magnetosphere and the ionosphere, and affects the way in which we observe many other features. Examples of magnetospheric phenomena caused, wholly or in part, by solar wind pressure variations include cavity and field-line resonances (e.g., [1], [2]), travelling convection vortices [3], ring current particle diffusion [4], quasi-periodic emissions [5], and possibly substorm onsets (e.g., [6]; but see [7]) and magnetopause reconnection [8]. Examples of the way in which solar wind pressure variations affect the way we observe and interpret magnetospheric features include the structure of the magnetopause boundary layer [9] and possibly some FTE identifications [10], and similarly the structure of the magnetotail.

The understanding of many of these phenomena is hampered by being built on an incomplete foundation of time-dependent magnetospheric compression, upon which are constructed more or less sophisticated models of the coupling of magnetospheric compressional waves to other wave modes and particles.

At first sight, the compressional foundation might seem quite straightforward and secure, namely the generalised Chapman-Ferraro (GCF) model [11], [12], in which the instantaneous position and shape of the magnetopause is determined by the balance between the solar wind dynamic pressure and the magnetospheric magnetic field pressure (i.e., $p_{sw} = kB_{mp}^2/2\mu_0$, where for a blunt shape like the magnetosphere, k is of order unity). How-

J. Moen et al. (eds.), Polar Cap Boundary Phenomena, 15–26.

ever, a number of deficiencies in this model may be pointed out. First, the GCF model is not time-dependent in which case there are to be expected inertial and damping effects [13], as well as propagation effects such as different positions on the magnetopause responding at different times and thus the magnetopause shape departing from its equilibrium configuration [10]. Second, the GCF model is based upon a corpuscular model of the solar wind. Such a model does not account for the bow shock and the effect that this might have on the compression. MHD theory does explain a bow shock, but only steady-state numerical solutions to the self-consistent mhd theory in various approximations have thus far been studied.

The limitations of the GCF model can be shown by applying it to a data example of the type Chapman and Ferraro [11] wished to describe, namely, a Sudden Impulse (SI). In Figure 1, the solid line shows the variation in the north-south component of the magnetic field, ΔH, measured at Moshiri, Japan (37.61°N, 213.23°E in corrected geomagnetic coordinates [14]). The scale for the Moshiri data is on the left-hand axis. A sudden decrease in ΔH at Moshiri correlates with a sudden decrease in the solar wind dynamic pressure, p_{sw}, at the WIND spacecraft [15] at 46 min lag. The dotted line shows the square root of the lagged solar wind dynamic pressure measured by WIND at $\sim 175 R_E$ upstream of Earth. The scale for the Wind data is on the right-hand axis. The size of the change equates to a 10.5 nT magnetic field perturbation per 1 $(\text{nPa})^{1/2}$ solar wind pressure perturbation, in line with what is predicted by the GCF model [16]. However, the correlation is not perfect: We note that the magnetic field responds more gradually to the solar wind dynamic pressure change. This indicates that the magnetopause response is not instantaneous, but instead it moves at a finite speed, of order 100 km/s [17]. Also, we can see that the magnetic field overshoots its final equilibrium position. Such overshoots are often seen in SI signatures [18]. They indicate that the magnetopause oscillates about its new equilibrium position.

Such properties are likely to emerge from any reasonable time-dependent theory of magnetopause motion. We demonstrate this by first considering a simple time-dependent extension of the GCF model (section 2). Next, in section 3, we review the Freeman, Freeman, and Farrugia (FFF) theory of magnetopause motion [19]. This is an analytical, time-dependent, hypersonic hydrodynamic theory offering the most complete, albeit simplified, solution to date. In both theories, we begin by outlining the theoretical method, including its approximations, and then derive the resulting equation of motion of the magnetopause and examine its properties. In section 4, we discuss how the properties of the FFF solution differ from those of the GCF theory, how they relate to the observations, and what more needs to be done. Though the FFF and other magnetospheric compression the-

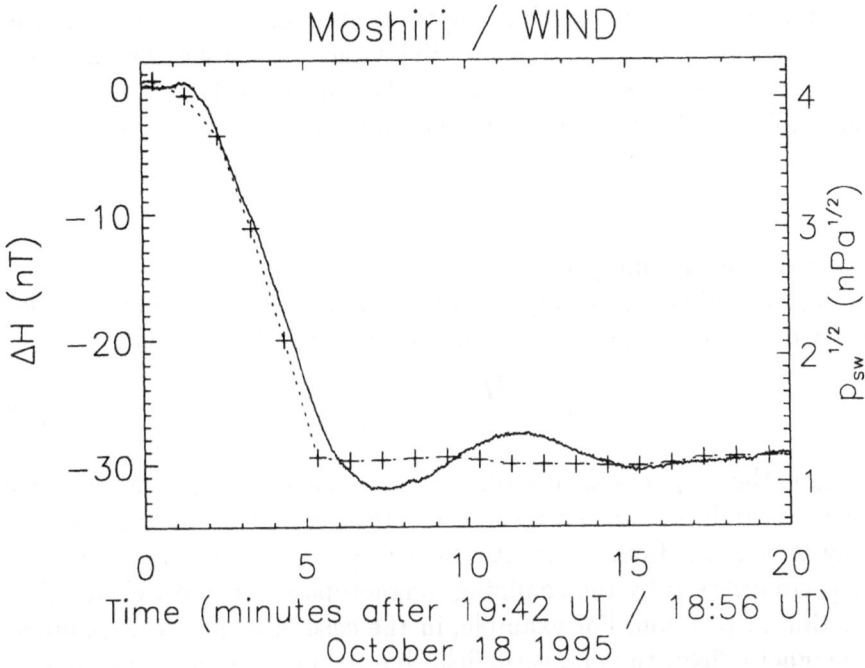

Figure 1. Comparison of the ground magnetic field perturbation at Moshiri (solid line) with the solar wind dynamic pressure variation at WIND (dotted line).

ories reveal some of the important physics to us, we are still far from the desired goal of a reliable foundation for time-dependent magnetospheric compression.

2. Time-dependent GCF model

2.1. STEADY STATE SOLUTION

The problem is to determine the free magnetopause boundary which separates the two distinct media of the magnetic field inside the magnetopause and the plasma outside it. In the GCF model the plasma outside the magnetopause comprises the solar wind particles which specularly reflect off the magnetopause surface.

In the GCF model the stresses exerted on the magnetopause surface are given by the rate of change of momentum of the solar wind particles scattering off the surface per unit area. If we assume that no energy is lost by the scattering particles and if the stress tangential to the surface is zero then the change in solar wind momentum normal to the surface is $2m\mathbf{v}_{sw}.\hat{\mathbf{x}}_1$, where $m\mathbf{v}_{sw}$ is the particle momentum and $\hat{\mathbf{x}}_1$ is the unit vector normal to the magnetopause. Assuming that the total number density of solar wind particles is n_{sw} (i.e., $n_{sw}/2$ incident particles and $n_{sw}/2$ reflected particles), then the particle number flux normal to the surface is $1/2 n_{sw} \mathbf{v_{sw}} . \hat{\mathbf{x}}_1$ and thus the total rate of change of solar wind momentum normal to the surface per unit area is given by:

$$p = n_{sw} m \left(\mathbf{v_{sw}} . \hat{\mathbf{x}}_1 \right)^2 \qquad (1)$$

where p is the Newtonian pressure exerted on the surface.

The internal pressure is that due to the magnetospheric magnetic field confined within the magnetopause surface. This can be written as:

$$p = \frac{\alpha^2 B_{eq}^2}{2\mu_0} \left(\frac{R_E}{r} \right)^6 \qquad (2)$$

where B_{eq} is the magnetic field strength on the earth's surface at the equator, R_E is the radius of the earth, and r is the geocentric distance. α is the factor by which the dipole magnetic field tangential to the magnetopause surface is compressed by the confining magnetopause current. Generally, α is a function of position. For example, in the case of a spherically confined dipole magnetic field, the magnetic field due to the magnetopause currents is uniform and equal to twice the dipole magnetic field strength at the equatorial magnetopause, i.e., $\alpha = 3 \cos \lambda$ where λ is the magnetic dipole latitude.

For a spherical geometry ($\hat{\mathbf{x}}_1 = \hat{\mathbf{r}}$), the external Newtonian pressure on the frontside magnetopause in the meridional plane varies as $\cos^2 \lambda$ which is the same variation as the internal magnetic pressure. However, elsewhere this is not so and thus a static spherical magnetopause cannot be in overall pressure balance. Instead, the magnetopause adopts a shape such that pressure balance is realised. This equilibrium shape is approximately ellipsoidal or parabolic [12], [20]. Much of the frontside magnetopause is approximately spherical but the very different shape behind means that the value of α at the subsolar point is reduced from $\alpha = 3$ for spherical confinement to $\alpha = 2.44$, with similar reductions elsewhere [12].

2.2. TIME DEPENDENT SOLUTION

If the solar wind dynamic pressure varies then the magnetopause pressure balance will change and we expect the magnetopause to move to a new equilibrium position of different size.

In general curvilinear coordinates, $\mathbf{r} = (x_1, x_2, x_3)$, moving with an axially symmetric ($\frac{\partial}{\partial x_3} \equiv 0$) magnetopause boundary, the boundary velocity, $\mathbf{v_b}$, is given by:

$$\frac{v_{b1}}{h_1} \frac{\partial \mathbf{r}}{\partial x_1} + \frac{v_{b2}}{h_2} \frac{\partial \mathbf{r}}{\partial x_2} = \left(\frac{\partial \mathbf{r}}{\partial t} \right)_{x_1, x_2} \qquad (3)$$

where h_1, h_2 are the curvilinear scale factors. In the boundary frame, the Newtonian pressure is:

$$p = n_{sw} m \left\{ (\mathbf{v_{sw}} - \mathbf{v_b}) \cdot \hat{\mathbf{x}}_1 \right\}^2 \qquad (4)$$

When the magnetosphere inflates or compresses, the strength of the dipole magnetic field tangential to the boundary changes and so also the strength of the magnetopause currents required to confine the dipole magnetic field. Thus the time-dependent magnetic pressure becomes:

$$p = \frac{\alpha^2 B_{eq}^2}{2\mu_0} \left(\frac{R_E}{r(t)} \right)^6 = \frac{\alpha^2 B_{eq}^2}{2\mu_0} \left(\frac{R_E}{r_i} \right)^6 \left(1 - 6 \frac{(r - r_i)}{r_i} \right) \qquad (5)$$

where r_i is the initial magnetopause position, and linearizing for small magnetopause displacements, $(r - r_i) \ll r_i$.

Equating the external Newtonian pressure to the internal magnetic pressure yields the equation for the geocentric distance to the subsolar magnetopause, $r(t)$:

$$\frac{dr}{dt} + 3 \sqrt{\frac{p_{swi}}{n_{sw} m}} \left(\frac{(r - r_i)}{r_i} \right) = \frac{\sqrt{p_{swi}} - \sqrt{p_{sw}}}{\sqrt{n_{sw} m}} \qquad (6)$$

In response to a solar wind dynamic pressure impulse, $p_{sw} = p_{swi} + (p_{swf} - p_{swi}) H(t)$, where $H(t)$ is the Heaviside unit step function, the magnetopause moves according to:

$$\frac{r - r_i}{r_i} = \frac{1}{3} \left(1 - \sqrt{\frac{p_{swf}}{p_{swi}}} \right) \left(1 - \exp \left(\frac{-t}{\tau} \right) \right) \qquad (7)$$

where the e-folding time, τ, on which the magnetopause moves asymptotically towards its final equilibrium, $r = r_f$ is given by:

$$\tau = \frac{r_i \sqrt{n_{swf} m}}{3 \sqrt{p_{swi}}} \qquad (8)$$

3. Time-dependent FFF model.

The FFF model [19] can be described as the hypersonic, hydrodynamic version of the GCF model described above. Like the GCF model it is the problem of determining the free magnetopause boundary which separates the two distinct media of the magnetic field inside the magnetopause and the plasma outside it. In contrast to the GCF model, the plasma outside the magnetopause is the shocked fluid of the magnetosheath, rather than the specularly reflecting corpuscles of the solar wind.

The FFF model solves the relevant hydrodynamic equations in the two limits of, first, Mach number, $M \to \infty$, then, second, ratio of specific heats, $\gamma \to 1$. In this, the Newton-Busemann approximation, a centrifugal term is added to the Newtonian pressure acting on the magnetopause [21]. This centrifugal term, which stems from the fact that particles in the shock layer attached to the body are moving around a curved surface, reduces the pressure on the body with respect to the Newtonian pressure. While this term seems well founded physically, it does lead to a well-known drawback, namely, that at some point, the pressure on the body drops to zero. Clearly, beyond that point the method breaks down [22]. If we confine our attention to the stagnation region then the Newton-Busemann approximation should be reasonably valid [19]. Thus we can say that the approximation is somewhat crude, but the theory is rational.

Since the Newton-Busemann solution for an ellipsoidal magnetopause geometry has been published elsewhere [19], we shall only outline the main steps of the method here.

3.1. STEADY STATE SOLUTION

In the FFF model we solve the hydrodynamic equations to find the external pressure exerted on the magnetopause surface by the shocked fluid, but the internal pressure is treated in exactly the same way as in the GCF model.

The steady state momentum equations in general curvilinear coordinates, $\mathbf{r} = (x_1, x_2, x_3)$, with axial symmetry ($\frac{\partial}{\partial x_3} \equiv 0$) and $v_3 = 0$ are

$$\frac{v_1}{h_1}\frac{\partial v_1}{\partial x_1} - \frac{v_2^2}{h_1 h_2}\frac{\partial h_2}{\partial x_1} + \frac{v_2}{h_2}\frac{\partial v_1}{\partial x_2} + \frac{v_1 v_2}{h_1 h_2}\frac{\partial h_1}{\partial x_2} = \frac{-1}{\rho h_1}\frac{\partial p}{\partial x_1} \tag{9}$$

$$\frac{v_2}{h_2}\frac{\partial v_2}{\partial x_2} - \frac{v_1^2}{h_1 h_2}\frac{\partial h_1}{\partial x_2} + \frac{v_1}{h_1}\frac{\partial v_2}{\partial x_1} + \frac{v_1 v_2}{h_1 h_2}\frac{\partial h_2}{\partial x_1} = \frac{-1}{\rho h_2}\frac{\partial p}{\partial x_2} \tag{10}$$

where h_1, h_2, h_3 are the curvilinear scale factors.

The magnetosheath and solar wind pressure, velocity, and momentum are related by the Rankine-Hugoniot relations:

$$\frac{v_{1s}}{v_{sw1}} = \frac{\rho_{sw}}{\rho_s} = \frac{(\gamma - 1)}{(\gamma + 1)} + \frac{2}{(\gamma + 1) M_{sw}^2} \underset{M_{sw} \to \infty}{=} \frac{(\gamma - 1)}{(\gamma + 1)} \tag{11}$$

$$\frac{v_{2s}}{v_{sw2}} = 1 \tag{12}$$

$$\frac{p_s}{p_{sw}} = \frac{2\gamma}{(\gamma + 1)} M_{sw}^2 - \left(\frac{\gamma - 1}{\gamma + 1}\right) \underset{M_{sw} \to \infty}{=} \frac{2\rho_{sw} v_{sw1}^2}{(\gamma + 1) p_{sw}} \tag{13}$$

$$M_{sw}^2 = \frac{\rho_{sw} v_{sw}^2}{\gamma p_{sw}} \tag{14}$$

Thus, in the limit of large solar wind Mach number, M_{sw}, the quantities v_{1s}/v_{sw1} and ρ_{sw}/ρ_s tend to $\epsilon \equiv (\gamma-1)/(\gamma+1)$, which is small as γ tends to unity. The quantities v_{2s}/v_{sw2} and $p_s/\rho_{sw}v_{sw1}^2$ are of order unity. Also, for large Mach number, the shock layer is thin and thus $(v_1/h_1)\partial/\partial x_1$ is of order unity, as is $(v_2/h_2)\partial/\partial x_2$. This is the Newton-Busemann approximation in which the momentum equation is expanded in terms of the small parameter, ϵ.

To lowest order the momentum equations become

$$\frac{v_1}{h_1}\frac{\partial v_2}{\partial x_1} + \frac{v_2}{h_2}\frac{\partial v_2}{\partial x_2} = 0 \tag{15}$$

$$\frac{v_2^2}{h_2}\frac{\partial h_2}{\partial x_1} = \frac{1}{\rho}\frac{\partial p}{\partial x_1} \tag{16}$$

The first equation implies that the magnetosheath velocity tangential to the magnetopause, v_2, is constant along the magnetosheath stream path. From equation (12), v_2 is equal to the solar wind velocity tangential to the bow shock at the point where the stream tube intercepts the shock.

The second equation gives the magnetosheath pressure gradient normal to the magnetopause at constant x_2. Integrating across the magnetosheath, the pressure exerted on the magnetopause is given by:

$$p = p_s + \int_{x_{1s}}^{x_1} \left(\frac{\rho v_2^2}{h_2}\right)\left(\frac{\partial h_2}{\partial x_1}\right) dx_1 \tag{17}$$

where p_s is the pressure behind the shock from equation (13) and is equal to the Newtonian pressure on the shock surface.

The second term is the Busemann correction [21]. This term represents the centrifugal force due to the fluid flowing around the magnetopause and acts so as to reduce the Newton-Busemann pressure relative to the Newtonian pressure. This pressure correction may be calculated using the relations at the shock and the hydrodynamic continuity equation [19]. At the stagnation point, the centrifugal correction is zero and the Newton-Busemann pressure reduces to the Newtonian pressure.

3.2. TIME-DEPENDENT SOLUTION

As in the time-dependent GCF model, we transform into a curvilinear coordinate frame moving with the magnetopause velocity, $\mathbf{v_b}$. The momentum equation in the moving frame (denoted by a prime) is:

$$-\nabla p' = \rho'\left(\frac{\partial'}{\partial t} + \mathbf{v'}.\nabla\right)\mathbf{v'} + \rho'\left(\frac{\partial'}{\partial t} + \mathbf{v'}.\nabla\right)\mathbf{v_b} \tag{18}$$

Using the Rankine-Hugoniot relationships in the moving frame, we neglect terms of order ϵ in the momentum equation to yield:

$$-\frac{1}{h_1}\frac{\partial p'}{\partial x_1} = \rho'\frac{\partial v_{b1}}{\partial t}' + g\rho'v_2'^2 \qquad (19)$$

where the last term is the centrifugal force (e.g., $g = 1/r$ for a spherical magnetopause).

Now we consider the magnetopause motion to be caused by a small perturbation in the solar wind density and speed, resulting in small perturbations in the magnetosheath quantities and magnetopause position. Linearizing and integrating the momentum equation across the shock layer, the Newton-Busemann pressure perturbation at the moving magnetopause is given by:

$$\Delta p' = \Delta p'_s - \int_{x_{1s}}^{x_1} \rho_i \left(\frac{\partial v_{b1}}{\partial t}\right)' h_1 dx_1 + I_c \qquad (20)$$

The first term on the right hand side is the perturbation pressure just inside the bow shock. From the Rankine-Hugoniot equations, this is equal to the perturbation in the solar wind dynamic pressure in the moving frame given by equation (4). The second term is an inertial term which has been computed for an ellipsoidal geometry [19]. The third term arises from the centrifugal term in the momentum equation and we neglect it.

Equating the external Newton-Busemann pressure to the internal magnetic pressure (again given by equation (5)) we find the linearized equation for the geocentric distance to the subsolar magnetopause, $r(t)$:

$$Kr_i\rho_{swi}\frac{d^2r}{dt^2} \quad + \quad 2\rho_{swi}v_{swi}\frac{dr}{dt} + 6\rho_{swi}v_{swi}^2\frac{(r - r_i)}{r_i} \qquad (21)$$

$$= \quad -\rho_{swi}v_{swi}^2\left(\frac{(\rho_{sw} - \rho_{swi})}{\rho_{swi}} + \frac{2\left(v_{sw} - v_{swi}\right)}{v_{swi}}\right) \qquad (22)$$

where K is a geometrical parameter of order unity for an ellipsoidal magnetopause [19]. Thus in the Newton-Busemann case, the magnetopause equation of motion describes a forced, damped simple harmonic oscillator. As a result, the magnetopause moves asymptotically towards its new equilibrium on an e-folding time of order:

$$\tau = K \left(\frac{r_i}{v_{swi}}\right) \qquad (23)$$

but also oscillates about the new equilibrium before settling there. The oscillation period is:

$$T = 2\pi\sqrt{\frac{K}{6}}\left(\frac{r_i}{v_{swi}}\right) \qquad (24)$$

4. Discussion

We have discussed two approaches towards the development of a theory of magnetopause motion due to solar wind dynamic pressure variations. The first was a simple extension of the GCF model, in which we estimated the external pressure exerted on a moving magnetopause by applying a Galilean transformation. In response to a solar wind dynamic pressure impulse, the magnetopause moved exponentially towards its new equilibrium position with an e-folding time of order the characteristic time scale of the system, r_i/v_{swi}. In the second approach, the FFF model, we estimated the external pressure exerted on a moving magnetopause by integrating the pressure gradient in the magnetosheath momentum equation from the bow shock to the magnetopause in the Newton-Busemann approximation. The e-folding time for the approach to equilibrium was of the same order as that in the GCF model, but, in addition, there was oscillatory motion about the new equilibrium position. We do not expect the applicability to the real magnetopause to be perfect, but the FFF model serves to demonstrate the importance of bow shock dissipation and system inertia in the magnetopause response.

That the magnetopause should respond on a characteristic time scale of the solar wind - magnetosphere system could be predicted by any reasonable dimensional analysis [13]. What is less predictable is the relative magnitudes of the damping and oscillation timescales. For example, Smit [13] estimated the magnetopause oscillation to be much ($\sim$ 100 times) more weakly damped than in the FFF model [19]. To test which estimate is more reasonable, we can compare the results derived here with the Sudden Impulse signature discussed in the Introduction. From equation (5), the magnetic perturbation due to a change in magnetopause position, $r - r_i$, is given by:

$$\Delta B = (\alpha - 1) B_{eq} R_E^3 \left(\frac{1}{r^3} - \frac{1}{r_i^3} \right) \approx -3 (\alpha - 1) \frac{B_{eq} R_E^3}{r_i^3} \frac{(r - r_i)}{r_i} \qquad (25)$$

In Figure 2, we plot ΔB for both models due to a solar wind dynamic pressure impulse of amplitude equal to that shown in Figure 1 and where, as before, the final equilibrium value of ΔB has been scaled according to 10.5 nT/(nPa)$^{1/2}$. Comparing Figures 1 and 2, it is evident that the Newton-Busemann theory agrees with the observations quite well. Nevertheless, there is much to improve on with the theory.

Here we have assumed that the magnetic field adjusts instantaneously to a change in the magnetopause position. This is a reasonable approximation for a vacuum magnetic field where communication is at the speed of light. In the real magnetosphere where the magnetic field is loaded with plasma

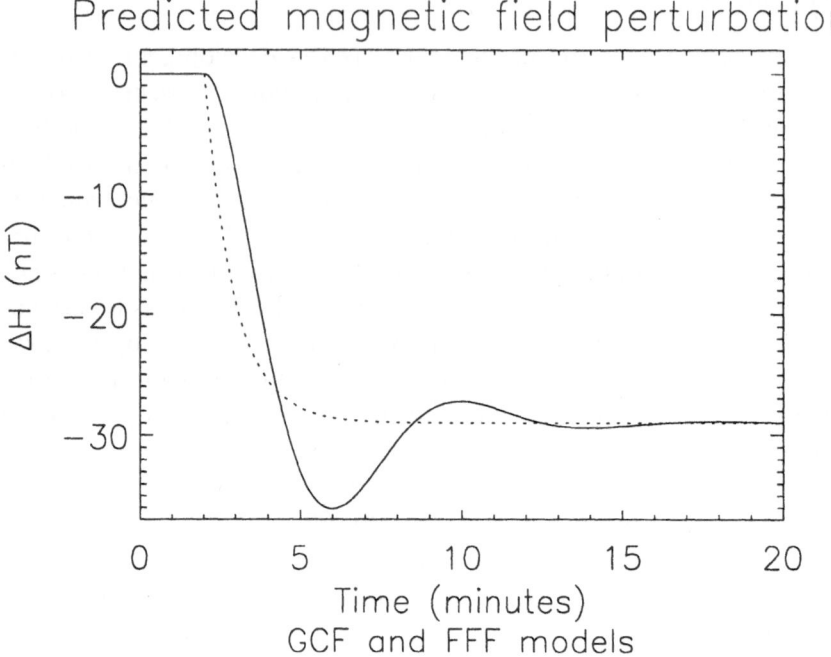

Predicted magnetic field perturbation

Figure 2. Magnetic field response to a solar wind dynamic pressure impulse predicted by the GCF model (dotted line) and the FFF model (solid line).

the communication speed is the fast mode which is of order a few hundred km/s in the outer magnetosphere. The effect of this on the magnetopause motion has been considered by Willis [23] whose equation of motion nevertheless has a similar asymptotic solution. Such a treatment is essential for describing magnetospheric cavity modes. Hitherto, cavity mode models assume the magnetopause to be a fixed reflecting boundary or to move in a simple manner. In reality, compressional waves are imperfectly reflected off the magnetopause and the reflection coefficient is likely a function of the magnetopause velocity. Thus the theory described here should be coupled to a cavity mode model. We expect that in this case the ability of the magnetosphere to sustain cavity modes would depend upon the degree to which the intrinsic magnetopause oscillation period due to magnetosheath inertia was matched to the magnetospheric cavity mode period.

Another effect to be modelled is that of the interplanetary magnetic field (IMF). Observations show a difference in the amplitude of magnetopause motion for northward and southward IMF [24]. It has been argued that the amplitude of the magnetopause fluctuations during northward IMF can be accounted for entirely by solar wind dynamic pressure variations but that the larger amplitude magnetopause fluctuations during southward IMF require additional non-compressional mechanisms to be operative. But it is at least conceivable that the different magnetopause response for northward and southward IMF might be explained by a compressional theory such as the Alfvénon theory [8] which exhibits different behaviour for northward and southward IMF conditions.

In summary, we have reviewed idealized models of the response of the magnetopause to solar wind dynamic pressure variations. The hydrodynamic FFF model revealed the importance of bow shock dissipation and magnetosheath inertia, but neglected the role of the magnetospheric cavity and the IMF emphasised in other models. All these models can only be considered as lowest order approximations to the actual magnetopause response. More detailed models need to be developed, coupling together these individual elements, in order to adequately describe this problem that is the source of many other magnetospheric phenomenon.

5. Acknowledgments

We would like to thank Prof. K. Yumoto and the 210° MM magnetic observation group for kindly providing data from the Moshiri station, Solar-Terrestrial Environment Laboratory, Nagoya University, as shown in Figure 1.

6. References

1. Allan, W., White, S.M., and Poulter, E.M. (1986) Impulse-excited hydromagnetic cavity and field-line resonances in the magnetosphere, *Planet. Space Sci.*, **34**, pp. 371–385
2. Kivelson, M.G., and Southwood, D.J. (1986) Coupling of global magnetospheric MHD eigenmodes to field line resonances, *J. Geophys. Res.*, **91**, pp. 4345–4351
3. Friis-Christensen, E., McHenry, M.A., Clauer, C.R., and Vennerstrom, S. (1988) Ionospheric traveling convection vortices observed near the polar cleft: a triggered response to sudden changes in the solar wind, *Geophys. Res. Lett.*, **15**, pp. 253–256
4. Walt, M. (1994) *Introduction to geomagnetically trapped radiation.* Cambridge University Press.
5. Coroniti, F.V., and Kennel, C.F. (1970) Electron precipitation pulsations, *J. Geophys. Res.*, **75**, pp. 1279–1289
6. Kokubun, S., McPherron, R.L., and Russell, C.T. (1977) Triggering of substorms by solar wind discontinuities, *J. Geophys. Res.*, **82**, pp. 74–86
7. Henderson, M.G., Reeves, G.D., Belian, R.D., and Murphree, J.S. (1996) Observations of magnetospheric substorms occurring with no apparent solar wind/IMF trigger, *J. Geophys. Res.*, **101**, pp. 10773–10791
8. Song, Y., and Lysak, R.L. (1994) Alfvénon, driven reconnection and the direct generation of the field-aligned current, *Geophys. Res. Lett.*, **21**, pp. 1755–1758
9. Hapgood, M.A., and Bryant, D.A. (1992) Exploring the magnetospheric boundary layer, *Planet. Space Sci.*, **40**, pp. 1431–1459
10. Sibeck, D.G. (1990) A model for the transient magnetospheric response to sudden solar wind dynamic pressure variations, *J. Geophys. Res.*, **95**, pp. 3755–3771
11. Chapman, S., and Ferraro, V.C.A. (1930) A new theory of geomagnetic storms,

Nature, **126**, p. 129

12. Mead, G.D., and Beard, D.B. (1964) Shape of the geomagnetic field solar wind boundary, *J. Geophys. Res.*, **69**, pp. 1169–1179

13. Smit, G.R. (1968) Oscillatory motion of the nose region of the magnetopause, *J. Geophys. Res.*, **73**, pp. 4990–4993

14. Yumoto, K., and the 210° MM Magnetic Observation Group (1996) The STEP 210° Magnetic Meridian Network Project, *J. Geomag. Geoelectr.*, **48**, pp. 1297–1309

15. Lepping, R.P., *et al.* (1997) The Wind magnetic cloud and events of October 18–20, 1995: Interplanetary properties and as triggers for geomagnetic activity, *J. Geophys. Res.*, **102**, pp. 14049–14063

16. Russell, C.T., Ginskey, M., Petrinec, S., and Le, G. (1992) The effect of solar wind dynamic pressure changes on low and mid-latitude magnetic records, *Geophys. Res. Lett.*, **19**, pp.1227–1230

17. Farrugia, C.J., Freeman, M.P., Cowley, S.W.H., Southwood, D.J., Lockwood, M., and Etemadi, A. (1989) Pressure-driven magnetopause motions and attendant response on the ground, *Planet. Space Sci.*, **37**, pp. 589–607

18. Russell, C.T., and Ginskey, M. (1993) Sudden impulses at low latitudes: Transient response, *Geophys. Res. Lett.*, **20**, pp.1015–1018

19. Freeman, M.P., Freeman, N.C., and Farrugia, C.J. (1995) A linear perturbation analysis of magnetopause motion in the Newton-Busemann limit, *Ann. Geophys.*, **13**, pp. 907–918

20. Roelof, E.C., and Sibeck, D.G. (1993) Magnetopause shape as a bivariate function of interplanetary magnetic field B_z and solar wind dynamic pressure, *J. Geophys. Res.*, **98**, pp. 21421–21450

21. Busemann, A. (1933) Flüssigkeits und gasbewegung, *Handwörterbuch der Naturwissenschaften, Vol. IV, 2nd edition*, p. 244

22. Hayes, W.D., and Probstein, R.F. (1966) *Hypersonic Flow Theory, Vol. 1*. Academic Press, New York.

23. Willis, D.M. (1964) The sudden commencement and first phase of a geomagnetic storm, *J. Atmos. Terr. Phys.*, **26**, pp. 581–602

24. Song, P., Elphic, R.C., and Russell, C.T. (1988) ISEE 1 & 2 observations of the oscillating magnetopause, *Geophys. Res. Lett.*, **15**, pp. 744–747

COMPARISON OF GASDYNAMICS AND MHD
PREDICTIONS FOR MAGNETOSHEATH FLOW

N.V. ERKAEV
Computing Center of Russian Academy of Sciences
660036, Krasnoyarsk, Russia

C.J. FARRUGIA
Institute for the Study of Earth, Oceans and Space, University
of New Hampshire
NH 03824, Durham, USA

H.K. BIERNAT
Space Research Institute, Austrian Academy of Sciences
A-8042, Graz, Austria

1. Introduction

Magnetohydrodynamics (MHD) has proved very useful in describing the behaviour of space plasmas, in particular the solar wind flow around the Earth's magnetosphere. The MHD flow problem is quite different from that in gasdynamics. This is because magnetic forces give rise to an asymmetric three-dimensional flow pattern. The structure of ideal MHD flow around obstacles has a special features caused by a frozen-in magnetic field, even though this is relatively small in the up-stream flow. The interplanetary magnetic field (IMF) is responsible for the boundary layer effects, which do not disappear regularly as the IMF is decreased.

Aim of the present paper is to analyze in details the structure of the MHD flow near the magnetopause, in particular the behaviour of different components of velocity along the normal and distribution of plasma parameters along the magnetopause. This numerical study is based on the magnetic barrier model described in papers [1 - 6].

27

J. Moen et al. (eds.), Polar Cap Boundary Phenomena, 27–40.

2. Advances in the studies of magnetosheath flow

During the last thirty years a substantial advance has been made in solving the MHD flow problem. We shall briefly summarize the results obtained by different authors concerning the flow problem.

Midgley and Davis [7] were the first to mention the possibility of compression of magnetic field in front of the magnetopause. Lees [8] calculated the flow parameters along the subsolar line without taking the magnetic tension into account. Stream lines were assumed to be symmetrical about the subsolar line, as in pure gasdynamics. Nevertheless, that simple model predicted the enhancement of magnetic field and simultaneous decrease of density along the subsolar line.

Spreiter et al. [9] used MHD model in a kinematic approximation to calculate the magnetosheath parameters. Kinematic approximation means that the terms of magnetic forces are removed from the momentum equations. In this approach, plasma parameters satisfy to the pure gasdynamic system of equations, and magnetic field is to be obtained from the equations of magnetic flux conservation and magnetic induction. They had predicted a number of magnetosheath features compared successfully with observations. They had also reproduced draping of magnetic field lines around the magnetosphere. This model was widely used as a basic theoretical instrument for interpretation of magnetosheath observations over the years. However, this kinematic approximation becomes unreliable near the magnetopause as the underlying assumption that the IMF is unimportant becomes invalid there. Magnetic field calculated in this approximation had a strong singularity at the streamlined surface. This implies that the magnetic field has a strong influence upon the flow near the magnetospheric surface. It is necessary to include the magnetic forces at the momentum equation to avoid the singularities of magnetic field. A full MHD treatment was done later by Spreiter and Rizzi [10] under special conditions: solar wind flow and magnetic field are parallel to each other.

Zwan and Wolf [11] presented the nonsteady model of magnetic flux tubes moving slowly from the shock to the magnetopause in the plane parallel to the IMF. They calculated the plasma parameters in the flux tube as a function of time, and showed the decreasing of plasma density and increasing of magnetic field while the tube was approaching the magnetopause. Zwan and Wolf [11] called the region of low density near magnetopause as "plasma depletion layer" (PDL). This is very important physical effect related to IMF and terminology (PDL) is widely used now. However, the distance from any point of the tube to the magnetopause was unknown in the model of Zwan and Wolf. Because of this, it was impossible to obtain the profiles of plasma parameters along any given normal to the magnetopause.

These profiles of decreasing plasma density and increasing magnetic field were determined only along the subsolar line under additional simplifying assumption. This assumption concerns the magnetic field tension which was assumed to be proportional to the gradient of total pressure tangent to the magnetopause. The latter was supposed to be equal to that obtained by Spreiter et al. [9] in gasdynamic calculations. The coefficient of proportionality was a constant free parameter. The profiles of plasma parameters and magnetic field were obtained along the subsolar line for different values of this coefficient which could not be determined in the model.

Pudovkin and Semenov [12] analyzed kinematically the behaviour of the ratio B/ρ for different flow topologies. They proposed the formation of two-dimensional stagnation line flow at the obstacle to avoid a singularity of the ratio B/ρ . The idea of a stagnation line was also proposed by Sonnerup [13].

It became clear that MHD effects in magnetosheath flow could be separated into two parts: large scale effects (variation of position of the shock and related phenomena) and boundary layer effects (decrease of density, increase of magnetic field near magnetopause). Large-scale MHD effects were analyzed by Pivovarov and Erkaev [14]. They predicted the influence of the IMF on the shock position: the increase of the stand-off distance of the shock is proportional to the inverse square of the Alfvén Mach number: $\sim 1/M_a^2 = B_\infty^2/4\pi\rho_\infty u_\infty^2$; the shock front position has an asymmetry related to IMF. The boundary layer effects were treated by Erkaev [1] as a nondissipative MHD boundary layer which exists not only near subsolar point but all over the magnetopause. The appropriate boundary layer equations were obtained and the analytical solution was examined in a case of MHD flow past a cone.

Another approach with a two-dimensional flow model near the subsolar line was proposed by Pudovkin et al. [15]. The flow velocity was assumed to be a linear function of distance from the magnetopause along the subsolar line. The electric field component along the prescribed stagnation line was assumed to be constant. At the subsolar point the normal component of the velocity was determined through the reconnection rate. The direction of reconnection line and value of reconnection rate were prescribed. The assumptions used in this model are difficult to justify, in particular, the two-dimensionality of the flow in the magnetosheath.

The boundary layer technique was further developed and the numerical solution of the MHD boundary layer equations was obtained [2, 4] for flow past the Earth's magnetosphere modelled as a paraboloid of revolution. In this last model the magnetic field tension was taken into account self-consistently without any further assumption. This nondissipative boundary layer was called "magnetic barrier" to emphasize the dominant role of the magnetic field there. The magnetic barrier was determined as a layer of strong magnetic field and low density adjacent magnetopause. It is physically equivalent to PDL introduced by Zwan and Wolf [11] near the subsolar point.

In parallel, the results of the global MHD simulation of the solar wind interaction with Earth started to appear. Wu [16] presented the results of the three-dimensional MHD simulation of the solar wind flow around magnetosphere, approximated as a sphere. In this model, the global scale features of the MHD flow around Earth and position of the shock are reproduced well. However, magnetic barrier (or PDL) is not reproduced properly in this simulation: plasma density has a little decrease ahead of the magnetopause.

MHD simulation of the position of the bow shock for low Alfvén Mach numbers was made by Cairns and Lyon [17]. They studied the variations of stand-off distance as a function of Alfvén Mach number and pointed out clear discrepancies between MHD predictions and gasdynamics: magnetotheath predicted by MHD model is thicker than that in pure gasdynamics. This effect is more pronounced for low Alfvén Mach numbers. The structure of the flow in the magnetosheath was beyond the scope of their work.

Thus the global MHD models cannot give a detailed description of magnetosheath properties near magnetopause, in particular, magnetic barrier. The latter can be reproduced better by the model based on the boundary layer approach.

In the strict sense, boundary layer technique is applicable for high Alfvén Mach numbers. But there are occasions when the Alfvén Mach number can be low. This happens, e.g., in magnetic clouds. To model these situations, the magnetic barrier theory was extended to low Alfvén Mach numbers [5]. In this approach the magnetic barrier need not to be thin but the total pressure is assumed to be prescribed function of coordinates. Both, linear [5] and quadratic [6] variations of the total pressure between the shock and the magnetopause, have been tried and compared with data. In the latter work, an extension of the Newtonian formula for the total pressure at the magnetopause was included.

Nondissipative MHD flow around the magnetosphere forms the background to the study of dissipative processes, magnetic field reconnection and instabilities at the magnetopause. Accordingly, the magnetic barrier solution was analyzed together with diffusion and reconnection layers for the northward IMF [4]. Magnetic barrier solution was also used to determine the boundary conditions for unsteady reconnection processes at the magnetopause [18].

3. Statement of the flow problem. Basic equations

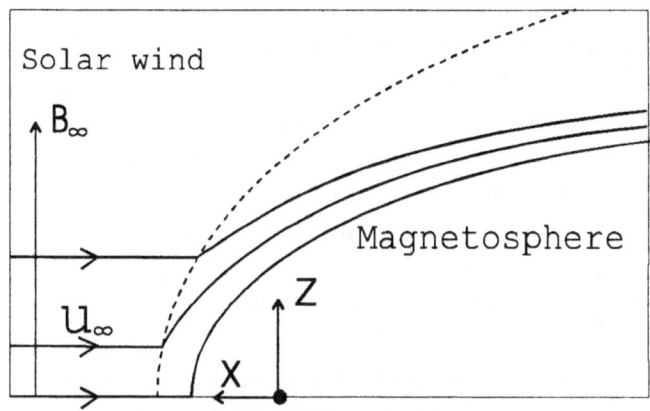

Figure 1. Coordinate system

An important characteristic of the problem is the generally supersonic nature of the flow:

$$p_\infty \ll \rho_\infty u_\infty^2, \qquad B_\infty^2 \ll 4\pi \rho_\infty u_\infty^2$$

where p_∞, ρ_∞, u_∞ and B_∞ are pressure, density, velocity and the magnetic field in an upstream (uniform) flow. For a perfectly conducting, inviscid gas we have the system of MHD equations without dissipation:

$$
\begin{aligned}
\rho\frac{\partial \mathbf{u}}{\partial t} + \rho(\mathbf{u}\cdot\nabla)\mathbf{u} + \nabla\Pi - \frac{1}{4\pi}(\mathbf{B}\cdot\nabla)\mathbf{B} &= 0 \\
\frac{\partial \rho}{\partial t} + div\,(\rho\mathbf{u}) = 0, \quad \frac{\partial S}{\partial t} + (\mathbf{u}\cdot\nabla)S &= 0 \\
\frac{\partial \mathbf{B}}{\partial t} - rot\,(\mathbf{u}\times\mathbf{B}) = 0, \quad div\,\mathbf{B} &= 0
\end{aligned}
\tag{1}
$$

Here Π is the total pressure: $\Pi = p + B^2/(8\pi)$, S is the entropy function: $S = p/\rho^k$, k is the adiabatic parameter.

For supersonic flow there exists a detached shock wave which separates the unperturbed solar wind from the magnetosheath. At the shock front we have the usual MHD Rankine–Hugoniot conditions:

$$
\begin{aligned}
[\![\rho u_n^2 + \Pi]\!] &= 0 \\
[\![\rho u_n \mathbf{u}_\tau - \frac{B_n}{4\pi}\mathbf{B}_\tau]\!] &= 0 \\
[\![\rho u_n]\!] = 0, \qquad [\![B_n]\!] &= 0 \\
[\![(\mathbf{u}\times\mathbf{B})_\tau]\!] &= 0
\end{aligned}
\tag{2}
$$

$$\llbracket \rho u_n \left[\frac{u^2}{2} + \left(\frac{\kappa}{\kappa - 1} \right) \frac{p}{\rho} \right] + \frac{u_n B^2 - B_n (\mathbf{u} \cdot \mathbf{B})}{4\pi} \rrbracket \;=\; 0$$

On the surface of the obstacle, which in this paper we model as a tangential discontinuity, we have the no-flow condition: $u_n = 0$. It is necessary to find a solution of the system (1) with the given parameters of an unperturbed flow and with the boundary conditions on the shock wave and the flow-around surface. The geometry of statement of problem is illustrated in Figure 1.

We introduce dimensionless variables as follows:

$$\mathbf{R} = \frac{\tilde{\mathbf{R}}}{R_m}, \qquad P = \frac{\tilde{P}}{\rho_\infty u_\infty{}^2}, \qquad \mathbf{B} = \frac{\tilde{\mathbf{B}}}{B_\infty}, \qquad \mathbf{u} = \frac{\tilde{\mathbf{u}}}{u_\infty}$$

Here, R_m is the distance from the Earth to subsolar point of the magnetopause. Substitution of the dimensionless set of variables into the initial system of MHD equations gives the set of MHD equations in dimensionless form:

$$\rho \frac{\partial \mathbf{u}}{\partial t} + \rho (\mathbf{u} \cdot \nabla) \mathbf{u} + \nabla \Pi - \epsilon (\mathbf{B} \cdot \nabla) \mathbf{B} \;=\; 0$$

$$\frac{\partial \rho}{\partial t} + div\,(\rho \mathbf{u}) \;=\; 0$$

$$\Pi = S \rho^k + \epsilon B^2 / 2, \quad \frac{\partial}{\partial t} S + (\mathbf{u} \cdot \nabla) S \;=\; 0 \qquad (3)$$

$$\frac{\partial \mathbf{B}}{\partial t} - rot\,(\mathbf{u} \times \mathbf{B}) = 0, \quad div\,\mathbf{B} \;=\; 0$$

Here ϵ depends on upstream parameters in the following fashion:

$$\epsilon = \frac{B_\infty^2}{4\pi \rho_\infty u_\infty^2} \qquad (4)$$

and is considered a small parameter.

For the description of ideal conducting plasma we use material coordinates, which are constant along the trajectories of the fluid particles and may be introduced by means of equations :

$$\frac{\partial \xi}{\partial t} + (\mathbf{u} \cdot \nabla)\xi = 0, \qquad \frac{\partial \eta}{\partial t} + (\mathbf{u} \cdot \nabla)\eta = 0, \qquad \frac{\partial \zeta}{\partial t} + (\mathbf{u} \cdot \nabla)\zeta = 0 \qquad (5)$$

These coordinates can be chosen so that they are proportional to Cartesian coordinates in the unperturbed flow:

$$\xi = z, \qquad \eta = y, \qquad \zeta = -x$$

We assume that the axis x is directed to the Sun and the direction of axis z is determined by the interplanetary magnetic field: the plane XZ

is parallel to the vector of IMF and the Z-axis is the IMF direction (see Figure 1). In this case, variables η and ζ are constant along magnetic field lines everywhere. Similar coordinates called as "frozen-in" were introduced by Pudovkin and Semenov [12]. In frozen-in coordinates the non-dissipative MHD equations can be written as follows:

$$\left(\frac{\partial^2 \mathbf{R}}{\partial t^2} - \epsilon \frac{\partial}{\partial \xi}\left(\rho \frac{\partial \mathbf{R}}{\partial \xi}\right)\right)_i + \frac{D(\Pi, R_k, R_\ell)}{D(\xi, \eta, \zeta)} = 0$$

$$S\rho^k + \epsilon \frac{\rho^2}{2}\left(\frac{\partial \mathbf{R}}{\partial \xi}\right)^2 = \Pi \qquad (6)$$

$$\frac{D(R_i, R_k, R_\ell)}{D(\xi, \eta, \zeta)} = \frac{1}{\rho}$$

Here the components of radius-vector $\mathbf{R}$ are Cartesian coordinates: $R_1 = x, R_2 = y, R_3 = z$; numbers (i, k, ℓ) are cyclic permutations of $(1, 2, 3)$.

The components of the velocity and magnetic field are given by the derivatives:

$$u_x = \frac{\partial x}{\partial t}, \qquad u_y = \frac{\partial y}{\partial t}, \qquad u_z = \frac{\partial z}{\partial t}$$

$$B_x = \rho \frac{\partial x}{\partial \xi}, \qquad B_y = \rho \frac{\partial y}{\partial \xi}, \qquad B_z = \rho \frac{\partial z}{\partial \xi}$$

The main assumption we use in the integration of these equations is that the total pressure is a known function of coordinates. As a first approximation, we use the so-called Newtonian formula for the total pressure along the magnetopause:

$$\Pi = \Pi_0 \cos^2(\psi) + \Pi_\infty \qquad (7)$$

where Π_0 is the pressure at the subsolar point of the magnetopause, and Π_∞ is the pressure in the upstream medium. The variation of total pressure between the magnetopause and bow shock is approximated as a quadratic function. The coefficients of this function are obtained from conditions at the bow shock (2) and at the magnetopause (7). We model the shape of the magnetopause as a paraboloid of revolution: $X = R_m - (Y^2 + Z^2)/(2L_0)$, where R_m and L_0 are, respectively, the distance from the Earth to the subsolar point of the magnetopause and the radius of curvature of the paraboloid at the subsolar point. We use the relation between R_m and L_0: $L_0 = 1.3 R_m$.

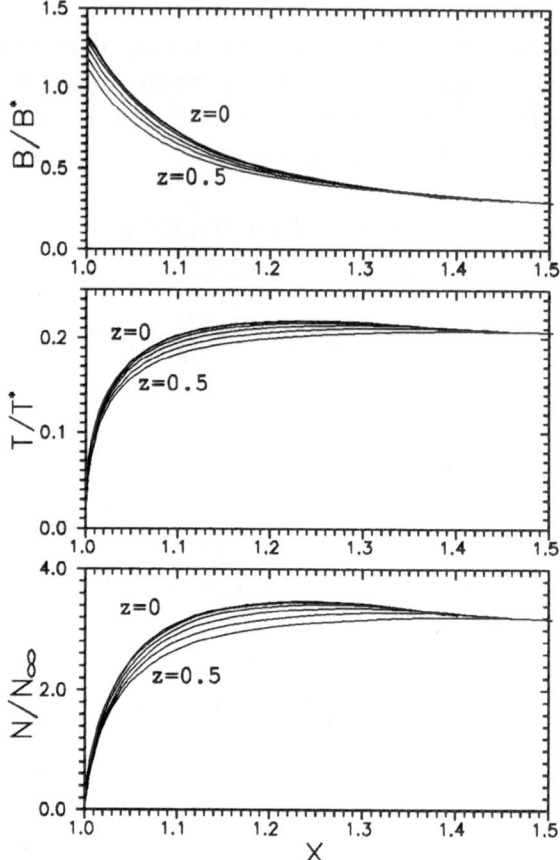

Figure 2. Profiles of magnetic field strength, temperature and density along the normal to the magnetopause at different distances from the subsolar point

4. Results of calculations

The results of the numerical integration of the magnetic barrier equations for $M_a = 10$ are presented in Figures 2 - 7. In all examples, the IMF directed North (i.e. the Z-axis is defined as North).

Profiles of magnetic field strength, temperature and density between the shock front and the magnetopause are shown in Figure 2 for 5 different distances from the subsolar line ($y = 0$): $0 \leq z \leq 0.5 \, R_m$. Step of variation of z is equal to $0.1 \, R_m$. Temperature and magnetic field strength are normalized to the following dimension parameters: $T^* = m_p u_\infty^2 / k$, $B^* = (4\pi\rho_\infty)^{0.5} u_\infty$. Here k is the Boltzman constant, m_p is the proton mass. While the magnetic field increases, the density, temperature and gas-kinetic pressure diminish to zero.

Figure 3 shows the profiles of z component of velocity in the magne-

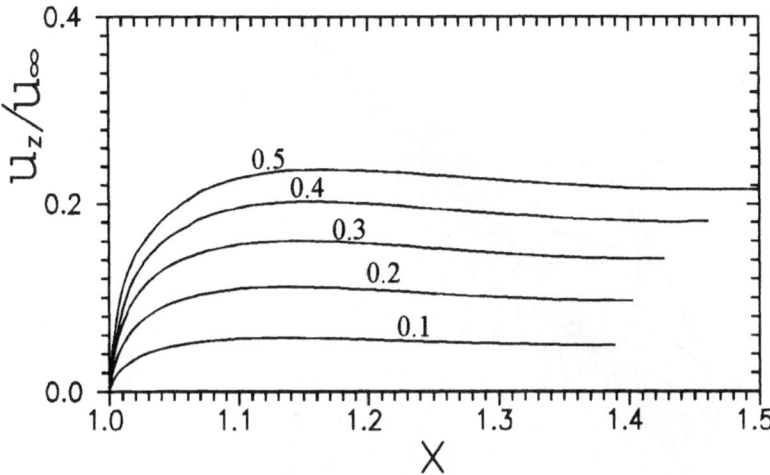

Figure 3. Profiles of u_z (North) component of velocity in the plane XZ at different distances from the subsolar line

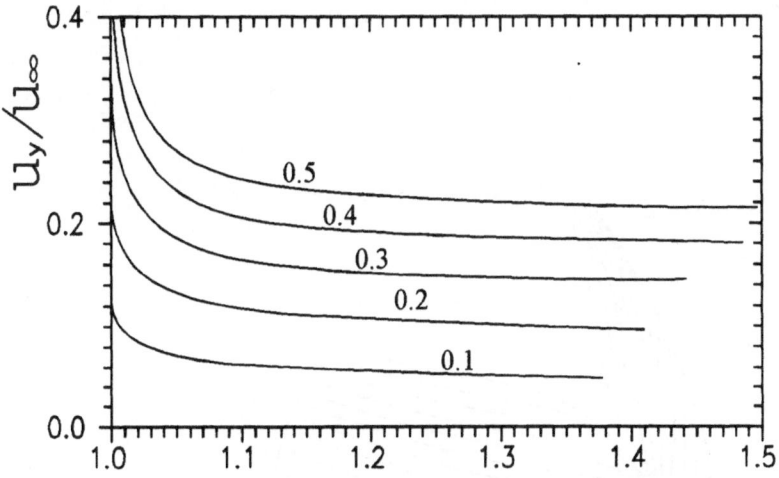

Figure 4. Profiles of u_y component of velocity in the plane XY at different distances from the subsolar line

tosheath at the different distances from the subsolar line ($0.1 \leq z \leq 0.5\, R_m$) in the plane XZ ($y = 0$). We see that the field-aligned component of velocity decreases to zero at any constant distance from the subsolar line.

Figure 4 shows the profiles of y component of velocity in the magnetosheath at the different distances from the subsolar line ($0 < y < 0.5 R_m$) in the plane XY ($z = 0$). The velocity component perpendicular to magnetic field increases because the magnetic field tension accelerates the plasma.

The calculated magnetic field and flow stream lines on the surface of

36

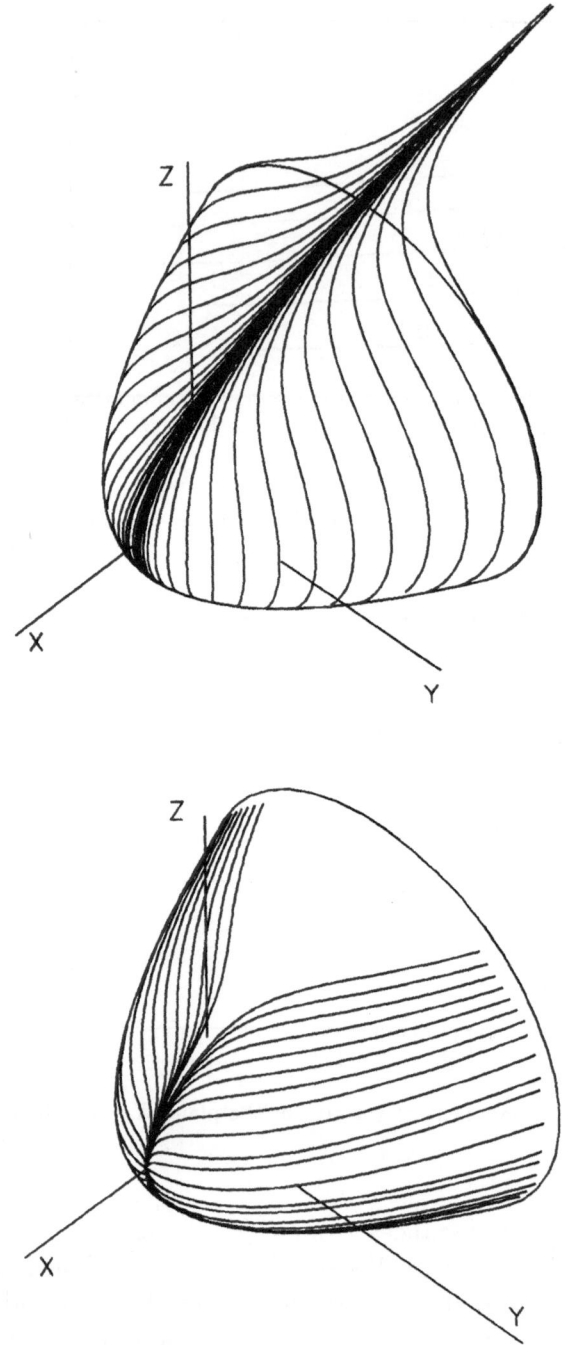

Figure 5. Magnetic field lines (top) and flow stream lines (bottom) on the boundary of the magnetosphere, modelled as a paraboloid of revolution

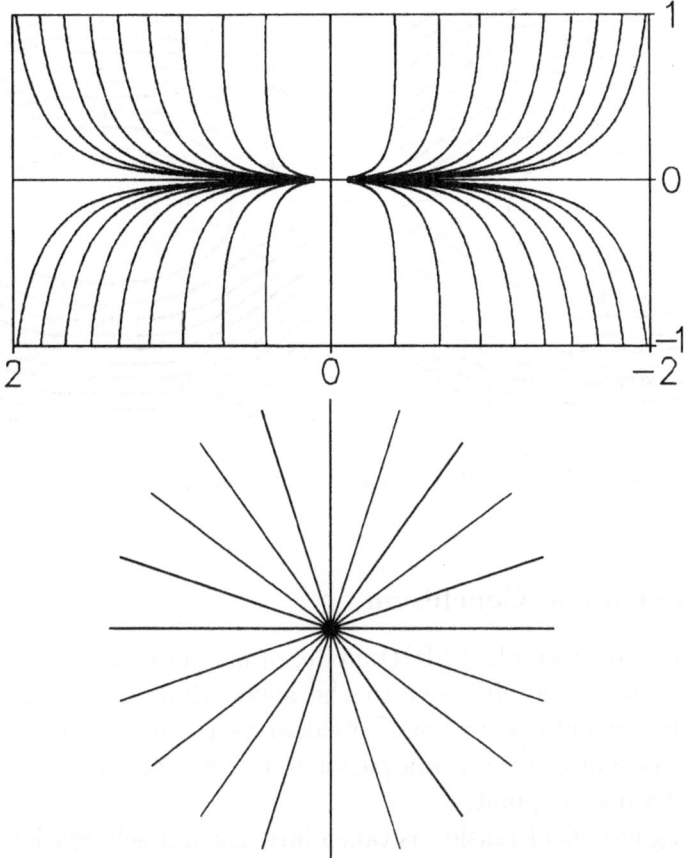

Figure 6. MHD (top) and gasdynamic (bottom) stream lines at the dayside magnetopause (view from the Sun)

the magnetosphere are shown in Figure 5. There is a stagnation line in the plane XZ when IMF points due to North. When the IMF rotates away from North, the stagnation stream line changes direction, remaining in the plane defined by the IMF and solar wind velocity. The angle between magnetic field and velocity tends to be $\pi/2$ near the magnetospheric boundary.

Figure 6 shows the stream lines at the magnetopause viewed from the Sun in magnetohydrodynamics (top) and gasdynamics (bottom). In the first case, the MHD stream lines are not radial ones, and each stream line touches the stagnation line directed along the IMF. In the second case, the gasdynamic stream lines go radially away from the subsolar point.

The contours of density and plasma pressure near the magnetopause are shown in Figure 7. The gradients are least in direction of IMF and greatest in the direction perpendicular to IMF.

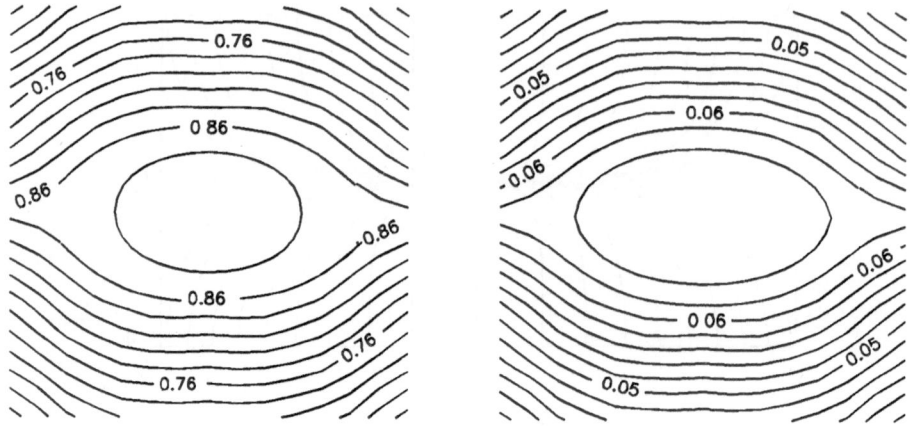

Figure 7. Contours of density (left) and plasma pressure (right) near the magnetospheric boundary

5. Discussion and Conclusion

We have described an ideal MHD model of magnetic barrier in the Earth's magnetosheath (in application to the solar wind flow around magnetosphere). Our model can be used to calculate the distribution of magnetic field, velocity and other plasma parameters over the entire magnetopause, not only at subsolar point.

The magnetic field tension is taken into account self-consistently. In our model the total pressure is a known function which can be improved iteratively [6]. The effects of the IMF are mainly concentrated inside the magnetic barrier, where the magnetic field eventually starts to exert a strong influence on the flow. The thickness of the layer depends on the small parameter ϵ, the inverse square of the Alfvén Mach number.

We have calculated the flow pattern at the obstacle with the stagnation stream line oriented along the IMF. We have shown the different behaviour of field-aligned and perpendicular components of velocity along a given normal to the obstacle. While the parallel velocity decreases towards the magnetopause, the perpendicular velocity increases. Explanation of this is following. A gradient of the plasma pressure along magnetic field lines is responsible for increasing of a field-aligned component of velocity. But inside the magnetic barrier the gradient of plasma pressure along magnetic field decreases. This is a reason for the small acceleration of plasma along magnetic field in the magnetic barrier. On the other hand, the magnetic tension accelerates plasma in direction perpendicular to magnetic field lines, resulting in an increase of the transverse component of velocity near the magnetopause.

The results of calculation show that the density, temperature and plasma pressure decrease along a normal to the obstacle. The frozen-in magnetic field lines stretch while the plasma flow goes towards the magnetopause. Because of this, the ratio $B/\rho = I$ tends to ∞. Magnetic field can not increase to infinity, because magnetic pressure is limited by the dynamic pressure of the solar wind. Hence, a density has to decrease to zero while the parameter I increases to infinity.

Stream lines tend to be orthogonal to magnetic field lines near magnetopause. At the magnetopause, stream lines have a specific structure with a stagnation line oriented parallel to IMF. Contours of plasma parameters are not axisymmetrical lines near the magnetopause: all parameters have minimal gradient in direction of IMF. The gradients of all parameters along the normal to the obstacles grows up with increasing of Alfvén Mach number.

The idea of "squeezing" of plasma along the magnetic field was proposed by Zwan and Wolf [11], and this terminology is very widespread. It means that, as the magnetic flux tubes approach the magnetopause, the plasma escapes along the tubes. According to solution presented here, the plasma velocity along the magnetic field lines is much smaller than that in gasdynamics. Along the magnetic field lines there is no force pushing plasma outward, only perpendicular to the tubes there is. The main point is that the plasma is escaping but not along the magnetic field lines. It is escaping perpendicular to them.

There is considerable interest in a phenomenon called "slow mode transition". This terminology was introduced to describe a local density peak sometimes observed in the magnetosheath density profile just before the decrease due to the depletion process takes place [19]. A theoretical explanation of this phenomenon was proposed by Southwood and Kivelson [20] who suggested the presence of an MHD slow shock wave standing inside the magnetosheath just ahead of the magnetopause. The question is: does a "slow mode transition" exist in the steady-state, ideal MHD flow around magnetosphere, modelled as a blunt body with a no-flow condition? Our steady-state MHD solution with a no-flow condition at the magnetopause does not have this structure. It may be related to nonsteady conditions or reconnection effects at the magnetopause, which we have not discussed here.

6. Acknowledgements

This work is partially supported by grant 97-05-64065 from the Russian Foundation of Basic Research, by the Austrian "Fonds zur Förderung der wissenschaftlichen Forschung", Project P09431-TEC, and by NASA grant NAG5-2834.

40

7. References

1. Erkaev, N.V. (1981) Effect of magnetic barrier in non-dissipative magnetohydrody-namics, *VINITY*, **3253-81 dep.**, Moscow, 1-55.
2. Erkaev, N.V. (1986) Peculiarities of the MHD flow of magnetosphere in the neigh-borhood of stagnation point, *Geomagnetism and Aeronomy*, **26**, 595-598.
3. Erkaev, N.V. and Mezentsev, A.V. (1992), The flow of the solar wind around the magnetosphere and the generation of electric fields, *Solar Wind - Magnetosphere In-teraction*, edited by M.F. Heyn, H.K. Biernat, V.S. Semenov, R.P. Rijnbeek, Vienna, Austria, 43-66.
4. Erkaev, N.V., Mezentsev, A.V., Denisenko, V.V., Zamai, S.S. and Troshichev, O.A. (1993) Electric field generation at the magnetospheric boundary for northward IMF, *J. Atmos. Terr. Phys.*, **56**, 153-166.
5. Farrugia, C.J., Erkaev, N.V., Biernat, H.K. and Burlaga, L.F. (1995) Anomalous magnetosheath properties during Earth's passage of an interplanetary cloud, *J. Geo-phys. Res.*, **100**, 19,245-19,257.
6. Farrugia, C.J., Erkaev, N.V., Biernat, H.K., Lawrence, G.R. and Elphic R.C. (1997) Plasma depletion layer model for low Alfvén Mach number: comparison with ISEE observations, *J. Geophys. Res.*, **102**, 11315-11324.
7. Midgley, J.E. and Davis, L.J. (1963) Calculation by a moment technique of the perturbation of the geomagnetic field by the solar wind, *J. Geophys. Res.*, **68**, 5111-5123.
8. Lees, L. (1964) Interaction between the solar plasma wind and the geomagnetic cavity, *AIAA J.*,**2**, 1576-1582.
9. Spreiter, J.R. and Alksne, A.Y. (1967) Plasma flow around the magnetosphere, *Rev. Geophys. and Phys.*,**7**, 11-68.
10. Spreiter, J. R. and Rizzi, A. W. (1974) Aligned magnetohydrodynamic solution for solar wind flow past the earth's magnetosphere, *Acta Astonautica*, **1**, 15-35.
11. Zwan B.J. and Wolf R.A. (1976) Depletion of the solar wind plasma near a planetary boundary,*J. Geophys. Res.*, **81**, 1636-1648.
12. Pudovkin, M.I. and Semenov, V.S. (1977) Stationary frozen-in coordinate system, *Ann. Geophys.*, **33**, 423-427.
13. Sonnerup, B.U.O., (1974) The reconnecting magnetopause, *Magnetospheric Physics*, edited by B.M. McCormac, D. Reidel Publishing Company, Boston, 23-33.
14. Pivovarov, V.G. and Erkaev, N.V. (1978) *Solar wind interaction with the Earth's magnetosphere*, "Nauka", Novosibirsk, 1-108.
15. Pudovkin, M.I., Heyn, M.F. and Lebedeva, V.V. (1982) Magnetosheath's parame-ters and their dependence on intensity and direction of the solar wind magnetic field, *J. Geophys. Res.*, **87**, 8131-8136.
16. Wu, C.C. (1992) MHD flow past an obstacle: large-scale flow in the magne-tosheath,*Geophys. Res. Let.*, **19**, 87-90.
17. Cairns, I.H. and Lyon, J.G. (1995) MHD simulations of Earth's bow shock at low Mach numbers: standoff distances, *J. Geophys. Res.*, **100**, 17173-17180.
18. Biernat, H.K., Bachmaier, G.A., Kiendl, M.T., Erkaev, N.V., Mezentsev, A.V., Far-rugia, C.J., Semenov, V.S. and Rijnbeek, R.P. (1995) Magnetosheath parameters and reconnection: A case study for the near-cusp region and the equatorial flank, *Planet. Space Sci.*, **43**, 1105-1120.
19. Song, P.C., Russel, C.T. and Thomsen. M.F. (1992) Slow mode transition in the frontside magnetosheath, *J.Geophys. Res.*, **97**, 8295-8305.
20. Southwood, D.J. and Kivelson M.G. (1995), Magnetosheath flow near the subsolar magnetopause: Zwan-Wolf and Southwood-Kivelson theories reconciled, *Geophys. Res. Let.*, **22**, 3275-3278.

TIME–VARYING RECONNECTION

Quantitative Analytical Results

H.K. BIERNAT
Space Research Institute, Austrian Academy of Sciences,
Lustbühelstr. 46, A-8042–Graz, Austria

V.S. SEMENOV AND O.A. DROBYSH
Institute of Physics, State University St. Petersburg,
St. Petergof 198904, Russia

C.J. FARRUGIA
Institute for the Study of Earth, Oceans and Space,
University of New Hampshire, Durham NH 03824, USA

N.V. ERKAEV
Computing Center, Russian Academy of Sciences,
660036 Krasnoyarsk 36, Akademgorodok, Russia

1. Introduction

Magnetic field line reconnection is a process which results in a change in the field topology, release of magnetic field energy, and associated acceleration and heating of plasma (see, e.g., [1], [2], [3]). This energy conversion process occurs in astrophysical, solar, space, and laboratory plasmas [4]. In space physics, reconnection has been investigated in analytical studies based on generalizations of the Petschek model ([5], [1], [6], [7]); in numerical simulations ([8], [9]); and in the study of in–situ data (e.g., [10]).

In this paper we present a reconnection model to illustrate the present level of understanding of this process and its application to magnetospheric physics. In particular, we use a generalization of the analytical shock–type Petschek model [5] in a rather general geometry which incorporates many features observed at the magnetopause, such as skewed magnetic fields, velocity shear, density jumps, and, in particular, time variations of the reconnection rate.

41

J. Moen et al. (eds.), Polar Cap Boundary Phenomena, 41–50.

2. The Model

The Petschek–type reconnection model has been described many times in the literature (e.g., [1], [6], [7]) and we can therefore be brief. Reconnection is initiated in a localized region of a current sheet, commonly referred to as the diffusion region. The current sheet itself is modelled as a tangential discontinuity, separating two magnetized plasma regions in which the behaviour is approximated by ideal magnetohydrodynamics (MHD). Since reconnection necessarily involves a breakdown of the ideal MHD approximation, we have to introduce some kind of a dissipative process. We model this processes globally through the introduction of a tangential electric field component $\mathbf{E}^*(\mathbf{r}, t)$ in the small–sized diffusion region. The reconnection–associated disturbances spread into the system and this results in the formation of an outflow region of reconnected, accelerated plasma along the current sheet embedded inside the region of reconnected flux. The latter region is defined by magnetic field lines which cross the plane of the initial current sheet and is bounded by separatrices, i.e., magnetic field lines which are in the process of being reconnected. The formation and subsequent development of the outflow region is determined by the outward propagation of MHD waves, including surface waves.

In time these MHD waves may steepen into large–amplitude waves or discontinuities. As a result, a reconnection layer forms ([6], [11]), bounded on each side by Alfvén discontinuities which rotate the magnetosheath and the magnetospheric fields to a common direction inside the layer. To achieve a matching of the field strength across the reconnection layer, the Alfvén discontinuities are followed by slow shocks. Finally, a contact discontinuity is needed to match any remaining asymmetries in the density.

If at some stage reconnection stops, then no more MHD waves are generated at the reconnection site so that the outflow region (region of accelerated plasma) and the separatrices detach from the reconnection site. The outflow regions then propagate in opposite directions away from the reconnection site. During this inactive phase of reconnection, the MHD waves created during the active phase continue to propagate through the system.

The reconnection electric field $\mathbf{E}^*$ provides us with a quantitative measure of the reconnection rate $\varepsilon(\mathbf{r}, t)$, which is given by $\varepsilon \equiv E^*/(v_{A0}B_0)$, where B_0 and v_{A0} are characteristic values of the magnetic field strength and Alfvén speed. We introduce $\mathbf{E}^*$ along a line segment of length $2l_0$, which corresponds to the reconnection line. We define this line to lie along the M–axis of a coordinate system in which $\hat{\mathbf{N}}$ is the unit vector pointing perpendicularly outward from the current sheet, and $\hat{\mathbf{M}}$ and $\hat{\mathbf{L}}$ are tangential vectors in so–called boundary normal coordinates [12]. In addition, we restrict ourselves to small values of the reconnection rate (weak reconnection), $\varepsilon \ll 1$, so that the outflow region can be treated as a thin boundary layer.

3. Model Results for Typical Magnetopause Conditions

The analytical Petschek–type reconnection model described above requires as input: (1) The plasma and magnetic field parameter values which specify the initial current sheet configuration, and (2) the reconnection electric field $\mathbf{E}^*(\mathbf{r}, t)$, which specifies the location where reconnection is initiated and its subsequent space–time variations.

For simplicity we consider in this chapter the incompressible limit, which means that the sound speed tends to infinity. In this case, the Alfvén wave and the slow shock propagate at the same speed, and merge to form a single structure, which is commonly called a Petschek shock.

To clarify the main features of our reconnection model in the context of the magnetopause, we consider an initial state consisting of a current sheet separating two plasma regions, the magnetosheath (sh) and magnetosphere (sp). The respective values for the density n, magnetic field $\mathbf{B}$, and velocity $\mathbf{v}$ are

$$n^{sh} = 10 \text{ cm}^{-3}, \ (B_L^{sh}, B_M^{sh}) = (-35, -20) \text{ nT},$$
$$(v_L^{sh}, v_M^{sh}) = (200, 100) \text{ km/s} \tag{1}$$

$$n^{sp} = 4 \text{ cm}^{-3}, \ (B_L^{sp}, B_M^{sp}) = (50, -20) \text{ nT}, \ (v_L^{sp}, v_M^{sp}) = (0, 0) \text{ km/s}. \tag{2}$$

The reconnection electric field is stipulated as follows

$$E^*(t, M) = |B_0 v_{A0}| \cos^2\left(\frac{\pi M}{2 l_0}\right) \sin^2\left(\frac{\pi t}{t_0}\right) \tag{3}$$

for $-1 \le M/l_0 \le 1, 0 \le t/t_0 \le 1$. We choose $t_0 = 1$ min and $l_0 = 2R_E$.

Figure 1 shows the propagating large–scale structures which result from reconnection, corresponding to the time $t = 2.5$ min. The original current sheet (magnetopause) is indicated by the shaded rectangle, and the upper half space corresponds to the magnetosphere. The heavy line in the center is the reconnection line. It is seen that the current sheet resolves into the outflow regions (region of accelerated plasma flow), which evolve into different directions above and below the position of the original current sheet. The uppermost and lowermost surfaces (heavy lines) are the separatrix surfaces, which envelop the reconnected tubes. The parts of the boundaries of the outflow regions indicated by the dark shaded areas consist of Petschek shocks, where plasma is energized and the magnetic field lines turn towards the original current sheet to establish a magnetic connection across the sheet. The remaining part of the outflow region boundary consists of a

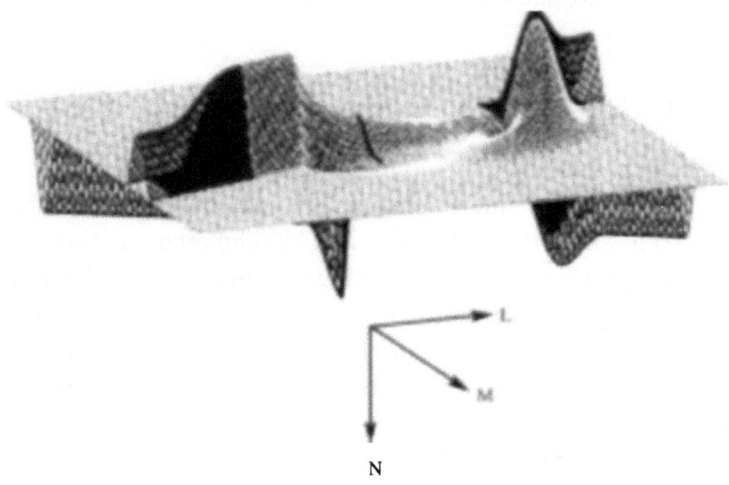

Figure 1. Reconnecting current sheet. Shown are the outflow regions, the Petschek shocks (dark), the reconnection line (heavy line in the center), and the separatrices (heavy lines).

tangential discontinuity to which the magnetic field lines run parallel. For the time shown, reconnection has just ceased, so that a new current sheet develops at the former reconnection site. The new current sheet is, however, deformed by surface waves.

Next we calculate, at special positions in space, the values of both the plasma parameters and magnetic field as a function of time. We identify these points with the position of a "test spacecraft". Since spacecraft or magnetopause boundary motions are generally at least one order of magnitude smaller than the local Alfvén speed, it seems safe to neglect them in a first approximation. All variations we observe in the model configuration are, therefore, purely the result of space–time variations in the reconnection rate, and the resulting propagation of disturbances along the current sheet.

Figures 2a and b show the plasma and magnetic field behaviour at two different points. The upper panels show the distance from the current sheet of: (1) the spacecraft, indicated by the dashed line, (2) the upper and lower surfaces of the outflow region (which bound the region of accelerated plasma flow), indicated by the dotted line, and (3) the separatrix, indicated by the long–dashed line. The next panels show two parameters: the parameter S indicates in which region the spacecraft is located. This parameter takes the value 0 or 2 if it is on the magnetospheric or magnetosheath side of the reconnected flux region, bounded by the separatrices, and 1 if it is inside the region of reconnected flux. This parameter is intended to mimic

a b

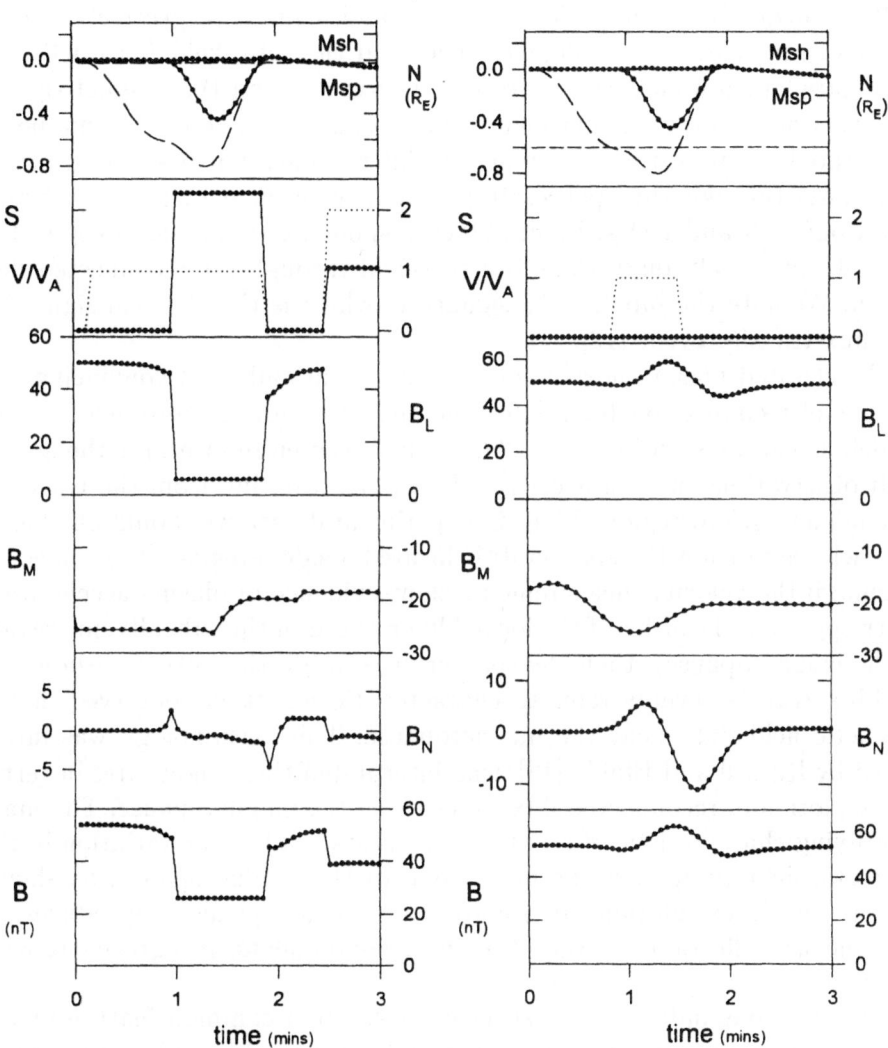

Figure 2. Calculated variation of the discontinuities, the plasma velocity and the magnetic field in time for two different points. a: For a point at distance $N = -0.02R_E$ from the magnetopause. b: For a point at $N = -0.6R_E$. The upper panels show the N coordinates of: (1) the spacecraft (dashed line). (2) the positions of the upper and lower surfaces of the outflow region (dotted line), and (3) the position of the separatrix (long–dashed line). Next panels show parameter S, which indicates in which region the spacecraft is located. This parameter takes the value 0 or 2 if it is on the magnetospheric or magnetosheath side of the reconnected flux region, and 1 indicates the region inside. In addition, in these panels, we show the plasma velocity V in multiples of $V_A = l_0/t_0 = 213$ km/s. Panels 3, 4, 5 show the L, M, N components of the magnetic field in nT, and the lowest panel shows the total magnetic field B.

the corresponding observations of energetic particle densities at the magnetopause. In addition, these panels show the plasma velocity V in multiples of $V_A = l_0/t_0 = 213$ km/s. Panels 3, 4, 5, and 6 show, respectively, the L, M, N components of the magnetic field, and the total field B (in nT).

Figure 2a shows a case, where the spacecraft is on the magnetosphere side, very near to the magnetopause. The second panel shows plasma being accelerated to twice the Alfvén speed. All reconnection–associated structures pass through the spacecraft position, as seen, e.g., in the variable behaviour of S and V/V_A. Figure 2b corresponds to a location deep in the magnetosphere, although there is a traversal through the reconnected flux region. We note the bipolar B_N signature, which is the classical signature of a flux transfer event [12].

Accelerated plasma flows are one of the main and most convincing signatures of reconnection (e.g., [10]) and they are indeed reproduced in our model, as can be seen in Figure 2a. We may therefore interpret the spacecraft observations of such accelerated flows as resulting from the traversal through an outflow region. Thus, in experimental data we should be able to infer whether or not the spacecraft is located inside a region of reconnected plasma; if the velocity measurements show evidence of plasma acceleration (with speeds of the order of the local Alfvén speed in the sub–alfvénic region of the magnetopause), then the spacecraft is inside the outflow region.

Flux transfer events refer to characteristic variations observed in the magnetic field data near the magnetopause. This terminology was introduced by Russell and Elphic [12], who interpreted these signatures in terms of the motion of reconnected flux tubes along the magnetopause. The main identifying characteristic of flux transfer events is a bipolar variation in the magnetic field component perpendicular to the magnetopause, as shown by our model calculations in Figure 2b. The model also reproduces simultaneous deflections in the other field components, in agreement with observations.

Observations indicate that surface waves are a common feature at the dayside magnetopause. Various source mechanisms have been suggested, including the Kelvin–Helmholtz instability and pressure pulses originating in the solar wind. The wave–like perturbations around the reconnection line, as shown in Figure 1 indicate that Petschek–type reconnection may be a further source of surface waves.

4. Observations on October 29, 1979, and Model Results

ISEE 2 plasma and magnetic field data during an outbound crossing of the dayside magnetopause at small northerly latitudes on October 29, 1979, between 01:15 UT and 01:45 UT, provide evidence of a reconnection layer.

Figure 3a shows the data (after Walthour et al. [13]).

Walthour et al. [13] interpret this event as a rotational discontinuity moving back and forth over the spacecraft and a slow shock at the magnetosheath side. The authors draw their results by checking the jump conditions separately at each discontinuity. The discontinuities appear, however, in reverse order, a fact, which is accounted for by plasma pressure anisotropy [13].

The data example has, however, a number of features which suggest a more detailed interpretation. We do this within the framework of the generalized Petschek–type reconnection model for a compressible plasma as discussed above. In line with Walthour et al. [13] we use an (X, Y, Z) coordinate system, where X is the sheet normal. We model the magnetopause as a tangential discontinuity.

Instead of hypothesizing a back and forth motion of a rotational discontinuity combined with steady–state reconnection (as done in [13]), it seems appropriate that two pulses of reconnection with a duration of ~ 3 minutes occur. To determine the initial conditions we have to follow Walthour et al. [13], their Table 1 and 2. We take the data for the magnetospheric side from Table 2, i.e., for $t < t_4$. Thus we estimate the density, the y and z components of the magnetic field and velocity, and plasma pressure p to be

$$n^{sp} = 3 \, \text{cm}^{-3}, \, (B_y^{sp}, B_z^{sp}) = (9.2, 36.7) \, \text{nT},$$
$$(v_y^{sp}, v_z^{sp}) = (-246, 181) \, \text{km/s}, \, p^{sp} = 0.48 \, \text{nPa}. \tag{4}$$

Similarly, we take for the magnetosheath values the boundary conditions for time t_6 from Table 1 in [13]

$$n^{sh} = 3.5 \, \text{cm}^{-3}, \, (B_y^{sh}, B_z^{sh}) = (0.25, -41.4) \, \text{nT},$$
$$(v_y^{sh}, v_z^{sh}) = (-116, 49) \, \text{km/s}. \tag{5}$$

The magnetosheath plasma pressure needs not be prescribed from data since it follows from the conservation of total pressure.

For the final input, we have to choose an appropriate spacecraft trajectory. In order to reproduce the data (see [13]) the spacecraft initially has to be in the magnetosphere, then enter the reconnection active region, before re–entering the magnetosphere. After that, the spacecraft again penetrates the reconnection layer, before it emerges into the magnetosheath.

Figure 3b shows the plasma and magnetic field behaviour, as measured along the trajectory (i.e., as a function of time). The spacecraft trajectory is shown starting from $\sim 0.09 R_E$ inside the magnetopause ~ 0.3 min before the first discontinuity is met, to $\sim 0.05 R_E$ inside the magnetosheath at a time ≈ 0.3 min after having exited the last reconnection signature. The

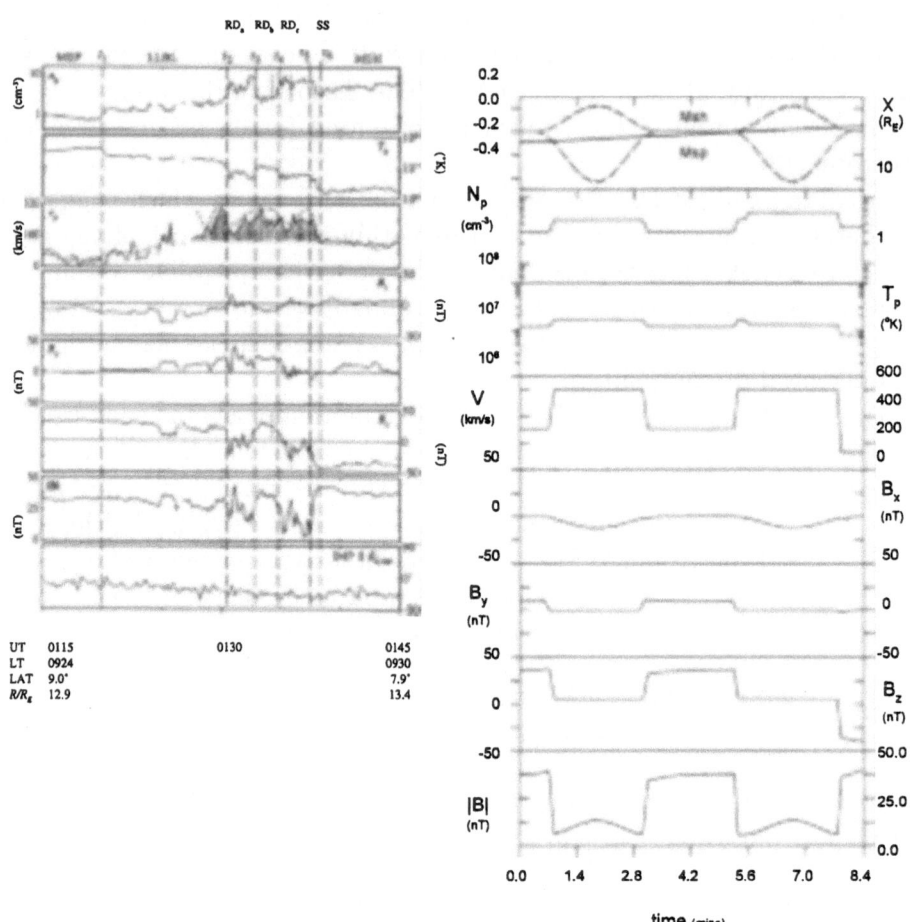

Figure 3. a: ISEE 2 observations during the magnetopause crossing on October 29, 1979. The panels are (from top to bottom): proton number density (n_p), proton temperature (T_p), magnitude of 2D (solid line) and 3D (dashed line) plasma velocity (v_p), the three components of the magnetic field (B_x, B_y, B_z), in the shock coordinate system (X, Y, Z), the total field (B), and the latitude angle of the IMP 8 magnetic field measurements (IMP 8 θ_{GMS}). (After Walthour et al. [13]) b: Calculated variation of the discontinuities, spacecraft trajectory, plasma and magnetic field quantities in time. The upper panels show the X coordinates of: (1) the spacecraft (heavy line), while it crosses the magnetopause from the magnetosphere (Msp) to the magnetosheath (Msh), and (2) the positions of the Alfvén discontinuities and the slow shocks, which virtually coincide. The next three panels show the plasma parameters density N_p in cm^{-3}, temperature T_p in kelvin (K), and the velocity V in km/s. The lower four panels show the components of the magnetic field B_x, B_y, B_z, and the modulus $|B|$ in nanoteslas.

whole time interval is ~ 8 min, in accordance with the corresponding interval shown in the paper of Walthour et al. [13]. The uppermost panel shows the X coordinates of: (1) the fictitious spacecraft (heavy straight line) while it crosses the magnetopause from the magnetosphere (Msp) to the magnetosheath (Msh) side, and (2) the positions of the upper and lower Alfvén discontinuities and slow shocks at virtually the same position, bounding the outflow region (accelerated plasma). The next three panels show the density N_p in cm^{-3}, temperature T_p in K, and the velocity V in km/s. Panels 5–8 show the X, Y, Z components of the magnetic field and the total magnetic field $|B|$ in nT.

Now we compare Figure 3b with Figure 3a. The spacecraft trajectory chosen shows a spacecraft which starts in the magnetosphere. The spacecraft then enters into the reconnection layer, crossing a discontinuity which is a combination of an Alfvén discontinuity and a slow shock. After about 2 min, the spacecraft enters the magnetosphere again, but shortly after that re–enters the reconnection region. Here it crosses the contact discontinuity (indicated by the increase of the density at time ≈ 5.6 min) and finally enters the magnetosheath region at the end of the interval shown.

One can see that the behaviour of the density N_p is very well modelled, with a larger increase during the second crossing of the reconnection region. Despite of the fact that the magnetosphere and magnetosheath densities are practically equal, our model predicts an increase in density inside the reconnection layer, in agreement with the observations. The increase in temperature as a result of plasma heating at the shocks, obtained in the frame of our model, is not seen in the data. The next panel (V) shows the acceleration of the plasma due to reconnection. Because of the two penetrations into the reconnection region we observe two regions of accelerated plasma. The data also show two bursts of acceleration, though not as strong as in the model. This discrepancy may be due to kinetic effects, which are, as yet, not included in the model. The B_x behaviour of the magnetic field is of minor importance because it is small compared to the total field, in both the model and in the data. While the model underestimates B_y, the behaviour of B_z reproduces the measured data very closely. Whenever the reconnection region is entered this is marked by a clear decrease, as in the data, with the final decrease to negative values upon entering the magnetosheath. The modelled field strength agrees with the measurements well: Every time the reconnection region is entered, the magnetic field clearly decreases both in the model and the data. Even the "humps" in the magnetic field strength in the middle of the reconnection region, as reproduced by the model at times ≈ 2 minutes and ≈ 6.5 minutes, might be identified in the data measured by ISEE 2.

In conclusion, we can state that a fair amount of progress has already been achieved in bringing together theory and observations of reconnection at the dayside magnetopause. This analysis suggests in particular that the time–varying generalized Petschek model is useful in interpreting magnetopause data to a considerably degree of accuracy.

5. Acknowledgements

We thank R.P. Rijnbeek, G.R. Lawrence, M.T. Kiendl, and D.F. Vogl for useful discussions. This work is partially supported by the Austrian "Fonds zur Förderung der Wissenschaftlichen Forschung", Project P09431-TEC, by the Russian Foundation of Basic Research, grant 97-05-64065 and by the program "Intergeophysics", by NASA grant NAG5-2834, and by grant 95-05-14192 from the Russian Foundation of Basic Research.

6. References

1. Vasyliunas, V.M. (1975) Theoretical models of magnetic field line merging, 1, *Rev. Geophys.*, **13**, 303–336.
2. Pudovkin, M.I. and Semenov, V.S. (1985) Magnetic field reconnection theory and the solar wind–magnetosphere interaction: A review, *Space Sci. Rev.*, **41**, 1–89.
3. Priest, E.R. (1985) The magnetohydrodynamics of current sheets, *Rep. Prog. Phys.*, **48**, 955–1090.
4. Hones, E.W., Jr. (1984) *Magnetic Reconnection in Space and Laboratory Plasmas*, Geophysical Monograph 30, AGU, Washington DC.
5. Petschek, H.E. (1964) Magnetic field annihilation, *NASA Spec. Publ.*, **SP–50**, 425–439.
6. Biernat, H.K. (1993) Reconnection at the dayside magnetopause: Theory and comparison with data, *Trends in Geophys. Res.*, **2**, 535–651.
7. Semenov, V.S., Lebedeva, V.V., Biernat, H.K., Heyn, M.F., Rijnbeek, R.P., and Farrugia, C.J. (1995) Time–varying reconnection: Implications for magnetopause observations, *J. Geophys. Res.*, **100**, 21,779–21,789.
8. Scholer, M. (1989) Asymmetric time–dependent and stationary magnetic reconnection at the dayside magnetopause, *J. Geophys. Res.*, **94**, 15,099–15,111.
9. Otto, A. (1990) 3–D resistive MHD computations of magnetospheric physics, *Comput. Phys. Commun.*, **59**, 1985.
10. Sonnerup, B.U.Ö., Paschmann, G., Papamastorakis, I., Sckopke, N., Haerendel, G., Bame, S.J., Asbridge, J.R., Gosling, J.T., and Russell, C.T. (1981) Evidence for magnetic field reconnection at the Earth's magnetopause, *J. Geophys. Res*, **86**, 10,049–10,067.
11. Semenov, V.S. (1997) Acceleration processes in the course of dayside solar wind–magnetopause interaction, *Adv. Space Res.*, in press.
12. Russell, C.T. and Elphic, R.C. (1978) Initial ISEE magnetometer results: Magnetopause observations, *Space Sci. Rev.*, **22**, 681–715.
13. Walthour, D.W., Gosling, J.T., Sonnerup, B.U.Ö., and Russell, C.T. (1994) Observation of anomalous slow–mode shock and reconnection layer in the dayside magnetopause, *J. Geophys. Res.*, **99**, 23,705–23,722.

AN EMPIRICAL MODEL OF THE MAGNETOPAUSE FOR BROAD RANGES OF SOLAR WIND PRESSURE AND Bz IMF

S. N. KUZNETSOV AND A.V. SUVOROVA
Skobeltsyn Institute of Nuclear Physics, Moscow State University,
Moscow, Russia,119899

Abstract. A dayside magnetopause model is presented as an alternative to the well-known Roelof and Sibeck model. Model parameters are the same: solar wind dynamic pressure and B_z-component of the interplanetary magnetic field. The main differences and features of the model are the larger statistics and wider effective range of model parameters and the different method of the data treatment. The first is due to geosynchronous satellite magnetopause crossings added to the basic data on high-apogee crossing. The second is based on a physical approach to the analysis of the data. An asymmetric dayside magnetopause shape during disturbed solar wind condition and the pressure balance change manner during strong negative B_z were derived from geosynchronous data analysis.

1. Introduction

We present a magnetopause model which allows estimation of the size and shape of the dayside magnetosphere and near-Earth depending on the solar wind conditions.

It has been established, that the main solar wind parameters controlling the magnetopause position are dynamic pressure of solar wind (SW) and negative B_z component of the interplanetary magnetic field (IMF). The MHD-theory predicts that the dayside magnetopause position can be described as a function of the SW dynamic pressure [1].

$$k \cdot p \cdot \cos^2 \alpha = \frac{\left(2 f B_0\right)^2}{2 \mu_0 R_{mp}^6} \tag{1},$$

where p is the solar wind dynamic pressure, which is $m_p n$ v^2 (n and v are the solar wind plasma density and velocity, m_p is the proton mass), with $p \cdot \cos^2 \alpha$ corresponding to the local dynamic pressure exerted on the magnetopause (MP); α is the angle between an outward normal to the MP and the Earth-Sun line; μ_0 is permittivity of vacuum; $B_0 =$ 3.03·10^4 nT (for the epoch 1990) is the equatorial Earth surface magnetic field strength; R_{mp} is the distance (in R_E) to a certain point of the MP. The coefficient f accounts for the contribution of the large-scale magnetospheric current system to the magnetic field on

J. Moen et al. (eds.), Polar Cap Boundary Phenomena, 51–61.

the magnetopause. The coefficient k is equal to the ratio of the magnetosheath plasma pressure to the SW dynamic pressure in the interplanetary space [1]. The coefficient k must always be <1 [1], while the coefficient f must depend on the position of the crossing point on the MP and can be >1 [2].

Many statistical studies of the magnetopause size and shape and their dependence on the SW parameters were made based on MP crossing data from single or several satellites [3-14]. In earlier work the parameters of the average shape of the MP surface, fitted by an ellipsoid or paraboloid of revolution about Earth-Sun line, were obtained for average pressure p about 1.5 nPa and orientations of the IMF when B_z component is positive and negative [3-5].

Recently Roelof and Sibeck [7] have created the empirical magnetopause model (R-S model) based on the largest statistics of MP crossings (1821) by high-apogee satellites during 1967-1986. As any model it has merits and problems. This model is the first attempt to describe the magnetopause shape as bivariate function of p and B_z. The effective range of the R-S model is limited because it does not cover extreme values of either the pressure and B_z component (the input data being limited to $p<8$ nPa; $|B_z|<7$ nT). The main shortcoming of the R-S model is an artificial MP dependence on the positive B_z which contradicts observations. The reason of this invalid result is the use of an unsuitable approximation function.

Alternative studies of the same data set [11,12] and of a smaller dataset [8-10] have showed the different MP dependence on p and B_z. We combine the results of Kuznetsov and Suvorova [12] with the results of an analysis of MP crossings by geosynchronous satellites [13,14] and develop the empirical dayside magnetopause model [15].

2. The Data Set

The empirical model of the magnetopause as a function of solar wind parameters was inferred from the measurements on 12 high-apogee satellites and 5 geosynchronous satellites made during 1967-1993 .

Dr. D.Sibeck kindly supplied us with the data on MP crossings by the high-apogee orbit satellites. Some of the data on the crossings near the geosynchronous orbit (113 occurrences) were borrowed from [16-19]. 59 geosynchronous magnetopause crossings were found from the GOES-7 magnetic measurements. The mean-hourly SW parameters and the GOES-7 magnetic data were obtained from the databases of the United States NSSDC and NGDC Scientific Centres. The dynamic SW pressure p and the IMF B_z component were taken to be the SW parameters that are responsible for magnetopause position. The MP crossings at high-apogee orbits were observed at geocentric distances of >7 R_E, with the mean-hour solar wind dynamic pressure not exceeding 10 nPa, and mostly, the IMF B_z exceeding -7 nT. The crossings at geosynchronous orbit (6.6 R_E) were observed either under much higher pressures (>15 nPa) or during intensive negative IMF B_z (<-6 nT). The data array used includes 886 magnetopause crossings when hourly averages of pressure and B_z varied from 0.5 nPa to 50 nPa and from -28 to +20 nT , respectively. The model is restricted to the domain of distances from -10 R_E to 12 R_E along x-axis. Within this domain, the model is of the highest statistical strength supported by the solar wind pressures from 0.5 nPa to 25 nPa and by the IMF B_z values above -15 nT .

3. Discussion of the Results

The results of the analysis performed in [10-15] can be summarized as following. We used a rotating paraboloid for approximation of the magnetopause shape

$$x = X_0 - g\rho^2 \qquad (2),$$

where $\rho^2 = y^2 + z^2$ and x,y,z are the solar-ecliptic coordinates of the MP position, X_0 is a standoff distance of the MP. The x axis directs from the center of geomagnetic dipole to the sun, the z axis is perpendicular to the ecliptic equator and the y axis a right-hand coordinate system.

If we introduce angles λ and φ (λ is a latitude of a given point of the MP, and φ is the angle between the direction to the sun and the direction to the projection of the given point of the MP to the xy plane) than $\cos\theta=\cos\lambda\cos\varphi$ (θ is an angle between directions to the Sun and to the point of magnetopause from dipole's center). For the nose part of the parabola, for $\theta<\sim40°$ parameters X_{01} (the MP standoff distance) and g_1 differ from parameters X_{02} and g_2 for the lateral part of parabola [10,11,13,15].

The most important part of the analysis, including analysis of closest positions of the MP to the Earth, was made for the equatorial plane. Therefore we used the 2D (x,y or R, φ) model, but this result can be easily extended to the 3D model by rotating the parabola to generate a paraboloid.

While changing of the B_z-component from north- to south-directed (i.e. $B_z>0$ is changed to $B_z<0$) the magnetopause occurs at the same distance under considerably lower pressure [16,11-13].

When the magnetopause is observed at $R=6.6$ R_E near the equatorial plane, a sharp dependence of the probability of magnetopause registration on φ is found [14,15]. This indicates the fact that in the equatorial plane, the best approximation is a shifted paraboloid toward the dusk sector.

$$x = X_0 - g(\rho - \rho_0)^2 \qquad (2')$$

Similar asymmetry in frequency occurrence of the magnetopause crossings was noted by Rufenach et al. [16].

In previous work [11, 12] a relationship between R (distance up to magnetopause) and p (solar wind pressure) was studied (of form $p\sim R^{-n}$) for different θ in the narrow range of $\Delta\theta\sim10°$. It was found that the coefficient of correlation (r) between R and p is a periodic function of θ. For r of about 1, the index $n\sim5$-6, at $r\sim0$ the index n increases.

In previous work [15], based on empirical analysis for X_{01} an expression was suggested, which does not depend on B_z for $B_z>0$ and depends on B_z for $B_z<0$. These two expressions can be integrated into one:

$$X_{01} = 12.1 \cdot p^{-0.19} - 0.058 \cdot \left(\frac{|Bz| - Bz}{2} \right)^{1.93} \cdot p^{-0.343} \qquad (3)$$

Obviously, this expression can be used if the second term on the right is significantly less than the first one. Because inward magnetopause motion is caused by squeezing of magnetosphere with enhanced p and with magnetic field's erosion caused by reconnection when $B_z < 0$ [14,16], expression (3) can be considered expanded as following:

$$X_{01} = 8.6 \cdot \left(1 + 0.407 \cdot \exp \frac{-(|Bz| - Bz)^2}{200 \, p^{0.15}} \right) \cdot p^{-0.19} \qquad (3')$$

In Figure 1 the contours of constant X_{01} are presented in as a function of p and B_z. On Figure 1 results from [7, 15] are also presented.

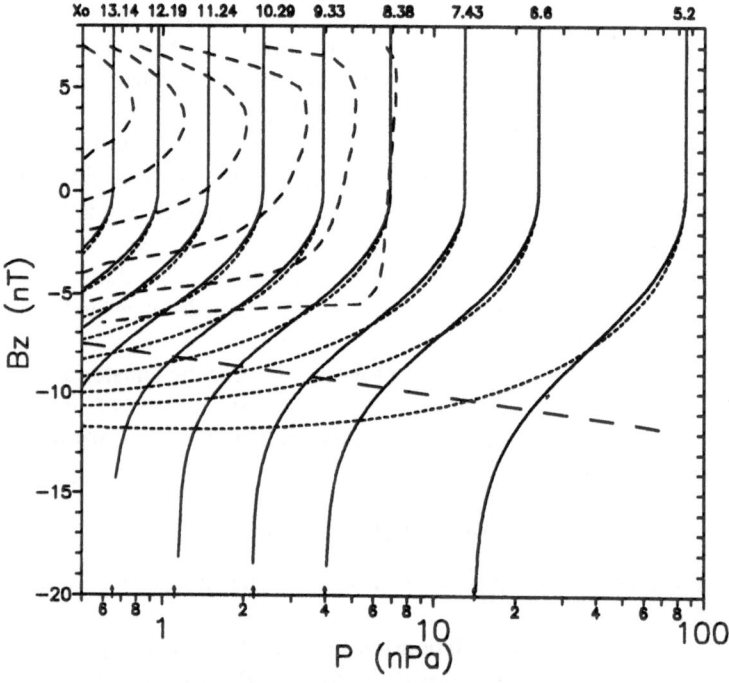

Figure 1. Fitted contours of X_{01} in (p,B_z) coordinates: the solid curve is the approximation suggested in the current paper; the dotted curve - approximation according to [15], the dashed curve [7], the dashed line - limitation of the model [15].

Similar dependence of X_{01} from B_z can also be explained as a result of surface currents' changings at the dayside. These currents' effect is described by f parameter. We can suppose, that

$$f = \left(a + b \cdot \exp \frac{-\left(|Bz| - Bz\right)^2}{F(p)} \right)$$

In this case with account for experiment's accuracy we derive (3') for describing the standoff distance X_{01}.

Let's analyze how a parabolic coefficient g in (2) influences the magnetopause shape. Obviously, if a dependence of R on pressure is the same as the dependence of the standoff distance X_0 on pressure, then $g = C/X_0^{-1}$ (C is constant). In the equation of pressure balance (1), in addition to known quantities (B_0) and measured quantities p, R and α theoretical quantities such as f^2 and k are present. They can change along magnetopause and can be functions of p and B_z.

Let's analyze equation (1) for a rotating paraboloid (2). Considering $\alpha = 90° - \psi$, where ψ can be determined from expression $\tan\psi = \partial\rho/\partial x$, we can obtain that $\cos^2\alpha = 1/(1 + 4g^2\rho^2)$. Further it is easy to obtain, that

$$\frac{\left(f^2/k\right)\big|_\varphi}{\left(f_0^2/k_0\right)\big|_\varphi} = \frac{R^6}{X_0^6 \cdot (1 + 4g^2\rho^2)} \tag{4}$$

We can try to determine conditions under which the value of $\left(f^2/k\right)\big|_\varphi$ does not change near $\varphi = 0°$. For this we need to set the derivative of expression (4) equal to zero.

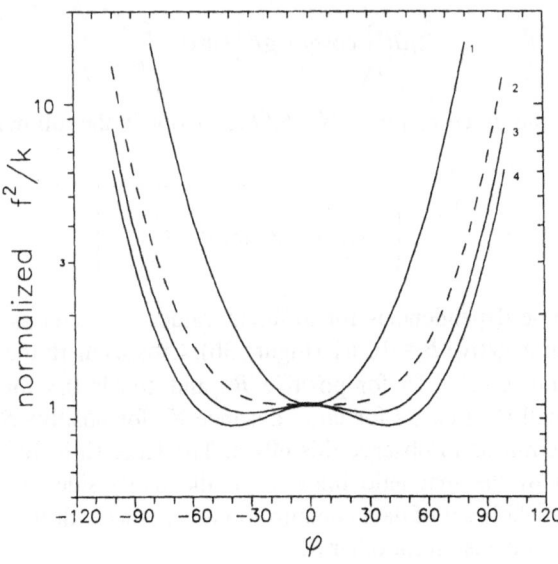

Figure 2. The dependence normalized f^2/k on φ for different values ($g*X_0$): 0.242 for curve labeled as 1; 0.396 for curve labeled as 2; 0.44 for curve labeled as 3; 0.48 for curve labeled as 4.

As a result we obtain:

$$g = \frac{\sqrt{21} - 3}{4X_0} \approx \frac{0.3956}{X_0}$$

In [15] the magnetopause was approximated by two parabolas. If we use only one parabola, we will decrease the position of magnetopause in φ-range $25°-45°$ systematically by 10%. In this case we have $C = 0.48 - 0.0018 \cdot (|B_z| - B_z)$. The dependence f^2/k on φ is shown on Figure 2 for different value of C. The value f^2/k increases when φ increases from $\varphi=0°$ for $C<0.3956$ and f^2/k decreases when φ increases (up to $\sim30°-50°$) for $C>0.3956$. In the case of two parabolas we have obtained $C<0.3956$ in the nose part and $C>0.3956$ in the flank part for any B_z. Therefore a jump down of the value f^2/k exists at the point of sewing parabolas (for example, transition from curve 1 onto curve 4 on Figure 2 at $\varphi\sim30°$ for $p\sim1.5$ nPa and $B_z>0$). This means, that surface currents on nose part differ from flank currents. Note that using single parabola we average the currents in the φ-range $0°-60°$.

Analysis of the magnetopause crossings at $R=6.6R_E$ [14] shows, that the probability of the magnetopause observation at $R=6.6R_E$ on the dawn side is more than on the dusk side. We describe this effect by shifting the parabola to the dusk side. We can calculate the relation of p and φ for $R=6.6R_E$. For the equatorial plane we can write

$$R\cos\varphi = X_0 - gR^2\left(\sin\varphi - \frac{\rho_0}{R}\right)^2$$

and compute the pressure corresponded to X_0.

$$p = \frac{(2B_0)^2}{2\mu R^6\left(\cos\varphi + gR\left\{\sin\varphi - \frac{\rho_0}{R}\right\}^2\right)^6}$$

If we propose, that p_0 corresponds $X_0=6.6R_0$, we obtain the following expression:

$$\frac{p}{p_0} = \frac{1}{\left(\cos\varphi + gR\left\{\sin\varphi - \frac{\rho_0}{R}\right\}^2\right)^6}$$

We present these dependencies for different values of C and $\rho_0/6.6$ for positive B_z (Figure 3a) and for negative $B_z=-10$ nT (Figure 3b). Crosses mark the experimental data. A value $\rho_0/6.6$ equals $0.1-0.15$ for positive B_z, and equals 0.3 for negative B_z. It is possible that a small shift exists for large distance X_0 for positive B_z but experimental errors does not permit us to observe this effect. The large shift for $B_z=-10$ nT testifies that reconnection of the magnetic lines is on the dawn side of the magnetopause. Unfortunately, the whole set of data contains too few points with $B_z \leq -10$ nT and small p, so as to verify this conclusion for other R.

Let us discuss the consequences of the dependence of the correlation coefficient r between R and p on φ. Figure 4 shows the dependence of r on φ for $B_z>0$ (upper panel); $0<B_z<-2$ nT. (middle panel) and $B_z<-2$ nT. We considered that this dependence on R

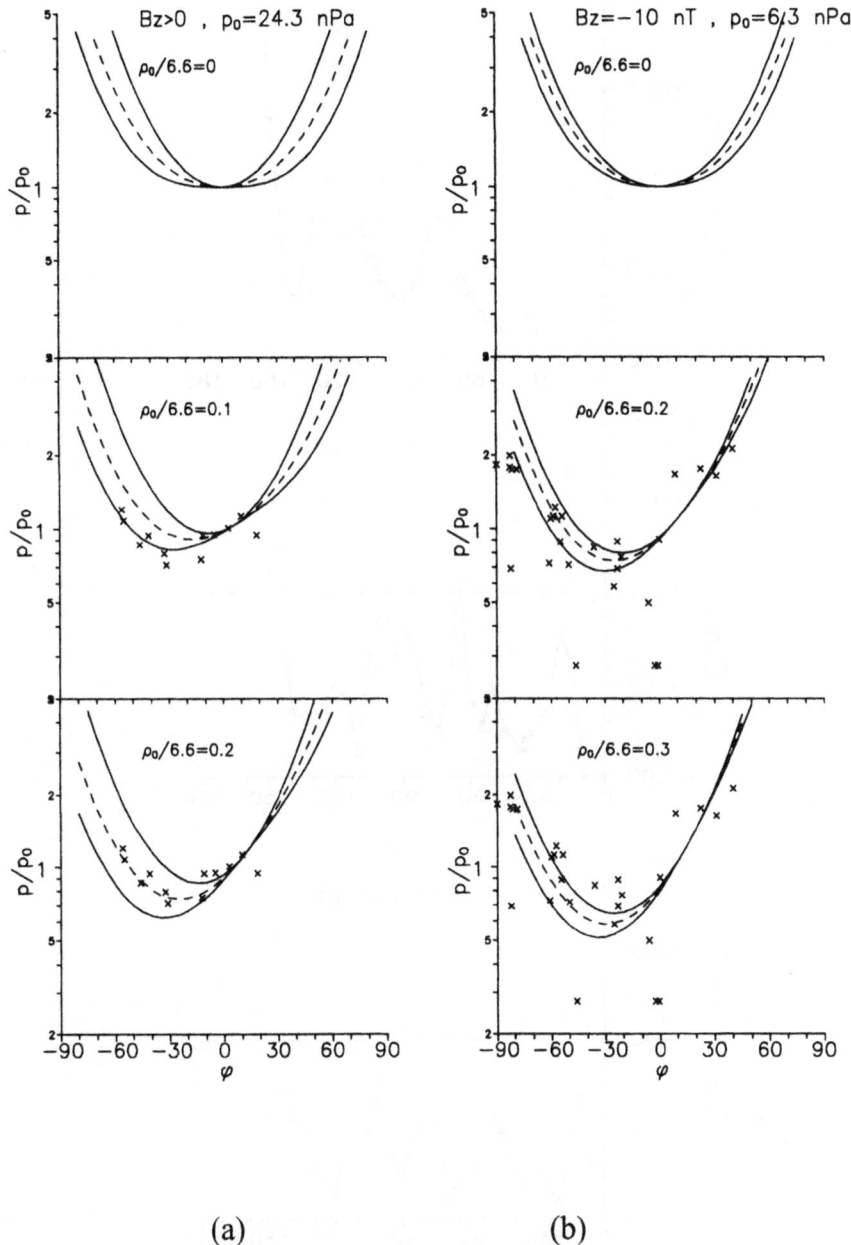

(a)　　　　　　　　(b)

Figure 3. The dependence p/p_0 on φ for different values of g^*X_0 and comparison with experimental data (crosses): (a) for $B_z>0$ and $g^*X_0=0.2885$; 0.3956; 0.48. (b) for $B_z=-10$ nT and $g^*X_0=0.3585$; 0.3956; 0.444.

58

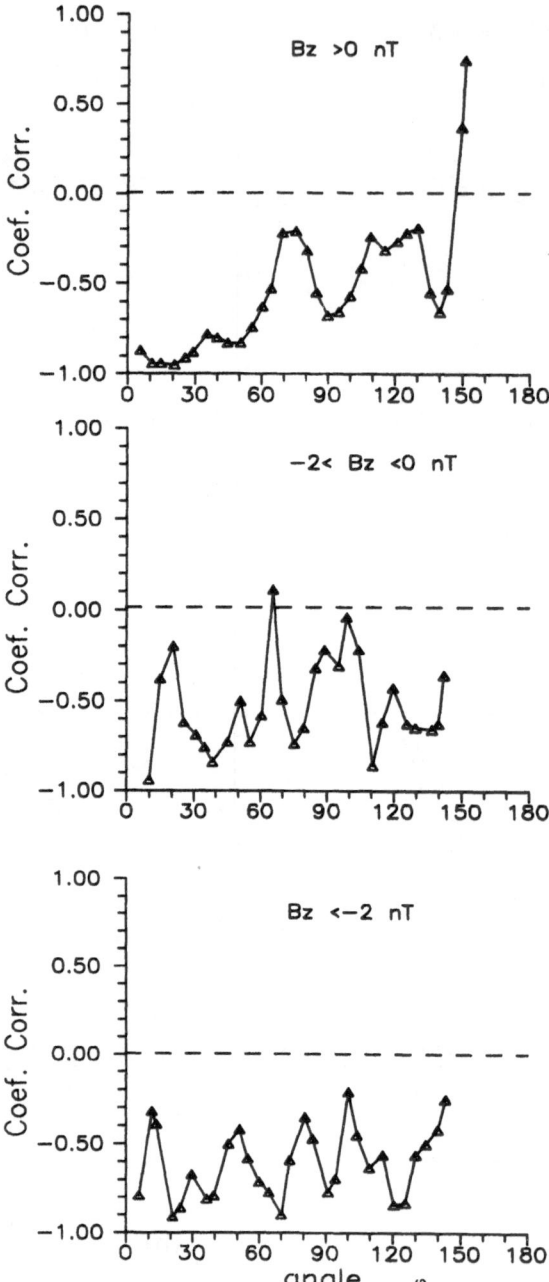

Figure 4. Dependence of the correlation coefficient *r* between *R* and *p* on φ for $B_z>0$, $0>B_z>-2$ and $B_z<-2$nT

and φ may be due to a consequence of the occurrence of proper oscillations of the magnetosphere and the existence of standing waves. The regions with large r would then be the regions of small amplitudes of standing waves and regions with smaller r correspond to region of large amplitude standing waves. In this particular case a membrane, which is fastened through the center may serve as a analogue of the oscillating magnetosphere.

4. Conclusions

Based on an analysis of experimental measurements of the magnetopause position on the dayside at distances from 5.2 R_E to 14 R_E under varying conditions of the solar wind, the following analytical semi-empirical approximation by a paraboloid of revolution with shift $\rho_0=y_0$ towards the dusk side is suggested:

$$x = X_0 - g(\rho - \rho_0)^2$$

$$X_{01} = 8.6 \cdot \left(1 + 0.407 \cdot \exp\frac{-\left(|Bz| - Bz\right)^2}{200\,p^{0.15}}\right) \cdot p^{-0.19}$$

$$g = \left(0.48 - 0.0018\{|Bz| - Bz\}\right)/X_0$$

$$\rho_0 \leq 0.66 \div 1 R_E \qquad \text{for} \quad X_0{\sim}6.6\,R_E, \quad B_z{>}0$$

$$\rho_0 \approx 2 R_E \qquad \text{for} \quad X_0{\sim}6.6\,R_E, \quad B_z{<}\text{-6 nT}$$

Comparison of this approximation with experimental data is shown in Figure 5. The change in the sign of the B_z component of the interplanetary magnetic field leads to changes in the mode of solar wind flowing around the magnetopause and to changes in the structure of surface currents.

Possibly, the magnetosphere is quite frequently in an oscillating mode, and in a certain dispersion in the experimental data is associated with this.

5. Acknowledgments

We gratefully acknowledge J.King and others at Goddard Space Flight Center for maintaining the OMNI online data base of solar wind parameters and D.Sibeck for giving us the high-apogee magnetopause data. This work was supported by the Russian Fund of Fundamental research under grant 96-05-65075.

60

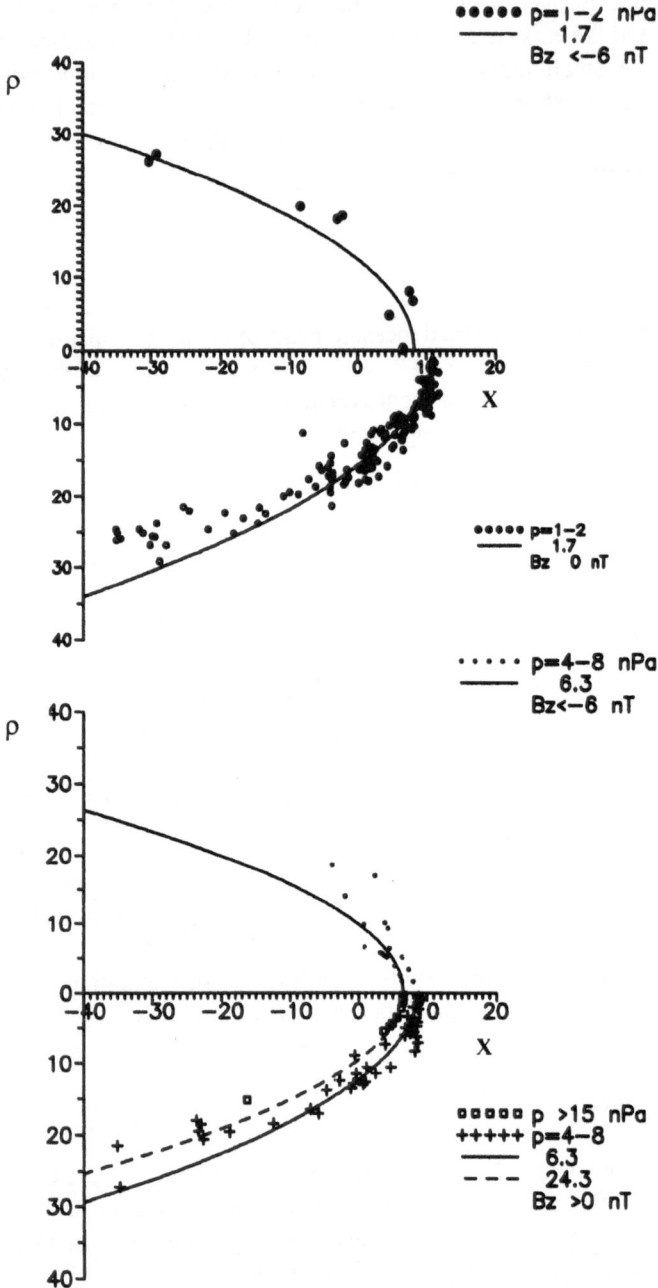

Figure 5. Comparison of the shape of magnetopause according to the model for different *p* with experimental data points.

6. References

1. Spreiter, J.R., Summers, A.L., and Alksne, A.Y. (1966) Hydromagnetic flow around the magnetosphere, *Planet.Space Sci.* **14**, 223-253.
2. Schield, M.A. (1969) Pressure balance between solar wind and magnetosphere, *J.Geophys. Res.* **74**, 1275-1286.
3. Fairfield, D.H. (1971) Average and unusual locations of the Earth's magnetopause and bow shock, *J. Geophys. Res.* **76**, 6700-6716.
4. Holzer, R.E, and Slavin, J.A. (1978) Magnetic flux transfer associated with expansions and contractions of the dayside magnetosphere, *J. Geophys.Res.* **83**, 3831-3839.
5. Formisano, V., Domingo, V., and Wenzel, K.-P. (1979) The three-dimensional shape of the magnetopause, *Planet.Space Sci.* **27**, 1137-1149.
6. Sibeck, D.G., Lopez, R.E., and Roelof, E.C. (1991) Solar wind control of the magnetopause shape, location, and motion, *J. Geophys. Res.* **96**, A4, 5489-5495.
7. Roelof, E.C., and Sibeck, D.G. (1993) The magnetopause shape as a bivariate function of IMF Bz and solar wind dynamic pressure, *J. Geophys. Res.* **98**, A12, 21421-21450.
8. Petrinec, S.P., Song, P., and Russell, C.T. (1991) Solar cycle variations in the size and shape of the magnetopause, *J. Geophys. Res.* **96**, A5, 7893-7896.
9. Petrinec, S.P., and Russell, C.T. (1993) External and internal influences on the size of the dayside terrestrial magnetosphere, *Geophys.Res.Lett.* **20**, 339-342.
10. Kuznetsov, S.N., Zastenker, G.N., and Suvorova, A.V. (1992) Correlation between interplanetary conditions and the dayside magnetopause, *Cosm.Res. (USA)* **30**, N6, 466-471.
11. Kuznetsov, S.N., Suvorova, A.V., Zastenker, G.N., and Sibeck, D.G. (1994) Solar wind control of the geomagnetopause position, *Proceeding of 1992 STEP Symposium, Cospar Colloquium Series* **5**, pp.293-295.
12. Kuznetsov, S.N., and Suvorova, A.V. (1996) Solar wind control of the magnetopause shape and location, *Radiation measurement, special issue "Space Radiation Environment: empirical and physical models, Dubna,1993"* **26**, N3, 413-415.
13. Kuznetsov, S.N., and Suvorova, A.V. (1994) Influence of solar wind to some magnetospheric characteristics, *Proceeding of 3rd Conference of Doctoral Students, WDS'94. Section "Physics of plasmas and ioni-ed media"*, Charles University, Prague 19 - 23 Sept.1994, pp.116-123.
14. Kuznetsov, S.N., and Suvorova, A.V. (in press) Solar wind magnetic field and plasma during magnetopause crossings at geosynchronous orbit, *Proceeding of 31st Cospar Scientific Assembly,14-21 July 1996, (Adv.Res).*
15. Kuznetsov, S.N., and Suvorova, A.V. (1996) Empirical model of the dayside magnetopause, *INP MSU Preprint No96-37/444*, Moscow.
16. Rufenach, C.L., Martin, R.F.,Jr., and Sauer, H.H. (1989) A study of geosynchronous magnetopause crossings, *J. Geophys. Res.* **94**, A11, 15125-15134.
17. McComas, D.J., Elphic, R.C., Moldwin, M.B., and Thomsen, M.F. (1994) Plasma observations of magnetopause crossings at geosynchronous orbit, *J. Geophys. Res.* **99**, 21249-21255.
18. Russell, C.T. (1976) On the occurrence of magnetopause crossings at 6.6 Re, *G. Res.Lett.* **3**, N10, 593-596.
19. Hoffman, K.A., Cahill, C.J., Anderson, R.R.Jr., et.al. (1975) Explorer 45(S3A) observations of the magnetosphere and magnetopause during August 4-6 1972 magnetic storm period, *J. Geophys.Res.* **80**, 4287-4296.

SOLAR WIND He^{2+} AND H$^+$ DISTRIBUTIONS IN THE CUSP FOR SOUTHWARD IMF

S. A. FUSELIER, E. G. SHELLEY, W. K. PETERSON,
and O. W. LENNARTSSON
Lockheed Martin Advanced Technology Center
Dept. H1-11 Bldg. 252; 3251 Hanover St.
Palo Alto, CA 94304 USA

Abstract. For southward Interplanetary Magnetic Field (IMF), the He^{2+}/H$^+$ density ratio in the cusp can be several times higher than the ratio in the solar wind. Proceeding poleward from the equatorial edge of the cusp, the ratio first increases to well above the solar wind value, then decreases to values at or below that in the solar wind. Superposed on this overall change are more rapid (2-3 minute) variations in the density ratio. A model for the magnetosheath ion distributions is used to explain both the overall change and the variations in the cusp density ratio.

1. Introduction

Solar wind ions (e.g., H$^+$, He^{2+}, O^{6+}) in the Earth's cusps are evidence for plasma transfer from the magnetosheath across the magnetopause and into the magnetosphere. For southward IMF, there is often a distinct energy-latitude signature for these cusp ions. Highest energy solar wind ions propagating earthward define the equatorial edge of the cusp. Proceeding poleward from this edge, successively lower energy ions are observed propagating earthward. At high enough altitudes and latitudes, a population propagating away from the Earth can also be seen. These are ions that have mirrored at lower altitudes and are propagating along the magnetic field toward the magnetotail. When solar wind ions are seen flowing predominately tailward along with ionospheric ions from the dayside ionosphere, this region is usually called the mantle. The energy-latitude signature in the cusp and mantle has been successfully reproduced by models which incorporate near-subsolar magnetic reconnection and velocity dispersion (or velocity filtering) of ions precipitating in the cusp.

Consistent with previous spacecraft observations at low and mid-altitudes, the energy-latitude signature is often observed in the ion data from the POLAR spacecraft. This spacecraft is in a 2 x 9 R$_E$ orbit with apogee over the north geographic pole. A typical nearly noon-midnight orbit for this spacecraft is shown in Figure 1 (for 24 April 1996). At about 1648 UT and an altitude of about 3.7 R$_E$, the spacecraft encountered the equatorial edge of the cusp. It remained in the cusp and mantle for about 45 minutes.

Figure 2 shows observations from the Toroidal Imaging Mass Angle Spectrograph (TIMAS) Shelley *et al.* [1] on the POLAR spacecraft for the time period from 1645 to 1710 UT on 24 April 1996. The top two panels show omni-directional fluxes of H$^+$ and

J. Moen et al. (eds.), Polar Cap Boundary Phenomena, 63–72.

He^{2+}, respectively from 0.016 to 27 keV/e. These fluxes were obtained by averaging 4 fully 3-dimensional phase space distributions (1 distribution is obtained every 6 s spin). The third panel shows the He^{2+}/H$^+$ density ratio from the cusp (solid line) and simultaneously from the solar wind. The solar wind density ratio was obtained from the Solar Wind Experiment (SWE) on the WIND spacecraft Ogilvie *et al.* [2]. These data were lagged by 21 minutes to allow for the plasma travel time from the WIND spacecraft (70 R$_E$ upstream from the Earth) to the cusp. Finally, the bottom panel shows the maximum flux for H$^+$ and He^{2+} obtained from the top two panels, respectively. For the maximum H$^+$ flux, the higher energy branch of the flux distribution in the top panel was used. The lower energy branch is from the ionosphere because it lacks a substantial He^{2+} component (compare the fluxes of H$^+$ and He^{2+} in the top two panels, respectively) but has a similar O$^+$ component (not shown).

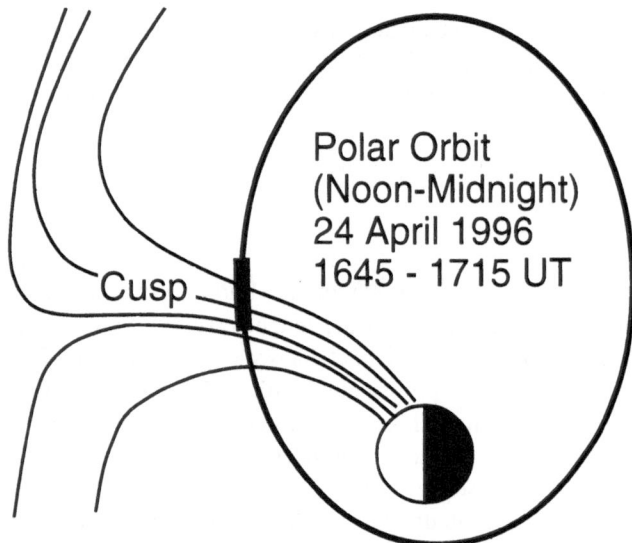

Figure 1. POLAR noon-midnight orbit for 24 April 1996. The equatorial edge of the cusp was encountered at about 1648 UT and the spacecraft moved poleward through the cusp/mantle until about 1715 UT.

The He^{2+}/H$^+$ density ratio shows two features which are the focus of this paper. First, the overall density ratio rises rapidly from a value below that in the solar wind at the equatorial edge of the cusp (1648 UT). From 1648 to 1655 UT, the ratio remains above that in the solar wind and after 1655 UT, it dips slightly below that in the solar wind. Second, relatively large fluctuations with a period of 2-3 min are superposed on this overall change (for example the maxima and minima between 1650 and 1656 UT).

Comparing the changes in the density ratio and the maximum flux in Figure 2, it is apparent that the large changes in the density ratio are the result of relatively large changes in the He^{2+} flux and comparatively smaller changes in the H$^+$ flux. Both maximum fluxes increase and then decrease over the period from 1648 to 1700 UT, but the He^{2+} flux peaks earlier than that for H$^+$. A similar trend in the He^{2+} and H$^+$ flux as a function of position in the cusp was observed previously Shelley *et al.* [3]. In Section

2, we develop a model for the magnetosheath ions that precipitate in the cusp. The overall changes in the He^{2+}/H^+ density ratio in the cusp are readily explained using these magnetosheath ion distributions. More rapid fluctuations in the density ratio require a time dependent precipitation mechanism, as discussed in Section 3.

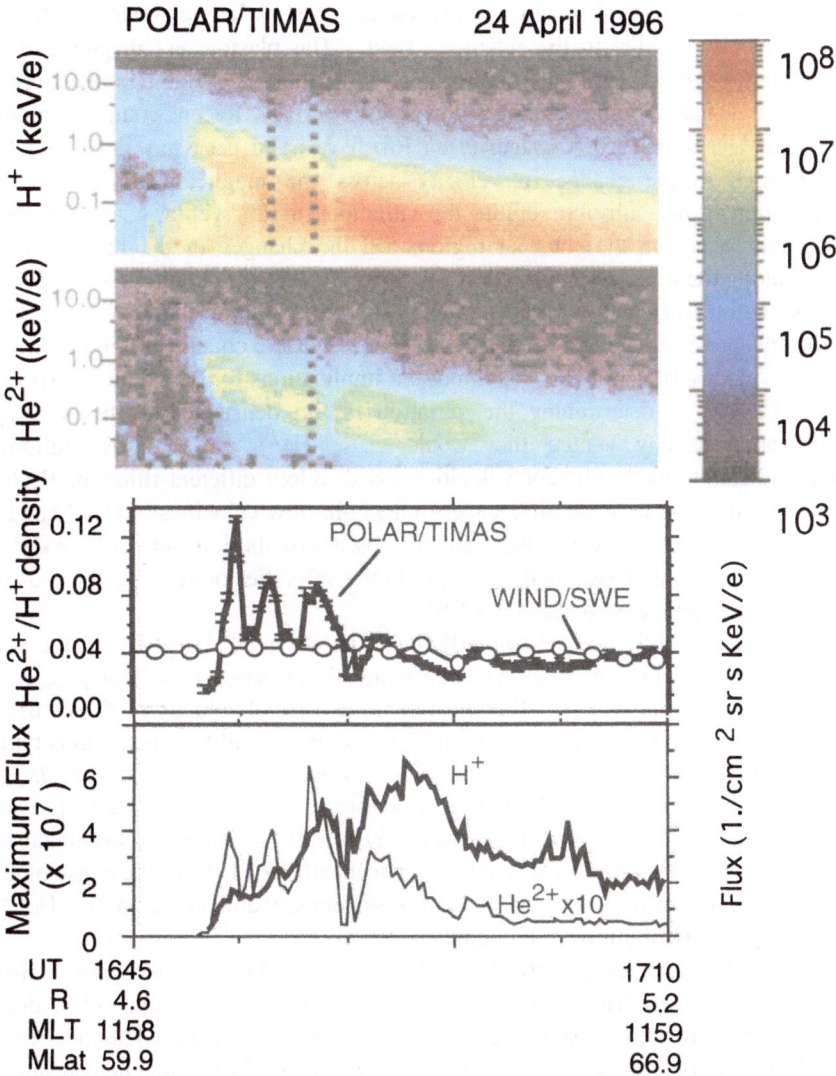

Figure 2. Omni-directional flux of H^+ and He^{2+}, the He^{2+}/H^+ density ratio and the maximum flux for He^{2+} and H^+. The cusp edge is at 1648 UT. He^{2+} shows a distinct energy-time (latitude) dispersion. H^+ also displays this dispersion with an additional component below 100 eV from the ionosphere. The He^{2+}/H^+ density ratio increases to values above the solar wind ration and later decreases to values at or below the solar wind ratio. This increase is attributed to a peak in the He^{2+} maximum flux prior to that for H^+.

2. Model

For our model of cusp ion precipitation, we use the very basic cusp model originally proposed by Rosenbauer et al. [4]. This basic model assumes that the cusp energy-latitude signature is the result of velocity filtering along the cusp field lines. Plasma is injected across the dayside magnetopause and into the low latitude boundary layer (LLBL) with a velocity parallel to the magnetic field. The plasma and magnetic field in the LLBL convect poleward. This convection causes the energy-latitude signature as faster ions arrive at low altitudes equatorward of slower ions. Recent computer models (e.g., Lockwood and Smith [5]; Onsager et al. [6]) have added necessary details to this basic model, such as the change in velocity across the magnetopause due to magnetic reconnection in the subsolar region, the variation in this velocity for plasma entry at points removed from the subsolar region, and the changes in the magnetosheath population along the magnetopause. Here, we focus on the properties of the magnetosheath ion population and illustrate that, with only the basic assumptions of the original model by Rosenbauer et al. [4], we can explain the density ratio changes in Figure 2.

The velocity filter process has important implications for different ion species in the cusp. Critical to determining the variation in the density ratio in the cusp is the identification of any feature that would cause He^{2+} and H^+ distributions in the magnetosheath to have different velocities and therefore different times of flight. These features could be systematic differences in the bulk flow velocities of He^{2+} and H^+ in the magnetosheath, differences in the velocity space distributions of He^{2+} and H^+ in the magnetosheath, and/or systematic changes in the velocities or velocity space distributions across the magnetopause and in the LLBL.

There can be bulk flow velocity differences between He^{2+} and H^+ in the solar wind. Solar wind He^{2+} can flow up to one solar wind Alfven speed (~50 km/s) faster than H^+. However, this difference is small compared to the thermal velocity of the distributions in the magnetosheath (several hundred km/s), is not present all the time, and is typically not observed in the magnetosheath (e.g., Peterson et al. [7]; Fuselier et al. [8]). No systematic differences in the bulk flow velocities of He^{2+} and H^+ are observed in the LLBL (e.g., Paschmann et al. [9]; Fuselier et al. [10]). Thus, we eliminate bulk flow velocity differences in the solar wind, magnetosheath, or LLBL as the cause of changes in the density ratio in the cusp and focus on systematic differences in the He^{2+} and H^+ velocity space distributions in the magnetosheath.

Figure 3 illustrates the shock processes that produce H^+ and He^{2+} distributions downstream from the super-critical, quasi-perpendicular bow shock. (The distributions downstream from a quasi-parallel bow shock are similar with the addition of an energetic (>10 keV/e) ion component (e.g., Gosling et al. [11]). In the magnetosheath, the solar wind H^+ and He^{2+} distributions can be divided into two components, identified here as a "core" and "shoulder" (e.g., Sckopke et al. [12]; Fuselier et al. [8]).

The bulk of the H^+ solar wind distribution (~80%) is transmitted directly across the shock and heated somewhat greater than adiabatic heating would imply. This core distribution can be reasonably well represented by a Maxwellian distribution (e.g., Sckopke et al. [12]). Unfortunately, there have been no systematic studies of the temperature of this core component and its relation to the upstream parameters such as the upstream solar wind thermal velocity. In the absence of a systematic study of the core component, we used data from Fuselier and Schmidt [13] to estimate that $V_{tcore}/V_{tsw} \sim 5.5\pm2$. However, this is probably an overestimate of the core temperature since the instrument used for this estimate lacked the time resolution to adequately resolve the core component in the magnetosheath. Little or no detail on the anisotropy of the core distribution is known. Therefore, we assume that it is isotropic.

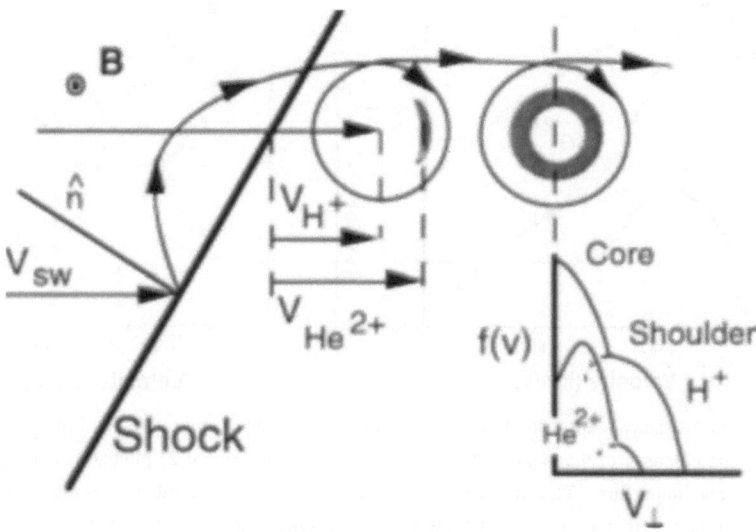

Figure 3. Processes at the high mach number, quasi-perpendicular bow shock and the resulting H^+ and He^{2+} distributions.

The shoulder H^+ component is more extensively studied. This component is the result of specular reflection of a portion of the incident solar wind off the quasi-perpendicular shock and subsequent gyration of this component into the downstream region (Figure 3) (e.g., Gosling and Robson [14]). For high mach number shocks like the bow shock in the subsolar region, the specularly reflected component is ~20% of the incident component. As the specularly reflected ions gyrate into the downstream region, they scatter in both pitch angle and energy, forming a shell in velocity space (e.g., Sckopke et al. [12]). The initial radius of this shell is $V_{in} \sim 2.15\ V_{sw} \cos \vartheta_{Vn}$, where ϑ_{Vn} is the angle between the shock normal and the incident solar wind velocity V_{sw} ($\vartheta_{Vn} \sim 0$ in the subsolar region). Although the initial radius of the shell is large, after energy scattering, the distribution better resembles a shoulder on the H^+ core distribution which we approximate with a Maxwellian with thermal speed $V_{tshoulder} = 1.2\ V_{sw}$ (e.g., Fuselier and Schmidt [13]). Although this distribution is anisotropic with $T_\perp/T_\parallel \sim 1.5$, for the initial model here, we assume that the Maxwellian is isotropic.

The bulk of the solar wind He^{2+} distribution behaves nearly as a test particle across the bow shock (e.g., Fuselier and Schmidt [15]). As a result, the He^{2+} distribution is slowed less than the H^+ distribution across the discontinuity. In the subsolar region for a quasi-perpendicular shock, this differential slowing occurs approximately perpendicular to the magnetic field in the downstream region (Figure 3). The initial He^{2+} distribution in

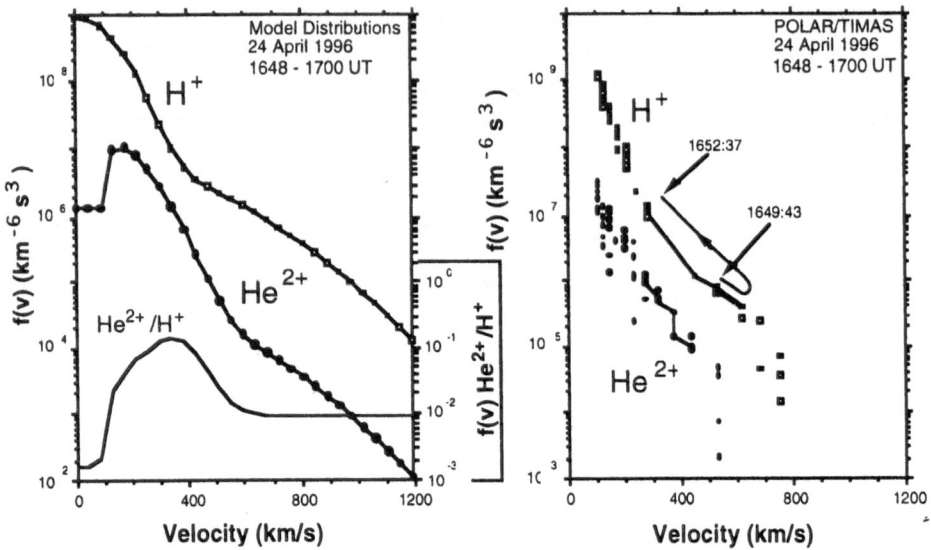

Figure 4. Model magnetosheath distributions consistent with the shock processes in Figure 3 (left hand panel). Phase space density of the maximum fluxes in the cusp (from Figure 2) plotted as a function of the velocity of the maximum flux. The model reproduces the observed H^+ distribution reasonably well and, because of the higher effective He^{2+} temperature, it reproduces the observed change in the He^{2+}/H^+ density ratio in the cusp.

the downstream region is bunched in both velocity and phase angle relative to the magnetic field. It rapidly scatters in phase angle forming a ring-beam distribution and further pitch angle scattering produces a shell distribution centered on the H^+ flow velocity. Slower energy scattering causes the shell to become filled at lower velocities, often forming a somewhat flat-topped distribution. For the high mach number perpendicular shock in the subsolar region, the shell radius $V_{shell} \sim 0.4\ V_{sw}$ (Motschmann and Raeder [16]; Fuselier and Schmidt [15]). This shell distribution represents the bulk of the He^{2+} temperature downstream from the shock. Although it is anisotropic with $T_\perp / T_{||} \sim 2.0$, for the initial model here, we assume that the distribution is isotropic.

We use a relatively simple model of the shell distribution consisting of the difference of two Maxwellian distributions. The first has $V_t = 0.4\ V_{sw}$ and 96% (+~24%) of the total He^{2+} density in the downstream region. The second Maxwellian is subtracted from the first and it has half the thermal speed, and 25% the density of the first distribution. Phase space densities for those velocities below the shell radius of about $0.4\ V_{sw}$ would be negative. However, we set these phase space densities equal to the first positive density outward from zero velocity. This approximates the filling of the shell at low velocities due to energy scattering.

In addition to the shell distribution, there is a shoulder component similar to the H^+ shoulder (Gloeckler et al. [17]; Fuselier et al. [8]). The origin of this component is not well known. However, it is similar to the H^+ shoulder except that, for a strong shock, the He^{2+} shoulder is only ~4% of the total He^{2+} density (Fuselier and Schmidt [13]). Therefore, we model the He^{2+} shoulder the same way as the H^+ shoulder (a Maxwellian with thermal velocity = 1.2 V_{sw}) except for the density difference.

To specify the H^+ and He^{2+} distributions in this simple model, we need the solar wind velocity, V_{sw}, the solar wind H^+ density, n_H, the solar wind He^{2+}/H^+ density ratio, n_{He}/n_H, and the solar wind H^+ thermal speed, V_{tsw}. In addition, we must specify the core temperature, or equivalently the change in the solar wind thermal speed across the shock (between 3.5 and 7.5 V_{tsw}) and we must specify the shock compression ratio.

For the interval in Figure 2, the WIND spacecraft observed the following: V_{sw} = 390 km/s, n_H = 5.7 cm^{-3}, V_{tsw} = 30 km/s, and n_{He}/n_H = 3.9%. For these solar wind parameters, a H^+ core temperature of 4.9 V_{tsw}, and a compression ratio of 4 (strong shock), the model magnetosheath velocity distributions in the rest frame of the H^+ distribution are shown in the left hand panel of Figure 4. The third curve is the ratio of these phase space densities (see the scale on the right hand side of the panel).

For velocities >600 km/s, the He^{2+}/H^+ phase space density ratio is constant at about 0.08%. This ratio is the result of a factor of 5 difference in the magnetosheath shoulder/core density ratio for He^{2+} and H^+ (4% for He^{2+} compared to 20% for H^+) and the nearly 4% He^{2+}/H^+ total density ratio in the solar wind. From 600 km/s to 400 km/s, the density ratio rises sharply from 0.08% to over 10%, more than 3 times the solar wind value. From 400 km/s to 0 km/s, the ratio decreases first gradually and then sharply to values below that in the solar wind.

The peak in the He^{2+}/H^+ density ratio occurs right at the transition between H^+ core and shoulder or at about 400 km/s (1 keV/e for H^+). The peak ratio occurs here because the core H^+ temperature is lower than the effective temperature of the He^{2+} shell. Obviously, the magnitude of the peak depends strongly on the H^+ core temperature and the He^{2+} shell radius. The radius of the He^{2+} shell is a weak function of the upstream mach number (Fuselier and Schmidt [15]). However, the solar wind mach number is typically high so that the shell radius is typically ~0.4 V_{sw}. Since the solar wind speed is typically ~400 km/s, the shell radius is typically ~160 km/s.

As discussed above, the H^+ core temperature is less well established. We estimate that this core temperature can be between 3.5 and 7.5 V_{tsw}. This could produce He^{2+}/H^+ peak density ratios from ~4% (i.e., the same ratio as that of the solar wind) to ~80%. The peak ratio between 10 and 20% in the left hand panel of Figure 4 represents somewhat typical solar wind conditions and approximately average H^+ core temperatures.

3. Comparison of Magnetosheath Model Distributions and Observations

Translating the model magnetosheath distributions in the left hand panel of Figure 4 into distributions observed in the cusp is relatively simple using the assumptions of the basic cusp model proposed by Rosenbauer et al. [4]. In this model, the velocity filter effect in the cusp will result in precipitation of the highest energy ions near the equatorial edge of the cusp and successively lower energy ions poleward of this boundary. Thus, by plotting the velocity space distributions (obtained from the maximum flux in Figure 2) as a function of velocity instead of time, the magnetosheath distribution should be reproduced. This plot is shown in the right hand panel of Figure 4.

In the cusp observations (right hand panel) and the model magnetosheath distributions (left hand panel), there are similarities and differences. For H^+, the shoulder component in the cusp is seen at velocities >400 km/s, with the transition between the hotter shoulder and cooler core components occurring somewhere between 300 and 400 km/s. For He^{2+}, the shoulder distribution is less well resolved. The He^{2+} distribution is clearly hotter than the H^+ core distribution. This temperature difference is most pronounced in the region between 300 and 400 km/s (~0.8 - 1 keV/e for H^+), where the He^{2+}/H^+ density ratio differs from the solar wind ratio by its largest amount (see Figure 2). Differences between the observations in the cusp and the model magnetosheath distributions include a cooler than predicted H^+ core temperature, higher than predicted phase space densities, and the lack of a discernible He^{2+} shell.

In summary, the overall change in the He^{2+}/H^+ density ratio in the cusp is caused by systematic differences in the velocity distributions of the source populations in the magnetosheath. In particular, it is caused by a magnetosheath He^{2+} distribution that is hotter than the magnetosheath H^+ core distribution. In this static model of cusp precipitation, the velocity of the precipitating ions decreases as the spacecraft moves from equatorward to poleward along its orbit. Since this velocity is a function only of position in the cusp (or time in Figure 2), the He^{2+}/H^+ density ratio should rise rapidly above the solar wind ratio and then decrease to values at or below that in the solar wind.

While this static model explains the overall change in the density ratio, it does not explain the large variations in the density ratio with a period of ~2-3 min in Figure 2. To understand these variations, a time variation must be included in the simple model of cusp precipitation. By allowing the position of the spacecraft relative to the equatorial edge of the cusp to vary on a timescale faster than the motion of the spacecraft through the cusp region, the velocity of the precipitating ions will change and, since the density ratio is a function of the ion velocity, this ratio will also change.

This variation in velocity is illustrated in Figure 4 (right hand panel). The H^+ (and He^{2+}) peak phase space densities observed between 1649:43 and 1652:37 UT are connected by lines. This time interval is between the first and second peaks in the He^{2+}/H^+ density ratio in Figure 2. As seen in Figure 4 (right hand panel), the large variation in the He^{2+}/H^+ density ratio is caused by variation in the H^+ and He^{2+} velocities over a range that includes the transition from the H^+ shoulder and core components.

Figure 5 illustrates two possible scenarios for variation of the spacecraft position relative to the equatorial edge of the cusp. In the first scenario, the position of the spacecraft in the cusp is changed by an inward motion of the magnetopause caused by an increase in the reconnection rate (increased erosion). If the convection electric field is changed in this process, the spacecraft position will appear to move poleward, the velocity of the precipitating ions will decrease, and the density ratio will change. As the reconnection rate decreases, the opposite will happen.

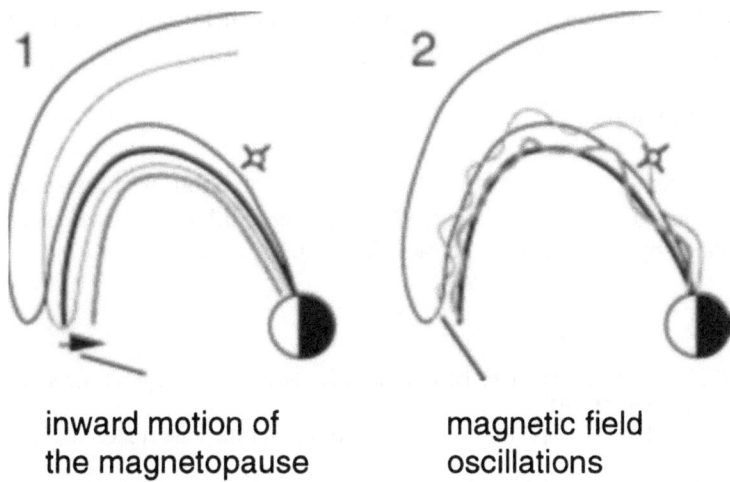

inward motion of   magnetic field
the magnetopause   oscillations

Figure 5. Two possible scenarios for changing the effective location of the equatorward edge of the cusp relative to a fixed spacecraft position. The first way is to increase the reconnection rate an move the cusp location equatorward and the second way is to have large amplitude waves in the cusp.

In the second scenario, the position of the spacecraft in the cusp changes due to waves in the cusp region. In this case, the spacecraft will periodically move equatorward and poleward in the cusp as the waves propagate, the velocity and the density ratio of the precipitating ions will also show periodic variations. Here, we make no distinction between these two scenarios. However, in the future, we will investigate further the variations in the density ratio shown in Figure 2 in order to distinguish between the two (and other) possible scenarios.

4. Summary and Conclusions

The model in this paper lacks some details important for the general interpretation of cusp observations. We have not traced precipitating ions in a realistic magnetic field, we have not included changes in the magnetosheath populations along the magnetopause from the subsolar region to high latitudes, and we have not included a change in the velocity across the magnetopause that occurs during magnetic reconnection.

Despite theses simplifications, we are able to show that a simplified model of the magnetosheath distributions can account for observed features in the cusp. In particular, we identify a high velocity shoulder on the H^+ and possibly the He^{2+} distributions in the cusp as the shoulder produced by the interaction of the solar wind distributions at the shock (Figure 4, right hand side). Also, we show that the change in the He^{2+}/H^+ density ratio in the cusp is the result of a relatively higher effective temperature of the He^{2+} magnetosheath distribution when compared to the core H^+ temperature in the same region. Short term variations in the cusp density ratio (Figure 2) require effective changes of the spacecraft position in the cusp on a timescale faster than that due to spacecraft motion. Figure 5 shows two ways this may be accomplished. The first is by a change in the reconnection rate (and convection electric field) at the magnetopause and the second is by the action of waves in the cusp.

5. Acknowledgments

The TIMAS instrument is the result of many years of effort by a large number of dedicated scientists and engineers. Research at Lockheed Martin Advanced Technology Center was funded by NASA under contract NAS5-30302.

6. References

[1] Shelley, E. G., et al. (1995) The toroidal imaging mass-angle spectrograph (TIMAS) for the polar mission, *Space Sci. Rev.* **71**, 497.

[2] Ogilvie, K., et al. (1995) A comprehensive plasma instrument for the wind spacecraft, *Space Sci. Rev.* **71**, 55.

[3] Shelley, E. G., Sharp, R. D., and Johnson R. G. (1976) He^{2+} and H^+ flux measurements in the dayside cusp: Estimates of convection electric field, *J. Geophys. Res., .81*, 2363.

[4] Rosenbauer, H., et al. (1975) Heos 2 plasma observations in the distant polar magnetosphere: The plasma mantle, *J. Geophys. Res.* **80**, 2723.

[5] Lockwood, M. and Smith, M. F. (1992) The variation of reconnection rate at the dayside magnetopause and cusp ion precipitation, *J. Geophys. Res.* **97**, 14,841.

[6] Onsager, T. G., Kletzing, C. A., Austin, J. B., and MacKiernan, H. (1993) Model of magnetosheath plasma in the magnetosphere: Cusp and mantle particles at low-altitudes, *Geophys. Res. Lett.* **20**, 479.

[7] Peterson, W. K., et al. (1979) H^+ and He^{++} in the dawnside magnetosheath, *Geophys. Res. Lett.* **6**, 667.

[8] Fuselier, S. A., Shelley, E. G., and Klumpar, D. M. (1988) AMPTE/CCE observations of shell-like He^{2+} and O^{6+} distributions in the magnetosheath, *Geophys. Res. Lett.* **15**, 1333.

[9] Paschmann, G. Fuselier, S. A., and Klumpar, D. M., (1989) High speed flows of H^+ and He^{++} ions at the Earth's magnetopause, *Geophys. Res. Lett.*, **16**, 56.

[10] Fuselier, S. A., Shelley, E. G., and Klumpar, D. M. (1993) Mass density and pressure changes across the dayside magnetopause, *J. Geophys. Res.* **98**, 3935.

[11] Gosling, J. T., Thomsen, M. F., Bame, S. J., and Russell, C. T. (1989) Ion reflection and downstream thermalization at the quasi-parallel bow shock, *J. Geophys. Res.* **94**, 10,027.

[12] Sckopke, N., et al. (1983) Evolution of ion distributions across a nearly perpendicular bow shock: Specularly and non-specularly reflected-gyrating ions, *J. Geophys. Res.,* **88**, 6,121.

[13] Fuselier, S. A., and Schmidt, W. K. H. (1994) H^+ and He^{2+} heating at the Earth's bow shock, *J. Geophys. Res.* **99**, 11,539.

[14] Gosling, J. T., and Robson, A. E. (1985) Ion reflection, gyration, and dissipation at super-critical shocks, in B. T. Tsurutani and R. G. Stone (eds.), *Collisionless Shocks in the Heliosphere: Reviews of Current Research, Geophys. Monogr. Ser. vol. 35*, AGU, Washington D. C., p. 153.

[15] Fuselier, S. A., and Schmidt, W. K. H. (1997) Solar wind He^{2+} ring-beam distributions downstream from the Earth's bow shock, *J. Geophys. Res.* **102**, 11,273.

[16] Motschmann, J., and Raeder, J. (1992) A simulation study of multiple ion wave generation downstream of low mach number quasi-perpendicular shocks, *Geophys. Res. Lett.* **19**, 1619.

[17] Gloeckler, G., et al. (1986) Solar wind carbon, nitrogen, and oxygen abundances measured in the Earth's magnetosheath with AMPTE, *Geophys. Res. Lett.* **13**, 793.

IDENTIFYING THE OPEN-CLOSED FIELD LINE BOUNDARY

M. LOCKWOOD
Rutherford Appleton Laboratory, Chilton, Oxfordshire,
OX11 0QX, United Kingdom.

Abstract: Of all the various definitions of the polar cap boundary that have been used in the past, the most physically meaningful and significant is the boundary between open and closed field lines. Locating this boundary is very important as it defines which regions and phenomena are on open field lines and which are on closed. This usually has fundamental implications for the mechanisms invoked. Unfortunately, the open-closed boundary is usually very difficult to identify, particularly where it maps to an active reconnection site. This paper looks at the topological reconnection classes that can take place, both at the magnetopause and in the cross-tail current sheet and discusses the implications for identifying the open-closed boundary when reconnection is giving velocity filter dispersion of signatures. On the dayside, it is shown that the dayside boundary plasma sheet and low-latitude boundary layer precipitations are well explained as being on open field lines, energetic ions being present because of reflection of central plasma sheet ions off the two Alfvén waves launched by the reconnection site (the outer one of which is the magnetopause). This also explains otherwise anomalous features of the dayside convection pattern in the cusp region. On the nightside, similar considerations place the open-closed boundary somewhat poleward of the velocity-dispersed ion structures which are a signature of the plasma sheet boundary layer ion flows in the tail.

1. Introduction

The large number of phenomena which have been explained using the open magnetosphere model and, in particular, the large number of genuine predictions that have later been verified in experimental data, leave no reasonable doubt that magnetic reconnection takes place in the magnetopause current sheet and in the cross-tail current sheet of the tail, as initially proposed by Dungey in 1961 [1]. Indeed, it is also now clear that this is the dominant process in magnetospheric morphology and dynamics. Reconnection in the dayside magnetopause current sheet generates open flux (which threads the magnetopause boundary) whereas reconnection in the cross-tail current sheet destroys open flux by converting it back to closed flux (which never threads the magnetopause). The cycles of accumulation and loss of open flux, caused

73

J. Moen et al. (eds.), Polar Cap Boundary Phenomena, 73–90.

respectively by dominant magnetopause and tail reconnection, are central to the basic substorm cycle of magnetospheric behaviour. The distribution of boundary-normal field at the magnetopause depends on a large number of factors, including the prevailing and previous solar wind and interplanetary magnetic field (IMF) conditions and the past history of the rate and location of the reconnection in the magnetopause and tail current sheets. When the IMF points northward, lobe reconnection can rearrange that distribution by changing the point at which open field lines thread the boundary: that point can instantly be changed from the far tail to the dayside. At low altitudes, the accumulation of all the open flux in the system is here referred to as the "polar cap" which is usually a single region, but may sometimes be bifurcated during northward IMF, or if the IMF B_y component has changed polarity [2]. The boundary between the open flux and closed flux is here referred to as the "open-closed boundary" (OCB).

There are a variety of other definitions of the polar cap boundary which have been employed and which, in some senses, are more practical as they are more readily detected than the OCB. However, these are usually dependent on the type and sensitivity of the measurements and do not have the physical significance of the OCB. Indeed, locating the OCB is of crucial importance in magnetospheric physics. It is frequently necessary to determine if a radar or satellite is observing open or closed field lines and the interpretation of any phenomenon will usually depend critically on this. In addition, recent work [3, 4, 5] has shown that the motion of the boundary is crucial to the excitation of ionospheric flow. Lastly, the size of the open field line regions is a key indicator of the disturbance state of the magnetosphere, the energy stored in the tail and the prevailing and immanent geomagnetic and auroral activity levels.

2. Magnetopause reconnection

The left hand schematics in figure 1 show magnetic field lines relative to noon-midnight cross-sections of the magnetosphere, as viewed from dusk such that the sun is to the left. Open, closed and interplanetary field lines are labelled o, c and i, respectively, and a particular class of open field line called "over-draped lobe" is labelled ol. The dashed line is the magnetopause (MP) and dots marked X are active reconnection sites. The right hand schematics are views of the northern hemisphere polar cap, viewed from above with the sun to the top, dawn to the right and dusk to the left: convection flow streamlines are marked with arrows, a solid line is a non-reconnecting segment of the open-closed boundary and dashed lines map to a reconnection X-line. Figure 1a is for southward IMF, whereas 1b and 1c are two possibilities for northward IMF. In all cases, we show steady-state situations in the ionospheric response for simplicity, but the considerations can readily be generalised for non-steady cases by allowing for the flows associated with moving boundaries [6]. The bulk of the flows shown in figure 1 are reconnection-driven. However, flow cells driven by non-reconnection (unspecified, "viscous-like") processes acting on closed field lines are also included for completeness and are labelled v. In practice, the voltage associated with such processes is very low (less than about 5 kV), much of the viscous-like voltage thought in the past to come from such processes is likely to be associated

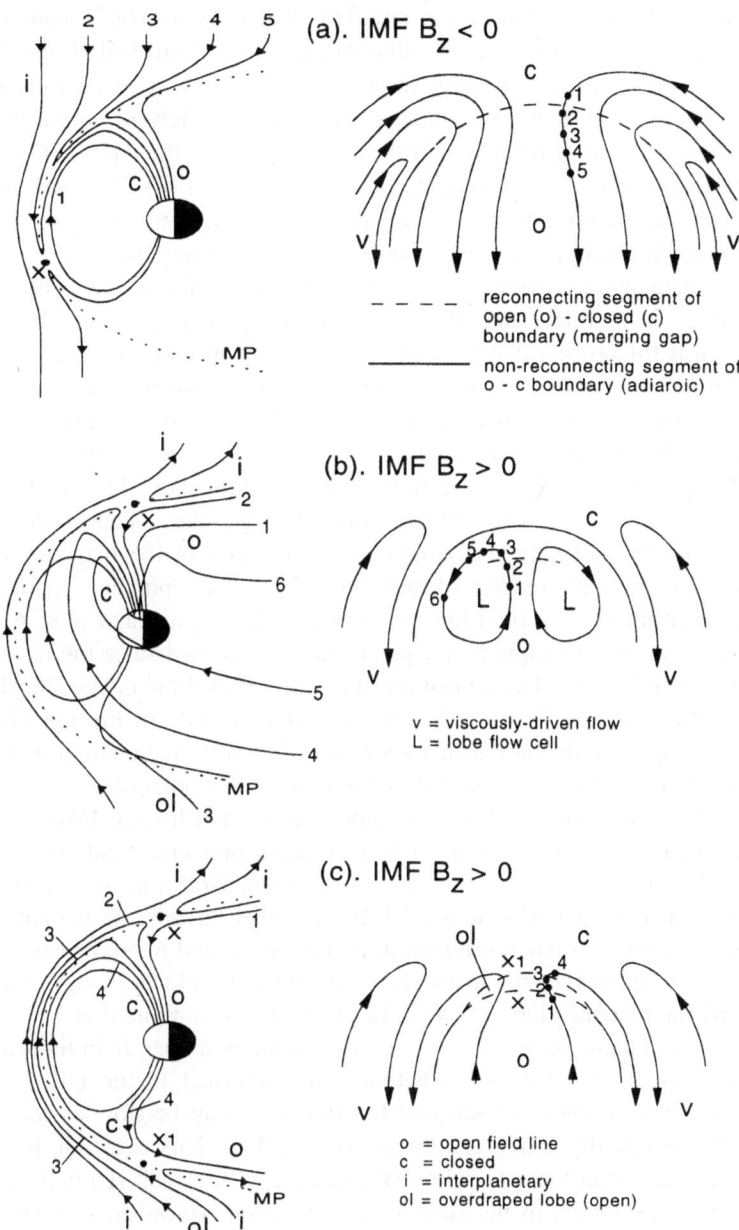

Figure 1. Schematic illustrations of (left) the evolution of reconnected field lines in the magnetosphere, as seen from the dusk flank and (right) the corresponding flow in the northern hemisphere ionosphere for magnetopause reconnection during (a) southward IMF and (b and c) northward IMF, with By ≈ 0. Field lines and regions that are open, closed, interplanetary and overdraped lobe are labelled, respectively, o, c, i, and ol.

with continued field line closure in the tail. This was indicated by Wygant et al. [7] who showed that the range of residual voltages during northward IMF dropped steadily with time since the northward turning, showing that is was associated with open flux produced by the prior period(s) of southward IMF. How poleward contraction of the polar cap (due to closure of open flux in the tail) can mimic the dayside flow patterns expected for a viscous-like mechanism was demonstrated by Lockwood and Cowley [8] and this effect was confirmed by Fox et al. [9] who showed examples where the apparent viscous-like voltage increased during polar cap contractions.

As mentioned above, magnetopause reconnection taking place between closed magnetospheric field lines and the shocked interplanetary field lines of the magnetosheath is the process that generates open flux. This can only take place where closed field lines are exposed to the magnetopause current layer. Adjacent to most of the tail lobe magnetopause is "old" open flux generated by prior reconnection and so this opening of flux is generally restricted to latitudes below the magnetic cusps, as in figure 1a. The positions numbered 1-5 in the left hand plot show the evolution of each field line from the closed field line region, through the reconnection site and subsequently into the tail lobe under the combined action of the magnetic curvature ("tension") force and the magnetosheath flow. The corresponding motion of the ionospheric footprint of such field lines is shown in the right hand plot. A cusp ion dispersion is seen as the precipitation depends on time elapsed since the reconnection. By the time a field line reaches a position like 5, the bulk flow of the sheath plasma crossing the magnetopause is towards the tail and very little of the ion distribution function is moving towards the Earth. As a result, the cusp precipitation will decay in flux and energy and will become classed as mantle and then polar cap.

Figure 1b shows one possible situation during northward IMF, when lobe reconnection takes place between an interplanetary magnetic field, draped in the magnetosheath, (i) and the open flux of the tail lobe. Note that the open flux was produced by a prior period of southward IMF and thus threads the boundary further down the tail and has a different orientation in the sheath and interplanetary space to i [10]. In this case, often referred to as "lobe stirring", the old open field lines like 1 evolve toward the reconnection site (to 2) before being reconfigured at X so that they thread the dayside magnetopause (like 3: these reconfigured open field lines have been called "over-draped lobe" flux [11]) before being returned to the tail lobe by the magnetosheath flow. Evolution back into the tail lobe may be slow as the magnetic curvature force is initially acting against the sheath flow. This case can, in principle, exist in steady state with lobe flux circulating around a flow cell (L) within the open field line region, as shown in the right hand schematic. However, in reality, this is unlikely to be a steady-state phenomenon and flux is very unlikely to circulate all the way around these cells as their lifetime is shorter than the circulation time. A reversed cusp ion dispersion precipitation is seen sunward of the projection of X (the dashed line), whereas the precipitation seen on the old open flux poleward of this (field lines 1 and 2) will classed as either mantle or polar cap. Because some of the overdraped lobe field lines can evolve only very slowly, due to the competing effects of the sheath flow and tension force, some of the cusp could be relatively stagnant and the precipitation dispersionless. Figure 1 is drawn for small IMF $|B_y|$ and so the lobe convection cells are roughly equal in size whereas one of these would dominate for large $|B_y|$ [4, 10]. Note that if the viscous-like flow is actually driven by continuing tail field closure, the streamlines of these v convection cells will cross the contracting polar cap boundary and can merge with the lobe circulation [8] to give a distorted 2-cell configuration, as has been inferred from satellite passes [12], as opposed to the 4 cells shown here for steady state. From observations, it appears that when IMF clock angles are such that B_z is only weakly positive, that some lower-latitude reconnection, as in 1a, can continue at the same time as the lobe reconnection [13, 14]. Observations of ion distribution functions and flows at the dayside magnetopause [15, 16] also imply that this can be the case and that this may be made possible by distortion of the magnetosheath field and intensification in the plasma depletion layer that forms when IMF $B_z > 0$ [17].

The lobe convection cells L are predominantly a summer phenomenon [18], which can be explained because the dipole tilt towards the sun favours lobe reconnection in the summer hemisphere. In the opposite hemisphere to X (the southern hemisphere in 1b), the lobe reconnection has no effect [13, 18, 19]. The IMF B_x component is also likely to be relevant as the draping of the field lines will favour lobe reconnection in one hemisphere over the other ($B_x<0$, $B_z>0$ favours the northern hemisphere, whereas $B_x>0$, $B_z>0$ favours the southern). As a result it may well be possible for lobe reconnection to take place in both hemispheres. One possibility is that the reconnection in the two hemispheres independently reconfigure the old open lobe flux and patterns like that shown in figure 1b are independently driven in the two hemispheres. However, the overdraped lobe field lines connected to the two hemispheres are likely to intertwine to restrict this. Another alternative is shown in figure 1c where the overdraped lobe flux produced by X (like 3) is itself reconnected at X1 in the other hemisphere to produce a closed field line (like 4). This is similar to the original suggestion for $B_z > 0$ by Dungey, recently invoked by Song and Russell [20] as a way of producing a closed field-line low-latitude boundary layer containing magnetosheath plasma, but is more realistic in that field lines are not simultaneously reconnected at both sites. In the case shown in 1c, a cusp precipitation, with reverse or weak dispersion, is seen in the northern hemisphere only. Note that the flow cells with sunward polar cap flow now cross between the open and closed field line regions and do not remain in the region of open flux as in 1b. If the reconnection rate at X1 exceeds that at X, the northern hemisphere overdraped lobe flux will decay, the width of the northern cusp would shrink to zero. After that time, X1 would act to expand an overdraped lobe connected to the southern hemisphere.

3. Locating the dayside open-closed boundary

Identifying the OCB is complicated by the variety or reconnection possibilities shown in figure 1. The easiest case may well be that in figure 1b because there is no flow across the OCB. The frozen-in approximation means that there is no significant mixing of the plasma populations on the two sides of the boundary and there will be no velocity-filter dispersion of particles and waves in the direction normal to the boundary. Thus the OCB would be marked by changes in the particle and wave characteristics, irrespective of their field-aligned velocity. As discussed below, this is certainly not the case when reconnection is driving boundary-normal flow. Thus in figure 1b, the closed field lines would have magnetospheric plasma sheet particles (of all energies) at latitudes equatorward of the OCB, whereas at all latitudes poleward of the OCB we would see either the reverse-dispersion/dispersionless cusp particles of the overdraped lobe or the mantle/polar cap precipitation of old open flux (note the polar cap in the opposite hemisphere to X would only contain the latter). The only confusion may be that gradient-B and curvature-B drifts which are breakdowns of frozen-in and which will allow some particles to appear on the opposite side of the boundary. These non-ideal MHD drifts are most significant for energetic magnetospheric particles and electrons/ions drift east/west, they will penetrate onto open field lines on the dawn/dusk flank, respectively. Otherwise the OCB would be clear and sharp in this case and reasonably well-defined in plasma characteristics.

The same is true for non-reconnecting segments of the boundary in figures 1a and 1c. There may be poleward flow streamlines crossing such boundaries if those boundaries are in motion (i.e. non steady-state cases), but this boundary-normal plasma motion is at the same speed as the motion of the boundary itself and no field lines or plasma cross the boundary. A likely cause is poleward flow and poleward boundary motion as tail reconnection closes open flux and contracts the polar cap [3, 4, 5, 8, 9]). In these cases, the boundary-normal motion does not cause a mixing of the plasmas on the two sides of the boundary, nor a dispersive spatial separation of particles. Rather, it moves all frozen in particles with the magnetic flux. Thus the OCB can still be recognised in a relatively abrupt change in the plasma characteristics.

However, in figure 1c the situation is not so clear. Consider a field line 4 which has recently been re-closed by lobe reconnection. On reconnection at X1, the source of magnetosheath plasma for middle and low altitudes is shut off and so the loss cone is no longer constantly replenished by the magnetosheath source, as it had been before. This might be noted almost immediately at lower altitudes in faster, higher-energy electrons for which travel times are short. However after the closure, ions already injected but yet to reach the satellite will continue to be seen for some time; the closing of the field line being noted first in higher energy and field-aligned ions for which transit times are shorter. Even in sheath electrons, the boundary may not be clear. This is because outside the loss cone, fluxes will remain high as the particles are trapped between two mirror points on the now-closed field lines. Even inside the loss cone, pitch angle scattering (possibly associated with the need to maintain quasi-neutrality) may maintain fluxes of sheath plasma for a while on closed field lines. In the polar cap local to X1, the flux of mantle/polar cap flux on the old open field lines will be shut off by the reconnection. Re-population of re-closed field lines with magnetospheric plasma sheet population would be by gradient and curvature drifts and may be mainly achieved by the next nightside injection event [21].

The remainder of this section will deal with the southward IMF case shown in 1a. The opening of the field line allows the escape of magnetospheric plasma to begin and the injection of solar wind plasma also to start. The resulting evolution of the particle populations has recently been modelled for a variety of conditions [22-27]: these include steady-state [25] and time-varying [22, 24] conditions and at the magnetopause

[28], middle altitudes [27] or low altitudes [25], and for satellites moving both normal to and along the OCB. The observed energy-observing time and pitch angle-energy dispersions have both been reproduced [26]. These models include some basic physical effects. The point where each newly-opened field line threads the magnetopause evolves with time elapsed since reconnection as the newly-opened field line moves toward the tail lobe under the influence of magnetic tension and the magnetosheath flow. Because of the spatial variations of plasma density and temperature adjacent to the boundary in the shocked magnetosheath (in the above models non self-consistent gas-dynamic predictions have generally been used), the characteristics of the source sheath population vary with time elapsed since reconnection. In addition, the acceleration of particles on crossing the boundary (which is significant for the ions) varies with this elapsed time as $\underline{J} \cdot \underline{E} > 0$ below the magnetic cusp turns to $\underline{J} \cdot \underline{E} < 0$ on the tail lobe boundary. The third consideration is the time-of-flight of particles to the satellite. This applies to injected sheath particles transmitted through any one point on the magnetopause and to magnetospheric particles which either escape from each initial location on the field line by transmission through the magnetopause or are reflected back towards the satellite by their interaction with the magnetopause. Only particles with a flight time less than the elapsed time since reconnection will be effected by the fact that the field line has been opened. Thus under these three main influences, the populations on newly-opened field lines at a given satellite altitude will evolve with elapsed time since the reconnection. In general, we should not expect a satellite to see this evolution in full as the reconnection may be pulsed and/or spatially patchy and even in steady-state, the satellite will not exactly follow the convection flow streamlines.

As a general point of principle we can note that only waves and particles with an infinite velocity of field-parallel motion could have reached the satellite at the OCB and so there can be no information that the field line has been opened. Thus the OCB which maps to an active reconnection site must lie within a region and cannot lie at any detectable feature which could define a boundary between two regions: all detectable changes will be somewhere downstream of the OCB.

Figure 2a shows the conventional view of the precipitation regions in the dayside ionosphere and their relation to the OCB. The precipitation regions are broadly as deduced from the statistical survey of Newell and Meng [29] except that the catch-all "void" classification, seen equatorward of the low-latitude boundary layer (LLBL) near noon, has here been included as part of the dayside BPS which is the other classification seen at such a location. From the mapping of precipitations originally proposed by Vasyluinas [30], the dayside BPS and LLBL have usually been thought of as being on closed field lines, whereas the cusp, mantle and polar cap precipitations are thought to be on open field lines [e.g. 31]. A point that needs making about these precipitation patterns is that they are deduced from low-altitude observations, i.e. from particles inside the loss cone. Thus, for example, the mechanism proposed by Song and Russell [20] would not be a direct source of the LLBL precipitation seen at low altitudes: this is because the re-closure of the field line would cut off the supply of particles for the loss cone and one would need to invoke a scattering mechanism of the subsequently trapped particles into the loss cone. In figure 2a, the equatorward edge of

80

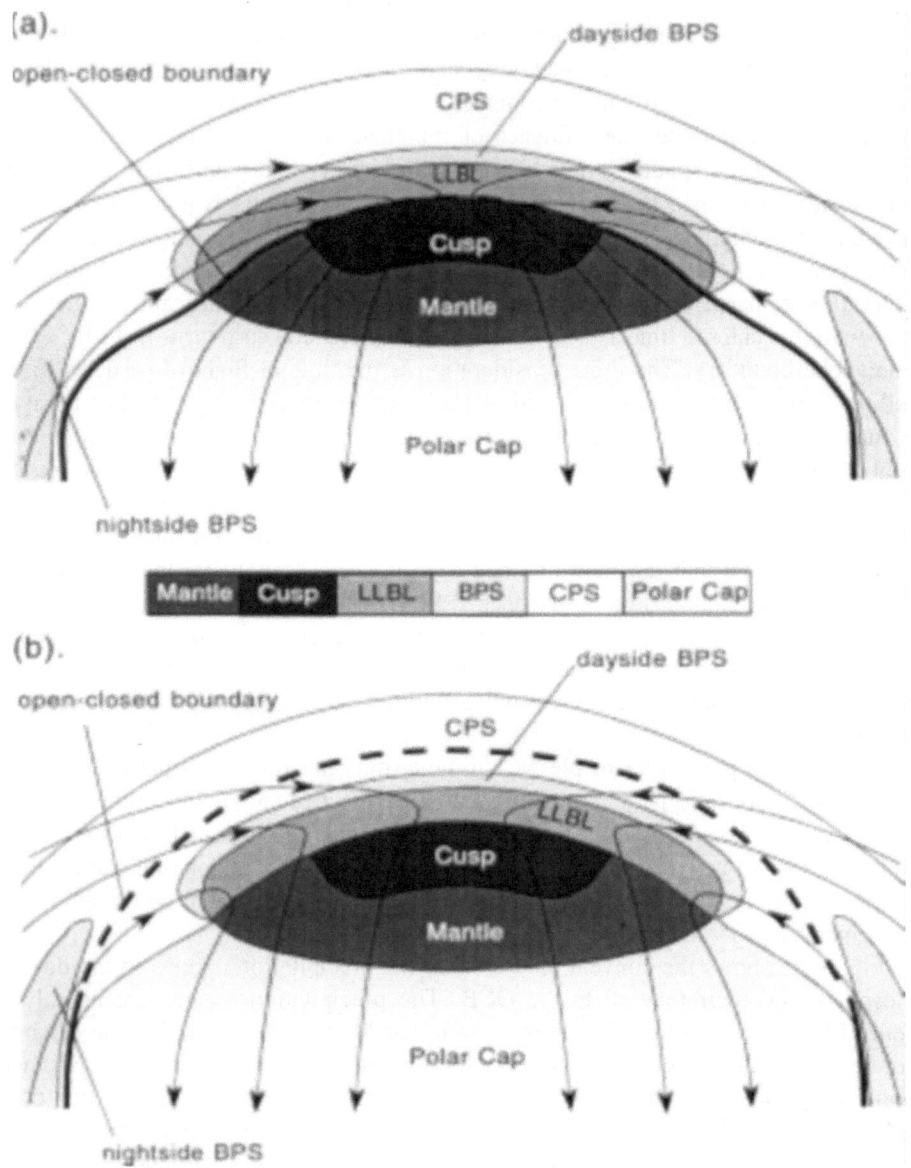

Figure 2. Schematic illustrations of proposed relationships of the open-closed boundary (OCB) to the dayside precipitation regions and the pattern of convective flow during southward IMF. In (a) the low-latitude boundary layer (LLBL) and dayside boundary plasma sheet (BPS) are on closed field lines, whereas in (b) they are on open field lines. Other precipitation regions are central plasma sheet (CPS), cusp, mantle and polar cap, all being shaded according to the key.

the cusp is the projection of the magnetopause reconnection X-line, this being reflected in the fact that it longitudinal extent is the same as that of the merging gap (the reconnecting segment of the OCB, shown as a dashed thick line; the "adiaroic", non-reconnecting segments of the OCB being shown as solid thick lines)

There are a number of problems with figure 2a. Firstly, because of ion flight times in the presence of convection, it is not possible that the OCB is on the boundary of two regions in the merging gap, where there is poleward flow across the OCB. This is because, as discussed above, the magnetosheath cusp particles take time to reach low altitudes along the newly-opened field lines, during which interval the field lines have convected poleward. Thus the poleward flow means that the OCB must lie equatorward of the equatorward cusp boundary. In addition, the constricted "throat" flow geometry shown in figure 2a is not seen, in radar data [32, 33], nor in satellite data surveys [34, 35], nor in flow patterns inferred from ground-based magnetometer networks [31, 19]. This problem can be quantified. For southward IMF, magnetopause reconnection can produce voltages up to about $V \approx 150$ kV in substorm growth phases, this is seen in transpolar voltages [36, 37, 38, 19] and in polar cap expansion rates [8]. In figure 2a, this is applies across the longitudinal extent of the cusp, which is statistically about 3 hours of MLT (i.e. a distance of $L \approx 1000$ km) wide, although it has been observed to be wider in specific cases [e.g. 35]. Given the ionospheric field is $B_i \approx 5 \times 10^{-5}$, these figure require poleward flow through the cusp at a speed of $V/(B_i L) \approx 3$ km s^{-1}, which is a factor of about 6 faster than is typically observed [31-35]. Other problems with figure 2a have been discussed by Lockwood [40].

Recently, it has been proposed that both the BPS and LLBL precipitations shown in figure 2 can be produced by reflecting magnetospheric ions off the two Alfvén waves produced by the magnetopause reconnection site and standing in the inflow regions (the outer one of which is the magnetopause, the inner one being the inner edge of the open LLBL) [41]. This idea has been shown to produce good fits to observed moments and partial moments of the ion gas, both at low altitudes [41, 42] and at the dayside magnetopause [28], and to also reproduce the observed ion spectra and pitch-angle distributions [26]. The key point is that this places both the BPS and LLBL precipitations on open field lines and thus the OCB lies close to the poleward edge of the magnetospheric Central Plasma Sheet (CPS) precipitation, as shown in figure 2b. This solves the anomalies discussed above [40]. It is consistent with the velocity filter effect of time-of-flight dispersion, because the OCB sits within a precipitation region and not on the boundary between two regions. In addition, the energisation of the dayside BPS over CPS energies is explained. The cusp longitudinal width is set by the longitudinal variation of density in the magnetosheath [40] and the reconnection merging gap can be considerably greater in its local time extent, as for example modelled by Crooker et al. [43]. This can therefore solve the problem of the poleward flow speed in the cusp and we can infer that the reconnection merging gap can be up to six times longer than the cusp extent. The interior Alfvén wave launched by the reconnection site arrives in the middle of the LLBL region, causing the convection reversal to sit in that region, as is observed [44]. This model can also explain the sunward-flowing "circum-polar ion precipitation" classification of Nishida et al. [45] as being between the OCB and the interior Alfvén wave.

Newell and Meng [46], argued that the LLBL was all on closed field lines in response to Lockwood and Smith [47, 48] who argued that equatorward of the cusp it must, at least partially be on open field lines. Recently, the idea that there is both an open and a closed LLBL at low altitudes has gained some acceptance, Newell and Meng making a distinction that was not made in their original LLBL classification [29], namely that the open LLBL shows a time-of-flight cut-off energy which is absent from a closed LLBL closer to the flanks [49]. On the other hand, Lockwood [40] argues that there is no need to invoke such a fundamental difference within the LLBL region at low altitudes. Towards the flanks, the LLBL would be on field lines that have been open long enough for the cut-off energy to fall below the range covered by the ion instrument, but fluxes will be lower than in the cusp simply because magnetosheath densities are lower on the flanks. In this scenario, the LLBL at low altitudes is entirely open, the loss cone being constantly replenished by sheath plasma streaming along open field lines. It should be noted that the low-altitude LLBL is not much more extensive in MLT than the cusp [29] and is about the same in this dimension as the mantle (which is on open field lines). This argues against there being a significant closed LLBL at low altitudes on the dawn and dusk flanks. Note that at higher altitudes there could be a trapped LLBL population on closed field lines, but the low fluxes in the loss cone would not give fluxes large enough to be classed as LLBL at low altitudes.

A last point to be made is that both the auroral luminosity seen and the classifications of the dayside precipitation regions is very much influenced by the electron precipitation, which does not bear a simple relationship to the ion precipitations. There are some time-of-flight electron effects and so, for example, an electron edge (at which energetic magnetospheric electrons are lost) is often seen close to where entering magnetosheath electrons are detected for the first time (e.g., plate 1 of [50], plates 1 and 3 of [51]; plates 1a-1c of [52]; and Plates 1, 3 and 4 of [53]). This electron edge is the same as that is seen at the magnetopause [54] on the inner edge of the open LLBL. At low altitudes such an electron edge will be found poleward of the OCB because of the antisunward convection, but close to it because the electron flight times are low). However, the electron precipitation is also highly influenced by the ion precipitation and the mobility of the electrons effectively maintains the observed quasi-neutrality of the cusp plasma [55]. Thus the ion fluxes are a major influence on the electrons but, in addition, field-aligned currents (caused by the stresses associated with the newly-reconnected flux), electron mirror points and ionospheric electron outflows will also important factors in determining how the overall charge neutrality balance is achieved and thus the electron precipitation morphology. In addition, the loss of magnetospheric electrons may be a misleading indicator of where the OCB is located because trapped energetic magnetospheric electrons can gradient-B and curvature-B drift onto open field lines at low altitudes for a quarter of a bounce after the field line is opened. In addition, fluxes can be maintained by magnetic bottles on open field lines [56 - 58] because the field strength has a minimum at middle latitudes in the magnetic cusp.

4. Tail Reconnection

The low-latitude magnetopause reconnection illustrated in figure 1a generates open flux. Because there are no long term trends in the amount of open magnetospheric flux, we know that, on average, there must be destruction of this open flux in the tail at the same rate. It is known that in the growth phases of magnetospheric substorms open flux accumulates in the tail lobe [59] and most theories of substorms regard much of the open flux destruction to take place in the expansion phase (or, at least, in the later half of it) and in the recovery phase. The reconnection site or sites may, in principle, be anywhere within the cross-tail current sheet but topologically the situation is less complex than the magnetopause case discussed in section 2, in that there are fewer possibilities which we need to consider here. These are shown in figure 3a.

If the reconnection in the cross-tail current sheet is between two open lobe field lines, then it is responsible for the destruction of open flux and we will here call that a "far X-line" (although the term far is relative and this reconnection can take place anywhere in a huge range of X GSE coordinates). There is one model of substorms which explains the appearance of features called plasmoids in the mid- and far-tail plasma sheet (and their corresponding lobe signatures called travelling compression regions), called the Near-Earth Neutral Line (NENL) model. A NENL forms Earthward of the far X-line and thus reconnects field lines that are already closed. The plasmoid is the resultant loops of magnetic flux (or the flux rope if the B_y is large) between these two reconnection sites. If the reconnection rate at the NENL exceeds that at the far X-line, the OCB field line will approach the NENL until the instant when the plasmoid is "pinched off" (disconnected from the Earth) when the OCB separatrix passes through both reconnection sites. After this time the NENL, by the above definition, has become the a X-line and the old far X-line is disconnected from the Earth.

For the sake of completeness, it is worth a mention of one other topological possibility, which may occur on the flanks of the tail. Any closed magnetic flux which is dragged into the tail by a genuinely viscous-like process (such as Kelvin-Helmholtz waves) may be pinched together and reconnect with itself to produce extended magnetic loops in the flank boundary layer [60]. This reconnection site is somewhat different from a NENL, as described above, as it is not be Earthward of any active far X-line.

5. Identifying the nightside open-closed boundary.

Figure 3a shows a situation after substorm onset, after the formation of the NENL (it is a controversial point as to whether or not these are different times), but before the plasmoid is pinched off, so that the OCB is the magnetic separatrix of the far X-line. At the moment of plasmoid pinch-off, the OCB passes through both X-lines and thereafter it is the magnetic projection of the newer reconnection site. In figure 3a, the

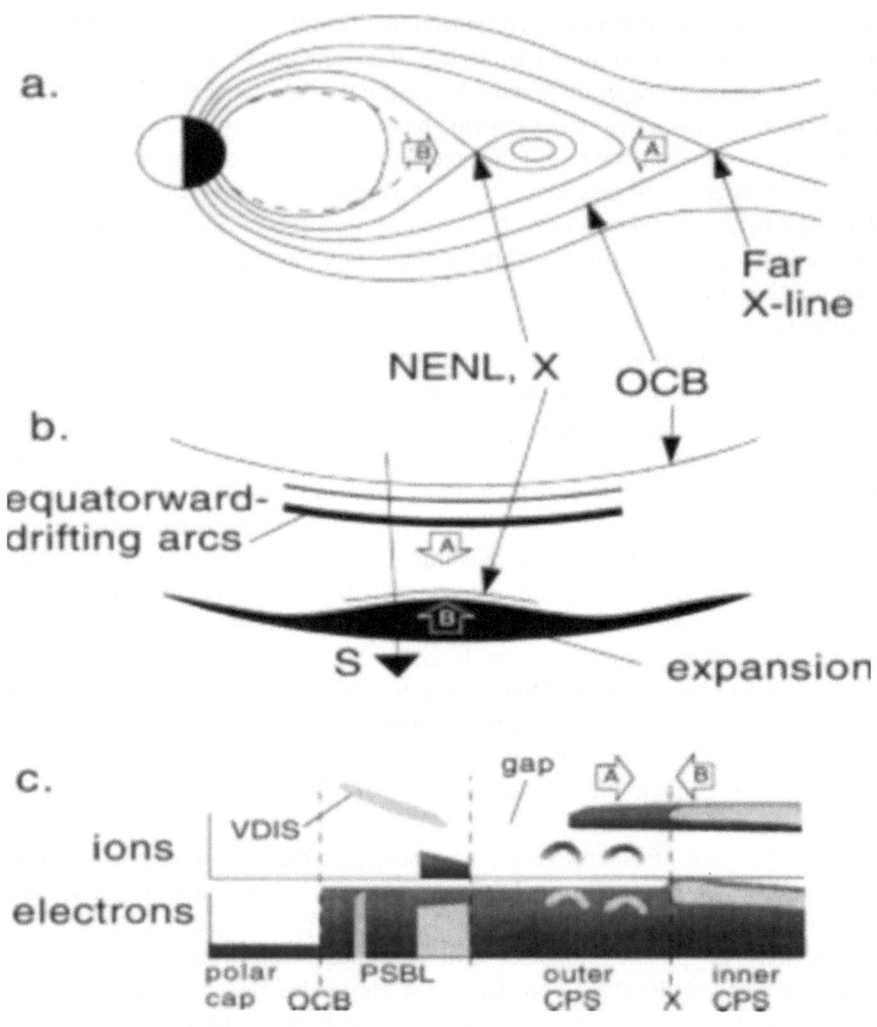

Figure 3. Schematic illustration of the nightside open-closed boundary (OCB): (a) the magnetic topology in the tail during a substorm expansion phase, showing a Near-Earth Neutral Line (NENL) and Far X-line, between which a plasmoid is growing; (b) a map of the ionospheric signatures; and (c) the form of the electron and ion precipitation spectrograms that would be seen by a satellite following the path S in part (b). The arrows A show the motion of equatorward-drifting arcs and the arrows B show the poleward expansion of the substorm aurora.

cross-tail current disruption is occurring in the Earthward outflow wedge of the NENL. It is a controversial point as to whether the current disruption is caused by the NENL, or the other way around [59].

At low altitudes near midnight, a frequently-observed precipitation structure is VDIS (velocity dispersed ion structure - somewhat analogous to the cusp), the "gap" and the central plasma sheet (CPS) [61-65], as shown schematically in figure 3c. Near the poleward edge of the VDIS are found the "boundary" field aligned currents [65] which are associated with a convection reversal [64]. A feature frequently seen in the electron precipitation are equatorward-drifting arcs [66-68], which appear in the VDIS/gap region [69], usually in the late growth phase and early expansion phase. As on the dayside, the question arises: "where is the OCB in relation to these regions?"

Onsager and Mukai [61, 62] have used a model which is directly analogous to the open magnetosphere dayside cusp models, to predict the ion precipitation on newly closed field lines produced by a far X-line. In their simulations, the VDIS, gap and plasma sheet are all reproduced and produced by field lines collapsing sunward after closure at a far X-line and the VDIS ions are the plasma sheet boundary layer ion flows seen at greater altitudes. The gap is a region of low density and appears as a gap in many datasets, only because of the threshold flux set by the instrument one-count level.

Figure 3b shows schematically the ionospheric footprint of these regions, with the NENL near the poleward edge of the expanding substorm electrojet, which maps to the region of disrupted cross-tail current. The OCB lies somewhere poleward of where the equatorward-drifting arcs first form, and poleward of the VDIS. This is analogous to the position of the dayside OCB inferred in figure 2b. How far poleward the OCB can be has recently been shown by Shirai et al [70]. They observed the dispersed disappearance of polar cap strahl electrons, which are seen only when the IMF B_x component connects that polar cap directly to the sun. The source of these electrons is shut off when the field line is closed by the far X-line and closer to Earth this is seen first in the higher energy electrons. This dispersed electron cut-off is continuous on a particle velocity - arrival time plot with the low-energy cut-off of the VDIS. These observations not only confirm the VDIS is on newly-closed flux but identify the nightside OCB (to within the small uncertainty associated with antisunward convection in the flight time of the strahl electrons to the satellite).

The surprising feature of these observations by Shirai et al. is how far poleward of the VDIS the OCB is, because of the large distances (and hence large particle travel times) between the far X-line and the Earth. In this example, the OCB is about 1.5° poleward of the VDIS. Close to the OCB, some very weak plasma sheet electron fluxes commence but, presumably because of the maintenance of quasi-neutrality, significant electron fluxes do not arrive until some time later, by when the field line has convected equatorward to near the poleward edge of the VDIS. Thus the OCB is roughly 200 km poleward of where an imager (or a satellite electron detector with just a slightly lower sensitivity) would place the poleward boundary of the auroral oval. It is also this far poleward of the poleward-most field-aligned currents [65]. Such a distance does not, at first, sound very significant. However, if we consider a circular polar cap with a dayside (cusp) OCB location at 80°, taking the poleward auroral boundary would give us a polar cap flux of 4.5×10^8 Wb, whereas the true OCB position gives us 4.0×10^8 Wb.

Lastly we note that there appears to be little flow signature in the ionosphere associated with the NENL [71]. However, as soon as the plasmoid is pinched off the higher reconnection rate associated with the NENL causes a discontinuous increase in the rate of field line closure and polar cap contraction [8] and the enhanced flows associated with this are then seen.

6. References

1. Dungey, J. W. (1961) Interplanetary magnetic field and the auroral zones, *Phys. Rev. Lett.*, *6*, 47.
2. Newell, P.T., D. Xu, C.I. Meng, and M.G. Kivelson, Dynamical polar cap: a unifying approach (1997) *J. Geophys. Res.*, *102*, 127-140.
3. Lockwood, M., S.W.H. Cowley, and M.P. Freeman (1990) The excitation of plasma convection in the high latitude ionosphere, *J. geophys Res.*, *95*, 7961-7971.
4. Cowley, S.W.H., and M. Lockwood, Excitation and decay of solar-wind driven flows in the magnetosphere-ionosphere system (1992) *Annales Geophys.*, *10*, 103-115.
5. Cowley, S.W.H. (1997) Excitation of flow in the Earth's magnetosphere-ionosphere system: observations by incoherent scatter radar, *this volume*.
6. Lockwood, M. (1994) Ionospheric signatures of pulsed magnetopause reconnection, *in "Physical signatures of magnetopause boundary layer Processes", ed. J.A. Holtet and A. Egeland, NATO ASI Series C, Vol. 425*, Kluwer, 229-243.
7. Wygant, J. R., R. B. Torbert and F. S. Mozer (1983) Comparison of S3-2 polar cap potential drops with the interplanetary magnetic field and models of magnetopause reconnection, *J. Geophys. Res.*, *88*, 5727.
8. Lockwood, M., and S.W.H. Cowley (1992) Ionospheric Convection and the substorm cycle, *in "Substorms 1, Proceedings of the First International Conference on Substorms, ICS-1", ed C. Mattock, ESA-SP-335*, 99-109, European Space Agency Publications, Nordvijk, The Netherlands.
9. Fox, N.J., M. Lockwood, S.W.H. Cowley, M.P. Freeman, E. Friis-Christensen, D.K. Milling, M. Pinnock and G.D. Reeves (1994) EISCAT observations of unusual flows in the morning sector associated with weak substorm activity, *Annales Geophys. 12*, 541-553.
10. Russell, C.T. (1972) The configuration of the magnetosphere, in *Critical Problems of Magnetospheric Physics*, edited by E.R. Dyer, p.1, Nat. Acad. Sciences, Washington.
11. Crooker N.U. (1992) Reverse convection, *J. Geophys. Res.*, *97*, 19,363-19,372.
12. Maynard, N.C., W.J. Burke, D.R. Wiemer, F.S. Mozer, J.D. Scudder, W.K. Peterson, R.P. Lepping and C.T. Russell (1997) Polar observations of cusp electrodynamics: evolution from 2- to 4- cell convection patterns, *this volume*.
13. Freeman, M.P., C.J. Farrugia, L.F. Burlaga, M.R. Hairston, M.E. Greenspan, J.M. Ruohoniemi, and R.P. Lepping (1993) The interaction of a magnetic cloud with the earth: ionospheric convection in the northern and southern hemispheres for a wide

range of quasi-steady interplanetary magnetic field conditions, *J. Geophys. Res., 98,* 7633-7655.

14. Sandholt, P.E., C.J. Farrugia, J. Moen and B. Lybekk (1997) The dayside aurora and its regulation by the interplanetary magnetic field, *this volume.*

15. Fuselier, S., B.J. Anderson, and T,G, Onsager (1997) Electron and ion signatures of field line topology at the low shear magnetopause, *J. Geophys. Res.,* in press.

16. Fuselier, S., B.J. Anderson, and T,G, Onsager (1995) Electron and ion signatures of field line topology at the low shear magnetopause, *J. Geophys. Res.,100,* 11805-11814.

17. Anderson, B.J., T.D. Phan, and S.A. Fuselier (1997) Relationships between plasma depletion and subsolar reconnection, . *J. Geophys. Res., 102.,* 9531-9542.

18. Crooker, N.U., and F.J. Rich (1993) Lobe-cell convection as a summer phenomenon, *Geophys. Res., 98,* 13,403-13,407.

19. Knipp, D.J., et al. (1993) Ionospheric convection response to slow, strong variations in a northward interplanetary magnetic field: A case for study for January 14 1988, *J. Geophys. Res.,* 98, 19,273-19,292.

20. Song, P., and C.T. Russell (1992) Model of the formation of the low-latitude boundary layer for strongly northward interplanetary magnetic field, *J. Geophys. Res., 97,* 1411-1420.

21. Hall, A.M., M. Lockwood, C.H. Perry, M. Grande, B. Kellet, M. Lester, G. Reeves, H.E. Spence, J. Woch and J. Fennell (1997) Dayside Polar observations of dispersed, substorm-associated particle injection features in the vicinity of the cusp, *Annales Geophys.,* in press.

22. Lockwood, M., and M.F. Smith (1994) Low- and mid-altitude cusp particle signatures for general magnetopause reconnection rate variations: I - Theory, *J. Geophys. Res., 99,* 8531-8555.

23. Onsager T.G., C.A. Kletzing, J.B. Austin, and H. MacKiernan (1993) Model of magnetosheath plasma in the magnetosphere: Cusp and mantle particles at low-altitudes, *Geophys. Res. Lett., 20,* 479-482.

24. Lockwood, M. (1995) The location and characteristics of the reconnection X-line deduced from low-altitude satellite and ground-based observations: 1. Theory, *J. Geophys. Res., 100,* 21791-21802.

25. Lockwood, M., and C.J. Davis (1996) On the longitudinal extent of magnetopause reconnection bursts, *Annales Geophys., 41,* 865-878.

26. Lockwood, M. (1997) Energy and pitch angle dispersions of LLBL/cusp ions seen at middle altitudes: predictions by the open magnetosphere model, *Annales Geophys.,* in press.

27. Lockwood, M., C.J. Davis, T.G. Onsager, and J.A. Scudder (1997) Modelling signatures of pulsed magnetopause reconnection in cusp ion dispersion signatures seen at middle altitudes, *Geophys. Res. Lett.,* in press.

28. Lockwood, M., and M.A. Hapgood (1997) How the Magnetopause Transition Parameter Works, *Geophys. Res. Lett., 24,* 373-376.

29. Newell, P.T. and C.-I. Meng (1992) Mapping the dayside ionosphere to the magnetosphere according to particle precipitation characteristics, *Geophys. Res. Lett., 19,* 609-612.

30. Vasyliunas, V.M. (1979) Interaction between the magnetospheric boundary layers and the ionosphere, in *Proceedings of the Magnetospheric Boundary Layers Conference, Alpbach*, pp.387-394, ESA SP-148, ESA, Paris.

31. Baker, K., A.S. Rodger and G. Lu (1997) HF radar observations of dayside magnetic merging rate: A Geospace Environment Modelling boundary layer campaign study, *J. Geophys. Res., 102*, 9601-9618.

32. Jørgensen, T.S., E. Friis-Christiansen, V.B. Wickwar, J.D. Kelly, C.R. Clauer, and P.M. Banks, P.M. (1984) On the reversal from "sunward" to "antisunward" plasma convection in the dayside high latitude ionosphere, J. *Geophys. Res. Lett., 1*, 887-890.

33. Etemadi, A., S. W. H. Cowley, M. Lockwood, B. J. I. Bromage, D. M. Willis, and H. Lühr (1988) The dependence of high-latitude dayside ionospheric flows on the north-south component of the IMF, a high time resolution correlation analysis using EISCAT "POLAR" and AMPTE UKS and IRM data, *Planet. Space Sci., 36*, 471.

34. Hairston, M.R. and R.A. Heelis (1990) Model of the high-latitude ionospheric convection pattern during southward Interplanetary Magnetic Field using DE 2 data, *J. Geophys. Res., 95*, 2333.

35. Heppner, J. P. and N. C. Maynard (1987) Empirical high-latitude electric field models, *J. Geophys. Res., 92*, 4467.

36. Reiff, P. H. and J. G. Luhmann (1986) Solar wind control of the polar cap voltage, in *'Solar Wind-Magnetosphere Coupling'*, edited Y. Kamide and J.A. Slavin, p. 453, Terra Scientifica, Tokyo.

37. Cowley, S.W.H. (1984) Solar wind control of magnetospheric convection, in *Achievements of the international magnetospheric study, IMS*, pp483-494, ESA SP-217, ESTEC, Noordwijk, The Netherlands.

38. Boyle, C.B., P.H. Reiff, and M.R. Hairston (1997) Empirical polar cap potentials, *J. Geophys. Res., 102*, 111-125.

39. Maynard, N.C., E.J. Weber, D.R. Wiemer, J. Moen, T. Onsager, R.A. Heelis and A. Egeland (1997) How wide in magnetic local time is the cusp? An event study, *J. Geophys. Res., 102*, 4765-4776.

40. Lockwood, M. (1997) The relationship of dayside auroral precipitations to the open-closed separatrix and the pattern of convective flow, *J. Geophys. Res., 102*, 17475-17487.

41. Lockwood, M., S.W.H. Cowley and T.G. Onsager (1996) Ion acceleration at both the interior and exterior Alfvén waves associated with the magnetopause reconnection site: signatures in cusp precipitation, *J. Geophys. Res., 101*, 21501 - 21515.

42. Lockwood, M., and J. Moen (1996) Ion populations on open field lines within the low-latitude boundary layer: theory and observations during a dayside transient event, *Geophys. Res. Lett., 23*, 2895-2898.

43. Crooker, N.U., F. Toffoletto, and M.S. Gusenhoven (1991) Opening the cusp, *J. Geophys. Res., 96*, 3497-3503.

44. Newell, P.T. W.J. Burke, E.R. Sanchez, C-I. Meng, M.E. Greenspan, and C.R. Clauer (1991) The low-latitude boundary and the boundary plasma sheet at low altitude: prenoon precipitation regions and convection reversal boundaries, *J. Geophys. Res., 96*, 21,013-21,023.

45. Nishida, A., T. Mukai, H. Hayakawa, A. Matsuoka, and K. Tsuruda (1993) Unexpected features of the ion precipitation in the so-called cleft/low-latitude boundary layer region: association with sunward convection and occurrence on open field lines, *J. Geophys. Res., 98,* 11,161-11,176.

46. Newell, P.T. and C.-I. Meng (1993) Reply, *Geophys. Res. Lett., 20,* 1739-1740.

47. Lockwood, M., and M.F. Smith (1993) Comment on "Mapping the dayside ionosphere to the magnetosphere according to particle precipitation characteristics" by Newell and Meng, *Geophys. Res. Lett., 20,* 1739-1740.

48. Smith, M.F. (1994) Transient reconnection and its effects on the ionosphere, *"Physical signatures of magnetopause boundary layer Processes", ed. J.A. Holtet and A. Egeland, NATO ASI Series C, Vol. 425,* Kluwer, 275-289.

49. Newell, P.T., and C.-I. Meng (1997) Open and closed low-latitude boundary layer *this volume*.

50. Watermann, J., O. de la Beujardiére and H.E. Spence (1993) Space-time structure of the morning aurora inferred from coincident DMSP-F6,-F8, and Sondrestrom incoherent scatter radar observations, *J. atmos. terr. phys.,* **55,** 1728-1739.

51. Newell, P.T. W.J. Burke, E.R. Sanchez, Ching-I. Meng, M.E. Greenspan, and C.R. Clauer (1991) The low-latitude boundary and the boundary plasma sheet at low altitude: prenoon precipitation regions and convection reversal boundaries, *J. Geophys. Res., 96,* 21,013-21,023.

52. de la Beaujardiere O., P. Newell, and R. Rich (1993) Relationship between Birkeland current regions, particle participation, and electric fields, *J. Geophys. Res., 98,* 7711-7720.

53. Ohtani, S.-I., T.A. Potemra, P.T. Newell, L.J. Zanetti, T. Iijima, M. Wantanabe, M. Yamauchi, R.D. Elphinstone, O de la Beaujardiére and L.G. Blomberg (1995) Simultaneous prenoon and postnoon observations of three field-aligned current systems from Viking and DMSP-F7, *J. Geophys. Res., 100,* 119-136.

54. Gosling, J.T., M.F. Thomsen, S.J. Bame, T.G. Onsager and C.T. Russell (1990) The electron edge of the low-latitude boundary layer during accelerated flow events, *Geophys. Res. Lett., 17,* 1833-1836.

55. Burch, J. L. (1855) Quasi-neutrality in the polar cusp, *Geophys. Res.Lett.,12,* 469-472.

56. Cowley, S.W.H. and Z.V. Lewis (1990) Magnetic trapping of energetic particles on open dayside boundary layer flux tubes, *Planet. Space Sci., 38,* 1343.

57. Scholer, M., P.W. Daly, G. Paschmann, and T.A. Fritz (1982) Field line topology determined by energetic particles during a possible magnetopause reconnection event, *J. Geophys. Res., 87,* 6073.

58. Daly, P.W. and T.A. Fritz (1982) Trapped electron distributions on open field lines, *J. Geophys. Res., 87,* 6081.

59. Lockwood, M. (1997) Testing Substorm Theories: The Need For Multipoint Observations, *Adv. in Space Res.,* 883-894.

60. Lockwood, M., S.W.H. Cowley, H. Todd, D.M. Willis and C.R. Clauer (1998) Ion flows and heating at a contracting polar cap boundary, *Planet. Space Sci., 36,* 1229-1253.

90

61. Onsager, T.G. and T. Mukai (1995) Low altitude signature of the plasma sheet boundary layer: Observations and model, *Geophys. Res. Lett. 22*, 855-858.
62. Onsager, T.G., and T. Mukai (1996) The structure of the plasma sheet and its boundary layers, *J. Geomag. Geoelect., 48*, 687-698.
63. Bosqued J.M., et al. (1993) Dispersed ion structures at the poleward edge of the auroral oval: low-altitude observations and numerical modelling, *J. Geophys. Res., 98*, 19,181-19,204.
64. Senior C., D. Delcourt, J.-C. Cerisier, C. Hanuise, J.-P. Villian, R.G. Greenwald, P.T. Newell and F.J. Rich (1994) Correlated observations of the boundary between polar cap and nightside auroral zone by HF radars and the DMSP satellite, *Geophys. Res. Lett., 21*, 221-224.
65. Fukunishi, H., Y. Takahashi, T. Nagatsuma, T. Mukai and S. Machida (1993) Latitudinal structures of nightside field-aligned currents and their relationships to the plasma sheet regions, *J. Geophys. Res., 98*, 11,235-11,255.
66. Persson, M.A.L., A.T. Aikio and H.J. Opgenoorth (1994) Satellite-groundbased coordination: Late growth and early expansion phase of a substorm, in *"Substorms 2"*, Proc. 2nd. Int. Conf. on substorms, 421-428, Geophysical Institute, Fairbanks, Alaska.
67. Persson, M.A.L., et al. (1994) Near-earth substorm onset: A co-ordinated study, *Geophys. Res. Lett. 21*, 1875-1878.
68. Gazey, N.G.J., M. Lockwood, M. Grande, C.H. Perry, P.N. Smith, S. Coles, A.D. Aylward R.J. Bunting, H. Opgenoorth and B. Wilken (1996) EISCAT/CRRES observations: nightside ionospheric ion outflow and oxygen-rich substorm injections, *Annales Geophys., 14 ,* 1032-1043
69. de la Beaujardière, O., L.R. Lyons, J.M. Ruohoniemi, E. Friis-Christensen, C. Danielsen, F.J. Rich and P.T. Newell (1994) Quiet-time intensifications along the poleward auroral boundary near midnight, *J. Geophys Res., 99 ,* 287-298.
70. Shirai, H., K. Maezawa, M. Fujimoto, T. Mukai, T. Yamamoto, Y. Saito, S. Kokubun and N. Kaya (1997) Drop-off in the polar rain near the plasma sheet boundary, *J. Geophys. Res., 102*, 2271-2278.
71. Lester, M., Lockwood. M, T.K. Yeoman, S.W.H. Cowley, H. Lühr, R. Bunting and C.J. Farrugia (1995) The response of ionospheric convection in the polar cap to substorm activity, *Ann. Geophys., 13*, 147-158.

OPEN AND CLOSED LOW LATITUDE BOUNDARY LAYER

PATRICK T. NEWELL AND CHING-I. MENG
Johns Hopkins University Applied Physics Laboratory
Johns Hopkins Rd., Laurel, Maryland, 20723

Abstract. We review and summarize the evidence that a closed LLBL exists, at least away from noon. We also emphasize the observation that just equatorward of the LLBL, overlapping magnetosheath and magnetospheric electrons exist on sunward convecting field lines which have no low-energy ion cutoff (i.e., closed field lines). Therefore it is inescapable to conclude that sheath plasma is introduced onto closed field lines. The dropoff of the high-energy electrons is identified not necessarily with the open/closed boundary but rather with the convection reversal boundary, since these electrons originate on the nightside and convect towards the dayside (hence they cannot exist on anti-sunward convecting field lines, even if the latter are closed). A model is discussed wherein diffusion introduces sheath plasma onto closed field lines, but merging removes these same field lines from the dayside. The competition between these two effects explains why the LLBL is thinnest at noon, where open LLBL signatures dominate (perhaps exclusively), and why the LLBL is thicker for northward IMF.

1. Introduction

The high-latitude magnetospheric boundary layer -- the plasma mantle -- was first reported by Rosenbauer et al. [1]. Observations of a permanent magnetospheric boundary layer on the frontside at lower latitudes was first reported by Eastman et al. [2]. However because these latter authors did not realize that the boundary layer had radically different characteristics at high and low latitude, they simply termed their discovery the "magnetospheric boundary layer". The term "Low-latitude boundary layer", was apparently

J. Moen et al. (eds.), Polar Cap Boundary Phenomena, 91–101.
© 1998 *Kluwer Academic Publishers.*

introduced by Haerendel *et al.* [3] to distinguish the very different properties observed at latitudes below about 50-60° on the magnetopause surface from the plasma mantle found above this latitude. These original researchers believed that the plasma mantle is open, and the LLBL closed.

Eastman *et al.* [4] discussed a number of reasons for believing that the LLBL is closed. They stressed that the flow speeds in the LLBL are typically slower than in the magnetosheath, rather than accelerated as merging might predict depending on the precise geometry (however these authors did find some cases of accelerated flows in high magnetic shear situations). The details of the energetic particles distribution functions also suggested trapped particles (they briefly ruled out magnetic mirroring at the magnetopause as an adequate explanation). The flow speeds were most often found to be aligned with the magnetosheath flow direction.

Further research continued to support the idea that much of the LLBL was closed. Sánchez and Siscoe [5] found that at large values of |z| the magnetotail boundary was generally a rotational discontinuity (RD, indicating field lines merged with the IMF) while for small |z| a tangential discontinuity (TD, indicating an open/closed transition) was often found. Mitchell *et al.* [6] found that the details of the LLBL energetic particles (tens of keV) distributions were consistent with the LLBL being typically closed for northward IMF, but that for southward IMF part or all of the LLBL could be on open field lines. These authors also reported the interesting result that the LLBL was thicker for northward IMF, and thinnest at the nose of the magnetopause.

Recently stress has been placed on the open LLBL [7, 8]. Notice however that these reports of an open LLBL seem concentrated near noon, while reports away from noon in the ionosphere and on the flanks of the magnetosphere continue to imply closed LLBL field lines. While there is no doubt that the field lines immediately equatorward of the cusp are "open LLBL" [9] -- for example they include the low-energy ion cutoff signature [10] -- the evidence for a closed LLBL away from noon remains strong.

In Section 2 we present an example clearly demonstrating that magnetosheath plasma can penetrate onto closed field lines. The open LLBL signature is discussed in Section 3, while the discussion in Section 4 presents a model which attempts to reconcile the varying observations.

2. Closed LLBL: A Clear Example

Figure 1 present an example of an LLBL crossing around 10 MLT as observed by the DMSP F7 satellite. This particular crossing was selected because simultaneous Sondrestrom data are available (courtesy of C. R. Clauer); the convection reversal boundary is shown on Figure 1. Notice the region of overlapping soft magnetosheath plasma with high energy magnetospheric electrons, which overlap region lies equatorward of the

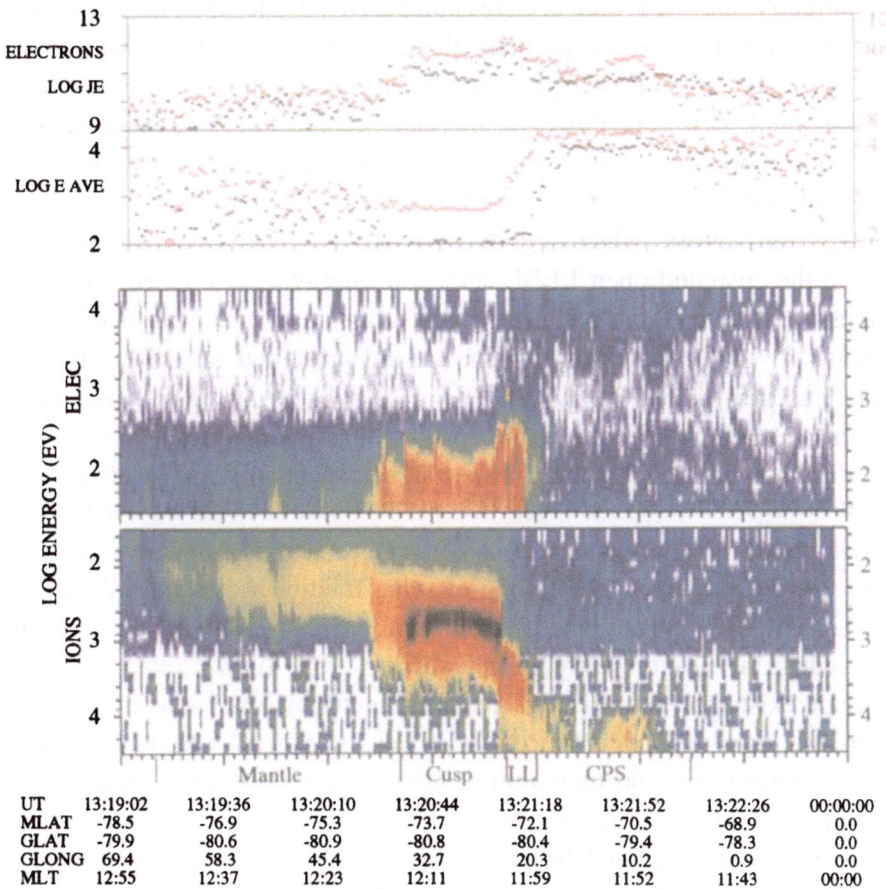

Figure 1. A DMSP F7 spectrogram from March 16, 1984 around 12 UT and 10 MLT. The CRB is indicated by an arrow. Note the region of overlapping magnetosheath and magnetospheric plasma which is convecting sunward.

convection reversal boundary [cf 11]. Let us consider several of the properties of the closed LLBL.

First, note that the average electron energy lies in the range 200-300 eV. Because this value depends on the range measured (i.e., it would be lower if DMSP measured electrons below 32 eV) it is helpful to report fitted electron temperatures; these are generally in the range 70-100 eV for this instance. Like other LLBL crossings at high-altitude [12] and low-altitude [13], these values are higher than typical of the magnetosheath (or cusp). The reason for this slightly elevated temperature compared to the sheath/cusp is not known; but since the ions are partially thermalized in the LLBL (with lower flow velocities but higher average energies), it is possible the electrons have acquired a small portion of this extra thermal energy. Another explanation is that diffusion works better for higher energy particles with larger gyroradi.

Next, note that the ions are spectrally complete. In particular, there is no low-energy ion cutoff, which signifies recently opened field lines. By contrast in the cusp and open LLBL such low-energy cutoffs exist, implying field lines merged so recently that most of the magnetosheath plasma has not yet reached low-altitude (individual spectra are presented in [13]). Indeed, from the presence of ions below 100 eV in the appropriate abundance, these field lines cannot have merged within about the last 10 minutes.

Consider now the region in which the magnetosheath plasma overlaps with the electrons of several keV energy. These electrons are moving many R_E/s, and hence would escape from the magnetosphere within a few seconds after the field line became open (depending on what fraction was reflected on each encounter with the magnetopause). Certainly this overlap region must be closed; indeed it lies on sunward convecting field lines. Inescapably a mechanism exists for introducing magnetosheath plasma onto closed field lines. In fact time-dependent injections of magnetosheath plasma on closed LLBL field lines in the morning region have been reported by *Clemmons et al.* [14] in several rocket flights.

Next, consider the region labeled "LLBL" in Figure 1 (convecting anti-sunward). No significant high energy electrons exist in this region, suggesting that it might be open. This hypothesis is however contradicted by many observations. First, the ion density is about a factor of 5 lower than in the cusp/magnetosheath. This is generally true of the LLBL at both high and low altitudes [3, 4, 10]. Although the density profile of the LLBL can be reproduced in a merging model by time-of-flight effects, this does not work when the ion spectra show no low energy cutoff (as in the spectra presented by Haerendel et al. and by us). Similarly, the fact that the electrons are slightly more energized in the LLBL than in the sheath (or cusp, or mantle) is difficult to account for in the merging model (if the merging did the energizing, these other regions would not be less energetic).

It is now possible to reproduce cusp/mantle observations in detail using a simple merging model [15, 16]. A field line which is open does not go through any stage in which its precipitation resembles that shown in Figure 1. Indeed, if an open field line maps to the frontside, its ion intensity will soon match the magnetosheath, as will its electron temperature. If it maps downstream from the magnetic cusp, the ions will be de-energized in entering the magnetosphere, and only weak soft fluxes will reach low-altitude. There does not appear to be any merging mechanism which would allow open field lines to have 1/5 the density of magnetosheath plasma, with slightly elevated electron temperatures, and no low-energy ion cutoffs. Incidentally, a plasma depletion layer cannot be appealed to as an explanation of the lower density, since (i) the same lower density is observed for southward IMF as northward; and (ii) the dropoff in density is observed in flank passes from the sheath into the LLBL.

3. The Open LLBL

Next we consider a fairly clear case of open LLBL, namely Figure 2 (a DMSP F7 pass from December 10, 1983). The open LLBL region, lying immediately equatorward of the cusp, is indicated by the two arrows on the spectrogram. Notice the narrow region of electrons at approximately the same energy but lower density in the open LLBL. This comes about because of charge quasineutrality: the majority of ions have not yet reached the ionosphere, so that a retarding potential keeps out the majority of the electrons. Although it has not been demonstrated, it is probable that this potential is concentrated in a narrow region near the magnetopause, simply because both theory and laboratory experience shows that sheath potentials are common but do not penetrate far into a plasma. Imposing charge quasi-neutrality reproduces this region quite nicely in model spectrograms [16].

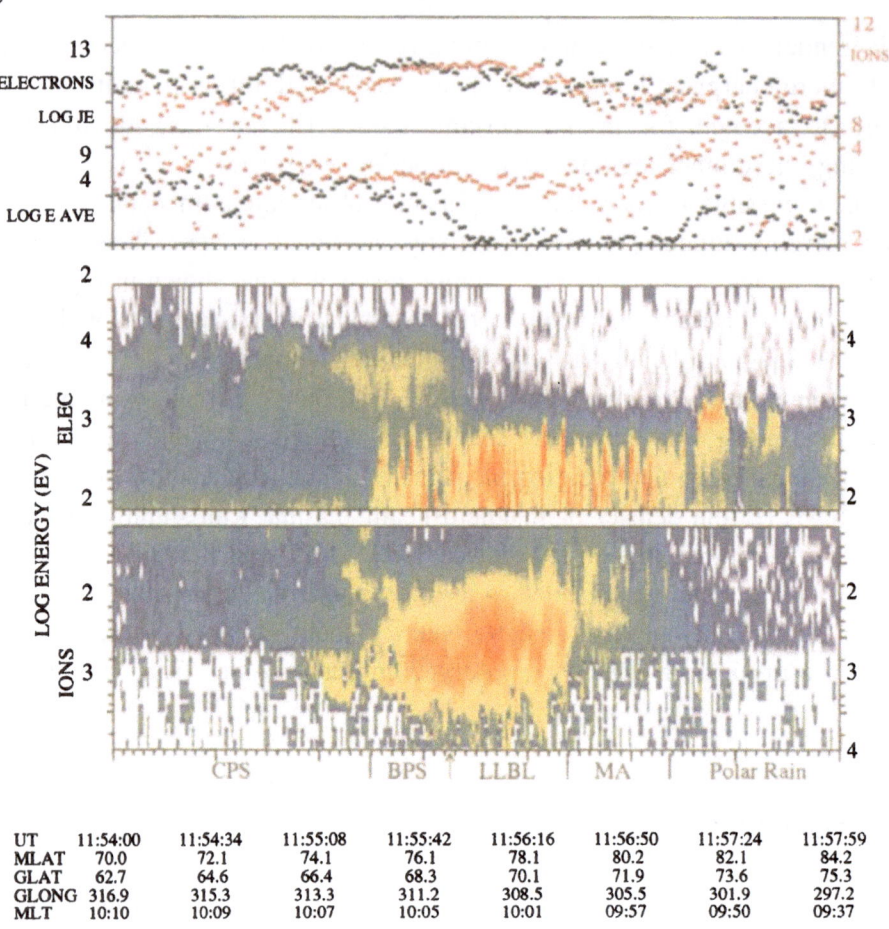

Figure 2. An open LLBL with equatorward high energy ions.

The precipitation of high energy ions extends well equatorward of this region of reduced density magnetosheath origin electrons. It is doubtful if such an extended region of ion precipitation could all be considered recently opened; and indeed the lack of accompanying electron precipitation (albeit at a reduced density) strongly argues against this interpretation. Therefore another explanation is needed for the extended region of high energy ion precipitation (the automated identification scheme introduced by Newell *et al.* [17] would attribute most of this to "CPS").

Three different recent studies have clarified the interpretation of this region. Alem and Delcourt [18] studied the non-adiabatic (full 3-d kinetic) motion of ions in a model magnetic field; they found that magnetospheric ions are led to precipitate equatorward of the cusp -- indeed, the region of magnetospheric ion precipitation extended several tenths of a degree into the cusp itself in their model. Similarly, Sergeev *et al.* [19] used various Tsyganenko magnetic field models and followed particle trajectories to conclude that a region of high energy ion precipitation should exist equatorward of the cusp from pitch angle scattered magnetospheric ions (when the magnetic field of curvature approaches about 1/8 that of the radius of curvature of the field lines). Notice that since ion scattering depends on the ion gyroradius, there is a latitude energy dispersion which has the same sense as that normally associated with the southward IMF cusp.

Grande *et al.* [20] have used Polar satellite measurements of energetic particles with composition data to find that the solar wind origin cusp dispersion curve does indeed smoothly join a dispersed magnetospheric origin ion population. The combination of the theoretical and observational work just described seems to nicely account for the extended region of high energy ion precipitation with any magnetosheath electrons. Indeed, the identification of this region as "CPS" appears to be the correct one.

A remainding question is the relationship of the "open LLBL" observed by Fuselier *et al.* [7] under northward IMF conditions near the subsolar point to the low-altitude cusp. They observed evidence of merging poleward of the cusp (in agreement with the reverse ion dispersion curves seen at low-altitude for Bz >0 [21]) which supplied magnetosheath plasma into the low-latitude magnetosphere. Fuselier *et al.* [7] and subsequent work present distribution functions on an extremely compressed scale, so it is not possible to readily infer whether they are observing essentially full magnetosheath density or not. If so, it seem logical to identify the northward IMF stagnating open LLBL of Fuselier *et al.* [7] with the northward IMF cusp as identified at low altitude. The latter has full magnetosheath density, electrons of magnetosheath energy (not heated), but spectrally complete ions.

4. Summary and Discussion

Let us briefly recall the following observations which need to be satisfied: The closed LLBL has lower density which is variable, but typically about 5 times below that of the magnetosheath. The electrons are similar to, but slightly higher temperature than that in the sheath. The LLBL has been

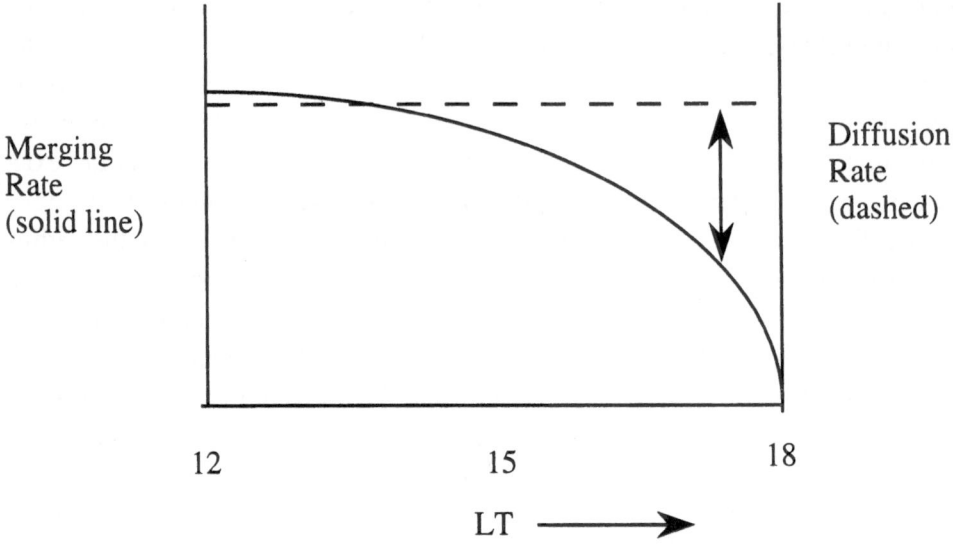

Figure 3. The merging rate depends on local time while the diffusion rate is constant, resulting in the LT-dependent thickness of the closed LLBL. No closed LLBL signatures are expected near noon under normal conditions.

reported from high altitude observations to be thicker away from noon and for northward IMF. Low-altitude observations of a closed LLBL (with spectrally complete ions) occur only away from noon. Low-altitude instances of the open LLBL occur only immediately equatorward of the cusp. Finally, as previously shown [13] and emphasized here, magnetosheath plasma can overlap with high energy electrons; this region is sunward convecting, and without the low-energy ion cutoff which indicates recent merging. Of course high energy ions can exist throughout the LLBL [3].

Figure 3 illustrates an effort to reconcile these observations. Simple kinetic calculations have demonstrated the ease with which magnetosheath plasma can cross onto closed field lines [e.g. 22]. However the merging rate can be expected to peak near 12 LT and decline moving away from noon. Thus two competing processes are at work: diffusion of magnetosheath plasma onto closed field lines, and the stripping away of these field lines followed by their convection into the magnetotail. Only where the merging rate drops well below the diffusion rate does a field line remain exposed to diffusion long enough for a closed boundary layer to accumulate. The boundary layer widens for northward IMF, and moving away from noon, for the same reason.

Next consider convection and the effects on high energy electrons. The calculations of *Omidi and Winske* [22] illustrate how diffusion introduces antisunward flow onto closed field lines. This antisunward flow damps away moving inward from the magnetopause, eventually reversing.   The high energy electrons such as shown in Figure 1 originate from nightside injections followed by subsequent convection around the Earth. Therefore these can be found only in sunward convecting regions. Thus it is that we consistently find that a dayside region containing these high energy electrons is convecting sunward, regardless of whether magnetosheath plasma overlaps the region (and whether the field line is open or closed).

5. Conclusions

Single particle kinetic calculations demonstrate that sheath plasma can cross the magnetopause onto closed field lines. The observation of sheath plasma deep inside the magnetosphere convecting sunward demonstrates that this mechanism is indeed operational, as appears unavoidable from theory. It cannot be argued that these field lines have merged too recently for an Alfvén wave to change the direction of convection, since the sheath ions do not exhibit the appropriate low-energy cutoff.

More generally, the signature of spectrally complete magnetosheath ions at a lower density but slightly higher electron temperature can only be understood as a closed LLBL. Although the open LLBL, the cusp, the mantle, and the polar rain can now all be modeled successfully using a reconnection model, this approach does not produce anything which resembles the closed LLBL signature. In particular, the closed LLBL is not a mixture of magnetosheath and magnetospheric plasmas based on time-of-flight effects; among other reasons the absence of a low-energy ion cutoff in the LLBL sheath-origin population rules this out.

The simple idea of a competition between diffusion, which builds up sheath plasma on closed field lines, and merging, which strips away those field lines, allows for an understanding of a number of otherwise puzzling observations about the LLBL. It explains why the LLBL is thinner at the subsolar point, and why "closed" LLBL signatures are rarely seen near noon at low-altitude. It explains why the LLBL is thicker for northward IMF, and how the density in the LLBL can be ~5 times lower than in the sheath, despite the absence of a low-energy ion cutoff. The superior ability of larger gyroradi electrons to cross the magnetopause might explain the somewhat higher electron temperatures observed in the LLBL. The high-altitude observation of lower bulk flow velocities which are however aligned with the magnetosheath flow becomes readily understandable. Finally the inability of the "Onsager-class" quantitative cusp simulation models to reproduce a "closed-LLBL" type signature is of course predictable.

6. References

1. Rosenbauer, H., H. Grunwaldt, M. D. Montgomery, G. Paschmann, and N. Sckopke (1975) *J. Geophys. Res., 80,* 2723.
2. Eastman, T. E., E. W. Hones, S. J. Bame, and J. R. Asbridge (1976) *Geophys. Res. Let., 3,* 685.
3. Haerendel, G., G. Paschmann, N. Sckopke, H. Rosenbauer, and P. C. Hedgecock (1978) *J. Geophys. Res., 83,* 3195.
4. Eastman, T. E., B. Popielawska, and L. A. Frank (1985) *J. Geophys. Res., 90,* 9519.
5. Sánchez, E. R., and G. L. Siscoe (1990) *J. Geophys. Res., 95,* 20771.
6. Mitchell, D. G., F. Kutchko, D. J. Williams, T. E. Eastman, L. A. Frank, and C. T. Russell (1987) *J. Geophys. Res., 92,* 7394.
7. Fuselier, S. A, B. J. Anderson, and T. G. Onsager (1995) *J. Geophys. Res., 100,* 11805.
8. Lockwood, M., and J. Moen (1996) *Geophys. Res. Lett., 23,* 2895.
9. Lockwood, M., W. F. Denig, A. D. Farmer, V. N. Davda, S. W. H. Cowley, and H. Lühr (1993) *Nature, 361,* 424.
10. Newell, P. T., and C.-I. Meng (1995) *J. Geophys. Res., 100,* 21943.
11. Yamauchi, M., J. Woch, R. Lundin, M. Shapshak, R. Elphinstone (1993) *Geophys Res. Lett., 20,* 795.
12. Formisano, V. (1980) *Planet. Space Sci., 28,* 245.

13. Newell, P. T., W. J. Burke, E. R. Sánchez, C.-I. Meng, M. E. Greenspan, and C. R. Clauer (1991) *J. Geophys. Res., 96,* 21013.

14. Clemmons, J. H., C. W. Carlson, and M. H. Boehm (1995) *J. Geophys. Res., 100,* 12133.

15. Onsager, T. G., C. A. Kletzing, J. B. Austin, and H. MacKiernan (1993) *Geophys. Res. Lett., 20,* 479.

16. Wing, S., P. T. Newell, and T. G. Onsager (1996) *J. Geophys. Res., 101,* 13155.

17. Newell, P. T., S. Wing, C.-I. Meng, and V. Sigillito (1991) *J. Geophys. Res., 96,* 5877.

18. Alem, F. and D. C. Delcourt (1995) *J. Geophys. Res., 100,* 19321.

19. Sergeev, V. A., G. R. Bikkuzina, and P. T. Newell (1997) *Annales. Geophys.*

20. Grande, M., J. Fennell, S. Livi, B. Kellett, C. H. Perry, P. Anderson, J. Roeder, H. Spence, T. Fritz, and B. Wilken (1997) *Geophys Res. Lett., 24.*

21. Woch, J., and R. Lundin (1992) *J. Geophys. Res., 97,* 1421.

22. Omidi, N., and D. Winske (1995) *J. Geophys. Res., 100,* 11935.

THE DYNAMIC MAGNETOSPHERE

T. A. POTEMRA

The Johns Hopkins University
Applied Physics Laboratory
Laurel, Maryland 20723

Abstract. The shape of the magnetosphere is defined by a complicated system of electric currents which flow within it in response to the interaction of the Earth's magnetic field with the solar wind. Changes in the solar wind pressure, such as occur during SSCs (storm sudden commencements), alter these complicated current systems and excite a variety of magnetohydrodynamic waves. The magnetic fields associated with these phenomena have been observed on the ground for over a century. For example, Stewart [1] reported on "pulsations" in the geomagnetic field observed during a major magnetic storm in 1859, and Birkeland [2] is credited as being the first to report on magnetic field oscillations later called "giant pulsations" [3]. The periods of these oscillations, in almost all cases, are shorter than about 10 min. Examples have been found in the Viking magnetic field measurements in which the total magnetic field varied in a quasi periodic manner with periods from 7 to 23 min (with an average of 12.2 min). Similar long-period fluctuations were also observed by a global network of ground-based magnetic observatories during two of the Viking examples. These long-period oscillations are interpreted as the periodic compression and relaxation of the magnetosphere, a long-period "breathing mode" of the magnetosphere. These long period variations are difficult to detect because their amplitudes are small (about 4 nT at Viking) and consequently may be overwhelmed by the higher frequency and stronger amplitude ULF waves often present in ground-based and satellite observations. However, several examples were readily found in the Viking data set, due to the unusually quiet conditions that prevailed during Viking's lifetime. Therefore, these long-period breathing modes may not be uncommon. These observations provide the opportunity to study long-period "equilibrium" global changes of the magnetosphere uncomplicated by transient effects, which exist at periods shorter than about 10 min (longer than most known resonances of the magnetosphere).

J. Moen et al. (eds.), Polar Cap Boundary Phenomena, 103–114.

1. The Historical Significance of Svalbard to Space Science

The international race to claim the Earth's poles captured the world's attention at the end of the nineteenth century. In many ways, that quest was similar to the space race that began nearly a century later. The Norwegian scientist Fridtjof Nansen was a marine biologist, oceanographer, intrepid explorer, artist, poet, and humanitarian, as well as a founding father of his country and a Nobel laureate. An early competitor in the pole race, he conceived, planned, and promoted a scientific endeavor comparable in scope to modern spacecraft projects. He constructed a specially designed ship, intentionally froze her in the Arctic ice, and used her as a scientific platform, a kind of spacecraft of the time. From 1893 to 1896, he and his crew conducted biological, oceanographic, atmospheric, geomagnetic, and auroral observations. Nansen's ship, the *Fram*, left Oslo on 24 June 1893 with 13 crew, and sailed north along the coast of Siberia and entered the polar ice pack on 22 September 1893. As planned by Nansen, the *Fram* became locked in the ice and began to drift toward the west. But in the spring of 1895, he realized that the drift of the *Fram* would not after all take her across the North Pole (see the map of the "Route of the *Fram*" in Nansen [4] or as reproduced in Potemra [5]. On 14 March 1895, Nansen left the *Fram* with Frederick Hjalmar Johansen, the ship's stoker, in an attempt to reach the Pole by dogsled. They reached "farthest north" (the title of Nansen's famous book published in 1897) at 86° 14′ N on 7 April 1895. Nansen and Johansen then turned south. On 17 June 1896, having walked for a year and three months after leaving the *Fram*, they met the British explorer Frederick Jackson on Cape Flora in Franz Joseph Land—one of the most remarkable chance meetings in history. Nansen sailed aboard Jackson's relief ship to Hammerfest on 21 August 1896. The *Fram*, under the command of Otto Sverdrup, was released from the polar ice near Spitsbergen on the same day that Nansen had arrived on Norwegian soil again, after drifting nearly three years in the ice.

In the early 1890s, the discovery of x-rays and electrons inspired Kristian Birkeland to develop vacuum chambers to study the influence of magnets on cathode rays. He discovered that charged particles, later identified as electrons, were guided toward the magnetic poles of a miniature "terrella" in his chamber where they produced visible emissions. A drawing of his terrella experiment appears on the left front of the 200 Crown Norwegian bank note, which also includes Birkeland's portrait. Birkeland concluded that the aurora was produced in the same way. He conducted several expeditions to the high-latitude regions to collect magnetic field data in order to test his theories. He established a network of geomagnetic observatories in the auroral regions. His highest latitude station was set up at Axelöen (now called Akselöya) in Spitsbergen, as shown in the map on the frontispiece of his book on "The Norwegian Aurora Polaris Expedition 1902–1903" Birkeland [6], reproduced as Figure 1 here. Birkeland's station was about 60 km from Longyearbyen, the site of this conference.

Birkeland used the observations collected from his ground-based network to derive patterns of ionospheric currents and he discovered the important role that these currents had in auroral phenomena and solar activity. These currents are referred to today as "auroral electrojets." Bikeland [6] also provided a definition for "elementary magnetic storms" as follows. We consider it to be beyond doubt that the powerful storms in the northern regions, both those of long duration, and the short, well-defined storms that we have called elemen-

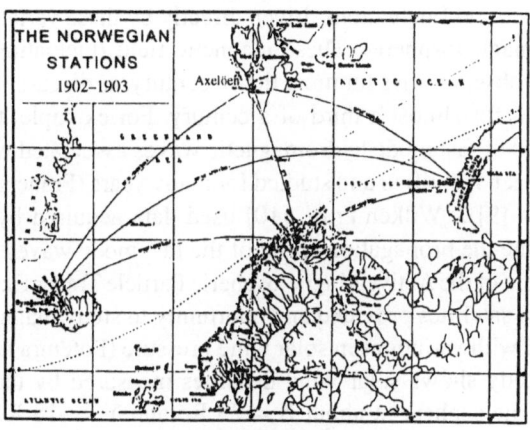

**THE NORWEGIAN
AURORA POLARIS EXPEDITION
1902—1903**

VOLUME I

ON THE CAUSE OF MAGNETIC STORMS AND
THE ORIGIN OF TERRESTRIAL MAGNETISM
BY

KR. BIRKELAND

SECOND SECTION

THE NORWEGIAN
STATIONS
1902–1903

CHRISTIANIA
H. ASCHEHOUG & CO.

LEIPZIG　　　　　**LONDON, NEW YORK**　　　　**PARIS**
JOHANN AMBROSIUS BARTH　　LONGMANS, GREEN & CO.　　C. KLINCKSIECK

Figure 1. The frontispiece of "The Norwegian Aurora Polaris Expedition 1902-1903, Vol. I. On the cause of magnetic storms and the origin of terrestrial magnetism," by Kr. Birkeland (1908, 1913). This shows the locations of the ground-based magnetic observatories, inlcuding the one in Spitsbergen established by Birkeland.

tary, are due to the action of electric currents above the surface of the Earth near the auroral zone. These currents, as far as the elementary storms are concerned at any rate, act, in the districts in which the perturbation is most powerful, as almost linear currents, that for a considerable distance are approximately horizontal."

He went on to say in the same book: "With regard to the further course of the current, there are two possibilities that may be considered. (1) The entire current system belongs to the Earth. The current-lines are really lines where the current flows upon the Earth's surface, or rather at some height above it. (2) The current is maintained by a constant supply of electricity from without. The current will consist principally of vertical portions. At some distance from the Earth's surface, the current from above will turn off and continue for some time in an almost horizontal direction, and then either once more leave the Earth, or become partially absorbed by the atmosphere." This was the first suggestion of the concept of field-aligned currents, which was argued and disputed over the following years, but only confirmed much later with satellite-borne magnetic field experiments. These field-aligned currents are now referred to as "Birkeland" currents, and are regarded as being an important component of the three-dimensional current system that defines the magnetosphere.

2. The Changing Shape of the Magnetosphere

It is now generally accepted that the shape of the magnetosphere is not static, but changes on time scales ranging from minutes to a solar cycle. Variations in the solar wind produce a wide range of magnetic field fluctuations driven directly by these variations and by resonant properties of the magnetosphere. These magnetic field fluctuations have been observed by ground-based observatories for more than a century, as discussed in the preceding section, and by satellites for almost a third of a century. For example, the ground-based magnetic field signatures of fast mode hydromagnetic waves associated with storm sudden commencements have been identified and studied for many years (Francis *et al.* [7]; Wilson and Sugiura [8]; Tamao [9]). Wilken *et al.* [10] used data acquired by a multi-satellite constellation to determine the propagation speed of the fast mode waves.

A fortunate alignment of the Active Magnetospheric Particle Tracer Explorers (AMPTE) CCE, Viking, and IMP 8 satellites provided an opportunity to study magnetic field fluctuations directly associated with variations in solar wind pressure (Potemra *et al.*, [11]). Figure 2, adopted from this study shows solar wind densities measured by the IMP 8 satellite, located outside the magnetosphere at about X = 29 Re and Y = −18 Re and the parallel magnetic field component from the AMPTE CCE satellite located in the equatorial plane at L = 8, just inside the magnetosphere at approximately 0800 MLT. The solar wind provided a "step" input, in the form a doubling of its pressure at about 0320 UT, and a periodic "driven" input in the form of two cycles of a 10 min wave beginning about 0410 UT. The magnetosphere response to these inputs, as measured by CCE, was a "high frequency" ringing from the step and a reproduction of the 10 min driven input. The later variations (beginning at 0410 UT) were observed, with almost identical characteristics, by the Viking satellite at about 2 Re altitude over the auroral region and on the ground by EISCAT magnetometer Cross (see Figures 4 through 9 in Potemra *et al.* [11]). Figure 2 suggests that long period variations may occur in the solar wind density which produce magnetic field variations with identical periods throughout the magnetosphere and on the ground. These variations, with periods longer than about 10 min, do not involve the excitation of resonant processes, but are the result of the magnetosphere slowly adjusting itself to "equilibrium" states in response to a slowly varying external driver.

In another multi-satellite study with IMP 8, GOES 5, GOES 6, and Viking, magnetic field perturbations due to solar wind dynamic pressure variations were investigated by Erlandson *et al.* [12]. Solar wind variations recorded at IMP 8 produced compressions and

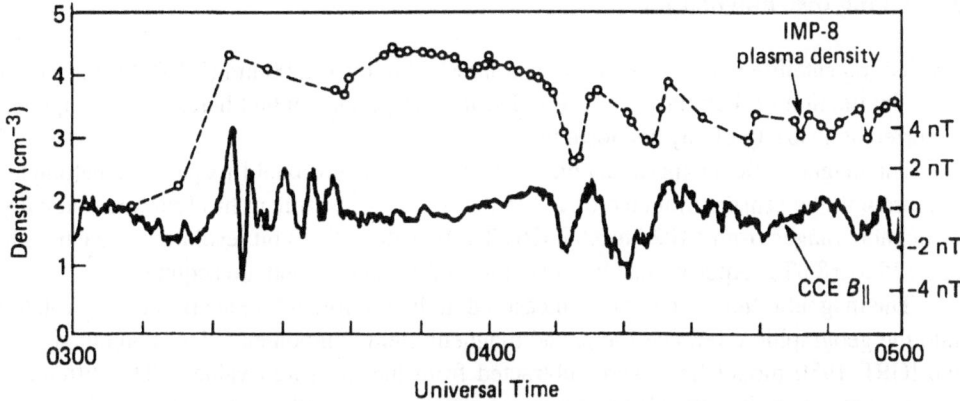

Figure 2. Solar wind plasma density observations from IMP 8 and parallel magnetic field component from AMPTE CCE during 0300 to 0500 UT on April 24, 1986 (adapted from Figure 3 of Potemra et al., 1989).

rarefactions of the magnetosphere as seen in GOES 5 and GOES 6 magnetic field observations at geostationary orbit. These same compressions and rarefactions were also recorded by Viking and on the ground at the Huancayo observatory. An example was provided in which two cycles of a quasi-periodic variation in the total field were observed by Viking with periods of 12 min and 14 min and amplitude of 3 nT (see Figure 8 of Erlandson *et al.* [12]).

In another multi-satellite study with GOES 5, GOES 6, Viking, and AMPTE/CCE and the ground-based Huancayo observatory, two cycles of a 10 min compressional wave were observed (Potemra *et al.* [13]). The amplitudes were about 20 nT at GOES 5 and GOES 6, 10 nT at AMPTE/CCE, 5 nT at Viking, and 25 nT at Huancayo. This study estimated that a solar wind pressure variation of $\Delta p/p = 33\%$ could produce the observed magnetic field variations.

A multi-satellite investigation was also conducted by Takahashi *et al.* [14] with data acquired by the SCATHA, GOES 5, GOES 6, and AMPTE/CEE satellites during a compression of the magnetosphere when the subsolar magnetopause remained near 7 Re for about 2 hours. Magnetic field and particle data were used to study ULF waves in the LLBL and their relationship to magnetic pulsations. The magnetic field exhibited both a 5–10 min irregular compressional oscillation and broad band primarily transverse oscillations with a mean period of about 50 s. The long period compressional oscillations were attributed to pressure variations in the solar wind.

Unique to almost all of these studies, but not specifically pursued, are the observations of quasi-periodic compressional variations of the geomagnetic field with amplitudes between 3 and 5 nT and periods ranging between 10 and 20 min.

3. Viking Instrumentation

The Viking satellite was launched into a polar 817 km by 13,500 km (3.1 Re) orbit with a 98.8° inclination on February 27, 1986. During the period studied here, Viking's apogee occurred at a variety of dayside local times.

The magnetic field experiment on Viking consists of a triaxial flux gate magnetometer system with the sensors mounted on a 2 m boom. This instrument has four automatically switchable ranges from ±1024 to ±65,536 nT full-scale with 13 bit resolution in each scale (±0.125 to ±8 nT). Approximately 53 vector samples per second are acquired.

The magnetic field components measured in the satellite reference frame were rotated into the geographic reference frame and magnetic field component values computed with the IGRF 1980 model field were subtracted from the measured values. The difference components were then transformed into eccentric dipole north, east, and parallel components.

The ground-based magnetic field data were assembled from 28 high- and mid-latitude stations by the Finnish Meteorological Institute for comparison with the Viking satellite during its lifetime from February 1986 to December 1986. The data used in this study were filtered to remove variations with periods shorter than 2 min.

4. Viking Data

Figure 3 shows the magnetic field measurements acquired by Viking on orbit 642 on June 18, 1986. This was a near polar orbit aligned close to the 1600 MLT meridian. The pass was almost directly over Kiruna, Sweden, and over the nearby magnetometer ground station at Abisko, Sweden. No solar wind data were available and the Kp sum was 16+ during this day and the three-hour Kp = 3+ during this pass. The perturbations in the magnetic southward, eastward, and parallel to the main field directions are shown in the top three panels. The perturbations of the total field are shown in the bottom panel. The gradients in the eastward component from about 1335 UT (72° magnetic latitude, MLAT) to 1350 UT (76° MLAT) and then to 1356 UT (78° MLAT) are due to a pair of region 2 and region 1 field-aligned Birkeland currents flowing into and away from the auroral ionosphere.

The bottom two panels display the component parallel to the geomagnetic field and the total magnetic field. These panels show long period variations in the total magnetic field that extend up to the geomagnetic pole. The largest amplitude is about 4 nT and the periods of the variations range from 10 to 13 min, corresponding to frequencies of 1.3 to 1.7 mHz.

Figure 4 shows the negative of the total field perturbation measured by Viking, shown in the bottom panel of Figure 3, in a stack plot with the horizontal magnetic field component variations measured by fourteen ground stations listed in Table 1. A compression of the magnetosphere will produce an increase in the horizontal component of magnetic field on the ground and a reduction of the total field at Viking's altitude in the north polar region. The ground-based data are arranged according to increasing longitude (local time) from top to bottom, with the data from Abisko (ABK) nearly directly below Viking's orbit, at the top. The variations observed by nearly every station are similar to that recorded by Viking. The peak variation at about 1356 UT at Viking occurs within the Alfvén travel time to Abisko,

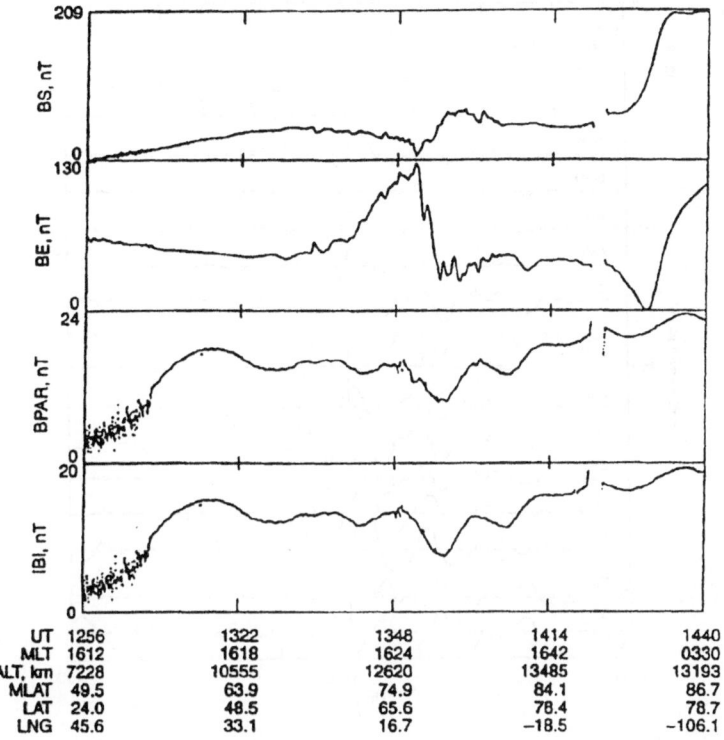

Figure 3. Magnetic field variations in the southward, BS, eastward, BE, and parallel to the main field, BPAR, and total magnetic field |B|, measured by Viking on June 18, 1986 (orbit 642).

and is delayed at stations located at larger local time separations from Viking's foot print. The longest delay is between Viking and Dixon Island (DIX), located at 80.6° east longitude, 63.9° east of Viking. The delays at stations further east than College (CMO) at 212.2° east longitude, half-way around the Earth from Viking, begin to decrease as the local time separation between the ground stations and Viking decreases. The GOES 5 satellite located near 0900 local time and nearly above Baker Lake (BLC), recorded magnetic field variations (not shown here) almost identical to those at Viking and with the same phase relationship. Since the same variation is observed at about 1600 MLT by Viking and 0900 MLT by GOES 5 and Baker Lake, it can be argued that the magnetospheric perturbation that caused them began almost exactly at noon and the effect propagated as a fast Alfvén wave mode east and west (toward the dawn and dusk flanks) similar to the Viking example studied by Erlandson *et al.* [12].

The top panel of Figure 5 shows the negative of the total field variations observed during a near-polar orbit of Viking on July 31, 1986. Viking was close to the 1300 MLT meridian and reached apogee at 2.1 Re altitude at 1600 UT where its magnetic field projection to the ground was located at 72.5° N and 299.3° E geographic latitude and longitude, respectively. Also shown in this plot are the horizontal geomagnetic field components measured at Barrow, BRW, (95.9° or 6.4 hours west of Viking's projected position), Cambridge

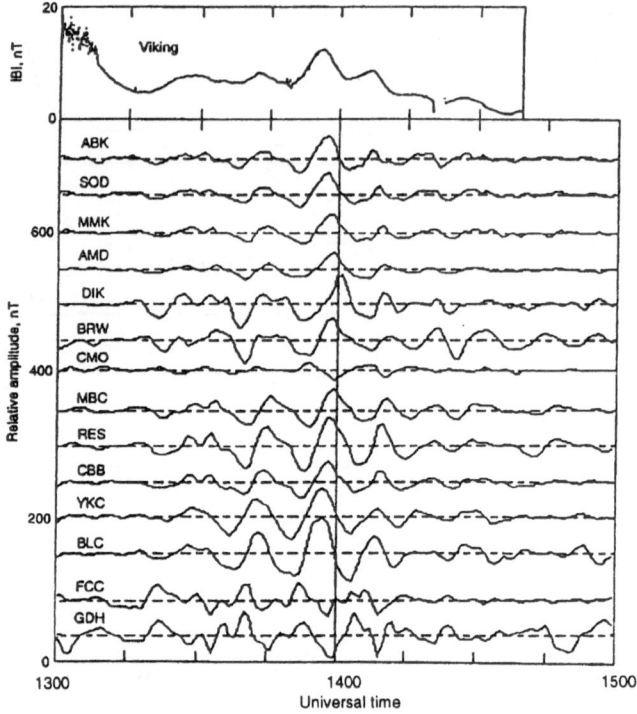

Figure 4. Negative of the variations in the total magnetic field measured by Viking (from Figure 1) with variations in the horizontal component measured by fourteen ground stations.

TABLE 1. Ground-based magnetometer stations

Code	Station Name	Geographic		Corr. Geomag.	
		Lat.°	Long.°	Lat.°	Long.°
ABK	Abisko	68.4	18.8	65.1	102.8
SOD	Sodankla	67.4	26.6	63.7	108.1
MMK	Murmansk	68.3	33.1	64.3	113.8
AMD	Amderma	69.5	61.4	64.7	137.9
DIK	Dixon Island	73.5	80.6	68.1	155.8
BRW	Barrow	71.3	203.4	69.8	248.8
CMO	College	64.9	212.2	64.9	261.8
MBC	Mould Bay	76.2	240.6	80.9	268.4
RES	Resolute Bay	74.7	265.1	83.9	313.4
CBB	Cambridge Bay	69.2	255.0	77.8	305.0
YKC	Yellowknife	62.4	245.5	69.7	297.9
BLC	Baker Lake	64.3	264.0	74.6	324.8
FCC	Fort Churchill	58.8	265.9	69.6	330.2
GDH	Godhavn	69.3	306.5	76.5	41.2

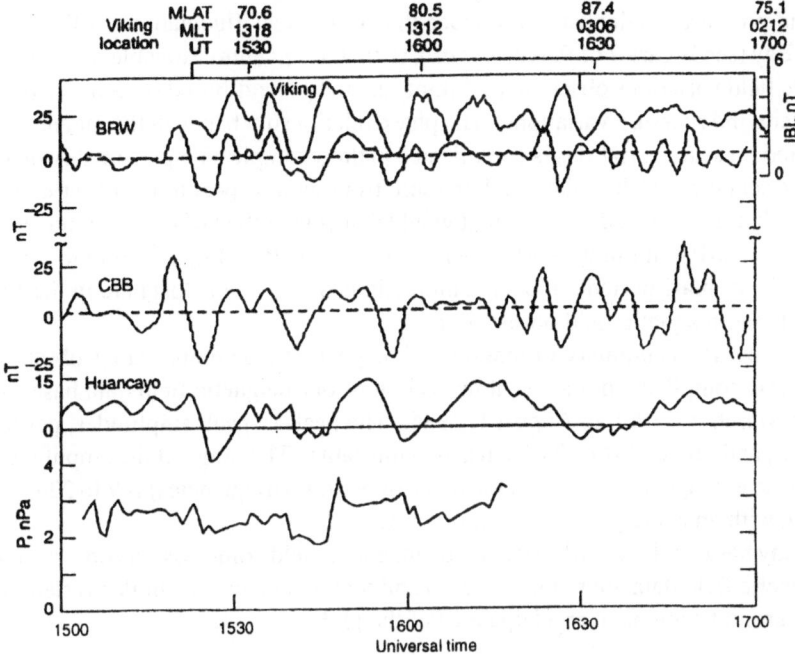

Figure 5. Negative of the variations in the total magnetic field measured by Viking on July 31, 1986 (orbit 879) with variations in the horizontal component measured at Barrow, Cambridge Bay, and the solar wind pressure derived from IMP 8 observations.

Bay, CBB, (44.3° or 3 hours west of Viking), and Huancayo (14.6° or 1 hour west of Viking). The Viking data trace is delayed by about 2.5 min in Figure 4 to provide the best correlation with the Barrow data. The total field variation measured by Viking is character-ized by a quasi-periodic variation with periods ranging from about 7 to 11 min and a peak of about 4 nT at 1546 UT. The variations in the Barrow horizontal component are similar to the Viking variations with a time delay of about 2.5 min, until about 1600 UT. A distinctive waveform can be identified in the Huancayo horizontal component which is somewhat similar to the Viking trace with an amplitude of about 15 nT and a relative peak at about 1552 UT. These phase delays may be due to a combination of the fast mode travel times (where the Alfvén speeds from Viking to the ground should be about 1000 km/s) and to the orientation of the solar wind pressure front.

Data were available from IMP 8 from about 1501 to 1615 UT on this day and the derived solar wind pressure is shown in the bottom panel of Figure 5. At 1600 UT on July 31, 1986, IMP 8 was located at X = −3.1 Re, Y= −36.7 Re, and Z = 2.2 Re. Conse-quently the spacecraft was in the solar wind, but at some distance from the Earth-Sun line in the early dawn hours. The IMF Bz was positive and steady at about 4 nT, the X component fluctuated between ±3 nT and the Y component varied between 2 and −4 nT. The solar wind pressure varied from 3 to 2 nPa and back to 3 nPa between 1505 and 1510 UT and from about 2.5 to 3 nPa and back again between 1520 and 1525 UT. There is a distinctive varia-tion consisting of a drop from 3 to 2 nPa at 1540 UT and an increase from 2 to 3.5 nT from 1545 to 1547 UT and a drop back to 3 nT at 1550 UT. A precise correspondence between solar wind pressure and magnetic field variations is not as clear in this case as it was in an

112

earlier event studied by Potemra *et al.* [11], shown here as Figure 1. However, a 50% change in solar wind pressure is adequate to produce the 30 to 50 nT fluctuations at Barrow and Cambridge Bay (see Figure 7 of Potemra *et al.*), so that it is not unreasonable to suggest that the magnetic field variations observed by Viking and the ground-based stations are due to a series of solar wind pressure variations. The phase relationship between the magnetic field and solar wind pressure variations is complicated in this example, and presumably involves the varying orientation of the solar wind pressure front with respect to the observing locations. Since Viking is located near noon, it would detect the effect of a solar wind pressure variation before IMP 8 did in the early dawn hours. The data in Figure 5 provide an additional example of quasi-periodic global compressions of the magnetosphere in the Viking and ground-based magnetic field observations.

Table 2 provides a summary of long-period magnetic field compressions observed by Viking that was compiled from a visual inspection of our magnetic field data base for the period from March 25, 1986 to August 1, 1986. The peak-to-peak amplitudes, periods of the variations, and range of IMF Bz are listed in this table. The range of the amplitudes is 2 to 5 nT, with an average of 3.7 nT, and the range of periods (frequencies) is 7 to 23 min (0.7 to 2.4 mHz), with an average of 12.2 min (1.4 mHz).

This study has concluded that long period magnetic field compressions observed in the Viking magnetic field data are not uncommon and that, in two cases, similar variations are observed at ground stations over a large area of the globe.

5. Discussion

The frequencies of the magnetic field compressions discussed here are within the range of ULF pulsations identified in the radar data by Ruohoniemei *et al.* [15], but it is difficult to argue that the two phenomena are the same because of the quasi periodic and compressional nature of the magnetic field variations observed by Viking. Our observations are consistent with the radar measurements in one respect however, in that the radar ULF pulsations show a packet structure as would be expected if they were triggered by a succession of solar wind pressure pulses, such as shown in Figure 4.

Table 2. Summary of global compressions observed by Viking

Date	Time, UT	ΔB_{pp}, nT	Period, min	IMF B_z, nT
3/25/86[1]	1700–1800	5	10,10	None
4/15/86	1340–1440	2	7,8,9,10,15	–1 to 2
4/23/86	2220–2350	5	14,22	~0
5/7/86	0053–0150	2	7,10,10,11	None
5/30/86	1946–2130	4	9,18	–5 to 0
6/12/86	1730–1900	4.5	19,20,23,23	–2 to 0
6/18/86	1330–1440	4	10,11,12,12,13	None
7/31/86[2]	1521–1630	4	6,7,7,8,9,11	~3
8/1/86	1710–1855	3	12,14	None

[1]Potemra *et al.* [13]
[2]Erlandson *et al.* [12]

Simulations have been performed on the responses of the magnetosphere to solar wind pressure fluctuations (Lee and Lysak [16], Lysak and Lee [17], and Lee [18]). One of the objectives of these simulations was to identify characteristics of toroidal and poloidal ULF wave modes. Lee and Lysak [16] were able to identify long-period compressional waves with a characteristic frequency equal to the source frequency of the solar wind pressure impulse used in the simulation. Inspired by these simulations, we suggest that the long-period quasi-periodic compressions identified here provide experimental confirmation of long period magnetospheric compressions due to solar wind pressure variations with the same periods. These compressions may be the source of the low frequency resonances identified in the radar and ground-based magnetometer observations discussed earlier, if there is some mechanism that can account for the stable frequencies in a dynamic magnetosphere.

6. Summary

This study has provided evidence for long-period quasi-periodic compressions of the magnetosphere at frequencies below almost all known ULF resonances within the magnetosphere (i.e., below the 1.3 mHz variations discovered in the radar and ground-based magnetometer data discussed earlier). These compressions produce, what has sometimes been referred to as a "breathing mode," of the magnetosphere with periods ranging from 7 to 23 min. These breathing modes are difficult to detect, because their amplitudes are small (about 4 nT at Viking) and may be overwhelmed by the higher frequency ULF waves often present in ground-based and satellite observations within the magnetosphere. However, several examples were readily found in the Viking data set, due possibly to the unusually quiet conditions that prevailed during Viking's lifetime. Therefore, these long-period breathing modes may not be uncommon. This finding supports the view that the magnetosphere may be constantly in motion and that it is difficult, and perhaps impossible, to provide a valid interpretation of magnetospheric phenomena within the framework of a static model.

Acknowledgments. The Viking magnetic field experiment was supported by the Office of Naval Research and the data analysis was jointly supported by the National Science Foundation and NASA.

7. References

1. Stewart, B. (1861) On the great magnetic disturbance which extended from August 28 to September 7, 1859, as recorded by photography at the Kew Observatory, *Philos. Trans. R. Soc. Landon* **151**, 423.
2. Birkeland, K. (1901) Expedition Norvegienne de 1899–1900 pour l'etude desaurores boreales, Resultants des recherches magnetiques, *Skr. Nor. Vidensk. Akad. Kl.* **1**, *Mat. Naturvidensk Kl.* **1**, 7–12.
3. Green, C.A. (1979) Observations of Pg pulsations in the northern auroral zone and at lower latitude conjugate regions, *Planet. Space Sci.*, **27**, 63.
4. Nansen, F. (1897, Vol. II) Farthest North, Being the Record of a Voyage of Exploration of the Ship "Fram" 1893–1896 and (1898, Vol. I) of a Fifteen Months' Sleigh Journey by Dr. Nansen and Lieut. Johansen, Harper and Brothers, New York.
5. Potemra., T.A. (1991) The Arctic Explorations of Fridtjof Nansen, Johns Hopkins APL Technical Digest, **12**, 275–283.

6.   Birkeland, K. (1908, 1913) The Norwegian Aurora Polaris Expedition 1902-1903, Vol. I. On the cause of magnetic storms and the origin of terrestrial magnetism, H. Aschehough & Co., Christiania (Oslo): First and Second Sections.

7.   Francis, W. E., M. I. Green, and A. J. Dessler (1959) Hydromagnetic propagation of sudden commencements of magnetic storms, *J. Geophys. Res.* **64**, 1643.

8.   Wilson, C. R. and M. Sugiura (1961) Hydromagnetic interpretation of sudden commencement of magnetic storms, *J. Geophys. Res.* **66**, 4097.

9.   Tamao, T. (1975) Unsteady interactions of solar wind disturbances with the magnetosphere, *J. Geophys, Res.* **80**, 4230.

10. Wilken, B., et al. (1982) The SSC on July 29, 1977 and it propagation within the magnetosphere, *J. Geophys. Res.* **87**, 5901.

11. Potemra, T. A., et al. (1989) Multisatellite and ground-based observations of transient ULF waves, *J. Geophys. Res.* **94**, 2543.

12. Erlandson, R. E., D. G. Sibeck, R. E. Lopez, L. J. Zanetti, and T. A. Potemra (1991) Observations of solar wind pressure initiated fast mode waves at geostationary orbit in the polar cap, *J. Atmos. Terr. Phys.* **53**, 231–239.

13. Potemra, T. A., L. J. Zanetti, R. Elphinstone, J. S. Murphree, and D. M. Klumpar (1992) The pulsating magnetosphere and flux transfer events, *Geophys. Res. Lett.* **19**, 1615–1618.

14. Takahashi, K., D. G. Sibeck, P. R. Newell, and H. E. Spence (1991) ULF waves in the low-latitude boundary layer and their relationship to magnetospheric pulsations: A multisatellite observation, *J. Geophys. Res.* **96**, 9503–9519.

15. Ruohoniemi, J. M., et al. (1991) HF radar observations of Pc 5 field line resonances in the midnight/early morning MLT sector, *J. Geophys. Res.* **96**, 15,697.

16. Lee, D.-H. and R. L. Lysak (1991) Monochromatic ULF wave excitation in the dipole magnetosphere, *J. Geophys. Res.,* **96**, 5811.

17. Lysak, R. L. and D.-H. Lee (1992) Response of the dipole magnetosphere to pressure pulses, *Geophys. Res. Lett.* **19**, 937.

18. Lee, D-H. (1996) Dynamics of MHD wave propagation in the low-latitude magnetosphere, *J. Geophys. Res.* **101**, 15,371.

IONOSPHERIC SIGNATURES OF MAGNETOPAUSE PROCESSES

A. S. RODGER,
British Antarctic Survey
Madingley Road, Cambridge CB3 0ET, U.K.

Abstract
Whilst it is well-accepted that short-duration changes to the Interplanetary Magnetic Field (IMF) and solar wind cause high-latitude ionospheric transients observed by magnetic, optical and radar techniques, a coherent explanation of the zoo of signatures remains illusive. The paper describes some of the problems involved in the interpretation of dayside transient features, all of which result from time- and spatially-varying field-aligned currents. A brief discussion of how internal magnetospheric processes can directly affect the rate at which energy is transferred across the magnetopause is discussed. Further progress will require carefully-coordinated measurements from ground-based networks of observatories with complementary measurements from space.

1. Introduction

Temporal changes of geospace occur over at least 16 orders of magnitude. On the millennium time-scale, changes result from alterations of the internal dynamo, and hence to the geomagnetic field. At the opposite end of the time spectrum, the electron gyro-frequency in the ionosphere is $\sim 10^{-6}$ s. The major source of the variability in the outer magnetosphere and high-latitude ionosphere on time-scales of minutes, results from the influences of the solar wind, and its embedded magnetic field (the IMF). Although the ionosphere is often described as the TV screen showing the solar wind-magnetosphere interaction programme, the unique interpretation of ionospheric signatures of the interaction of the solar wind on the closely-coupled magnetosphere-ionosphere system remains illusive. In this paper, some reasons for the difficulties in interpreting the transient ionospheric signatures of these interactions are discussed. They include the dynamic pressure effects and short-lived enhancements in reconnection at the dayside magnetopause. However the way in which the ionosphere and magnetosphere can actively influence magnetopause processes is also discussed briefly.

J. Moen et al. (eds.), Polar Cap Boundary Phenomena, 115–125.

116

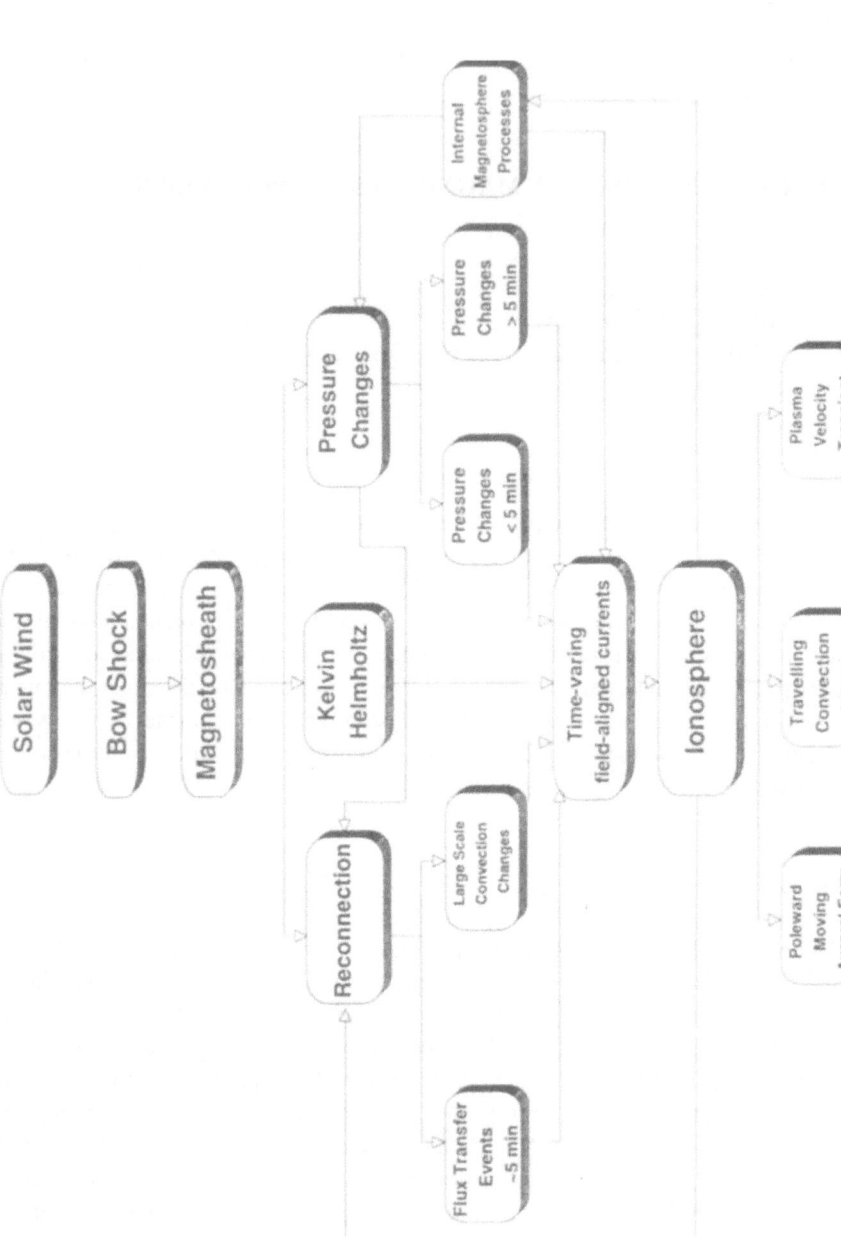

Figure1. Block diagram showing that changes both of the solar wind dynamic pressure and of the IMF result in time-varying field-aligned currents. These in turn cause poleward moving auroral forms, travelling convection vortices and plasma velocity transients in the ionosphere.

2. Disturbances at the magnetopause

The solar wind plasma is heated substantially as it crosses the bow shock. Each ion species is heated to a different extent. The resultant particle energy spectrum depends critically on the precise nature of the shock [8]. Ions reflected from the shock also contribute to the generation of pc3/4 geomagnetic pulsations (e.g. [21] and the references therein). These ULF waves provide an additional coupling process from the solar wind through to the inner magnetosphere and ionosphere. The magnetosheath too is a highly turbulent region [33]. For example, large mirror mode waves are observed between the shock and the magnetopause. However a detailed description of bow shock and magnetosheath processes lies outwith this paper. Here the focus is on two major classes of disturbance at the magnetopause arising from interactions of the solar wind and IMF with the closely-coupled magnetosphere and ionosphere (see Figure 1). These are the effects caused by changes of the dynamic pressure of the solar wind and the reconnection rate. Because Kelvin-Helmholtz waves have the greatest probability of occurrence on the flanks of the magnetosphere rather than near the sub-solar point, owing to the much higher plasma velocity shears in this region, they will not be discussed further here.

The magnetopause forms where the dynamic pressure of the shocked solar wind equals the magnetic pressure of the geomagnetic field. To a good approximation, the location of the magnetopause is give by equation 1 in which N_{sw} and V_{sw} are the concentration (cm^{-3}) and the velocity (km s^{-1}) of the solar wind plasma respectively.

$$L_{\text{MAGNETOPAUSE}} = 107.4(N_{sw} \times V_{sw}^2)^{-1/6} \tag{1}$$

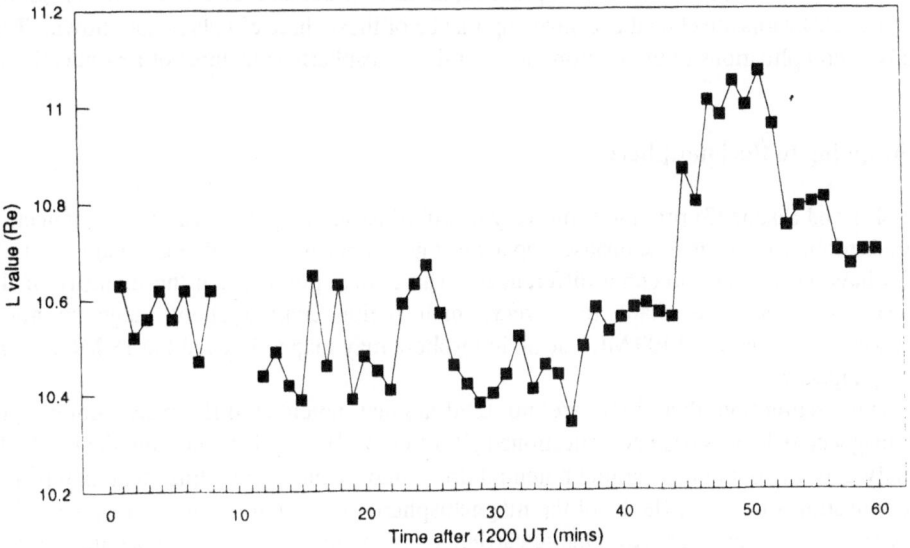

Figure 2 Location of the magnetopause on 15 May 1996 between 1200-1300 UT derived using equation 1

The dynamic pressure of the solar wind is very seldom stable. This is illustrated in Figure 2 which shows the magnetopause location for 1 h chosen at random, determined using equation 1. Over the course of the hour, the magnetopause moves first inward, then outward by about 0.8 Re. However on time-scales of a few minutes it can move ~0.5 Re (e.g. near 1245 and 1250 UT). To achieve this equilibrium, the magnetopause moves at speeds of 100s km s^{-1}, a speed comparable with the velocity of the plasma in the magnetosheath. These changes in location cause perturbations throughout the magnetosphere and ionosphere ([6], and the references therein). These variations are described further in section 4.

Reconnection is the physical process responsible for the majority of energy transfer from the solar wind/IMF into the magnetosphere. The concept of reconnection at the dayside magnetopause, introduced over 30 years ago [4], has been very successful in describing many of the large-scale features of the high-latitude ionosphere and magnetosphere. For example it can explain the influence of the IMF on the magnitude and direction of high-latitude ionospheric convection patterns, and the cusp ion plumes. There has been an intense debate as to whether reconnection is quasi-steady or highly transient (see [14], [22]). Cusp ion signatures have been found to be a very powerful diagnostic of pulsed reconnection (see [32]). Furthermore there is increasing evidence that reconnection can occur simultaneously at two or more sites on the magnetopause. Sometimes two sites near the sub-solar point may be active but on other occasions it might be a subsolar and a high-latitude location ([10], [26], [20]).

An increase in the rate at which open flux is created is usually regarded as having an associated enhancement of the reconnection electric field, which in turn causes increased plasma velocity. However two other equally-valid alternatives can occur. First there can be inward erosion of the magnetopause to satisfy flux conservation. Alternatively the length of the active x-line can be extended. In practice, all three processes are likely to occur but the geophysical factors affecting the relative importance of these three effects is not known. This leads to complications in the interpretation of the ionospheric signatures of reconnection.

3. Mapping to the ionosphere

Crooker and Siscoe [3] provide some very useful illustrations of how various reconnection sites distributed on the magnetopause map to the ionosphere, assuming that geomagnetic field lines have no parallel potential differences. The critical point is that the majority of the dayside magnetopause maps to a very small region centred about magnetic noon. Observations made near 1400 MLT in the ionosphere may map to beyond the 18 MLT at the magnetopause.

The assumption that there are no field-aligned potential differences along open geomagnetic field lines has been questioned (Reiff et al., 1996, private communication). To satisfy current continuity, parallel potential drops may occur on field lines that map from a reconnection site on the flanks of the magnetosphere to the high latitude ionosphere. This may arise because there are few charge carriers owing to the very low plasma concentration at the flanks. The parallel potential drop will cause particle acceleration, and the formation of auroral arcs at 557.7 nm on open field lines in the polar cap. Thus observations of discrete auroral arcs can no longer be taken as a definitive signature of closed field lines, as has previously been assumed.

Changes at the magnetopause are communicated to the polar ionosphere via field-aligned currents. Any time variation of dynamic pressure of the solar wind, or the reconnection rate at the magnetopause will cause a corresponding change of the field-aligned currents. Herein lies the heart of the problem, in that all significant processes on the magnetopause lead to time-varying currents in the ionosphere (see Figure 1). Thus the separation of their causes becomes complex. Often insufficient information is available from ground- and space-based instruments to allow unambiguous interpretation.

If time-varying currents near the cusp are carried mainly by magnetosheath ions, then there is a significant time delay between the causative process on the magnetopause and the first effects being observed in the ionosphere. A typical transit time for such ions from the magnetopause near 10 Re is of the order of 2-3 minutes. During this time, the ions will convect about 75 km in latitude from the flux open/closed field line boundary assuming modest reconnection electric field (equivalent to 25 mV m^{-1} at ionospheric altitudes). Minow [19] has shown that there is often a separation of this order between the most poleward 557.7 nm emission, attributed to ring current precipitation on the most poleward closed field line, and the most equatorward extent of the 630 nm emission caused by the precipitation of energetic plasma from the magnetosheath.

4. Transient signatures in the dayside ionosphere

There are three major classes of time-varying signature in the high-latitude ionosphere reported in the literature. These are travelling convection vortices (TCVs), plasma velocity disturbances and poleward moving auroral forms (PMAFs). TCVs were originally identified in magnetometer data by Friis-Christensen et al. [7] and initially were attributed to spatially-varying field-aligned currents associated with solar wind pressure pulses. Plasma velocity disturbances have been largely ascribed to the ionospheric signatures of short-duration periods of enhanced reconnection, termed flux transfer events (FTEs). PMAFs are short-duration optical features, observed mainly near magnetic noon. The relationship of these features to solar pressure changes and time-varying reconnection is discussed in this section.

4.1 SHORT-DURATION RECONNECTION SIGNATURES (<5 MINUTES)

FTEs are a major source of energy transfer from the solar wind into the magnetosphere. The search for the ionospheric signatures of FTEs has been intense in the last decade. FTE signatures in VHF radar [9], incoherent scatter radar [13], [12], optical [30] magnetometer [11] and HF radar data [25], [24] have all been reported.

From these observations, Cowley and Lockwood [2] proposed a time-sequence of events in response to enhanced reconnection at the magnetopause. First a protrusion starts to grow, expanding equatorward the ionospheric footprint of the open/closed field line boundary about 3 minutes after enhanced reconnection occurs at the magnetopause i.e. the Alfvén wave travel time from the reconnection site at the magnetopause to the ionosphere. If reconnection continues, then the protrusion continues to expand in latitude and longitude with time. If reconnection diminishes, then the protrusion relaxes to its original quasi-circular shape. Many authors have assumed that the open/closed field line boundary and the equatorward edge of the 630 nm auroral emission associated with the cusp precipitation are collocated but this is not so, as explained in section 3. Second, Cowley and Lockwood [2] suggest that the Alfvén wave carries with it the first information about reconnection at the magnetopause, and the orientation of the sheath magnetic field, a proxy for the IMF. Therefore the first dynamical response in the ionosphere, namely an increase in plasma velocity, would also be expected to be observed at the equatorward edge of the 630 nm emission region about 3 minutes after enhanced reconnection occurred at the magnetopause [34]. The relative importance of the boundary motion and the plasma velocity increase is not known.

Recent high-time resolution measurements from the PACE HF radar at Halley, Antarctica have been used to show that the first plasma velocity increase associated with flux transfer events (FTEs) occurs ~100 km equatorward of the region to which magnetosheath precipitation maps to the ionosphere [28]. These authors suggest that the velocity transients are initiated at the ionospheric footprint of the open/closed field line boundary. Such observations are not consistent with the current Cowley-Lockwood model. The velocity variations have rise times ~100 s and fall times of ~10 s, time-scales that the authors could not explain.

The increase in the plasma velocity at the open/closed field line boundary reduces the necessity for the magnetopause to move inward, and thus for its ionospheric signature to move equatorward. Very careful examination of the many published papers showing optical data from the ionospheric footprint of the cusp (e.g. [3] and the references therein) does not reveal any consistent equatorward motion of the red or green line auroral forms when each and every new PMAF is established, though some PMAFs do show an apparent equatorward motion of their boundary. Thus key elements of the FTE paradigm need revision to accommodate both the plasma velocity and the optical observations summarised above.

To date there has been no unique ground-based magnetometer signature of FTEs identified. Modelling of magnetic signature of FTEs was carried out by McHenry and Clauer [18]. As a result of a considerable body of new observations of FTEs published in the last decade, significant improvements to their model are now possible. For example, there are better estimates of the spatial-scale size of the ionospheric signatures of FTEs. Also a more realistic representation of the ionospheric conductivity should be possible

4.2 PRESSURE CHANGES

There is universal agreement that changes in the solar wind dynamic pressure on the magnetopause will lead to field-aligned currents, but there is no agreement on the number and direction of these currents e.g. [17]. In response to a sudden commencement, an extreme form of pressure pulse, Araki [1] has proposed that two pairs of oppositely-directed field-aligned currents form near noon. Then they propagate around the morning and afternoon polar cap boundary with speeds ~ 10 km s^{-1}. However recent observations (Moretto, private communication) show that field-aligned currents appear to form, intensify and then diminish near noon, rather than propagate. Engebretson (private communication) has shown some effects of pressure pulses in the nightside ionosphere occur almost immediately after the pulse hits the dayside magnetopause (i.e. propagation speed >1000 km s^{-1}).

Zesta [36] has investigated the ionospheric effects of some short-lived (<5 minutes) pressure pulses using the many magnetometer data at high latitudes in the northern hemisphere. These pressure pulses caused magnetic disturbances which propagate away from noon approximately along a constant magnetic latitude. Inversion of the magnetometer data gives patterns consistent with there being three field-aligned currents observed in the morning sector. The propagation speeds of each of the currents varies from less than 1 km s^{-1} to more than 6 km s^{-1}. In general, Zesta [36] found that propagation speeds varied through an event, and from one event to the next.

There has been a considerable body of evidence that has suggested that the travelling convection vortices (TCVs) (e.g. [7]) could be the magnetometer signature of pressure pulses. However Yahnin and Moretto [35] have shown that the majority of published examples of TCVs occur near the ionospheric image of the central plasma sheet rather than at the ionospheric footprint of the magnetopause. One interpretation is that the pressure variations on the magnetopause are the ultimate source of the energy, but they establish a short-lived field-line resonance region inside the magnetosphere, and that the effects of this resonance propagate away from noon.

To date no consistent ground-based optical signature of pressure pulses has been identified, although there have been a few candidate descriptions (e.g. [6]). If the scale size of the field-aligned currents systems associated with pressure pulses is confirmed to be ~500 km diameter or more, then it is difficult for such features to be resolved easily in ground-based all-sky camera data; the field of view of the camera at E -region altitudes being similar or smaller than the feature to be identified. Also if the current is evenly distributed across such a large area, then the change in intensity of the optical emissions will be small. Therefore networks of sensitive all-sky cameras are likely to be necessary for such studies. Data from such networks are only just becoming available from both hemispheres.

Another serious limitation to the further understanding of effects of pressure pulses is their identification in the solar wind. Recent evidence suggests that the solar wind is not in dynamic and thermal equilibrium. It is certainly not clear how short-lived concentration enhancements of a factor of 2 that last a couple of minutes can be generated in the solar wind. Also the spatial scale of such features is not known. Therefore unless spacecraft measurements are made both close to the Earth and directly on the Sun-Earth line, it will be difficult to know exactly what is impinging upon the dayside magnetopause. Cluster should offer unique possibilities to address these critical questions.

These examples illustrate how a consistent ionospheric response to pressure variations at the magnetopause is not observed. The fundamental reasons for this difficulty arise because the present approach over-simplifies the real situation. For example, most models assume that the observed magnetic perturbations arise from twin or triplet field aligned currents. This does not take account of many of the processes that contribute to the observed perturbations. These include the effects of the changing intensity of the Chapman-Ferraro current, the changing distribution of the ring current owing to its compression or rarefaction, spatial and temporal variations of ionospheric conductivity and a realistic distribution of field-aligned current. In the latter case a simple, wire-like field-aligned current is used but more complex current distributions are observed [36].

4.3 THE EFFECTS OF INTERNAL PROCESSES ON THE MAGNETOPAUSE

Field-line resonances (FLRs) may play an important role in key nightside processes such as the formation of auroral arcs and indeed in the triggering of substorms [29]. However less attention has been paid to their importance on the dayside. Most often FLRs are identified in data from magnetometer chains where magnetic disturbances often exceed 100 nT peak-to-peak variation, but they have been identified in HF radar data [5].

What effect do FLRs have if they occur on field-lines just inside the open/closed field line boundary, i.e. close to the magnetopause? It is suggested that changes in the geomagnetic field strength will lead to changes in the location of the magnetopause. Furthermore because of the spatial and temporal nature of the FLR, then similar variations will be imposed on the magnetopause. Thus it will no longer be the smooth, quasi-parabolic feature so often drawn in published literature but will have a 3-dimensional corrugated surface. The consequences of this corrugation may be very significant as it will both increase and decrease the velocity shears across the magnetopause. Also there will be changes in the orientation of the geomagnetic field relative to the magnetosheath field. Both effects will combine to make reconnection more likely in some locations and less likely in other locations on the dayside magnetopause, depending upon the detail of the changes locally.

There are some ground-based observations to support the above hypothesis. Øierset et al. (1996) show the magnetic, optical and plasma velocity transient signatures that are consistent with those of the ionospheric signature of a series of FTEs, forming and then propagating poleward over Svalbard. However the authors also show IMAGE magnetometer data from the mainland of Norway and Sweden. Clear magnetic variations with a poleward phase progression are observed at lower latitudes *before* the FTEs are formed. Indeed the timing is such that the FTEs appear to form when the maximum positive amplitude perturbation on the H component reaches the open/closed field line boundary, which for this occasion lies somewhere between mainland Norway and Svalbard.

The natural resonance period of the last closed field line under normal conditions (say L~9) is between about 7-9 minutes depending upon the assumptions made about the mass on the flux tube. This is very close to the supposed magic period for FTE occurrence that is often reported [27], although Lockwood and Wild [12] have shown that the mode in FTE occurrence is about 3 minutes. The data presented by Pinnock and Rodger [26] show that FTEs are essentially omni-present but large FTEs occur on average every 7-8 minutes. The similarity of the natural resonance period of the most poleward closed field lines and the occurrence of large FTEs suggests that there may be a causative link, a topic that requires further study. Perhaps the most important element of this discussion is an appreciation that the internal magnetosphere-ionosphere system need not necessarily be a passive respondent to the effects of the IMF and solar wind, but it may play an important role in determining both the location and the amount of energy transferred across the magnetopause.

5. Concluding remarks

The focus of this paper has been on the difficulties of interpreting the ionospheric signatures of short-duration changes to the IMF and solar wind. All of the changes cause time-varying field-aligned currents. Although there is a wealth of published data now on the three major classes of dayside ionospheric transients, TCVs, PMAFs and plasma velocity transients, there is no synthesised, unambiguous interpretation of them. Some of the reasons why difficulties still persist have been briefly described in the paper. The way forward on the topic requires high-spatial and temporal resolution measurements by multiple instruments of the common volume in which these transients occur. Suitable networks of optical, magnetometer and radar measurements now exist in both hemispheres. Therefore rapid progress in this topic should be expected soon but especially when CLUSTER data become available early in the next millennium.

Acknowledgements

I am most grateful to the organisers of this excellent meeting, and to NATO for its support. The work described in this paper has benefitted substantially from discussions with many scientists in recent years, particularly John Dudeney, Jeff Hughes, Mike Lockwood, Therese Moretto, Mike Pinnock and Eftyhia Zesta. Some of this work was carried out while I was supported by the Leverhulme Trust; their support is greatly appreciated.

References

1. Araki, T. (1994) A physical model of the geomagnetic sudden commencement, in M. J. Engebretson, K. Takahasi and M. Scholer (eds.), *Solar wind sources of magnetospheric ultra-low frequency waves, Geophysical Monograph 81,* American Geophysical Union, Washington DC, USA.
2. Cowley, S.W.H. and Lockwood, M. (1992) Excitation and decay of solar-wind driven flows in the magnetosphere-ionosphere system, *Ann. Geophys.* **10**, 103.
3. Crooker, N. U. and Siscoe, G.L. (1990) On mapping flux transfer events to the ionosphere, *J. Geophys. Res.,* **85**, 3795.
4. Dungey, J. W. (1961) Interplanetary field and the auroral zones, *Phys. Rev. Lett.* **6**, 47.
5. Fenrich, F.R., Samson, J.C., Sofko, G., and Greenwald, R.A. (1995) ULF high- and low-m field line resonances observed with the Super Dual Auroral Radar Network, *J. Geophys. Res.,* **100**, 21535.
6. Freeman, M.P. and Farrugia, C.J. (1998) Magnetopause motions in a Newton-Buseman approach, these proceedings.
7. Friis-Christensen, E., McHenry, M.A., Clauer, C.R., and Vennerstrom, S. (1988) Ionospheric travelling convection vortices observed near the polar cleft: A triggered response to sudden changes in the solar wind, *Geophys. Res. Lett.* **15**, 253.
8. Fuselier, S. A., Shelley, E.G., and Lennnartson, O.W. (1997) Solar wind composition changes across the Earth's magnetopause, *J. Geophys. Res.,* **102**, 275.
9. Goertz, C. K., Neilsen, E., Korth, A., Glassmeier, K.H., Haldoupis, C., Hoeg, P., and Hayward, D. (1985) Observations of a possible ground signature of flux transfer events, *J. Geophys. Res.,* **90**, 4069.
10. Kan, J R. (1988) A theory of patchy and intermittent reconnections for magnetospheric flux transfer events, *J. Geophys. Res.,* **93**, 5613.

11. Lanzerotti, L. J., Wolfe, A., Trevedi, N., Maclennan, C.G., and Medford, L.V. (1990) Magnetic impulse events at high latitude: magnetopause and boundary layer plasma processes. *J. Geophys. Res.* **95**, 97.

12. Lockwood, M. and Wild, M.N. (1993) On the quasi-periodic nature of flux transfer events, *J. Geophys. Res.*, **98**, 5935.

13. Lockwood, M., Cowley, S.H.W., Sandholt, P.E., and Lepping, R.P. (1990) The ionospheric signatures of flux transfer events and solar wind dynamic pressure changes, *J. Geophys. Res.*, **95**, 17113.

14. Lockwood, M., Denig, W.F., Farmer,A.D., Davda, V.N., Cowley, S.W.H., and Lühr, H. (1993) Ionospheric signatures of pulsed reconnection at the Earth's magnetopause, *Nature,* **361**, 424.

15. Lockwood, M., Cowley, S.W.H., and Smith, M.F. (1994) Comment on "By fluctuations in the magnetosheath and azimuthal flow velocity transients in the dayside ionosphere" by Newell and Sibeck, *Geophys. Res. Lett.*, **21**, 1819.

16. Lühr, H., Lockwood, M., Sandholt, P.E., Hansen, T.L. and T. Moretto. (1996) Multi-instrument ground-based observations of a traveling convection vortex event, *Ann. Geophys.*, **14**, 162.

17. Lysak, R. L., Song, Y., and Lee, D.-H. (1994) Generation of ULF waves by fluctuations in the magnetopause position, in M. J. Engebretson, K. Takahasi and M. Scholer (eds.), *Solar wind sources of magnetospheric ultra-low frequency waves, Geophysical Monograph 81*, American Geophysical Union, Washington DC, USA.

18. McHenry, M. A. and Clauer, C.R. (1987) Modeled ground magnetic signatures of flux transfer events, *J. Geophys. Res.*, **92**, 11231.

19. Minow, J. (1996) PhD Thesis, University of Alaska, USA.

20. Moen, J., Evans, D., Carlson, H.C., and Lockwood, M. (1996) Dayside moving auroral transients related to LLBL dynamics, *Geophys. Res. Lett.*, **23**, 3247.

21. Morrison, K. and Freeman, M. P. (1995) The role of upstream ULF waves in the generation of quasi-periodic ELF-VLF emissions. *Ann. Geophys.* **13**, 1127.

22. Newell, P.T. and Sibeck, D.G. (1993) Upper limits on the contribution of FTEs to ionospheric convection, *Geophys. Res. Lett.*, **20**, 2829.

23. Øierset, M., Lühr, H., Moen, J., Moretto, T., and Sandholt, D.E. (1996) Dynamical auroral morphology in relation to ionospheric plasma convection and geomagnetic activity: signatures of magnetopause x-line dynamics and flux transfer events, *J. Geophys. Res.*, **101**, 13275.

24. Pinnock, M., Rodger, A.S., Dudeney, J.R., Baker, K.B., Greenwald, R.A., and Greenspan, M. (1993) Observations of an enhanced convection channel in the cusp ionosphere, *J. Geophys. Res.*, **98**, 3767.

25. Pinnock, M., Rodger, A.S, Dudeney, J.R., Greenwald, R.A, Baker, K.B., and Ruohoniemi, J.M. (1991) An ionospheric signature of possible enhanced field merging on the dayside magnetopause, *J. Atmos. Terr. Phys.*, **53**, 201.

26. Pinnock, M., Rodger, A.S., Dudeney, J.R., Rich, F.,and Baker, K. (1995) High spatial and temporal observations of the ionospheric cusp, *Ann. Geophys.* **13**, 919.

27. Rijnbeck, R. P., Cowley, S.W.H., Southwood, D.J., and Russell, C.T. (1984) A survey of dayside flux transfer events observed by the ISEE 1 and 2 magnetometers, *J. Geophys. Res.*, **89**, 786.

28. Rodger A. S. and Pinnock, M. (1997) The ionospheric response to flux transfer events - the first few minutes, *Ann. Geophys.* **15**, 685-691.

29. Samson, J.C. , Wallis, D. D., Hughes, T.J., Creutzberg, F., Ruohoniemi J.M., and Greenwald, R.A. (1992) Substorm intensifications and field line resonances in the nightside magnetosphere, *J. Geophys. Res.*, **97**, 8495.

30. Sandholt, P. E., Lockwood, M., Lybbek,B., and Farmer, A.D. (1990) Auroral bright spot sequence near 1400 MLT: Co-ordinated optical and ion drift observations, *J. Geophys. Res.*, **95**, 21095.

31. Sandholt, P. E., and 13 other authors. (1994) Cusp/cleft auroral activity in relation to solar wind pressure and interplanetary magnetic field Bz and By, *J. Geophys. Res.*, **99**, 17323.

32. Smith, M. P. and Lockwood, M. (1996) Earth's magnetospheric cusps, *Rev. Geophys.*, **34**, 233.

33. Song, P. (1994) Observations of waves at the dayside magnetopause, in M. J. Engebretson, K. Takahasi and M. Scholer, (eds.), *Solar wind sources of magnetospheric ultra-low frequency waves, Geophysical Monograph 81,* American Geophysical Union, Washington DC, USA.

34 Southwood, D. J. (1987) The ionospheric signature of flux transfer events, *J. Geophys. Res.*, **92**, 3207.

35. Yahnin, A. G., and Moretto, T. (1996) Travelling convection vortices in the ionosphere map to the central plasma sheet, *Ann. Geophys.*, **14**, 1025.
36. Zesta, E. (1997) *Traveling convection vortices at cusp latitudes observed by the Magnetometer Array for Cusp and Cleft Studies*, PhD Thesis, Boston University, 1997.

EXCITATION OF FLOW IN THE EARTH'S MAGNETOSPHERE-IONOSPHERE SYSTEM: OBSERVATIONS BY INCOHERENT-SCATTER RADAR

S.W.H. COWLEY
Department of Physics & Astronomy, University of Leicester
University Road, Leicester LE1 7RH, United Kingdom

1. Introduction

The most important consequence of the coupling processes which occur at the magnetopause boundary of the Earth's magnetosphere and in the geomagnetic tail is the large-scale convective flow that pervades the magnetosphere-ionosphere system, which dominantly determines the structure and dynamics of both the magnetospheric and the high-latitude ionospheric plasmas. The detailed properties of these flows are, however, difficult to discern from sparse single-point observations available from high-altitude spacecraft, and most of what is known has been determined from studies of the flow at ionospheric heights. Up until the early 1980s the main source of such information was obtained from data returned by low-altitude polar-orbiting spacecraft, which provide ~10-min passes over the polar regions every ~90 min for a given hemisphere. The principal result of these studies was that the flows depend strongly on the direction of the interplanetary magnetic field (IMF), thus implying that reconnection plays a central, if not exclusive, role in their generation. By combining together the data from different orbits obtained under similar IMF conditions, the overall pattern of the flow was determined. Figure 1 provides a summary of the results, adapted from the work of Reiff and Burch [1]. The flow is represented as an IMF-dependent set of steady-state patterns, driven principally by balanced dayside and tail reconnection when IMF B_z is negative (upper row), and by steady single-lobe reconnection (which does not change the amount of open flux) when IMF B_z is positive (lower row). The additional possibility of double-lobe reconnection when the IMF points close to northward, which closes open flux on the dayside and produces a time-dependent "reversed" twin-vortical flow on both open and closed flux tubes, is not represented. In the figure the region of open flux is represented by the central circular areas, while the dashed lines map along the field to the sites of reconnection. The figures also represent the zonal flows and the dawn-dusk shifts of the open flux region which are produced by the field tension effects associated with IMF B_y (positive on the left, negative on the right). These flows and shifts are shown for the northern hemisphere, and are simultaneously opposite in sense in the southern hemisphere. The diagrams also include the possible presence of steady-state non-reconnection "viscously-driven" flows, indicated by the flow vortices located wholly on closed field lines at high latitudes.

While the representation of the plasma flows shown in Fig. 1 contains much essential physics, it is nevertheless deficient in not including any description of time-dependent effects. Temporal evolution of the flow will follow from time-dependent

127

J. Moen et al. (eds.), Polar Cap Boundary Phenomena, 127–140.
© 1998 *Kluwer Academic Publishers.*

128

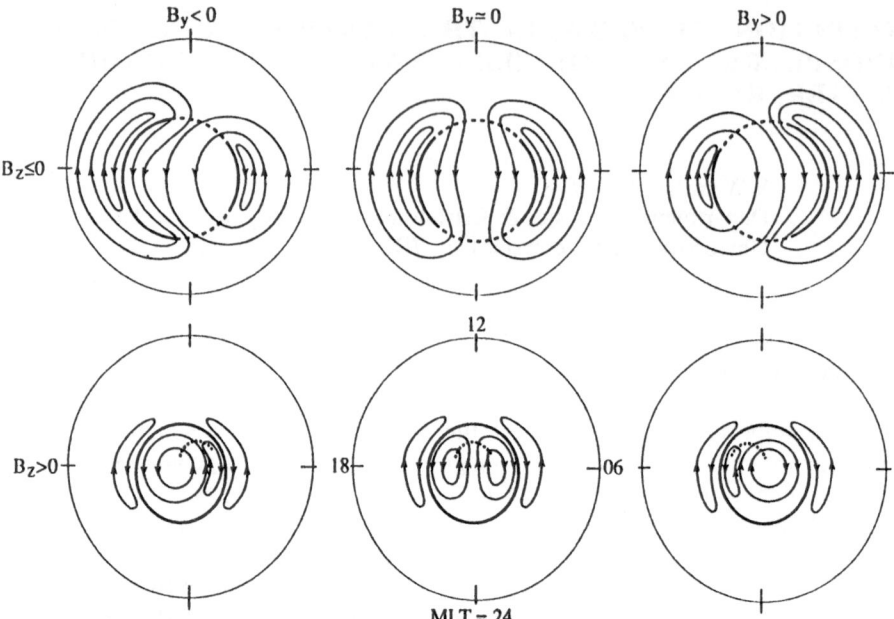

Figure 1. Sketches of the northern hemisphere ionospheric flow ordered according to the direction of the IMF. The arrowed solid lines show the plasma streamlines, while the centre circle represents the boundary between open and closed magnetic flux, the dashed portions of which in the upper row (negative IMF Bz) are the "merging gap" segments which map to the magnetopause reconnection sites. In the lower row (positive IMF Bz) the dashed lines also map to the lobe magnetopause reconnection sites, but do not lie on the open-closed field line boundary if reconnection for a given IMF field line takes place in one hemisphere only, as assumed here.

coupling processes at the magnetopause (e.g. flux transfer events, FTEs) and in the tail, from changes in the direction of the IMF, and from the intervals of unbalanced dayside or nightside reconnection which are inherent in the magnetospheric substorm cycle. Since the time-scales for these variations lie typically in the range 5-30 min, they are not amenable to investigation using low-altitude spacecraft data, but are readily accessible to suitably-placed radars, which can monitor the flow in a given region of the ionosphere with good time resolution.

2. Incoherent-Scatter Radar Observations of Dayside Flows

The first incoherent-scatter radar experiments which had sufficiently short cycle times to be able to resolve the time scales of flow variations in the high-latitude ionosphere were conducted in the mid-1980s using the EISCAT UHF radar. In these experiments the radar beam was pointed at low (21.5°) elevation to the north of the transmitter site at Tromsø, and cycled between two azimuths displaced 12° on either side of the magnetic meridian. Line-of-sight (l-o-s) velocities were then obtained at 15 s resolution in a number of range gates covering a range of latitudes, as the beam passed obliquely through the F-region ionosphere. With a dwell time at each azimuth of 2 min and a 30 s swing between, 2.5 min resolution vectors could also be derived (with appropriate assumptions) by pairwise combination of the l-o-s values along the beams [2]. Since

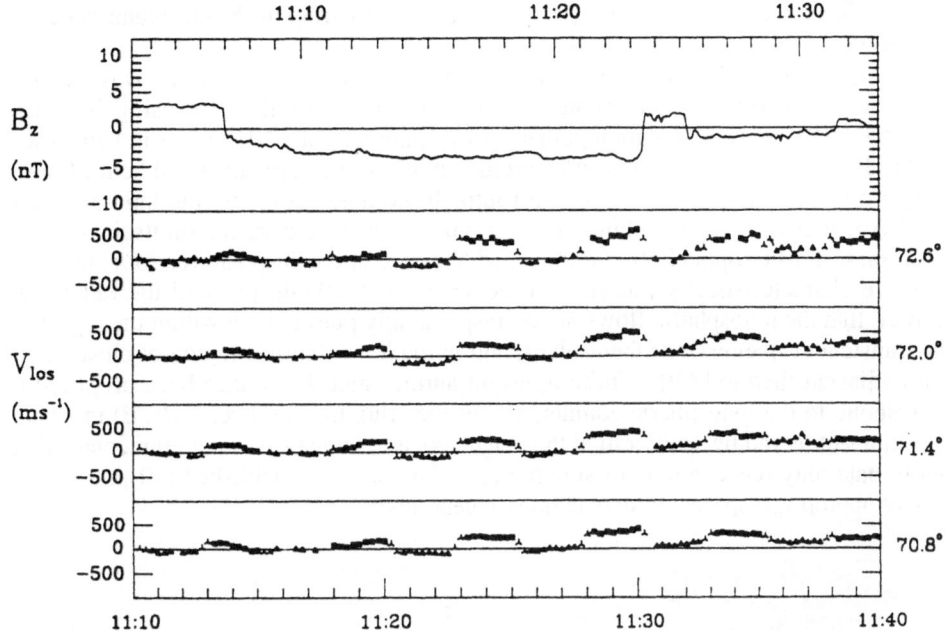

Figure 2. IMF Bz from AMPTE-UKS (top panel) and line-of-sight velocity from gates 1-4 of the EISCAT "Polar" experiment (lower panels) are plotted versus UT on 27 October 1984, with the former shifted with respect to the latter by 6.8 min to account for IMF propagation with the plasma from the spacecraft to the subsolar magnetopause. From Todd *et al.* [5].

the initial experiments were conducted at a time of solar minimum, the coverage in magnetic latitude was typically restricted to the range ~71°-73°, usually corresponding to the equatorward region of the high-latitude flow cells on closed field lines in the near-noon sector. The data showed that the flow in this regime responds strongly to the north-south component of the IMF, with the flow near noon appearing within ~5 min of southward fields arriving at the subsolar magnetopause [3,4]. An example is shown in Fig. 2, taken from the work of Todd *et al.* [5]. The top panel shows IMF Bz measured by the AMPTE-UKS spacecraft, which is shifted by 6.8 min relative to the EISCAT l-o-s velocity data shown in the lower panels to account for the propagation of the field with the interplanetary plasma from the spacecraft located upstream of the bow shock to the subsolar magnetopause. The radar data plotted as squares come from the "west" radar azimuth, and those as triangles from the "east" azimuth, with positive values indicating flow away from the radar. Positive values of the former and negative values of the latter indicate westward flow, corresponding to the dusk cell observed in the immediate post-noon sector (1340-1410 MLT for the data in the figure). An onset of flow was observed to occur coherently across the radar field-of-view at 11:19 UT, about 5 min after a switch from positive to negative IMF Bz was estimated to have occurred at the subsolar magnetopause. Similar results were found for the onset of flow reductions following the arrival of northward fields, with the overall time scale of the subsequent flow decay being ~15 min. Away from noon the flow response time was found to increase from ~5 min at noon to ~15 min near the dawn-dusk meridian, implying an east-west phase propagation speed away from noon of several km s^{-1}. For the case illustrated here, the phase speed deduced directly from the beam-swung radar ion temperature data was 2.6 km s⁻¹ [6].

Later, in the early 1990s, vector flows were derived in a similar way at 10 s resolution by using two simultaneous radar beams transmitted by using both the EISCAT UHF and VHF radars together, or by employing the VHF radar in split-beam mode. In addition, because solar maximum conditions then prevailed and hence higher F-region densities, observations could routinely be made covering the latitude range ~71°-76°, thus often approaching and entering the dayside cusp region itself on open field lines. Earlier optical observations had shown that the cusp auroras for southward IMF are characteristically pulsed on time scales of 5-10 min [7], and the radar data showed that the ionospheric flows are correspondingly pulsed, both within the region of the transient cusp emissions themselves, and in the lower latitude region of closed field lines adjacent thereto [8,9]. These transient auroras and flows have been proposed to correspond to the ionospheric counterparts of the "flux transfer event" (FTE) magnetic signatures observed by spacecraft at the magnetopause, indicative of pulsed reconnection, but to date only one example of simultaneous data has been published [10], due to the lack of appropriate spacecraft data in more recent times.

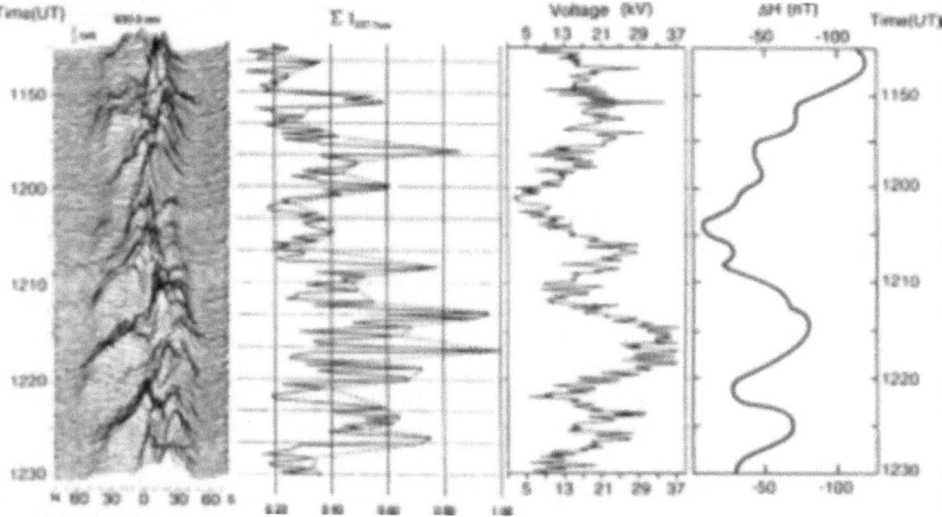

Figure 3. MSP 630 nm emission profiles, integrated 557.7 nm emission, dusk flow cell voltage determined from EISCAT CONV flow data, and H-component magnetic perturbation observed at Greenland West station ATU, on 7 January 1992. From Moen *et al.* [11].

An example of pulsed phenomena on the dayside is shown in Fig. 3, taken from the work of Moen *et al.* [11]. Here from left to right we show 630 nm meridian scanning photometer (MSP) scans versus zenith angle, the integrated 557.7 nm intensity (taken to be an indicator of region 1 current flow), the dusk cell voltage obtained at 10 s resolution by integrating the vector flow data obtained as described above from simultaneous EISCAT UHF and VHF radar data in the latitude range 71°-76°, and the H-component magnetic perturbation observed at Greenland West station ATU at 75°

magnetic latitude. The MSP and EISCAT data correspond to mid-afternoon local times, ~15 MLT, while ATU was in the mid-morning sector at ~10 MLT. The MSP data indicate the occurrence of six poleward-moving transient auroral forms during the 45 min interval, which simultaneous all-sky TV images show propagated eastward into the field-of-view of the MSP in the region just poleward of the radar observations, where westward flows on closed field lines were being observed. The first three transients were separated by intervals of ~5 min, and were associated with a single peak in the dusk cell voltage. The next three were separated by ~10 min and were associated with separated peaks in the flow. Corresponding variations were observed in the magnetic perturbations at ATU (and over much of the Greenland chain), indicative of simultaneously pulsed flows in the dusk convection cell, thus showing the large-scale nature of these transient effects.

3. Theoretical Implications

The results described above clearly require the development of theoretical ideas which go beyond the steady-state pictures shown in Fig. 1. Such ideas have been proposed by Cowley and Lockwood [12,13], following earlier partial concepts discussed by Siscoe and Huang [14] and Freeman and Southwood [15]. The Cowley-Lockwood picture is based on the idea of a "zero-flow equilibrium state" of the magnetosphere-ionosphere system. Suppose initially we have a convecting magnetosphere driven by some combination of dayside and tail reconnection, but that at some time all reconnection (and other coupling processes too) are switched off; what happens to the flow? We argue that under these circumstances all the flow in the near-Earth region must die away after some interval as equilibrium is approached, despite the continued presence of open flux in the system (which remains constant due to the cessation of all reconnection). The system overall is not in a steady zero-flow state, of course, because the open tubes which are present continue to be stretched downstream by the solar wind, but once the "last" open tubes have been stretched beyond a few tens of Earth radii (the scale size of the near-Earth system), and the near-Earth system has adjusted to their presence, no further evolution will take place in the latter region and the flow will die away. The time-scale for this decay of the flow is simply estimated to be 10-20 min on this basis. Subsequent reconnection at the magnetopause or in the tail then perturbs the near-Earth system away from this equilibrium state, and flows will then be excited which return the system towards equilibrium with the changed amount of open flux.

Based on these ideas, the flow response to a single pulse of dayside magnetopause reconnection in the presence of negative IMF By (corresponding to the data in Fig. 3) is shown in Fig. 4. For simplicity the open-closed field line boundary is shown as a segment of a circle in the initial zero-flow equilibrium, with the pulse of reconnection then creating a patch of new open flux which is appended to the boundary near noon as shown in sketch (a). Initially, therefore, the open-closed field line boundary jumps equatorwards, and with it a new patch of cusp emission will be produced due to precipitating accelerated magnetosheath plasma. The initial flow within the patch will be mainly eastwards due to the tension force associated with IMF By [sketch (b)], with an ionospheric closure which is mainly in a single vortex on closed field lines, since the motion of the open flux in the magnetosphere mainly perturbs closed dayside flux tubes as it propagates "sideways" over the magnetopause. These flows will dominate in the ionosphere for intervals of ~5 min, leading to displacements of the patch of open flux around the boundary of ~1000 km (~2 h of MLT) at speeds of 2-3 km s^{-1}. Eventually,

132

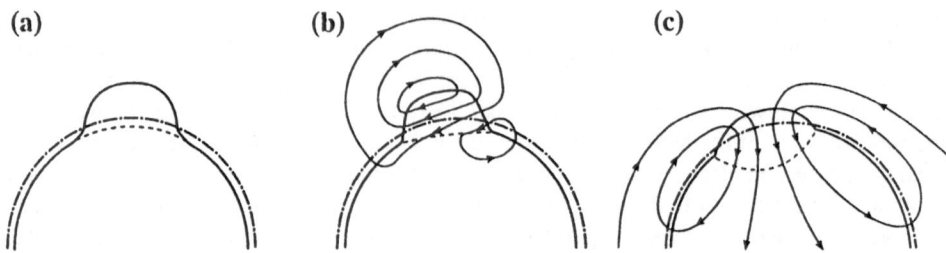

(a)   (b)   (c)

Figure 4. Sketches showing the evolution of the dayside flow in response to an isolated pulse of low-latitude magnetopause reconnection in the presence of negative IMF By and Bz.

however, the open tubes will have contracted over the dayside boundary and will then be pulled by the magnetosheath flow into the tail. The open flux will then move predominantly poleward with the excitation of twin-vortex flow [sketch (c)], while the cusp auroras will start to fade as the magnetosheath source to the ionosphere is shut off. Magnetosheath plasma still enters the magnetosphere along the open tubes, but it then principally flows down-tail in the plasma mantle rather than into the ionosphere. The twin-vortex flow will start in the vicinity of the open tube near noon, subsequently spreading out over the whole polar region at a phase speed of several km s⁻¹ (corresponding roughly to magnetospheric Alfvén speeds mapped along the field into the ionosphere). This flow will be excited, and will subsequently decay, over a 15-min time scale as the system returns toward equilibrium with the changed amount of open flux. The equilibrium position of the boundary is indicated in the figure by the dot-dashed line. The twin-vortex flows which are excited carry the system towards this equilibrium configuration, and then decay as it is approached.

Overall, this picture clearly provides a reasonable physical description of the observed behaviour of dayside optical and flow transients, with initial azimuthal motions which depend on the sense of IMF By, followed by poleward motion and fading. Rapid (~5 min) flow responses to reconnection on the dayside, followed by ~15 min expansions away from the vicinity of noon, and subsequent decay on such time scales are also inherent. However, the complete sequence of events in Fig. 4 could only be observed as depicted if the system remains uninfluenced by other activity over intervals of at least several tens of minutes before and after the reconnection pulse. More typically, however, dayside transients recur on time-scales of 5-10 min as in Fig. 3, such that the overall flow will be a combined effect produced by more than one evolving open flux region. Since the time-scale for overall excitation and decay of the flow is ~15-20 min, we may expect that when transients recur on a time-scale which is comparable with or longer than this a clearly pulsed flow will be observed, while if the recurrence time is much shorter, an essentially continuous flow will be generated. This expectation is clearly compatible with the flow data shown in Fig. 3.

The opposite limit to that illustrated in Fig. 4 for isolated pulses is that of steady reconnection, though the system will still generally be time-dependent unless the reconnection rates are exactly balanced between day and night, as first illustrated in the model calculations of Siscoe and Huang [14]. The Cowley-Lockwood picture applied to intervals of steady reconnection is illustrated in Fig. 5, where we also consider how the resulting flow will reconfigure due to rapid switches in the direction of the IMF. For definiteness we assume that the IMF switches direction from positive Bz and By to

(a)

(b)

(c)

(d)

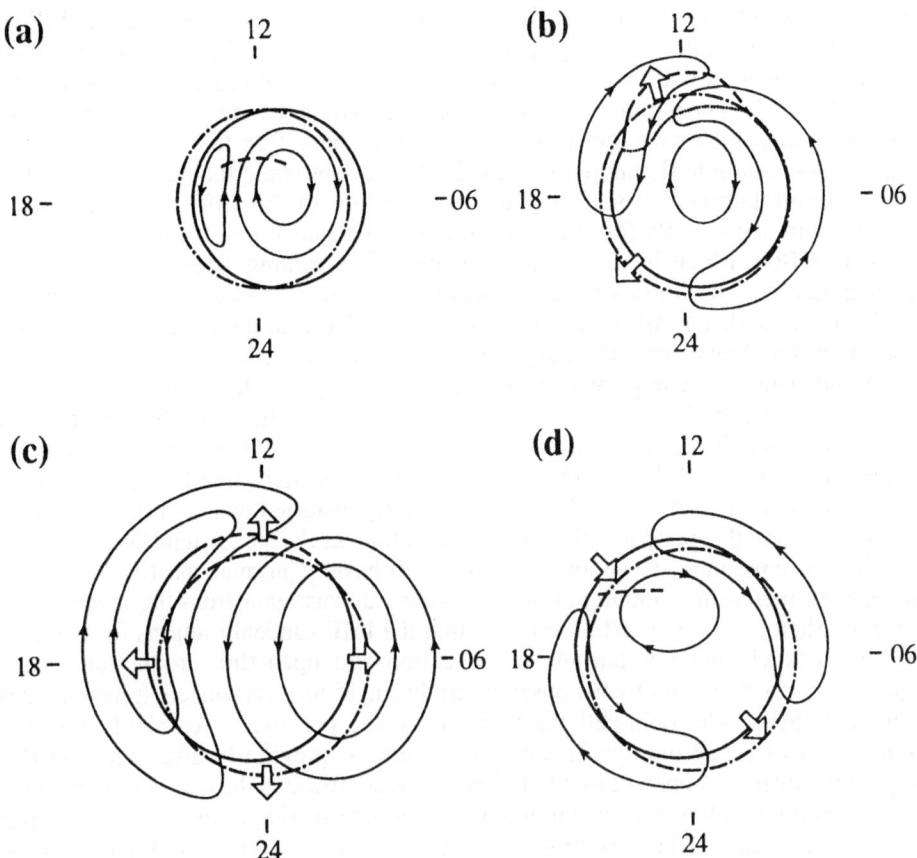

Figure 5. Sketches illustrating the reconfigurations of the dayside flow in response to sudden switches in the direction of the IMF from By and Bz both positive to both negative, and back again. Reconnection in each regime is assumed steady, and corresponds to single lobe reconnection for By and Bz positive, and low-latitude reconnection for By and Bz negative.

negative Bz and By across a sharp discontinuity, and then after a significant interval (long compared with the overall excitation and decay time constant of ~10-20 min) it switches back again. We consider only the effects of steady magnetopause reconnection, and do not include the further complications that may result either from "viscous" coupling at the magnetopause, or from tail processes. Sketch (a) shows the initial flow configuration assuming steady single-lobe reconnection for northward IMF. The flow pattern is steady both because the reconnection rate is assumed to be steady and because for single-lobe reconnection the amount of open flux is constant; the pattern is essentially that shown at the lower right in Fig. 1 with "viscous flows" omitted for simplicity. As in the latter diagram, the reconnection site projected along the field is shown by the dashed line, and a "reversed dispersion" cusp ion precipitation pattern (i.e. one in which the ion energy increases with increasing latitude) will be present downstream from this. The open-closed field line boundary is shown by the solid line, while the dot-dashed line shows the zero-flow equilibrium position of the latter which contains the same amount of open flux. When the directional change in the IMF (to negative Bz and By) arrives at the dayside magnetopause, low-latitude reconnection and open flux production will begin, while lobe reconnection will cease shortly thereafter. Nevertheless, lobe-stirring in the sense appropriate to the previous direction of IMF By (positive) will persist over intervals of ~15 min as those flux tubes evolve over the magnetopause and into the more distant tail. At the same time the newly-opened flux tubes at lower latitude respond to the new direction of the IMF, flowing initially eastward with negative By (as in Fig. 4) and then also generating a outward-spreading twin-vortex flow which increasingly surrounds the remaining "lobe-stirring" cells at higher latitudes. The flows which will occur during the ~15 min transitional interval are illustrated in sketch [b], where the broad arrows indicate the motions of the open-closed field line boundary. The cusp ion precipitation will consist of regions of both falling and rising ion energy with increasing latitude, generally giving rise to a "V" dispersion profile in the vicinity of noon. After ~15 min the new flow pattern will have become established appropriate to steady unbalanced dayside reconnection in the presence of negative IMF Bz and By. The open-closed field line boundary expands equatorward at all local times in the presence of a By-distorted twin-cell flow, as shown in sketch [c], and this pattern will continue for as long as the steady unbalanced dayside reconnection continues. Cusp ion dispersion will have a "normal" profile (falling ion energy with increasing latitude) at all local times downstream from the reconnection region (dashed line). Now let us suppose that the IMF suddenly returns to its former orientation, such that low-latitude reconnection and open-flux production ceases, followed shortly thereafter by the onset of steady single-lobe reconnection with the new northward IMF. The twin-cell flow driven by the previous interval of open flux production will decay over an interval of ~15 min as previously discussed, the flow carrying the displaced open-closed field line boundary towards the equilibrium position. During this interval lobe reconnection will also initiate stirring in the sense appropriate to the new direction of IMF By (positive) within the expanded region of open flux, the lobe reconnection site propagating into the polar cap from the open-closed boundary as described previously by Øieroset et al. [16]. The form of the flow during this ~15 min transition is shown in sketch [d], and the cusp ion dispersion will again assume a "V" profile. After this interval the twin-vortex flows will largely have died away, leaving a new steady lobe circulation flow and "reversed" cusp ion dispersion of the same form as that shown in sketch [a], but now lying within an expanded region of open flux.

In this section we have dealt with descriptions of the flow for two particular cases only, for an isolated pulse of reconnection with IMF By negative, and transitions in flow driven by steady reconnection due to switches in direction of the IMF from positive Bz and By to negative Bz and By, and back again. Clearly there are many other possibilities which could have been considered. However, our purpose has been mainly to illustrate how the thinking behind the Cowley-Lockwood picture can be used to understand the form and time scale of the flow response to time-dependent reconnection processes, at least in a qualitative way.

4. Incoherent-Scatter Radar Observations of Nightside Flows

While the above discussion indicates that the flow in the dayside high-latitude ionosphere can be reasonably accounted for in terms of the effects of magnetic reconnection at the magnetopause, the flow in the nightside auroral zone is influenced by a number of factors, and is far less well understood.

The first factor is that the flow is clearly influenced by the same magnetopause reconnection processes as discussed above, which propagate with significant delays into the nightside, as well as by nightside reconnection processes. The same dual influences must also in principle be present on the dayside, but it is evident from the dayside data studied to date that any nightside influences must be relatively more subtle than the analyses performed so far have been able to discern. On the nightside, however, the flows originating from dayside reconnection must play a substantial role in the adiaroic expansion of the auroral zone during the growth phase of substorms, and a significant issue then concerns the time response of this nightside flow component relative to changes in the direction of the IMF at the subsolar magnetopause. The results of Etemadi *et al.* [3] and Todd *et al.* [4] extend from the dayside to near the dusk-dawn meridian, where response times were found to have increased to ~10-15 min, as reported above. The response time in the auroral zone near midnight may thus be anticipated to be somewhat longer than this. However, no systematic studies of the nightside response delay have yet been reported, while the results of a few case studies have yielded a range of values between ~10 and ~40 min (see Taylor *et al.* [17] and references therein).

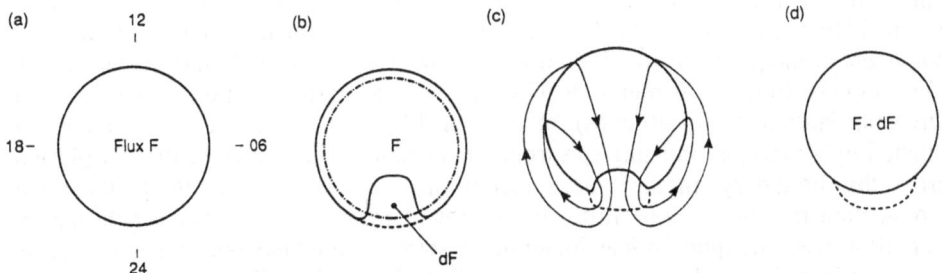

Figure 6. Ionospheric flow and open-closed field line boundary response to an impulse of tail reconnection. From Cowley and Lockwood [12]

The second factor is the flows driven by tail reconnection itself, by which we mean specifically the closure of open flux in the tail lobes. In the Cowley-Lockwood picture an impulse of nightside reconnection will produce an expanding twin-vortex flow in a corresponding manner to that already described above for open flux production at the magnetopause. This is illustrated in Fig. 6. Here the zero-flow equilibrium configuration with open flux *F* in sketch (a) is perturbed by a pulse of tail reconnection in (b) to cause the open-closed field line boundary to jump poleward on the nightside. The dot-dash line in (b) shows the zero-flow equilibrium boundary for the changed amount of open flux *F-dF*, and twin-vortex flow is excited, first near midnight, then expanding over the polar cap, which carries the system from its perturbed condition towards equilibrium (sketch (c)). As before, the excitation and decay time scale of the flow will be ~10-20 min, and after it has died away the newly closed flux (and region of hot plasma precipitation) will be located at the poleward edge of the nightside open-closed field line boundary (sketch (d)). Continuous unbalanced tail reconnection will similarly drive a continuous but time-dependent twin-vortex flow, associated with a continuously contracting open-closed field line boundary.

Ground-based observations related to tail reconnection have been made in a number of experiments using the Sondrestrom incoherent-scatter radar located on the west coast of Greenland, and related instrumentation [18-20]. In these experiments the radar was

136

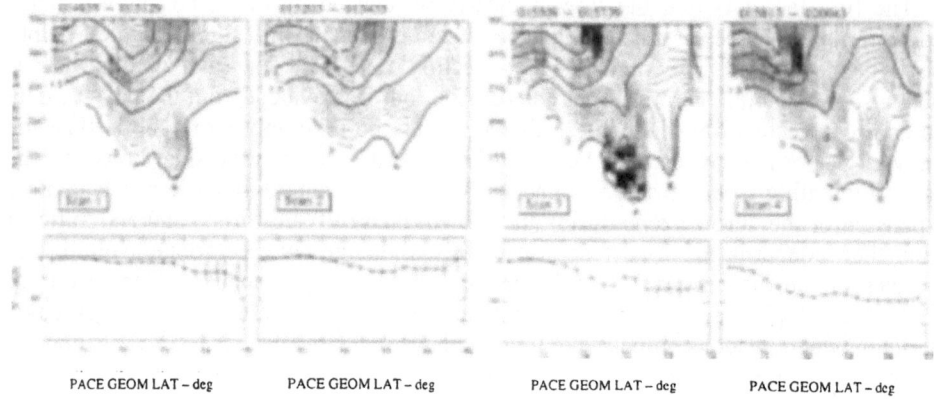

PACE GEOM LAT – deg   PACE GEOM LAT – deg   PACE GEOM LAT – deg   PACE GEOM LAT – deg

Figure 7. Meridional electron density contour plots and northward velocities for successive 3-min scans of the Sondrestrom radar on 13 January 1989. From de la Beaujardière *et al.* [19]

scanned in elevation, typically between 30° north and 30° south, with a 3-5 min cycle time, making measurements of the E- and F-region density structure, and the north-south component of the plasma drift. The position of the open-closed field line boundary was inferred either from the E-region density structure (a sharp reduction of density with increasing latitude at all altitudes), or from profiles of the 630 nm auroral emission obtained by a meridian imaging spectrograph, or both. The measured flow of plasma across this boundary, in the latter's rest frame, then relates directly to the local reconnection rate of tail lobe flux. In one set of measurements obtained during an interval of magnetic quiet, it was found that a series of impulsive reconnection events took place in which the boundary moved rapidly poleward by ~50-100 km, followed by the excitation of a ~15 min surge of equatorward flow and boundary relaxation [19], very reminiscent of the sequence shown Fig. 6. These reconnection impulses thus appear to represent a nightside counterpart of dayside FTEs. An example is shown in Fig. 7, taken from the work of de la Beaujardière *et al.* [19]. This shows meridional electron density profiles from four successive radar scans, together with the simultaneously measured north-south component of the plasma drift. Scans 1 and 2 show the presence of an E-region density enhancement at position A, which optical data show corresponds to an east-west aligned auroral arc. Plasma flows in its vicinity are equatorward at 100-200 m s^{-1}. Poleward of the arc plasma densities drop to much lower values, and satellite overpass data confirm that it lay just equatorward of the open-closed field line boundary via the presence of a velocity-dispersed ion signature (VDIS) on its poleward side, indicative of a mapping to the plasma sheet boundary layer. (The VDIS is just the nightside counterpart of the dispersed cusp ion signatures on the dayside.) In scan 3 this arc intensified in concert with the appearance of a new arc (B) on its poleward side, such that the precipitation region moved poleward by ~100 km. The equatorward flow simultaneously increased to ~300 m s^{-1}, though it was somewhat suppressed within intensified arc A itself. The poleward motion of the boundary combined with an enhancement of the equatorward flow indicates the occurrence of a sudden enhancement in the pre-existing tail reconnection rate. In scan 4 both arcs move equatorward in the presence of a ~500 m s^{-1} flow, which subsequent data show died away to smaller values (~200 m s^{-1}) over a ~10 min interval as the boundary moved back towards its former location. During the nightside interval studied, major events of this nature were found to occur about once each hour, with smaller events recurring on 10-20 min time scales. We may anticipate that in the near-Earth tail these events will be related to recurring outward expansions of the plasma sheet, typically by about half an Earth radius (RE), which subsequently relax over intervals of ~15 min. The expansions will propagate towards the Earth from the more distant tail, with the fastest earthward-directed ions being observed first. Thermal (few keV) ions will arrive at Earth ~15 min after the corresponding reconnection enhancement took place, for a reconnection region located somewhat beyond ~100 RE downtail. This is illustrated schematically in the sketches in Fig. 8, where the hatched region represents the region of hot thermal plasma, and the dotted region the boundary layer of energetic ions (VDIS region). The sketches show both the initial configuration for low reconnection rate (sketch a), the perturbed system just after the end of a ~5 min burst of enhanced reconnection (sketch b), and the arrival of the hot plasma near the Earth after a further ~10 min (sketch c).

(a)

(b)

(c)

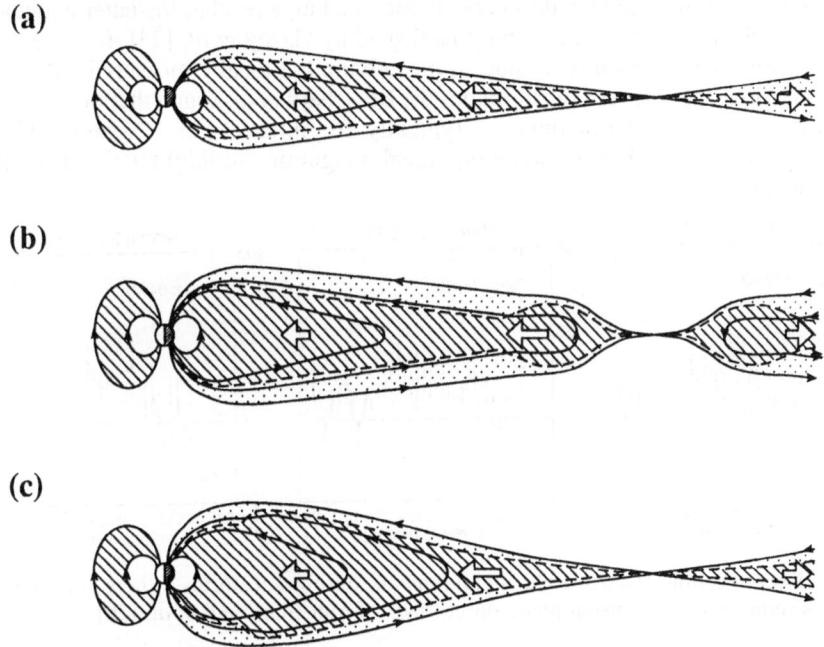

Figure 8. Sketches of the plasma sheet response to a pulse of distant tail reconnection.

Similar "equatorward drifting arcs" recurring on ~5-10 min times scales have also been observed by the EISCAT radar during the growth and early expansion phases of substorms, suggesting that pulsed reconnection in the more distant tail may be a typical phenomenon [21]. In the F-region these arcs are associated with increases of electron temperature indicative of soft electron precipitation, and decreases in the density indicative of ion outflow, but no increases in the ion temperature that might result from rapid differential flows and ion-neutral frictional heating. It has also been found that the equatorward motion of these arcs is uninterrupted both by the eventual formation of the substorm auroral bulge in the equatorward region, and its poleward progress towards the arcs, indicating that they are part of separate dynamical phenomena associated with the distant and near-Earth tail respectively.

138

The overall occurrence of tail reconnection has been studied statistically by Blanchard *et al.* [20], using data from the Sondrestrom radar obtained by the technique outlined above. They found that the majority of the overall flux transfer takes place within a 2 h-segment of local time on either side of midnight, and that the local transfer rate responds to southward fields in the interplanetary medium with a delay determined from cross-correlation of 1-1.5 h. It may initially seem reasonable to suppose that this time scale corresponds to the time scale for substorm expansion phase onset following intervals of southward fields in the interplanetary medium, since ~1 h response times are typical of past correlation analyses of the relation between the IMF and geomagnetic disturbance indices such as AL [22]. However, this is clearly not perfectly so, because substorm expansion onset occurs at the equatorward edge of the band of discrete auroras, often many hundreds of km equatorward of the open-closed field line boundary and thus deep within the closed flux tube region of the plasma sheet (most probably near its inner edge). A substantial interval may thus occur before the bulge reaches the latter boundary and open tube closure starts. In the study published by Gazey *et al.* [23], for example, this interval was ~10 min or more. Indeed, in studying the locally measured tail (open flux) reconnection rate after six substorm onsets, Blanchard *et al.* found that the rate was usually low at onset, and remained so typically for intervals of ~20 min. Three examples are shown in Fig. 9, spanning local magnetic midnight (02 UT at the Sondrestrom radar).

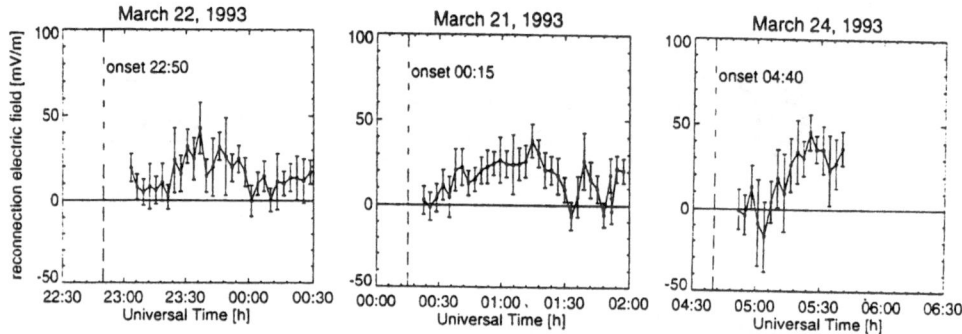

Figure 9. Reconnection electric field measurements during substorms, where the vertical dashed lines indicate the expansion phase onset times. From Blanchard *et al.* [20]

The third factor which influences nightside flows arises from the strong gradients in ionospheric conductivity which are produced by the spatially and temporally structured plasma precipitation from the geomagnetic tail [24,25]. For example, in the nightside polar cap (open field region) the height-integrated Pedersen and Hall conductivities are both typically less than 1 mho. Inside the growth-phase equatorward-drifting arcs these values increase to ~15 mho (Pedersen) and ~20 mho (Hall), while between the arcs they fall to ~3 and ~5 mho. Within the substorm auroral bulge, however, Pedersen conductivities rise to ~20-50 mho, and Hall conductivities to ~50-200 mho. Consideration must then be given to issues of current closure in the coupled magnetosphere-ionosphere system, and associated flow modification. Up until the present, however, very few such studies have been published, such that this aspect of nightside flows remains very ill-understood. The simplest initial picture which one may suggest, based e.g. on the EISCAT results of Kirkwood *et al.* [24], is one in which the flow tends to be suppressed within regions of very high conductivity, with corresponding flow deflections and enhancements in surrounding regions. However, the details relating to this picture remain highly uncertain from both experimental and theoretical viewpoints.

5. References

1. Reiff, P.H. and Burch, J.L. (1985) IMF By-dependent plasma flow and Birkeland currents in the dayside magnetosphere. 2. A global model for northward and southward IMF, *J. Geophys. Res.* **90**, 1595.
2. Willis, D.M., Lockwood, M., Cowley, S.W.H., van Eyken, A.P., Bromage, B.J.I., Rishbeth, H., Smith, P.R., and Crothers, S.J. (1986) A survey of simultaneous observations of the high-latitude ionosphere and interplanetary magnetic field with EISCAT and AMPTE-UKS, *J. Atmos. Terr. Phys.* **48**, 987.
3. Etemadi, A., Cowley, S.W.H., Lockwood, M., Bromage, B.J.I., Willis, D.M., and Lühr, H. (1988) The dependence of high-latitude dayside ionospheric flows on the north-south component of the IMF: a high time resolution correlation analysis using EISCAT "Polar" and AMPTE-UKS and -IRM data, *Planet. Space Sci.* **36**, 471.
4. Todd, H., Cowley, S.W.H., Lockwood, M., Willis, D.M., and Lühr, H. (1988) Response time of the high-latitude dayside ionosphere to sudden changes in the north-south component of the IMF, *Planet. Space Sci.* **36**, 1415.
5. Todd, H., Cowley, S.W.H., Etemadi, A., Bromage, B.J.I., Lockwood, M., Willis, D.M., and Lühr, H. (1988) Flow in the high-latitude ionosphere: measurements at 15 s resolution made using the EISCAT "Polar" experiment, *Planet. Space Sci.* **36**, 423.
6. Lockwood, M., van Eyken, A.P., Bromage, B.J.I., Willis, D.M., and Cowley, S.W.H. (1986) Eastward propagation of a convection enhancement following a southward turning of the interplanetary magnetic field, *Geophys. Res. Lett.* **13**, 72.
7. Sandholt, P.E., Deehr, C.S., Egeland, A., Lybekk, B., Viereck, R., and Romick, G.J. (1986) Signatures in the dayside aurora of plasma transfer from the magnetosheath, *J. Geophys. Res.* **91**, 10063.
8. Lockwood, M., Denig, W.F., Farmer, A.D., Davda, V.N., Cowley, S.W.H., and Lühr, H. (1993) Ionospheric signatures of pulsed reconnection at the earth's magnetopause, *Nature* **361**, 424.
9. Lockwood, M., Moen, J., Cowley, S.W.H., Farmer, A.D., Løvhaug, U.P., Lühr, H., and Davda, V.N. (1993) Variability of dayside convection and motions of the cusp/cleft aurora, *Geophys. Res. Lett.* **20**, 1011.
10. Elphic, R.C., Lockwood, M., Cowley, S.W.H., and Sandholt, P.E. (1990) Flux transfer events at the magnetopause and in the ionosphere, *Geophys. Res. Lett.* **17**, 2241.
11. Moen, J., Sandholt, P.E., Lockwood, M., Denig, W.F., Løvhaug, U.P., Lybekk, B., Egeland, A., Opsvik, D., and Friis-Christensen, E. (1995) Events of enhanced convection and related dayside auroral activity, *J. Geophys. Res.* **100**, 23917.
12. Cowley, S.W.H., and Lockwood, M. (1992) Excitation and decay of solar wind-driven flows in the magnetosphere-ionosphere system, *Ann. Geophysicae* **10**, 103.
13. Cowley, S.W.H. (1996) The auroral ionosphere and its coupling to the magnetosphere and solar wind, in H. Kohl, R. Rüster, and K. Schlegel (eds.), *Modern Ionospheric Science*, EGS Publishers, Katlenburg-Lindau, p.32.

140

14. Siscoe, G.L. and Huang, T.S. (1985) Polar cap inflation and deflation, *J. Geophys. Res.* **90**, 543.
15. Freeman, M.P. and Southwood, D.J. (1988) The effect of magnetospheric erosion on mid- and high-latitude ionospheric flows, *Planet. Space Sci.* **36**, 509.
16. Øieroset, M., Sandholt, P.E., Denig, W.F., and Cowley, S.W.H. (1997) Northward interplanetary magnetic field cusp aurora and high-latitude magnetopause reconnection, *J. Geophys. Res.* **102**, 11349.
17. Taylor, J.R., Lester, M., Yeoman, T.K., Emery, B.A., Knipp, D.J., Orr, D., Solovyev, S.I., Hughes, T.J., and Lühr, H. (1997) The response of the magnetosphere to the passage of a coronal mass ejection on March 20-21 1990, *Ann. Geophysicae* **15**, 1174.
18. de la Beaujardière, O., Lyons, L.R., and Friis-Christensen, E. (1991) Sondrestrom radar measurements of the reconnection electric field, *J. Geophys. Res.* **96**, 13907.
19. de la Beaujardière, O., Lyons, L.R., Ruohoniemi, J. M., Friis-Christensen, E., Danielsen, C., Rich, F.J., and Newell, P.T. (1994) Quiet-time intensifications along the poleward boundary near midnight, *J. Geophys. Res.* **99**, 287.
20. Blanchard, G.T., Lyons, L.R., de la Beaujardière, O., Doe, R.A., and Mendillo, M. (1996) Measurements of the magnetotail reconnection rate, *J. Geophys. Res.* **101**, 15265.
21. Persson, M.A.L., Aikio, A.T., and Opgenoorth, H.J. (1994) Satellite-groundbased coordination: late growth and early expansion phase of a substorm, in *Proc. Second Internat. Conf. on Substorms*, Geophysical Institute, Fairbanks, Alaska., p.157.
22. Bargatze, L.F., Baker, D.N., McPherron, R.L., and Hones, E.W., Jr (1985) Magnetospheric impulse response for many levels of geomagnetic activity, *J. Geophys. Res.* **90**, 6387.
23. Gazey, N.G.J., Lockwood, M., Smith, P.N., Coles, S., Bunting, R.J., Lester, M., Aylward, A.D., Yeoman, T.K., and Lühr, H. (1995) Development of substorm cross-tail current disruption as seen from the ground, *J. Geophys. Res.* **100**, 9633.
24. Kirkwood, S., Opgenoorth, H.J., and Murphree, J.S. (1988) Ionospheric conductivities, electric fields and currents associated with auroral substorms measured by the EISCAT radar, *Planet. Space Sci.* **36**, 1359.
25. Aikio, A.T. and Kaila, K.U. (1996) A substorm observed by EISCAT and other ground-based instruments - evidence for near-Earth substorm initiation, *J. Atmos. Terr. Phys.* **58**, 5.

LARGE-SCALE ELECTRIC FIELDS
IN THE DAYSIDE MAGNETOSPHERE

M.I. PUDOVKIN

Institute of Physics, St. Petersburg University
St. Petersburg, Petrodvorets, 198904, Russia

AND

A. EGELAND

Department of Physics, University of Oslo
P.O. Box 1048, Blindern, 0316, Oslo 3, Norway

Abstract. Observations of auroral dynamics in the polar cusp region is used to derive regularities of the time and space distribution of large-scale electric fields at the magnetopause and in the polar ionosphere. The interpretation of experimental data is based on the supposition that the magnetopause electric field (the reconnection electric field) consists of the potential and vortex components which have different sources in the magnetopause vicinity, and different signatures in the polar ionosphere.

Energy sources of magnetic field reconnection at the magnetopause are discussed; the distribution of electric currents in the magnetic barrier region and in the magnetopause vicinity are calculated. There is shown that the magnetopause electric currents are partially closed within the magnetic barrier region.

The relationship between the development of electric currents in the day- and night-side auroral zone is considered.

1. Introduction

Electric fields are known to play a fundamental role in the development of magnetospheric disturbances and in the energy balance of the magnetosphere. At the same time, the regularities of the space and time behaviour of those fields, as well as the mechanism of their generation are known rather vaguely as yet. In particular, the time evolution of the electric field dur-

J. Moen et al. (eds.), Polar Cap Boundary Phenomena, 141–156.
© 1998 *Kluwer Academic Publishers*.

ing magnetospheric substorms, as well as its distribution along the plasma sheet are studied in outline only.

On the other hand, there exist numerous publications in which various individual features of the magnetospheric electric field are revealed. In this paper, we summarize results of those studies, and try to construct on that base a general picture of the space and time variations of the large-scale magnetospheric electric field.

As the most abundant and detailed data on the magnetospheric electric fields are obtained from observation of the dynamics of aurorae, we shall concentrate our attention mainly on studies of that kind.

The interpretation of the data on the auroral dynamics is based on the following consideration. The velocity of energetic electrons with the energy of about 1–5 keV is $(2-4)\cdot 10^9$ cm·sec^{-1}, so that it takes for them about 2–3 sec to pass the way from the magnetopause or the plasma sheet to the ionosphere. This allows us to suggest that auroral arcs or other auroral forms locate practically at the same geomagnetic field lines at which their sources in plasma sheet or at the magnetopause are situated, and, correspondingly, are moving with them. In turn, sources of precipitating electrons may be either frozen in the ambient plasma, or move with respect to it. Among the sources of the first type may be plasma clots located at closed geomagnetic field lines [1]. The sources of the second type may be associated with some MHD waves propagating across the geomagnetic field lines, or magnetic diffusion waves in the vicinity of a reconnection line [2]. This variety and complexity of the expected relationship between the motions of the magnetospheric plasma and the sources of precipitating particles compels us to be very careful in the interpretation of the data on the dynamics of aurorae. In this connection, in the further analysis, we shall try to confirm the conclusions obtained from the data on the motion of aurorae by the data of independent observations.

Besides, we shall suppose that the magnetospheric electric field, as any other vector field, may be represented as the sum of a potential ($\mathbf{E}_p = -\nabla\varphi$) and a vortex ($\mathbf{E}_v = -\frac{1}{c}\partial\mathbf{A}/\partial t$) electric fields. Each of these fields has its own sources within the solar wind or in the magnetosphere, they develop in quite different ways, and differently manifest themselves in the ionosphere [3], [4], [5]; therefore, we shall consider these two components of the magnetospheric electric field separately.

2. Electric fields at the magnetopause and in the dayside magnetosphere

The magnetopause is a region where the solar wind contacts the geomagnetic field, and where the processes determining the state of the entire magnetosphere are developing. And a fundamental role among those processes is plaid by the magnetic field reconnection.

For many years, beginning with a pioneering paper by Petschek [6], magnetic field reconnection models have been developed under the supposition on the steady-state character of the process. However, even the earliest observations of auroral dynamics in the cusp region [7], [8], [9], [10] have shown that the steady-state situation is observed rather seldom (if ever), and most often, the magnetic field reconnection proceeds at the dayside magnetopause as a series of burst-like pulses. Recognition of this fact results in that many present-day models suppose the reconnection at the magnetopause to have no steady-state component on the whole. However, this supposition seems to be unsubstantiated, and we shall try to show that the reconnection at the dayside magnetopause proceeds as a superposition of pulses of the spontaneous reconnection and of a quasi-stationary slowly varying reconnection which does not cease between those pulses.

In Fig. 1 after Pudovkin *et al.* [4] there are presented contour plots of the red auroral emission in Ny Alesund on Jan. 10, 1985. One can see the aurorae to display on the day under consideration in a usual manner with a quasi-periodical recurrence of brightening and poleward motion of auroral arcs [10], [11]. Based on the analogy with the poleward leaps of aurorae during nightside breakups, these events have been termed as midday breakups [11], [12]. It has to be noted however that the poleward drift of aurorae in the cusp region is associated with a similar motion of the ionospheric plasma [11] and is caused correspondingly by the potential component of the magnetospheric–ionospheric electric field [3]. At the same time, the poleward jumps of aurorae during the night side breakups are not associated with the poleward drift of the ionospheric plasma [13] and are driven correspondingly by the vortex component of the electric field. Thus, the poleward motion of the auroral transients shown in Fig. 1 cannot be identified with auroral breakups [4].

At the same time, as is seen in the Figure, just before the appearance of each transient, there may be observed a rapid jump of the southward boundary of the auroral belt to the equator. As the poleward motion of auroral transients goes on existing also during the equatorward jumps of the separatrix, these jumps cannot be associated with the equatorward drift of the ionospheric plasma, and are driven by the vortex component of the magnetopause electric field [4]. This conclusion has been confirmed by Lockwood *et al.* [14] who have found on the base of radar observations that there are occasions when the separatrix boundary jumps equatorward without the corresponding motion of the ionospheric plasma. Thus, equatorward jumps of the southern boundary of the cusp region auroral belt may be considered as real day side breakups. However, characteristics of those breakups sig-

144

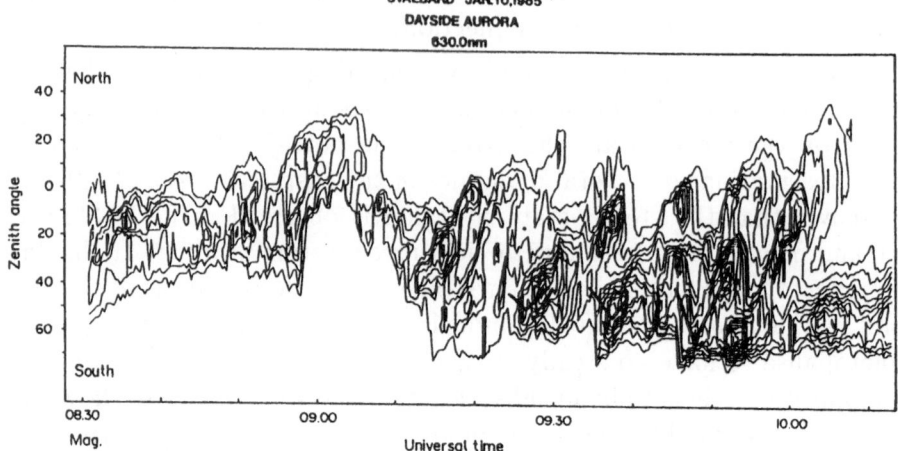

Figure 1. Contour plot of the red auroral emission line in Ny Ålesund on 10 Jan. 1985. Thick solid lines show auroral arc motion during breakups.

nificantly differ from the characteristics of the night side breakups, and not only in concern to the direction of the auroral arc motion. Indeed, these two types of breakups are distinguished also by the sense of physical processes associated with them: while the night side breakups result in closing previously open field lines, on the day side they result in opening field lines which were previously closed. Thus, the day side breakups may be considered as inversed (with respect to the night side phenomena) breakups.

Intensity of the both components of the magnetopause electric field may be estimated from the velocity of the auroral arc motion:

$$E_{mp} = \frac{1}{c} v_{iN} B_i \frac{\delta l_i}{\delta l_m} \tag{1}$$

$$E_{mv} = -\frac{1}{c} v_{iS} B_i \frac{\delta l_i}{\delta l_m} \tag{2}$$

here E_{mp} and E_{mv} are the potential and vortex components of the E_m field, positive eastwards, v_{iN} and v_{iS} are velocities of the poleward drift and equatorward jumps respectively, B_i is the geomagnetic field intensity in the ionosphere, and $\frac{\delta l_i}{\delta l_m}$ is a geometric factor which characterizes the divergence of the geomagnetic field lines [4].

Variations of the induced and potential electric field intensity on the day under consideration for $\frac{\delta l_i}{\delta l_m} = 30$ are presented in Fig. 2 after Pudovkin *et al.* [4].

It is seen in the Figure that the vortex component of E_m field really develops as a series of pulses with peak values of about 2–2.5 mV/m.

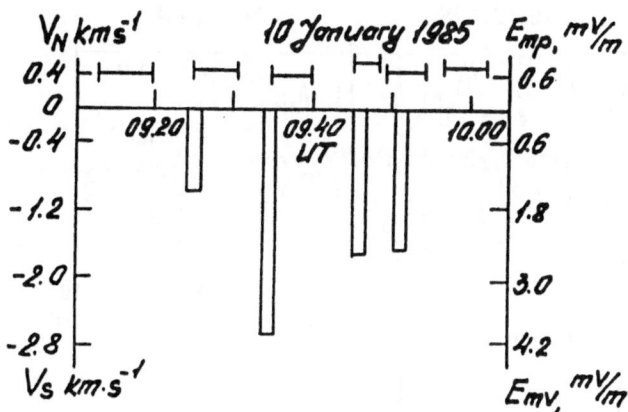

Figure 2. Variations of the velocity of the auroral arc motion to the pole and to the equator and corresponding values of the electric fields $E_{m,p}$ and $E_{m,v}$.

What concerns the potential component of the electric field, it is seen to exist during pulses of the E_{mv} as well as at intervals between them. Thus, the magnetic field reconnection at the magnetopause really proceeds as a superposition of pulse-like and of a quasi-stationary processes.

Another sample of the development of the breakups in the cusp region is presented in Fig. 3 after Pudovkin *et al.* [5]. Once again, one can see the reconnection electric field to consist of a slowly-varying quasi-stationary component E_{mp} superposed by numerous short duration pulses of the induced field E_{mv}. For convenience of comparison with E_{mp}, E_{mv} values are given in the Figure in units corresponding to the ionospheric level. In this connection, there has to be once more reminded that this recalculation is quite formal, and the E_{mv} field exists (in a motionless frame of reference) only at the magnetopause and in its vicinity. This field accelerates and heats the plasma in the reconnection region providing the brightening of aurorae in the cusp region; besides, it causes the motion of the separatrix in the ionosphere. Correspondingly, it contributes into the voltage along the reconnection line; at the same time, in the ionosphere the induced field exists only in the separatrix frame of reference, and hence does not influence the potential drop across the polar caps measured as a rule in the motionless frame of reference. This makes groundless any attempts to obtain the potential drop across the polar caps from the voltage produced by the total electric field along the reconnection line at the magnetopause, or from the voltage along the separatrix moving in the ionosphere.

A similar consideration of ionospheric signatures of the reconnection pulses at the magnetopause was carried out by Lockwood [15] and by Lockwood and Davis [16]. However, in contradiction to our model, the authors suppose that the reconnection electric field has a vortex character (that

146

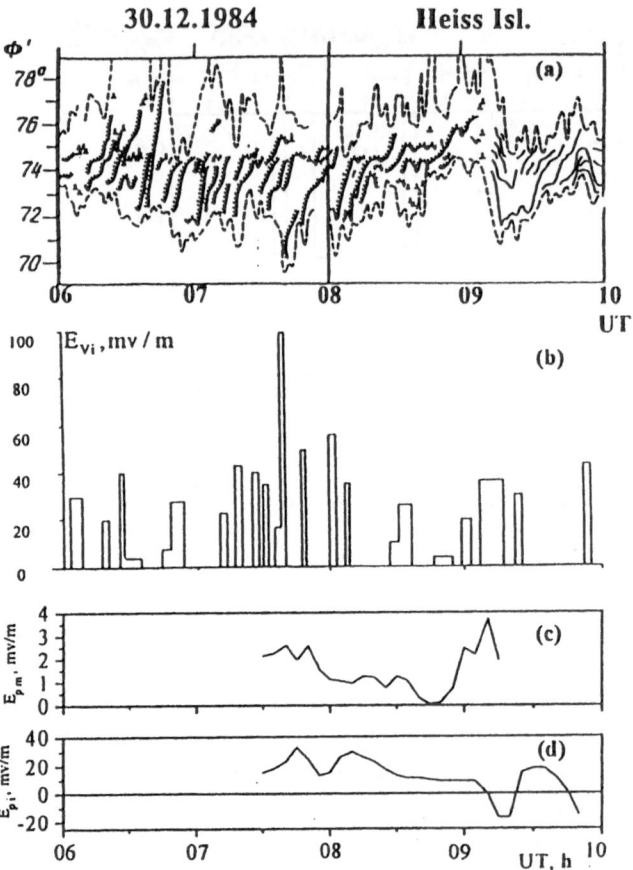

Figure 3. Variations of auroral and other geophysical indices on Dec., 30, 1984. a — Auroragram from the Heiss Island observatory; dashed line shows the boundaries of the auroral belt; solid lines with dashes and thin solid lines show rayed and homogeneous auroral arcs respectively. b — Intensity of the vortex electric field E_{vi}; c — intensity of the model solar wind electric field at the magnetopause (E_{pm}); d — ionospheric electric field $E_{pi} = \frac{1}{c} v_{iN} B_i$.

is $\partial \Phi_y / \partial s \neq 0$, where s is the distance along the reconnected field line) only during periods of about 1–2 min taken by an Alfvenic wave to propagate from the magnetopause to the ionosphere; afterwards, the reconnection electric field is mapped onto the ionosphere as a potential field and causes there bursts of the poleward motion of auroral transients and of the ionospheric plasma. Of course, such a possibility cannot be entirely excluded, and we shall check it by experimental data. In this connection we turn again to Fig. 3, in the lower panel of which variations of the ionospheric electric field E_{pi} derived from the arc motion velocities are shown. As is seen in the Figure, the value of E_{pi} varies very smoothly without any significant bursts which could be ascribed to the pulses of E_{mv}. This result seems to

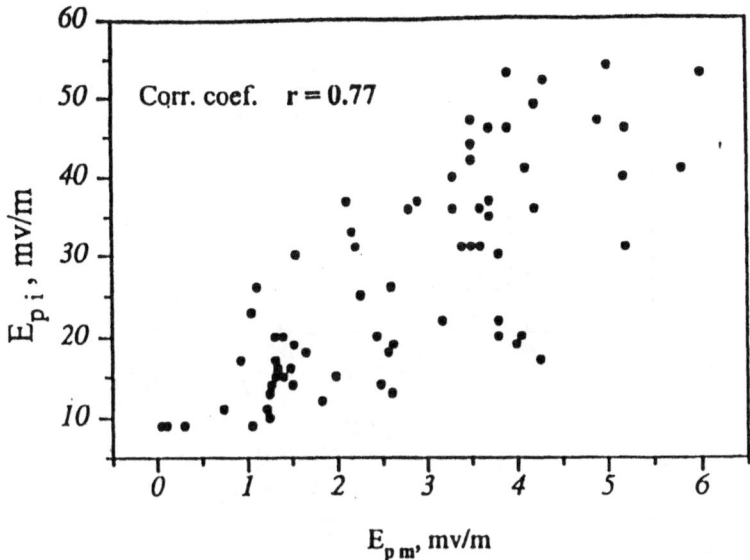

Figure 4. Ionospheric electric field intensity E_{pi} versus $E_{pm} = \frac{1}{c} B_T \sin^2 \frac{\Theta}{2}$.

confirm our supposition on the principally vortex character of the pulsed reconnection field.

Nevertheless, the intensity of the potential component E_{mp} is seen to vary in a rather wide range, and according to the model of the quasi-stationary reconnection these variations are determined by the variations of the solar wind electric field. To confirm this supposition, there are presented in Fig. 3c variations of the model electric field at the magnetopause: $E_{pm} = \frac{1}{c} v_w B_T \sin^2(\Theta/2)$ where v_w is the solar wind velocity. As is seen in the Figure, variations of the ionospheric (Fig. 3d) and magnetopause (Fig. 3c) electric fields at the period 07:30–08:50 are rather close, and the coefficient of correlation between them equals $r = 0.6$ at time delay $\tau = 15$ min; the last figure agrees with results by Horwitz and Akasofu [17] and Clauer and Banks [18].

In the same manner there were analyzed data on some more days, and the scatter plot of E_{pi} versus E_{pm} is given in Fig. 4. One can see that the correlation between E_{pi} and E_{pm} values is sufficiently close with $r = 0.77$ at $\tau = 15$ min.

Beside the meridional drift, aurorae in the cusp region experience even a more rapid motion in the azimuthal direction [11]. Velocity of this motion is supposed to be determined by the IMF B_y component [19], [20]. This supposition is convincingly confirmed by the data presented in Fig. 5 after Pudovkin *et al.* [21] in which there is shown intensity of the X-component of geomagnetic disturbances at Mould Bay ($\Phi = 80°56$) at 11–13 MLT on

148

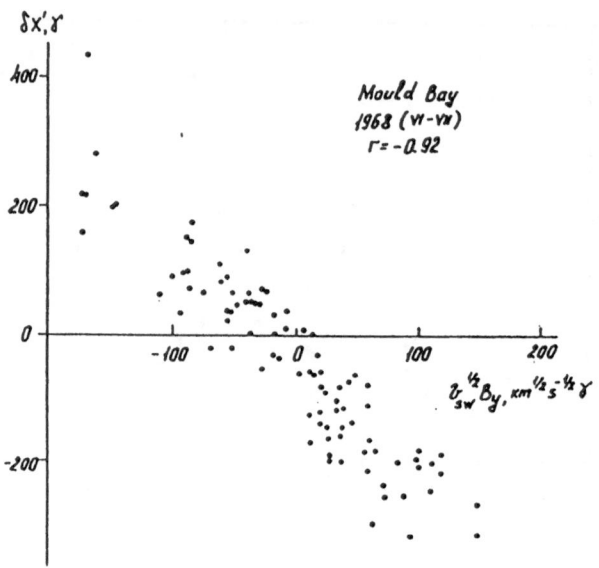

Figure 5. The dependence of geomagnetic disturbance vector (Mould Bay; 11.00-13.00 MLT; June–July 1968) projected onto the Earth–Sun line $\delta X'$ on $(B_y V_{sw}^{1/2})$.

June–July 1968 in dependence on $v_w^{1/2} B_y$. One can see a close correlation ($r = -0.9$) between the two variables, which also confirms the fundamental role of the solar wind electric field and, hence, the importance of the quasi-stationary magnetic field reconnection in the electrodynamics of the magnetosphere.

Another parameter characterizing the global intensity of the magnetospheric plasma convection is the potential drop across the polar caps ($\Delta \Phi$). According to the model by Pudovkin *et al.* [21] the value of $\Delta \Phi$ may be estimated as

$$\Delta \Phi \text{ (kV)} = \Delta \Phi_{vis} + 4.2 n^{-1/2} B_T^2 | \sin^3 \frac{\Theta}{2}| \tag{3}$$

where $\Delta \Phi_{vis}$ is a quasi-viscous term [22], n (cm^{-3}) is the solar wind plasma number density, and $B_r(\gamma) = (B_y^2 + B_z^2)^{1/2}$ is the IMF transversal component intensity. The value of $\Delta \Phi_{vis}$ may be presented as [23]:

$$\Delta \Phi_{vis} \text{ (kV)} = 13.6 + 0.07(v_w - 300) \tag{4}$$

where v_w is measured in km/sec.

It is interesting to note that in contrast to the common point of view, the reconnection term in Eq. (3) does not depend on the solar wind velocity v_w. And according to the model, this is explained by that while the electric field intensity at the magnetopause is proportional to the v_w, the reconnection line length L is inversely proportional to v_w so that in the product $E_m \cdot L$ the solar wind velocity terms cancel each other.

Values of the $\Delta\Phi$ predicted according to Eq. (3) versus the observed ones are given in Fig. 6. As is seen in the Figure, predicted values of $\Delta\Phi$ agree with experimental ones rather well, and the coefficient of correlation between them amounts the value $r = 0.9$. However, the amount of data presented in Fig. 6 is rather small. Because of that, data presented by Boyl et al. [24] seem to be of a great interest. ¿From the analysis of a great number of polar cap satellite crossings, the authors obtained the following empirical formula:

$$\Delta\Phi = 1.01 \cdot 10^{-4} v_w^2 + 11.7 |B \sin^3 \frac{\Theta}{2}|. \qquad (5)$$

It is not difficult to see that the structure of Eq. (5) is very close to that of Eq. (3); in particular, the second term in the right hand side of Eq. (5) does not contain the solar wind velocity either, and it is proportional to $|\sin^3 \frac{\Theta}{2}|$. This allows us to believe that in spite of some difference between Eq. (3) and Eq. (5), the model based on the supposition on the magnetic field reconnection adequately describes gross features of the solar wind – magnetosphere interaction.

Electric field transferred from the solar wind into the magnetosphere, cause within it a large scale plasma convection and generate there global systems of electric currents. To maintain these phenomena, a certain energy is needed. Of course, the primary source of that energy is the solar wind.

It is usually supposed that the solar wind energy enters the magnetosphere in the form of the Poynting vector flux. At the same time, there is known that within the solar wind the energy is concentrated mainly in the form of the kinetic energy of the solar wind plasma. This suggests that somewhere in the vicinity of the magnetopause, the solar wind kinetic energy transforms into the electromagnetic one. The transformation of the solar wind kinetic energy into the electromagnetic and thermal energy of the magnetosheath plasma, as well as the rate of the electromagnetic energy input into the reconnection region at the dayside magnetopause were considered by Pudovkin and Semenov [25]. There has been shown that the transformation of the solar wind kinetic energy takes place mainly at the bow shock and within the magnetic barrier in front of the magnetopause. The flux of the Poynting vector across the outer boundary of the reconnection region was estimated as:

$$\varepsilon = \frac{1}{4\pi(Ma)^{1/3}} \frac{\rho_m}{\rho_0} v B^2 Dm^2 \sin^4 \frac{\Theta}{2} \qquad (6)$$

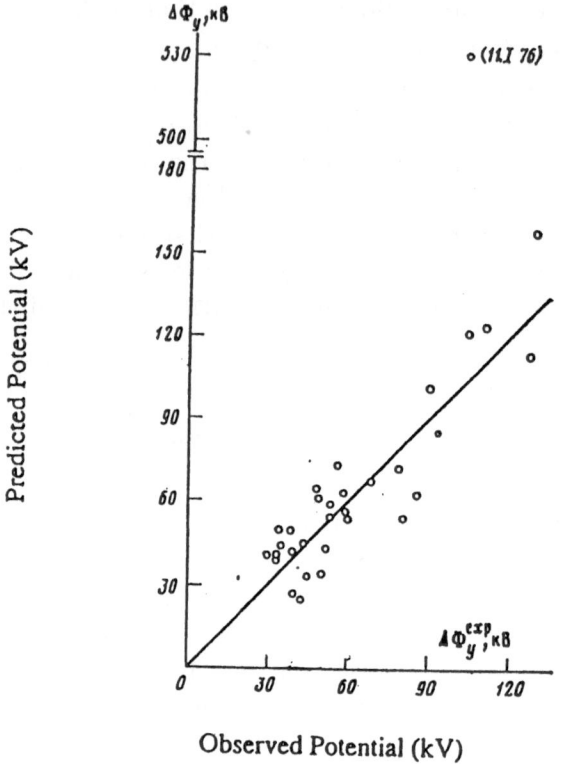

Figure 6. Calculated potential drop across the polar cap in dependence on the observed one.

where Dm is the magnetosphere's diameter, ρ_0 and ρ_m are the magnetosheath plasma density just after the bow shock crossing and at the magnetopause respectively, and Ma is the Alfvenic Mach number.

It is interesting to compare this result with the empirical formula by Perreault and Akasofu [26]:

$$\varepsilon = l_0^2 v B^2 \sin^4 \frac{\Theta}{2} \qquad (7)$$

with $l_0 = (7-8)\,R_e$. One can see that for $Ma \approx 10$, the model Eq. (6) and the empirical Eq. (7) practically coincide, which once more confirms validity of the theoretical model.

Thus, the model of the magnetic reconnection seems to describe adequately the energetics of the solar wind – magnetosphere interaction. However, beside the question on the energy sources, the problem on the solar wind – magnetosphere interaction includes also a question on the circuit connecting the reconnection region to the energy sources [27]. To clear up this question, we shall use results by Bernikov and Semenov [28] who have calculated in a kinematic approximation the magnetic field in a perfectly conductive fluid flowing around a sphere. As the result, the following expression for the components of the magnetic field in the vicinity of the leeward surface of the sphere were obtained:

$$b_r = \frac{2}{3} \frac{\sqrt{3(r-1)} \sin\Theta \sin\varphi}{(1+\cos\Theta)}; \quad b_\Theta = \frac{\sin\varphi}{\sqrt{3(r-1)}}; \quad b_\varphi = \frac{\cos\varphi}{\sqrt{3(r-1)}} \quad (8)$$

here $\mathbf{b} = \mathbf{B}/B_\infty$; r is the normalized (by the radius of the sphere) distance; Θ is the polar angle with respect to the polar axis pointed in $(-\mathbf{v}_\infty)$ direction; φ is the azimuthal angle calculated from the X-axis; and the Y-axis is directed along the magnetic field $\mathbf{B}_\infty$.

It has to be noted that solution (8) is asymptotic one, and is valid therefore only for a certain region where $(r-1)/r << 1$ and $-\pi/2 \leq \Theta \leq \pi/2$.

Using (8), we can calculate rot $\mathbf{b}$ and thereby the intensity of the electric currents:

$$j_r \frac{(\cos\Theta - 1)\cos\varphi}{r\sqrt{3(r-1)}\sin\Theta}; \quad j_\Theta = \frac{2}{3}\frac{\sqrt{3(r-1)}\cos\varphi}{\sqrt{3(r-1)}} - \frac{(3r-6)\cos\varphi}{2r(3(r-1))^{3/2}}$$

$$j_\varphi = \frac{(3r-6)\sin\varphi}{2r(3(r-1))^{3/2}} - \frac{2}{3}\frac{\sqrt{3(r-1)}\cos\Theta\sin\varphi}{r(1+\cos\Theta)} - \frac{2}{3}\frac{\sqrt{3(r-1)}}{r(1+\cos\Theta)^2} \quad (9)$$

The distribution of the currents density at the equatorial plane for a southward IMF is shown in Fig. 7. Data presented in the Figure show that in the magnetic barrier region currents flow westwards, so that the dot product $(\mathbf{E}\cdot\mathbf{j})$ is there negative. Thus, the magnetic barrier may be considered as an MHD generator in which the kinetic energy of the solar wind is transformed into the electromagnetic one. Besides, one can see that in spite of that the normal component of the magnetic field vanishes at the magnetopause $(r \rightarrow 1)$, the radial component of the current density goes on existing there, and the currents forming the magnetic barrier flow into the magnetopause and close along the latter. Thus, the magnetopause (DCF) currents indeed are connected to the energy source in the barrier region, which allows the magnetic reconnection process to proceed in a quasi-steady-state regime.

Till now, we have considered the interaction of the solar wind with the magnetosphere supposing the latter to be a passive object of that interaction. Of course, this supposition oversimplifies the problem, and we may expect the intensity of the electric field in the magnetosphere to depend

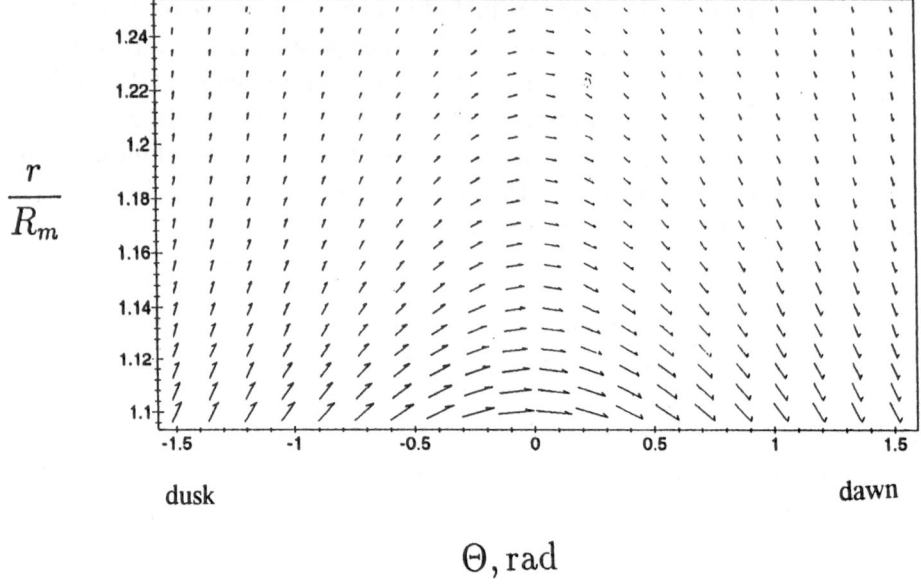

$$\frac{r}{R_m}$$

dusk   dawn

$$\Theta, \mathrm{rad}$$

Figure 7. The distribution of the currents density at the equatorial plane for a southward IMF.

not only on the solar wind parameters but also on the state of the magnetosphere.

And indeed, an effect of that kind may be seen in Fig. 8 after Starkov *et al.* [29] in which variations of the electric field intensity in the cusp region (at Spitzbergen) calculated from the velocity of auroral transients, and of the intensity of geomagnetic disturbances in the nightside auroral zone (obs. College) are presented. One can see in the Figure that at approximately 20 min before the onset of an auroral substorm in the nightside magnetosphere, the intensity of the potential component of the electric field in the cusp region and correspondingly at the magnetopause, significantly decreases.

To clear up the cause of this decrease, we shall consider the spatial distribution and temporal variation of the electric fields in the night side magnetosphere.

In Fig. 9 after Zverev *et al.* [30], there are shown variations of the auroral arc velocity in the auroral zone, during the initial phase of substorms; the figures at the curves show the values of IMF B_z during corresponding substorms.

As is seen in the Figure, in the case of a northward IMF ($B_z = 2.3 \gamma$), the auroral arc velocity and hence the potential electric field intensity in the plasma sheet are almost constant during the initial phase of substorms. However, with increase of the southern component of the IMF, there ap-

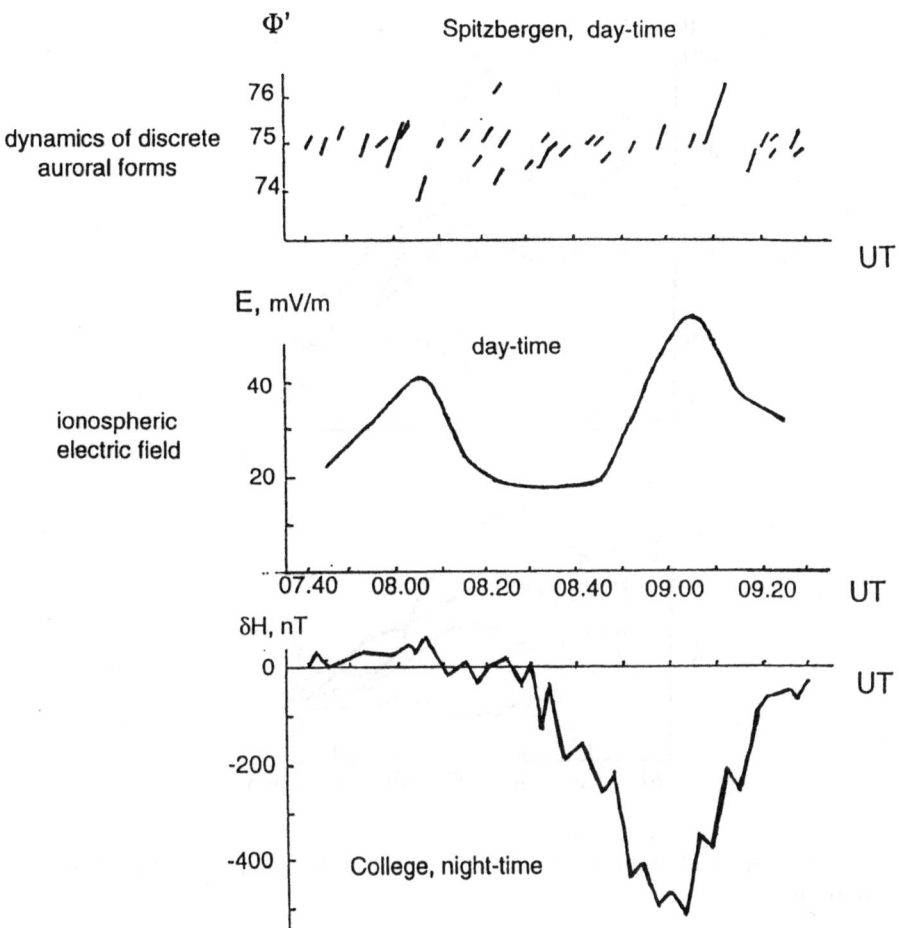

Figure 8. Development of auroral substorm 18.12.1981.

pears in the curves of $v(t)$ and becomes more and more evident a peculiarity consisted of a significant decrease of the arc velocity at $\sim$20–25 min before the beginning of the active phase of the substorm. Thus, the decrease of the electric field intensity is observed at the end of the initial phase not only in the day side of the magnetosphere but also in the night side of it. It is interesting to note that at the same period, the intensity of the induced electric field associated with stretching the geomagnetic field lines into the tail, is maximum [31]. This allows us to suppose that the observed decrease of the potential electric field is caused by the polarization of the magnetospheric plasma in the magnetotail.

3. Conclusions

All the regularities of the auroral arc motion and the electric field distribution in the day-side magnetosphere and ionosphere may be described in the

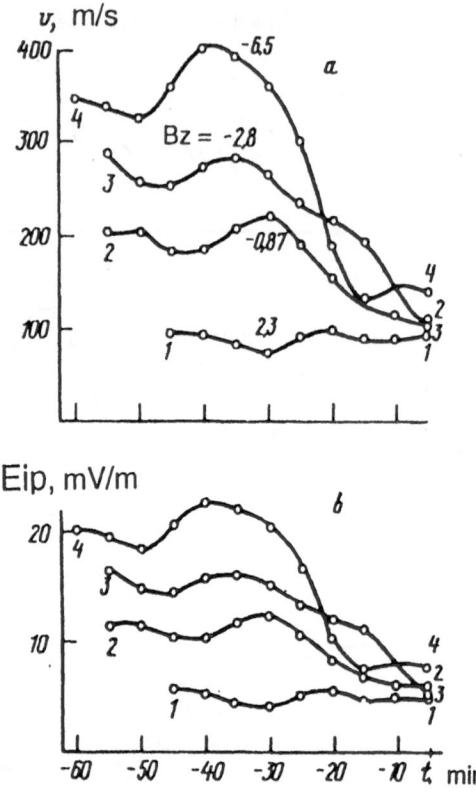

Figure 9. Variations of auroral arc velocity and potential ionospheric electric field E_{ip} at the inital phase.

frame of a model with the magnetic field reconnection at the magnetopause as follows.

1. Magnetic field reconnection at the magnetopause proceeds in a dual manner as a superposition of a quasi-steady-state reconnection and of burst-like pulses of the spontaneous reconnection [3], [4], [32], [33], [34].

2. The zonal component of the potential electric field in the polar ionosphere amounts the value of 10–60 mV/m and is determined by the solar wind electric field component parallel to the reconnection line with a time delay of about 15 minutes.

3. Intensity of the induced electric field during reconnection pulse peaks amounts a value several times higher than the background quasi-stationary electric field [3], [4], [34]. At the same time, this electric field, being local in its nature, does not contribute into the potential drop across the polar caps.

4. Energy sources of the reconnection processes locate within the magnetic barrier region which is connected to the magnetopause by a 3D electric current system.

5. Development of auroral substorms in the nightside magnetosphere results in a decrease of the electric field in the cusp region at 5–25 minutes before the breakup [29].

Acknowledgements: This work has been supported by the Russian Foundation for Fundamental Researches, Grants N 97-05-64458 and 96-05-00006G.

References

1. Pudovkin, M.I. and Golovchanskaya, I.V. (1989) On the formation of discrete auroral arcs, *Planet. Space Sci.* **37**, 783–793.
2. Kornilova, T.A., Pudovkin, M.I., and Starkov, G.V. (1990) Fine structure of aurorae in the vicinity of poleward boundary of the auroral bulge during breakups, *Geom. Aeron.* **30**, 250–254 (Russian Edition).
3. Pudovkin, M.I., Semenov, V.S., Starkov, G.V., and Kornilova, T.A. (1991) On separation of potential and vortex parts of the magnetospheric electric field, *Planet. Space Sci.* **39**, 563–568.
4. Pudovkin, M.I., Zaitseva, S.A., Sandholt, P.E., and Egeland, A. (1992) Dynamics of aurorae in the cusp region and characteristics of the magnetic reconnection at the magnetopause, *Planet. Space Sci.* **40**, 879–887.
5. Pudovkin, M.I., Zaitseva, S.A., Sandholt, P.E., and Egeland, A. (1996) Relationship between the cusp aurora poleward motion velocity and solar wind parameters, in: *Proc. Third International Conference on Substorms (ICS-3), Versailles, France, 12-17 May 1996*, ESA SP-389, pp. 737–742.
6. Petschek, H.E. (1964) Magnetic field annihilation, in: *AAS/NASA Symp. of the Physics of Solar Flares*, NASA Spec. Publ., SP-50, 425–439.
7. Vorobjev, V.G., Gustafsson, G., Starkov, G.V., Feldstein, Ya.I., and Shevnina, N.F. (1975) Dynamics of day and night aurora during substorms, *Planet. Space Sci.* **23**, 269–278.
8. Vorobjev, V.G., Zverev, V.L., and Leontiev, S.V. (1988) The structure of auroral luminosity in the midday sector, *Geom. Aeron.* **28**, 256–261 (Russian Edition).
9. Sandholt, P.E., Egeland, A., Holtet, J.A., Lybekk, B., Svenes, K., and Asheim, S. (1985) Large- and small-scale dynamics of the polar cusp, *J. Geophys. Res.* **90**, 4407–4414.
10. Sandholt, P.E., Deehr, C.S., Egeland, A., Lybekk, B., Viereck, R. and Romick, G.J. (1986) Signatures in the dayside aurora of plasma transfer from the magnetosheath, *J. Geophys. Res.* **91**, 10,063–10,079.
11. Sandholt, P.E., Lockwood, M., Oguti, T., Cowley, S.W.H., Freeman, K.S.C., Lybekk, B., Egeland, A., and Willis, D.M. (1990) Midday auroral breakup events and related energy and momentum transfer from the magnetosheath, *J. Geophys. Res.* **95**, 1039–1060.
12. Baker, K.B., Rodger, A.S., and Lu, G. (1997) HF-radar observations of the dayside magnetic merging rate: a Geospace Environment Modeling boundary layer campaign study, *J. Geophys. Res.* **102**, 9603–9617.
13. Kelley, M.C., Starr, J.A., and Mozer, F.C. (1971) Relationship between magnetospheric electric fields and the motion of auroral forms, *J. Geophys. Res.* **76**, 5269–5277.
14. Lockwood, M., Moen, J. Cowley, S.W.H., Farmer, A.D., Lovhaug, U.P., Lühr, H., and Davda, V. N. (1993) Variability of dayside convection and motions of the

cusp/cleft aurora, *Geoph. Res. Letters* **20**, 1011–1014.

15. Lockwood, M. (1994) Ionospheric signatures of pulsed magnetopause reconnection, in J. A. Holtet and A. Egeland (eds.), *Physical signatures of magnetospheric boundary layer processes*, NATO ASI series C, vol. 425, Kluwer Acad. Press, pp. 229–243.

16. Lockwood, M. and Davis, C.J. (1996) On the longitudinal extent of magnetopause reconnection pulses, *Ann. Geophys.* **14**, 865–878.

17. Horwitz, J.L., and Akasofu, S.-I. (1977) The response of the dayside aurora to sharp northward and southward transition of the interplanetary magnetic field and magnetospheric substorms, *J. Geophys. Res.* **82**, 2723–2734.

18. Clauer, C.R. and Banks, P.M. (1986) Relationship of the interplanetary electric field to the high-latitude ionospheric electric field and currents: observations and model simulation, *J. Geophys. Res.* **91**, 6959–6971.

19. Svalgaard, L. (1968) Sector structure of the IMF and daily variation of the geomagnetic field at high latitudes, prepr. Det Danske Met. Inst., Charlottenlund, 11 p.

20. Mansurov, S.M. (1969) New evidences on the relationship between the IMF and the geomagnetic field, *Geom. Aeron.* **11**, 115–118 (Russian Edition).

21. Pudovkin, M.I., Zaitseva, S.A., Bazhenova, T.A., and Andrezen, V.G. (1985) Electric fields and currents in the Earth's polar caps, *Planet. Space Sci.* **33**, 407–414.

22. Reiff, P.H., Spiro, R.W., and Hill, T.V. (1981) Dependence of polar cap potential drop on interplanetary parameters, *J. Geophys. Res.* **86**, 7639–7648.

23. Pudovkin, M.I., and Zaitseva, S.A. (1983) Electric fields in the polar cap, *Geom. Aeron.* **23**, 285–289 (Russian Edition).

24. Boyl, C.B., Reiff, P.H., and Hairston, M.R. (1997) Empirical polar cap potentials, *J. Geophys. Res.* **102**, 111—125.

25. Pudovkin, M.I., and Semenov, V.S. (1986) The flux and transformation of the solar wind energy in the magnetosheath, *Geom. Aeron.* **26**, 887–891 (Russian Edition).

26. Perreault, P. and Akasofu, S.-I. (1978) A study of geomagnetic storms, *Geoph. J. Roy. Astron. Soc.* **54**, 547–573.

27. Heikkila, W. (1997) The reconnection myth, *EOS* **78**, 153–156.

28. Bernikov, L.V. and Semenov, V.S. (1979) On the problem of an MGD flow around the magnetosphere. *Geom. Aeron.* **19**, 671–675 (Russian Edition).

29. Starkov, G.V., Pudovkin, M.I., Smith, P.R., and Rijnbeek, R.P. (1996) Dynamics of aurora distribution of electric fields on day side during substorms, in: *The 1st EGS Alfven Conference on Low-Altitude Investigation of Dayside Magnetospheric Boundary Processes. Abstracts*, Kiruna, Sweden, p. 41.

30. Zverev, V.L., Pudovkin, M.I., and Starkov, G.V. (1994) The aurora motion and the electric fields during the initial substorm phase, *Geom. Aeron.* **34**, 49–55 (Russian Edition).

31. Kornilova, T.A., Kornilov, I.A., Pudovkin, M.I., and Starkov, G.A. (1997) Velocities of aurora motion and electric field distribution during active phase of substorms, *Geom. Aeron.* (In press).

32. Pudovkin, M.I. (1994) A model of the magnetosheath and dayside magnetopause and their coupling to the cusp/cleft ionosphere, in J.A. Holtet and A. Egeland (eds.), *Physical signatures of magnetospheric boundary layer processes*, Kluwer Acad. Press, Netherlands, pp. 421–431.

33. Cowley, S.W.H., Freeman, M.P., Lockwood, M., and Smith, M.F. (1991) The ionosphere signature of flux transfer events, in C.I. Barron (ed.), *CLUSTER-dayside polar cusp*, ESA SP-330, ESA, Nordvijk, The Netherlands, pp. 105–112.

34. Lockwood, M., Cowley, S.W.H., Sandholt, P.E., and Lovhaug, U.P. (1995) Causes of plasma bursts and dayside auroral transients: An evaluation of two models invoking reconnection pulses and changes in the Y-component of the magnetic field, *J. Geophys. Res.* **100**, 7613–7626.

POLAR OBSERVATIONS OF CUSP ELECTRODYNAMICS: EVOLUTION FROM 2- TO 4-CELL CONVECTION PATTERNS

N. C. MAYNARD
Mission Research Corporation, Nashua, NH

W. J. BURKE
Phillips Laboratory, Hanscom Air Force Base, MA

D. R. WEIMER
Mission Research Corporation, Nashua, NH

F. S. MOZER
University of California, Berkeley, CA

J. D. SCUDDER
University of Iowa, Iowa City, IO

W. K. PETERSON
Lockheed Martin Space Sciences Laboratory, Palo Alto, CA

R. P. LEPPING
Goddard Space Flight Center, Greenbelt, MD

C. T. RUSSELL
University of California, Los Angeles, CA

1. Introduction

The dayside magnetosphere quickly responds to changes in the polarities of IMF B_Y and/or B_Z. Within a few minutes of the changes reaching the magnetopause, characteristic optical [1] and plasma convection signatures [2] appear in the ionospheric projection of the cusp. Global ionospheric convection patterns at high geomagnetic latitudes, however, represent a mixture of IMF conditions over the previous half hour. Maezawa [3] first reported observing magnetic perturbations during sustained periods of northward IMF whose explanation required sunward convection in the central polar cap. Electric and magnetic fields measured by the S3-2 [4], Atmospheric Explorer [5] and MAGSAT [6] satellites suggested that with IMF $B_Z > 0$

157

J. Moen et al. (eds.), Polar Cap Boundary Phenomena, 157–172.

and $B_Y \approx 0$, a four-cell convection patterns evolves. This convection pattern consists of two cells in the polar cap, whose polarity is opposite to the adjacent, standard negative potential (clockwise) afternoon and positive (counter-clockwise) morning cells. The polar cap convection cells are driven by magnetic merging at the poleward boundary of the cusp [7]. The residual, standard-polarity pair of cells at auroral latitudes are weak and probably are related to the low latitude boundary layer (LLBL) [8].

Heppner and Maynard [9] showed that during periods of northward IMF in which B_Y had large values, electric fields measured by the DE 2 satellite were consistent with the existence of two standard-polarity convection cells at high latitudes, distorted in shape. Flow (equipotential) lines had the same sense of rotation as those observed when $B_Z < 0$. Further analysis for B_Y positive showed that the dayside portion of the afternoon cell rotated into the prenoon sector and consisted of two parts: (1) equipotentials whose associated magnetic flux is always open (lobe cells), and (2) equipotentials whose associated flux is both open and closed [10]. The lobe cell was embedded within the afternoon cell and had the same sense of rotation. Recent SuperDARN observations showed dayside convection patterns evolved from two distorted cells into four cells as IMF B_Y decreased in magnitude relative to B_Z [11]. The observed evolution of dayside convection compared favorably with predictions of the empirical model of Weimer [12].

Satellite passes through the dayside high-latitude ionosphere encounter distinctive plasma characteristics. Newell *et al.* [13] identified the spectral properties of electrons and ions in the dayside ionosphere whose magnetospheric sources are the central plasma sheet (CPS), the boundary plasma sheet (BPS), the LLBL, the cusp and the mantle. In energy-versus-time spectrograms the cusp is marked by intense fluxes of low-energy (<100 eV) electrons and energy-dispersed ions. The latter signature is a time-of-flight, velocity-filter effect. During periods of northward (southward) IMF, the highest energy ions are detected near the poleward (equatorward) boundary of cusp precipitation [14]. Ionospheric plasma convection in the cusp has a sunward (poleward) component when the IMF has a northward (southward) component. The azimuthal component of convection is controlled by the polarity of B_Y. Besides the large scale, Region 1/Region 2 systems, the dayside ionosphere is marked by cusp and mantle field-aligned current (FAC) systems, referred to as Region 0 [15].

This paper extends a study by Maynard *et al.* [16] based on particle and field measurements taken during three Polar orbits at dayside, high latitudes throughout which the B_Z and B_Y were positive for extended periods of time. The observed energies/pitch angles of different ion species are used to improve identifications of the source regions for electrodynamic signatures detected by Polar. The T96 magnetic field model of Tsyganenko

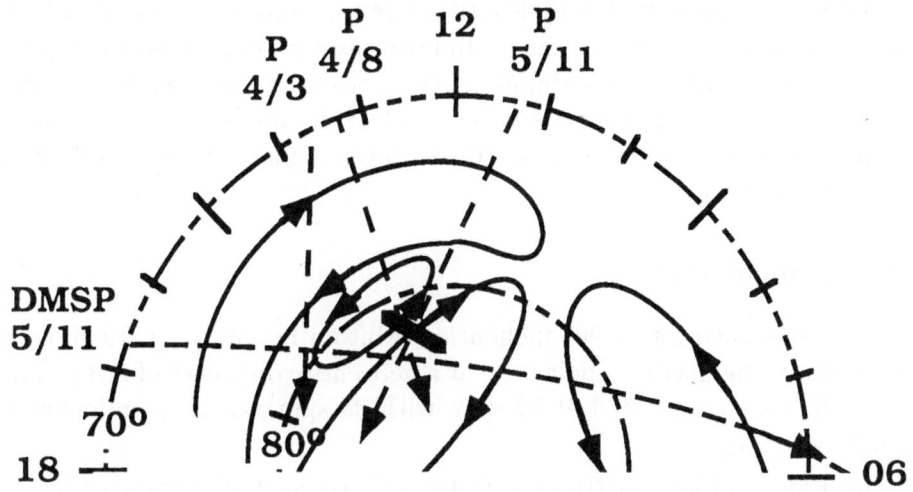

Figure 1. Schematic representation of high-latitude convection patterns encountered by Polar on April 3 and 8 and on May 11, 1996. Approximate orbital traces relative to convection patterns for Polar and DMSP F13 are indicated for reference [16].

[17] is employed to map from locations of Polar to the magnetosphere and the ionosphere. Results of Maynard *et al.* [16] are summarized in Figure 1 which shows three Polar orbits in the early afternoon to noon magnetic local time (MLT) sector. It also schematically represents the convection patterns they encountered. After leaving the region of plasma sheet precipitation, Polar detected quite different particle and field phenomena. During the April 3 orbit Polar skimmed along the projection of the LLBL where it sporadically detected standard ion dispersions and He^{++} fluxes. The April 8 orbit crossed the ionospheric projection of the merging line at the poleward boundary of cusp precipitation within a small positive potential (counterclockwise rotating) cell in the afternoon sector. The orbit on May 11 moved along a zero equipotential line between two lobe cells where it encountered multiple reverse ion-dispersion events. The interpretation of Polar data was confirmed in observations of a four-cell convection pattern detected during a simultaneous high-latitude pass of the Defense Meteorological Satellite Program (DMSP) F13 satellite.

The following sections contain brief descriptions of the Polar sensors used in this study, particle and field measurements taken at middle altitudes ($\sim 5R_E$) by Polar on April 27, and 29, 1996, and their interpretation in terms of encounters with middle altitude projections of previously identified magnetospheric plasma regimes. In both cases we were able to compare POLAR measurements with simultaneous particle and plasma drift observations of high-latitude convection by two DMSP satellites which confirm the interpretation that Polar made direct measurements of lobe-cell parts of four-cell convection patterns.

2. Instrumentation

Polar was launched into a 90° inclination orbit on February 24, 1996, with apogee above the northern polar cap at a geocentric distances of 9 R_E. The spacecraft is spin stabilized at 10 rpm, with its spin axis perpendicular to the orbital plane.

The Electric Field Instrument (EFI) [18] consists of three dipoles to measure vector electric fields from potential differences between three pairs of spherical sensors. Two of the sensor pairs are held at separation distances of 100 m and 130 m by wire booms that rotate in the spacecraft's spin plane. The third pair is held at a separation of 14 m by a pair of rigid booms, aligned with the spin axis. The two spin-plane components of the electric field are represented by the symbols E_{X-Y} and E_Z. E_{X-Y} is approximately the projection of the spin-plane component of the electric field onto the geocentric solar ecliptic (GSE) XY plane. It is positive whenever the unit vector has a component in the $-X_{GSE}$ direction. E_Z is positive toward the GSE north pole. The third component, called E_{56}, points along the spin axis, positive in the sense that completes an orthogonal, right-hand coordinate system (see Figure 1 of Maynard et $al.$ [16]. Thus, with POLAR orbiting along the noon meridian, components of the vector (E_{X-Y}, E_Z, E_{56}) are positive in the $(-X_{GSE}, +Z_{GSE}, +Y_{GSE})$ directions. Spin-fit measurements are presented for E_{X-Y} and E_Z every 6 s and are used to calculate the potential along the orbit and V_{56}, the velocity in the 5-6 direction. Measurements by the short booms of E_{56} are contaminated by differing levels of dc offsets [16]. During the two passes reduced-accuracy values of E_{56} could be determined and are used to estimate V_{X-Y}.

The Magnetic Field Experiment (MFE) [19] consists of two orthogonal, triaxial fluxgate magnetometers that are mounted on a nonconducting boom at separation distances from the nearest satellite surface of 5.97 m and 4.75 m. Here, we are concerned with magnetic perturbations produced by field-aligned currents (FACs) that couple the high-latitude to the magnetosphere or the magnetosheath. The positive-slope deflections in B_{56} with time (latitude) presented below are generated by FACs directed into the ionosphere.

The 3-dimensional electron and hot plasma instrument (HYDRA) [20] consists of two pairs of electron and ion/electron spectrometers, which are each mounted 180° apart on the spacecraft body. In this paper we only use ion and electron measurements from the Duo Deca Electron Ion Spectrometer (DDEIS), which consists of six pairs of 127° electrostatic analyzers looking in different directions outward on a unit sphere. The electron spectrometer measures fluxes in the 2 eV to 35 keV range.

The Toroidal Imaging Mass-Angle Spectrograph (TIMAS) [21] uses a first-order, double focusing system of ion optics that simultaneously measures the spectral characteristics of positively charged ions in the mass per charge range 1 - 32 AMU/q, and energy per charge from 15 eV/q to 32 keV/q. Here we are only concerned with fluxes of He^{++} ions which primarily come from the solar wind. They provide tracers for identifying direct (cusp/mantle) and indirect (LLBL) paths from the magnetosheath.

The DMSP F12 and F13 satellites are in sun-synchronous, circular polar orbits at an altitude of ∼840 km near the 1000 - 2200 and 0600 -1800 local time meridians. Both satellites carry up-looking spectrometers [22] to measure ion and electron fluxes in the energy range 30 eV to 30 keV, and drift meters [23] to monitor the vertical and cross-trajectory components of the ionospheric plasma's bulk motions.

3. Observations

In this section we present plasma and field measurements taken at dayside high latitudes by the Polar satellite between 1500 and 1800 UT on April 27, 1996, and between 1930 and 2130 UT on April 29, 1996. Subsidiary information about global conditions are provided from simultaneous, high-latitude passes of two DMSP satellites. In both cases the solar wind speed, measured by the Wind satellite, $X_{GSE} \approx 85$ R_E, was constant at ∼350 km/s. Thus, signal propagation times to the magnetopause are ∼25 minutes. Average values of IMF B_Z were ∼3.5 nT in both cases. On April 27, IMF B_Y was ∼0 until 1527 UT, then oscillated between +1 and -1 nT with a period of ∼38 minutes. On April 29, IMF B_Y was varied between +1 and 3.5 nT with a period of ∼50 minutes. We regard the magnetosphere as responding to nearly constant northward IMF conditions in which exact propagation times from Wind to the magnetosphere are not critical.

3.1. APRIL 27, 1996

Figure 2a contains electron (top) and ion (bottom) spectra acquired by Hydra between 1500 and 1800 UT on April 27, 1996, in standard energy-versus-time spectrogram formats. The color bars to the right of the spectrogram provide and the count-rate magnitudes. The DDEIS spectra give averages of 72 measurements at the given energy steps, accumulated over

13.8 s intervals. Data are presented as functions of universal time (UT), invariant latitude (ILT), magnetic local time (MLT) and geocentric distance in R_E. Values of the ILT and MLT were determined the using IGRF 95 model. TIMAS measurements are not available for this orbit.

On a purely empirical basis, we divide the particle measurements from the April 27 pass into three time intervals. (1) Prior to 1525 UT the highest ion count rates were at energies in the 8 to 10 keV/q range. The energy of peak counts was nearly constant with increasing ILT. Electron counts were ~100 at energies between 1 and 10 keV. After 1510 UT average electron energies decreased with increasing ILT. (2) The period between 1525 and 1550 UT is marked by an intense, but fairly constant flux of electrons with average energies near 70 eV. The flux of ions reaching HYDRA also increased during this interval. The average energies of ions increased with increasing ILT, characteristic of an inverse energy dispersion structure. Smaller scale structures also appear during this interval in the ion spectrogram, with peak counts at 1532, 1540 and 1550 UT. They show both standard and inverse ion dispersion characteristics. (3) After 1600 UT, DDEIS detected electron fluxes whose intensities and energies were significantly reduced, and ion count rates near background levels. Based on experience with the spectral characteristics of ion and electron fluxes observed at low altitudes [13], we identify the source regions encountered by Polar during the three intervals as (1) the CPS, (2) the LLBL (1525 – 1531 UT) and cusp (1531 – 1550 UT), and (3) the polar cap.

From top to bottom, Figure 3a shows the spin-plane components of the electric field, the electric potential distribution along the trajectory, two components of the plasma drift V_{X-Y} and V_{56} and magnetic field B_{56}. The plasma velocity components are roughly in the sun–Earth and dawn–dusk directions. The B_{56} trace has the T96 model field subtracted. The reduced-accuracy V_{X-Y} should only be used as a flow direction indicator. Potential values are given for the corotating (dashed line) and inertial (solid line) frames of reference for easy comparison with ionospheric patterns. In the presentation below we describe convection patterns in the inertial frame of reference.

Attention is directed to the following seven points. (1) Prior to 1522 UT, the absolute values and variability of electric field was <1 mV/m. After this the amplitudes of variations grew. (2) From 1520 to 1555 UT the amplitude of electric field variations assumed values <3 mV/m. The periods of the variations ranged from a few tens of seconds to a few minutes (Pc 1 to Pc 4). (3) Average (quasi dc) values of the electric field along the spacecraft velocity vector reversed polarity near 1547 UT (see E_Z and Φ traces). (4) The third panel shows that Polar crossed regions of positive (1525 – 1650 UT) and negative (after 1700 UT) potential. Viewed from above the north pole,

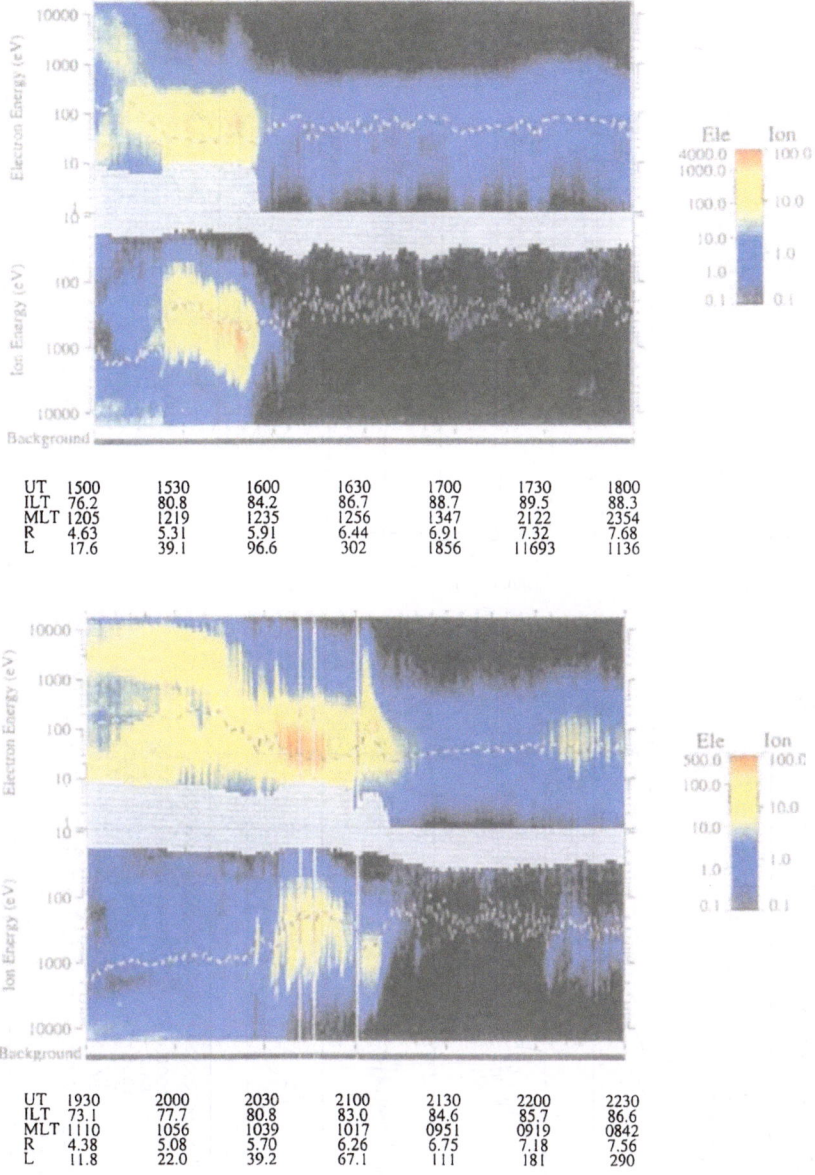

UT	1500	1530	1600	1630	1700	1730	1800
ILT	76.2	80.8	84.2	86.7	88.7	89.5	88.3
MLT	1205	1219	1235	1256	1347	2122	2354
R	4.63	5.31	5.91	6.44	6.91	7.32	7.68
L	17.6	39.1	96.6	302	1856	11693	1136

UT	1930	2000	2030	2100	2130	2200	2230
ILT	73.1	77.7	80.8	83.0	84.6	85.7	86.6
MLT	1110	1056	1039	1017	0951	0919	0842
R	4.38	5.08	5.70	6.26	6.75	7.18	7.56
L	11.8	22.0	39.2	67.1	111	181	290

Figure 2. HYDRA measurements from (A) 1500 to 1800 UT on April 27, 1996, and (B) 1930 to 2230 UT on April 29, 1996, in energy-versus time spectrograms. The top panels show omnidirectional averaged counts from 12 detectors that is proportional to differential energy flux for electrons with energies between 1 eV and 20 keV The bottom panels give count rates for ions with energies per charge between 10 eV/q and 20 keV/q. The dynamic ranges represented by color-bar scales are not the same for the two dates. The dashed lines on HYDRA spectrograms give mean energies derived from distribution functions associated with observed count rates, assuming that the positive ions are H^+.

164

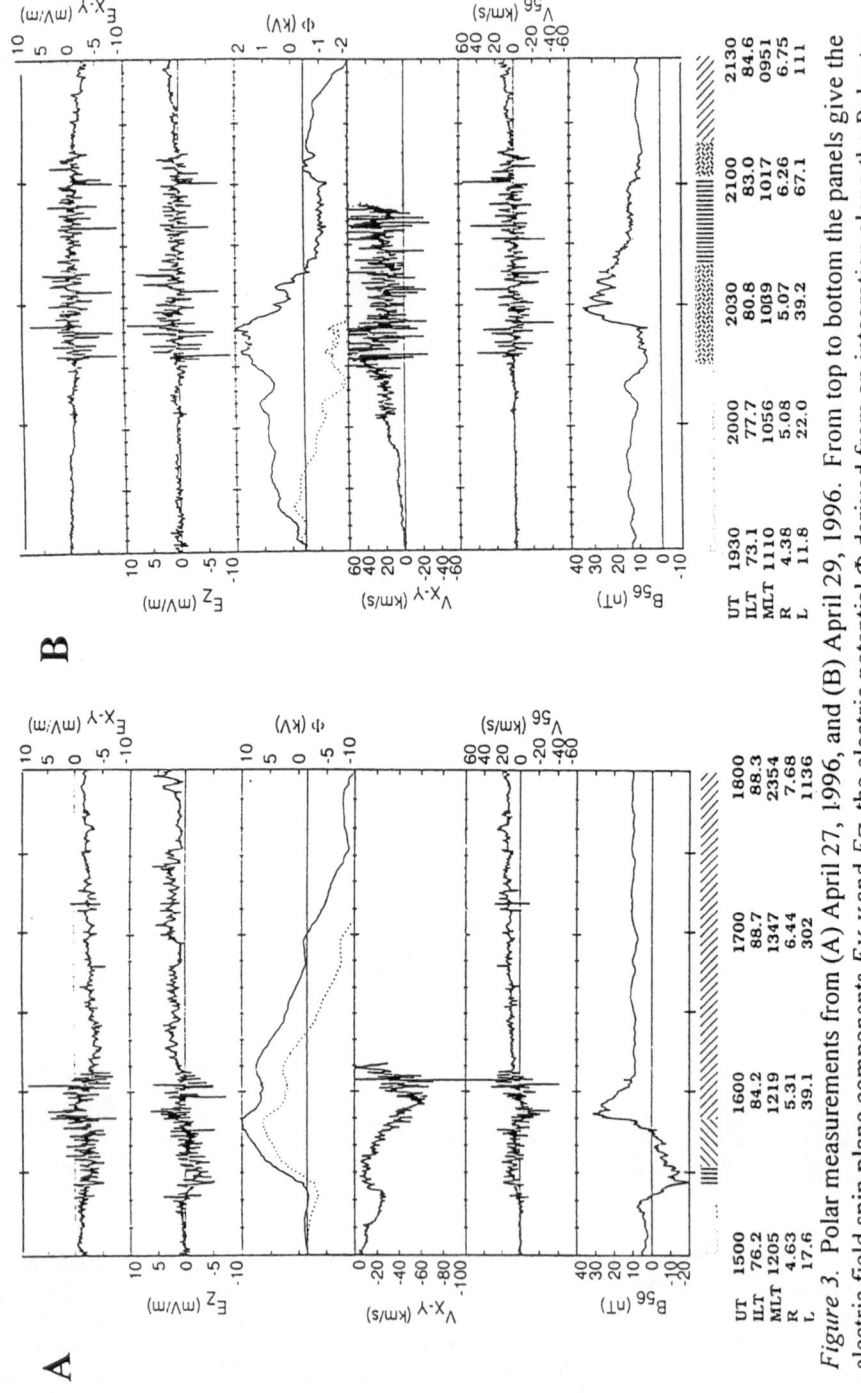

Figure 3. Polar measurements from (A) April 27, 1996, and (B) April 29, 1996. From top to bottom the panels give the electric field spin plane components E_{X-Y} and E_Z, the electric potential Φ derived from an integration along the Polar trajectory, the plasma drift components V_{56} and V_{X-Y}, and the magnetic field component B_{56} transverse to the orbital plane. the particle regions described in the text are marked below the figures.

the sense of rotation for plasma convection is clockwise/counterclockwise in regions of negative/positive potential. In this case, the potential of the counterclockwise rotating cell is ~10 kV. (5) Negative values of V_{X-Y} indicate that plasma convection had a sunward component in the regions of cusp and polar rain precipitation. (6) The dawn–dusk component V_{56} had low values prior to 1520 UT. From 1520 to 1550 UT Polar detected eastward flow (positive V_{56}), with an average value of ~15 km/s, and westward from 1550 to 1600 UT. (7) Variations in the trace of B_{56} mimic those of quasi-dc values of E_{X-Y} and E_Z. This indicates that Polar crossed several large-scale FAC sheets which close via Pedersen currents in the ionosphere. (8) Positive (negative) slopes in the B_{56} trace indicate FACs into (out of) the ionosphere. Consistent with a postnoon MLT trajectory, the negative (1505 – 1512 UT) slope in B_{56} appears to result from Polar crossing the dayside Region 1 system [15]. It corresponds in time to Polar detecting LLB fluxes. The afternoon Region 2 current system is either absent or very weak. The remaining FACs belong to the Region 0 system, associated with the cusp precipitation. The polarity of the main Region 0 current is opposite to that of the adjacent Region 1. A small region of upward current is located poleward of the main Region 0 current.

At middle altitudes the Polar satellite moves rather slowly across the dayside high-latitude region while measuring local particles and fields. To understand their significance, it is useful to place these measurements in a wider context whenever DMSP satellites cross the auroral/polar latitudes at nearly the same time. On April 27, 1996, the DMSP F13 and F12 satellites crossed the region between 70° invariant latitude in the dusk/evening sectors to 70° in the dawn/day sectors from 1605 to 1615 UT and 1602 to 1614 UT, respectively. Figure 4a shows the trajectories of the three satellites and the particle precipitation regions [13] that they encountered. The directions of plasma flow detected by DMSP F13 and convection cells observed by Polar are schematically indicated. Drift meter measurements from DMSP F12 were contaminated by sunlight and are not shown. A four-cell convection pattern is seen along the F13 trajectory. Flow is sunward in the auroral oval and in the central polar cap, and antisunward along its flanks. This global flow pattern is consistent with Polar's detection of a positive-potential lobe cell in the postnoon MLT sector. All three satellites detected CPS precipitation whose poleward boundary moved to higher geomagnetic latitudes on the day side. The latitudes for detecting cusp precipitation at the locations of Polar and F12 appear to be consistent. At this time BPS fluxes were detected near the dusk, evening and dawn MLT sectors, but not on the dayside. In their region of overlap, all of three satellites detected polar rain fluxes. The figure also indicates the polarities of FACs detected by Polar. The upward Region 1 current spans the latitudinally narrow strip in which HYDRA detected LLBL fluxes [24].

3.2. APRIL 29, 1996

Figure 2b contains electron (top) and ion (bottom) spectra acquired by HYDRA between 1930 and 2230 UT on April 29, 1996. Note that the dynamic ranges expressed by the color bars differ from Figure 2a. Again we divide the particle measurements empirically , this time into five intervals. (1) Prior to 2015 UT the highest ion count rates were again at energies in the 8 to 10 keV/q range and nearly constant with increasing ILT. Electron counts were ~100 at energies near 5 keV. Their average energies decreased slightly with increasing ILT. (2) From 2015 to 2040 UT electron fluxes became more irregular and their average energies decreased from 200 to 70 eV. The average energies of ions reaching Hydra also decreased with latitude. (3) Between 2040 and 2102 UT electron fluxes were most intense and varied smoothly. Average electron energies were ~70 eV. The spectrogram shows three ion injection structures, all with standard dispersion characteristics. Significant fluxes of He^{++} ions were detected by TIMAS only during this period. They were marked by three intensifications with standard energy-versus-latitude characteristics (4) During the interval 2102 to 2110 UT the flux of electrons and ions intensified and their average energies increased to values similar to those found in the second interval. (5) After 2110 UT, HYDRA detected electron fluxes whose intensities and energies were significantly reduced, and ion count rates near background levels. Applying the criteria of Newell *et al.* [13], we identify the source regions encountered by Polar during the intervals (1) and (5) as the central plasma sheet and the polar cap. We tentatively suggest that during intervals (2), (3) and (4) Polar cross field lines connected to the BPS, the LLBL, then returned to the BPS before entering the polar cap. We defer justification for these assertions to the discussion section.

Plasma and field measurements from the interval 1930 to 2130 UT are presented in Figure 3b. Attention is directed to four aspects of these data: (1) Prior to 2015 UT, the region of CPS precipitation, and after 2110 UT, the region of polar rain, measured electric fields were a few mV/m in magnitude and varied smoothly. Between these two intervals the magnitudes of electric field fluctuations more than doubled. (2) Prior to 2040 UT Polar crossed through ~2 kV of a positive potential cell. The potential turned negative (-1 kV), returned to ~0 kV near 2105 UT, then became increasingly negative. For later reference we note that the first and second negative potential excursions correspond to the second and fifth particle precipitation regions. The region in which the potential returned nearly to zero matches the particle flux intensification centered near 2004 UT. (3) Positive values

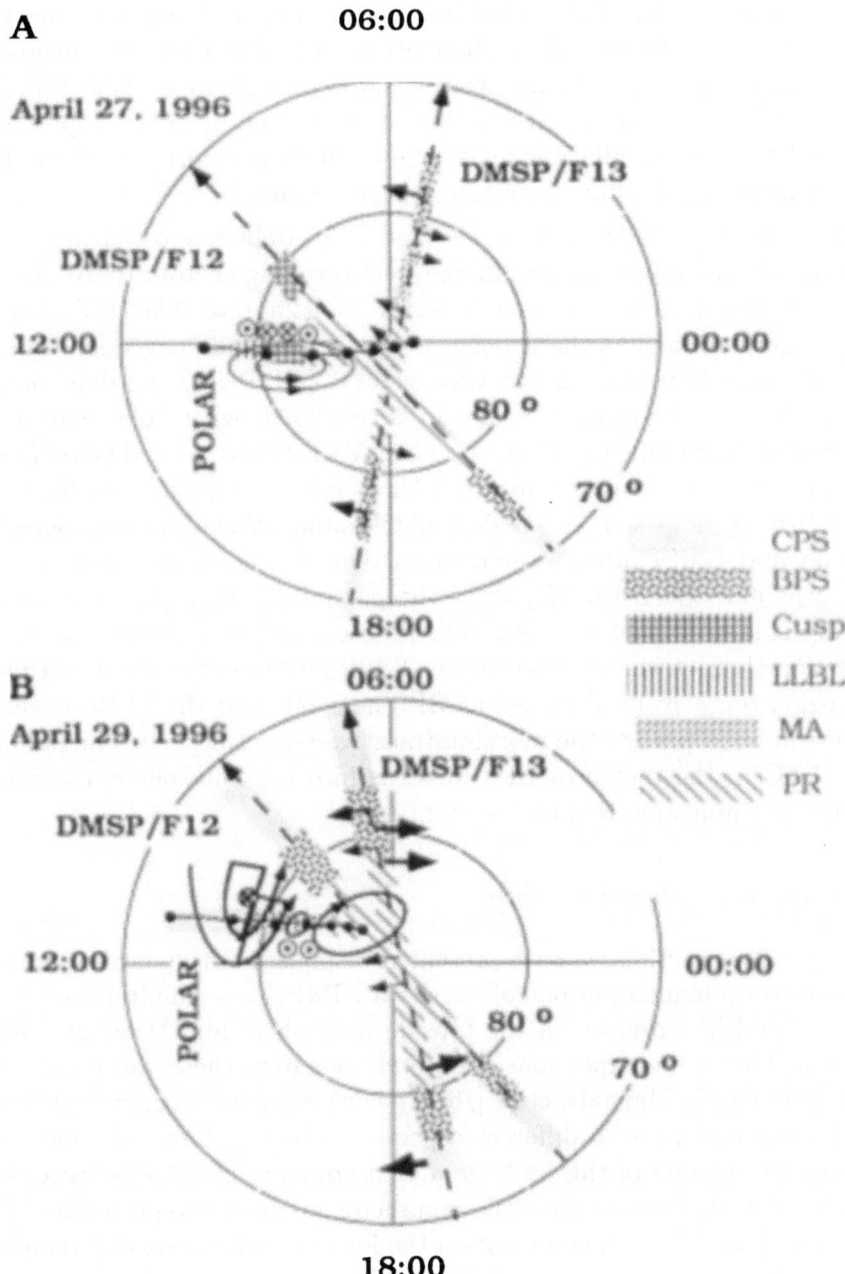

Figure 4. Schematic representation of high-latitude convection patterns and particle populations encountered by Polar on (A) April 27, 1996, and (B) April 29, 1996. Orbital traces for DMSP F12 and F13 along with particle populations are indicated. Arrows along the F13 trajectories show directions of plasma drift.

of V_{X-Y} indicate that the plasma flow had an antisunward component of drift throughout the period of observations. (4) The east-west magnetic perturbation B_{56} has a strong positive slope from 2025 to 2030 UT, followed by three oscillations in a region of large-scale negative slope (2030 to 2102 UT). The oscillations in B_{56} after 2030 UT correspond to clear intensifications in the flux of precipitating electrons.

On April 29, 1996 the DMSP F13 and F12 satellites crossed the region between 70° invariant latitude in the dusk/evening sectors to 70° in the dawn/day sectors from 2043 to 2054 UT and 2041 to 2052 UT, respectively. Figure 4b depicts the trajectories, velocities and convection cells. A four-cell convection pattern was traversed by DMSP F13. Its detection of sunward flow in the central polar cap is consistent with Polar entering a negative potential lobe cell in the prenoon MLT sector. As indicated in the figure, the DMSP satellites observed CPS, BPS and polar rain fluxes as their latitudes increased. The regions of CPS and polar rain fluxes detected by Polar and DMSP are mutually consistent. A significant difference between this and the April 27 pass is the partial detection by Polar of the morning side auroral convection cell. Its structure is consistent with the convection characteristics detected by DMSP F13 near the dawn meridian. The region tentatively identified as BPS near 80° and the LLBL occur in the antisunward part of the morning (positive potential) cell. The polarity of the FAC in the region of BPS fluxes is into the ionosphere, consistent with the morning side Region 1 system.

4. Summary and Discussion

In the previous section we have presented responses of the dayside magnetosphere to prolonged periods of northward IMF from simultaneous viewpoints of DMSP satellites in the topside ionosphere and Polar at middle altitudes. Through comparisons of observations from these viewpoints presented here and in Maynard et al. [16], we have come to recognize signatures of Polar encounters with different aspects of evolving four-cell convection patterns. In the rest of this section we first compare Polar observations on April 27 with those from the afternoon sector previously reported by Maynard et al. [16]. We then comment on the ion source location and temporal behavior. Lastly, we present a justification for our interpretation of measurements acquired during intervals (2), (3) and (4) on April 29 as encounters with BPS, LLBL and BPS particles and comment on its significance for understanding the structure of the magnetotail at this time.

Particle precipitation and convection patterns detected during the Polar orbits of April 8 and April 27 show repeatable elements. The universal times spanned by these two orbits are almost the same. Figures 1 and 4a show that since the Polar orbit precessed slowly, the MLT–versus–invariant latitude trajectories are similar. In both instances Polar crossed a positive potential lobe cell in the early afternoon sector. Within these cells HYDRA detected cusp electron and ion spectra. The ions had inverse dispersion characteristics in which higher energies were detected at higher latitudes. There are two striking differences. First, on April 8, but not on April 27, Polar detected an auroral zone negative-potential convection cell extending into the noon region, presumably driven by the LLBL. Secondly, the highest voltages sampled within the lobe cells were ~3 kV on April 8, and ~10 kV on April 27. In both cases the solar wind speed was moderate, in the 300 – 350 km/s range. However, the relative strengths of IMF B_Y to IMF B_Z were quite different. On April 8 the IMF clock angle was near 40°. On April 27, the clock angle was nearer 20°. We know that on the May 11 event, which had a similar clock angle to April 8, a 4-cell pattern was observed [16]. The 4-cell nature of the pattern on April 27 was confirmed by DMSP.

Using these results and those of Burke *et al.* [10] we can piece together a scenario for positive B_Y to evolve distorted 2–cell patterns into 4–cell patterns with decreasing clock angle. For IMF clock angles greater than 45° distorted 2-cell patterns dominate [9][10][12]with the large negative cell dominating into the noon sector. A negative lobe cell distorts the afternoon negative convection cell. As B_Y weakens the negative lobe cell moves prenoon and negative potential boundary-layer-driven effects no longer extend into the prenoon region. A positive lobe cell develops postnoon (April 8) for clock angles between 30 and 40°. For smaller clock angles that positive cell becomes stronger (April 27) leading toward a more symmetric 4–cell pattern for pure B_Z north. The afternoon negative cell no longer extends into the noon region as the positive cell strengthens.

The relation of particle signatures to convective flows combined with the location within the global context provided by the mapping of magnetic field lines passing through the Polar trajectory provide clues relative to the characteristics of the particle sources. In measurements from the slow moving Polar satellite, we consistently found that both large and small scale ion dispersions have standard and reverse structures in the antisunward flowing boundary layers and sunward moving cusp plasmas, respectively. We have also consistently detected fluxes of He^{++} ions only in regions where our analysis of HYDRA electron and ion data indicate that Polar was crossing the cusp or a boundary layer. It is through these two boundary regions where ions of solar wind origin have relatively direct access to middle altitudes and the ionosphere. The reverse dispersion is a signature of merging on the poleward edge of the cusp. The structure observed as the ion energy decreased in Figure 2a as well as the multiple injections observed in the May 11 event [16] indicate that the lobe merging is both temporally and spatially variable. Forward dispersion signatures with He^{++} ions on April 3, April 29, and May 11 which map to the boundary layers on the flanks indicate entry of magnetospheath fluxes on the flanks. The exact nature of the process which creates the observed forward dispersion signature remains an open question.

In our presentation of Polar measurements from the April 29 orbit, we tentatively identified the particle fluxes detected in regions (2), (3) and (4) as originating in the BPS (2012 – 2040 UT), the LLBL (2040 – 2102 UT), and the BPS (2102 – 2110 UT), respectively. Our interpretation of the HYDRA measurements for regions (2) and (3) appears to be a simple application of criteria established by Newell et al. [13]. Region (2) appears just poleward of the CPS precipitation, is characterized by structured fluxes of electrons whose average energies exceed 100 eV, and straddles the convection reversal of the morning side convection cell. On the other hand, electron fluxes detected in Region (3) are less structured and have average energies below 100 eV. Ion fluxes have standard dispersion characteristics, contain a significant He^{++} component, and appear amid antisunward convecting plasma. The persistent presence of a high energy electron spectral component (Figure 2b) suggests Polar was crossing the middle altitude projection of a boundary layer to which plasma sheet electrons had access, the LLBL, rather than the cusp. The electrons with energies >100 eV and ions near 1 keV measured between 2102 and 2110 UT are spectrally similar to those observed at invariant latitudes equatorward of the LLBL. Data contained in the third panel of Figure 3b shows that Polar detected these enhanced fluxes at locations where the electric potential was similar to that measured prior to entering the LLBL. These particle and field measurements force us to conclude that after leaving the LLBL and before entering the polar cap, Polar again sampled field lines threading the BPS. Figure 4b shows that DMSP F12 and 13 detected BPS ion and electron fluxes up to magnetic latitudes of ~83° near the dawn meridian but only to ~80° near dusk. This asymmetric intrusion of BPS fluxes to high latitudes on the dawn side is consistent with our interpretation of Polar data.

To help understand our Polar measurements and the source regions they imply, we have mapped its locations between 2010 and 2130 UT to the magnetosphere using the T96 model. This mapping suggests that the field lines crossed by Polar mapped close to the dawnside magnetospheric boundary. After 2030 UT the mapping was to the night side of the dawn terminator. Figure 5 represents a cross section of the magnetotail with the source regions of BPS, the LLBL and polar rain particle fluxes indicated. A schematic magnetic mapping of the Polar trajectory (dashed line) onto this plane suggests that an intrusion of the BPS to high latitudes on the morning side could indeed provide the sequence of electron and ion fluxes observed

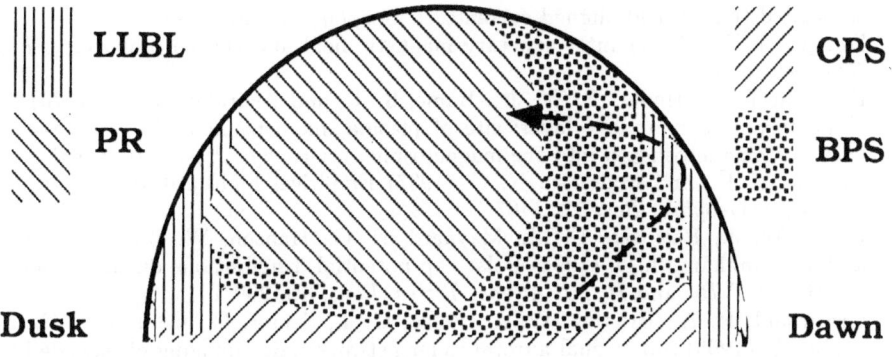

Figure 5. Schematic representation of a cross section for northern half of the magnetotail indicating sources of particles measured by HYDRA after 1530 UT on April 29, 1996. A qualitative magnetic mapping of the Polar trajectory onto this plane is indicated by the dashed line.

by HYDRA after 1530 UT on April 29, 1996. The IMF B_Y in this case was varying. Chang *et al.* [25] recently developed a model for the generation of theta aurora arcs due to shifts in IMF B_Y, while B_Z remains northward. The Polar and DMSP results are suggestive of a polar cap arc or theta bar starting to break away from the dawn side of the oval as envisioned in the Chang *et al.* model.

References

1. Sandholt, P. E. (1991) Auroral electrodynamics at the cusp/cleft poleward boundary durin northward interplanetary magnetic field, *Geophys. Res. Lett.* **18**, 805.
2. Clauer, C. R., and Friis-Christensen, E. (1986) High-latitude electric fields and currents during strongly northward magnetic field: observations and model simulations, *J. Geophys. Res.* **91**, 6959.
3. Maezawa, K. (1976) Magnetospheric convection induced by positive and negative Z components of the interplanetary magnetic field: Quantitative analysis using polar cap magnetic field records, *J. Geophys. Res.* **81**, 2289.
4. Burke, W. J., Kelley, M. C., Sagalyn, R. C., Smiddy, M., and Lai, S. T. (1979) Polar cap electric field structures with northward interplanetary magnetic field, *Geophys. Res. Lett.* **6**, 21.
5. Reiff, P. H., and Heelis, R. A. (1994) Four cells or two? Are four cells really necessary, *J. Geophys. Res.* **99**, 3955.
6. Iijima, T., and Shibaji, T. (1984) Global characteristics of northward IMF-

associated (NBZ) field-aligned currents, *J. Geophys. Res.* **89**, 2408.

7. Dungey, J. W. (1961) Interplanetary magnetic field and the auroral zones, *Phys. Rev. Lett.* **6**, 47.
8. Eastman T. E., Hones, E. W. Jr., Bame, S. J., and Asbridge, J. R. (1976) The magnetospheric boundary layer: site of plasma, momentum and energy transfer from the magnetosheath into the magnetosphere, *Geophys. Res. Lett.* **3**, 685.
9. Heppner J. P., and Maynard, N. C. (1987) Empirical high-latitude electric field models, *J. Geophys. Res.* **92**, 4467.
10. Burke, W. J., Basinska, E. M., Maynard, N. C., Hanson, W. B., Slavin, J. P., and Winningham, J. D. (1994) Polar cap potential distributions during periods of positive IMF B_Y and B_Z, *J. Atmos. Terres. Phys.* **56**, 209.
11. Greenwald, R. A., Bristow, W. A., Sofko, G. J., Senior, C., Cerisier, J.-C., and Szabo, A. (1995) Super dual auroral radar network radar imaging of dayside high-latitude convection under northward magnetic field: Toward resolving the distorted two-cell versus multicell controversy, *J. Geophys. Res.* **100**, 19,661.
12. Weimer, D. R. (1996) A flexible, IMF dependent model of high latitude electric potentials having "space weather" applications, *Geophys. Res. Lett.* **23**, 2549.
13. Newell, P. T., Burke, W. J., Sanchez, E. R., Meng, C.-I., Greenspan, M. E., and Clauer, C. R. (1991) The low latitude boundary layer and the boundary plasma sheet at low altitude: prenoon precipitation regions and convection reversal boundaries, *J. Geophys. Res.* **96**, 35.
14. Reiff, P. H., Burch, J. L., and Spiro, R. W. (1980) Cusp proton signatures and the interplanetary magnetic field, *J. Geophys. Res.* **85**, 5997.
15. Iijima, T., and Potemra, T. A. (1976) Field-aligned currents in the dayside cusp observed by TRIAD, *J. Geophys. Res.* **81**, 5971.
16. Maynard, N. C., Burke, W. J., Weimer, D. R., Mozer, F. S., Scudder, J. D., Russell, C. T., Peterson, W. K., and Lepping, R. P. (1997) Polar observations of convection with northward IMF at dayside high latitudes, *J. Geophys. Res.*, in press.
17. Tsyganenko, N. A. (1996) Effects of the solar wind conditions on the global magnetospheric configuration as deduced from data based field models, in *Third International Conference on Substorms (ICS-3)*, ESA SP-389, ESA Pub. Div., Noordwijk, The Netherlands, pp. 181–185.
18. Harvey, P., *et al.* (1995) The electric field instrument on the Polar satellite, in C. T. Russell (ed.), *The Global Geospace Mission*, Kluwer Academic Publishers, Dordrecht, pp. 583–596.
19. Russell, C. T., *et al.* (1995) The GGS/Polar magnetic fields investigation, in C. T. Russell (ed.), *The Global Geospace Mission*, Kluwer Academic Publishers, Dordrecht, pp. 563–582.
20. Scudder, J., *et al.* (1995) Hydra - A 3-dimensional electron and ion hot plasma instrument for the Polar spacecraft of the GGS mission, in C. T. Russell (ed.), *The Global Geospace Mission*, Kluwer Academic Publishers, Dordrecht, pp. 459–495.
21. Shelley, E. G., *et al.* (1995) The toroidal imaging mass-angle spectrometer (TIMAS) for the Polar mission, in C. T. Russell (ed.), *The Global Geospace Mission*, Kluwer Academic Publishers, Dordrecht, pp. 497–530.
22. Hardy, D. A., Gussenhoven, M. S., and Brautigam, D. (1989) A statistical model of auroral ion precipitation, *J. Geophys. Res.* **94**, 370.
23. Greenspan, M. E., Anderson, P. B., Pelagatti, J. M. (1986) Characteristics of the thermal plasma monitor for the Defense meteorological Satellite Program (DMSP) spacecraft S8-S10, *AFGL-TR-86-0227*, AFGL, Hanscom AFB, MA.
24. Siscoe, G. L., Lotko, W., Reiff, P. H., and Sonnerup, B. U. Ö. (1991) A high-latitude, low-latitude boundary layer model of the convection current system, *J. Geophys. Res.* **96**, 3487.
25. Chang, S.-W., *et al.*. (1997) A comparison of a model for the theta aurora with observations from Polar, Wind, and SuperDARN, *J. Geophys. Res.*, in press.

RELATIONSHIP BETWEEN LARGE-, MESO-, AND SMALL-SCALE FIELD-ALIGNED CURRENTS AND THEIR CURRENT CARRIERS

M. YAMAUCHI, R. LUNDIN, and L. ELIASSON
IRF-Kiruna
Box 812, S-98128 Kiruna, Sweden.

S. OHTANI
JHU/APL
Johns Hopkins Road, Laurel, MD 20723-6099, U.S.A.

J. H. CLEMMONS
NASA/GSFC
Greenbelt, MD 20771, U.S.A.

ABSTRACT. Carriers of the dayside large-scale field-aligned currents (FACs) are discussed. Since the gyro-radius of the current carriers are smaller than the size of small-scale FACs (a pair of upward and downward FACs associated with inverted-V potential structure as shown in Figure 1), the current carriers of large-scale FAC could be controlled by small-scale (and hence meso-scale) FACs. We restrict the discussion to only a few regions.
(1) Although the current carries are electrons in most cases, the framework of the large-scale FAC system is sometimes determined by positive ions, especially in the cusp.
(2) There is a dawn-dusk asymmetry in the relationship between the large-scale FACs and the current carriers. A substantial fraction of the Region-1 FAC is probably composed of many small-scale paired FACs which are associated with the inverted-V structure (see Figure 1), whereas the Region-2 FAC is carried by CPS electrons in the morning sector and by thermal electrons in the afternoon sector.
(3) Meso-scale FAC is formed by individual ion injections, but it has no relation to the large-scale FACs even inside the cusp region. Thus the large-scale cusp FACs (and the cusp itself) are formed by a steady mechanism but not by the meso-scale injections such is FTEs.. This rules out the FTE cusp model.

1. Introduction

The purpose of this study is to show low- and mid-altitude satellite observations of the field-aligned currents (FACs). Since Iijima and Potemra [1, 2] derived Region-1, Region-2, and cusp region (Region-0) large-scale FAC system from the satellite (TRIAD) magnetometer data, many observational efforts have been made on the subjects listed below. Some of them have been reviewed by Potemra [3].

1. Alternative means to derive FACs such as from ground geomagnetic disturbances. optical images, ionospheric convection patterns, and the electric fields [4-13; and references therein].
2. Distinguishing the wave and the real FACs [14-21; and references therein].
3. Seasonal variations [22-24; and references therein].
4. The current closure in the ionosphere [5, 8; and references therein].
5. Subdivision of the current system in terms of the source and behaviors (see review by Potemra [3] and references therein).

J. Moen et al. (eds.), Polar Cap Boundary Phenomena, 173–188.

6. The distribution (shape, alignment, size) of each FAC regions [7, 23, 25-33; and references therein].
7. Macroscopic current carriers, i.e., relation to BPS, CPS, PSBL, cusp, cleft, and mantle. [16, 27, 31, 34-39; and references therein].
8. Microscopic current carriers, i.e., potential-accelerated aurora particles, black aurora, bursts, low-energy particles, and thermal particles [31, 40-44; and references therein].
9. The relation to small-scale and meso-scale FAC systems [13-16, 19, 45-47; and references therein].
10. The short-term development and long-term behavior in response to the change of external conditions [13, 23, 48-51; and references therein].
11. The current closure (source) in the magnetosphere.

All questions are of course further examined for various solar wind conditions and geomagnetic activities. Here, the term "meso-scale" simply means anything smaller than "regions" and larger than the inverted-V potential structures. For example, individual electron/ion energy-time dispersion and multiple injections are all meso-scale phenomena [46, 51, 52]. Compared to the first six questions (*1-6*), the next four questions (7-10) are rather poorly investigated and the available results are controversial and puzzling. The last question (*11*) has a large weight on the theoretical aspect, and lies outside the scope of this paper.

Thus, identifying the current carriers and revealing the cross-scale relations are the most needed topics. Here, we should note that Alfvén waves are related to both the conduction current (real) and the induction current (wave), and it is often impossible to distinguish wave (small and meso scales) or conduction current (meso and large scales) for the stationary Alfvén structure [53, 54]. The same problem occurs for the electrostatic shocks and weak double layers which are associated with cavitons [31, 55]. Obvious questions related to the current carriers are: (a) carrier species (ion or electron); (b) energy of the carrier; and (c) their energization mechanisms, e.g., the parallel electric potential, waves (burst), thermalization, Fermi-acceleration, and J x B force. The answers to these questions depend on the location (MLT/latitude), altitude, and the scale-size of the FAC system. For example, the current carrier can be different between the morning side Region-1 FAC, near-noon downward FAC, afternoon Region-2 FAC, and substorm-related downward FAC, although it has been generally assumed that upward FACs are carried by auroral electrons whereas downward FACs are carried by ionospheric thermal electrons. It is impossible to cover all the aspects of current-carrier problem, and we show only a few observations of the dayside FAC system.

2. Relation Between Large-scale FACs and Small-scale FACs

The large-scale FACs often contain many upward and downward pairs of small-scale FACs [14, 15, 19]. It is a question whether these small-scale FACs are simply added to the background large-scale FACs, or the difference of the upward and downward small-scale FACs makes up the large-scale FACs. In the former case, the large-scale FACs are basically carried by the thermal electrons and are independent of the small-scale FACs of the Alfvén waves. In the latter case a substantial portion of the small-scale FACs must be a stationary structure (such as the electrostatic shocks and weak double layers [19]) and the current carriers are not necessarily the thermal electrons. It has been generally believed that small-scale FACs are induction currents of Alfvén waves and therefore they cannot form the large-scale FACs.

However, FREJA [31] and FAST [44] satellites revealed that many FAC spikes in the night-side sector are associated with electron bursts and field-aligned potential drops

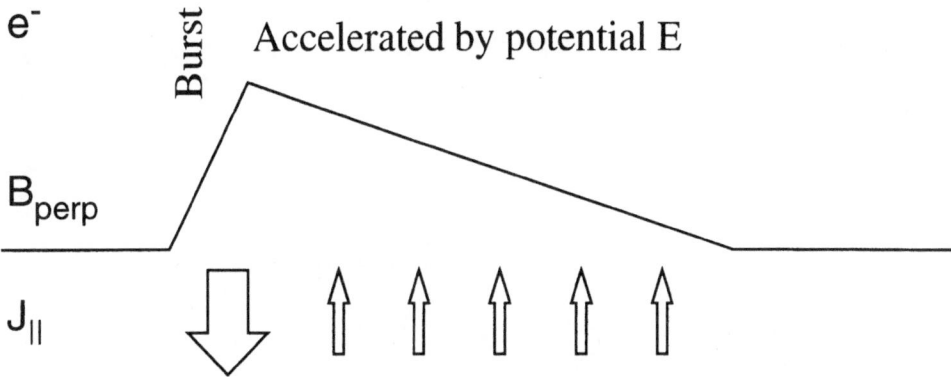

Figure 1. Relative strength and carriers of small-scale FACs in the night-side sector.

as shown in Figure 1. They are of the small-scale, but yet stationary structure associated with the ionospheric plasma cavities [31, 55]. This structure is often denoted the electrostatic shocks. The observations are made at a few thousand km altitude, and the picture is altitude-dependent [R. Ergun, private communication, 1997]. Burch *et al.* [42] using DE-1 data at mid-altitude concluded that 20-200 eV upward electrons are the major carriers of the downward FAC near the cusp. Although they concluded that electric potential causes an energization of upward electrons (carriers of downward FAC), the observation itself generally agrees with the above view, i.e., the energetic electrons are not necessarily accelerated by a potential electric field but by the waves. Therefore, we have to consider this type of FAC carriers (Figure 1) as well as the traditional thermal electrons and precipitating auroral electrons [40, 41].

Now we have the following questions: How should we understand these small-scale stationary FACs in the framework of large-scale FACs? What portion of the FAC is carried by energized electrons and what portion by thermal electrons? This is not a simple question because the same types of plasma population (CPS, BPS, LLBL, cusp, cleft, etc. [56-59; and references therein]) must explain the opposite sense of FACs between dawn and dusk. For example, we have to associate the CPS plasma population to both upward (dawn) and downward (dusk) Region-2 FACs, whereas the BPS or LLBL plasma population must explain downward (dawn) and upward (dusk) Region-1 FACs.

2.1. MORNING SECTOR

We start with the dawnside FACs. The small-scale FAC of Figure 1 has some similarity to Region-1 and Region-2 FACs: the downward FAC is associated with a bursty electron region (this is the characteristics of BPS and LLBL), whereas the upward FAC is associated with a weak potential drop (this is the characteristic of CPS because of the mirror force [60, 61]). Figure 2 shows one such example [43]. Nearly 40% of the Region-1 FAC is concentrated within a tiny region of a spike-like FAC at around 21:35 UT, which is accompanied by the upward electron burst of BPS.

On the other hand, 60% of the downward FAC are not associated with energized electrons. This need further investigation because VIKING could have missed the small-scale electron bursts if the scale size of the electron burst is smaller than the time resolution of VIKING which detects upward electrons only once every 20 seconds (spin

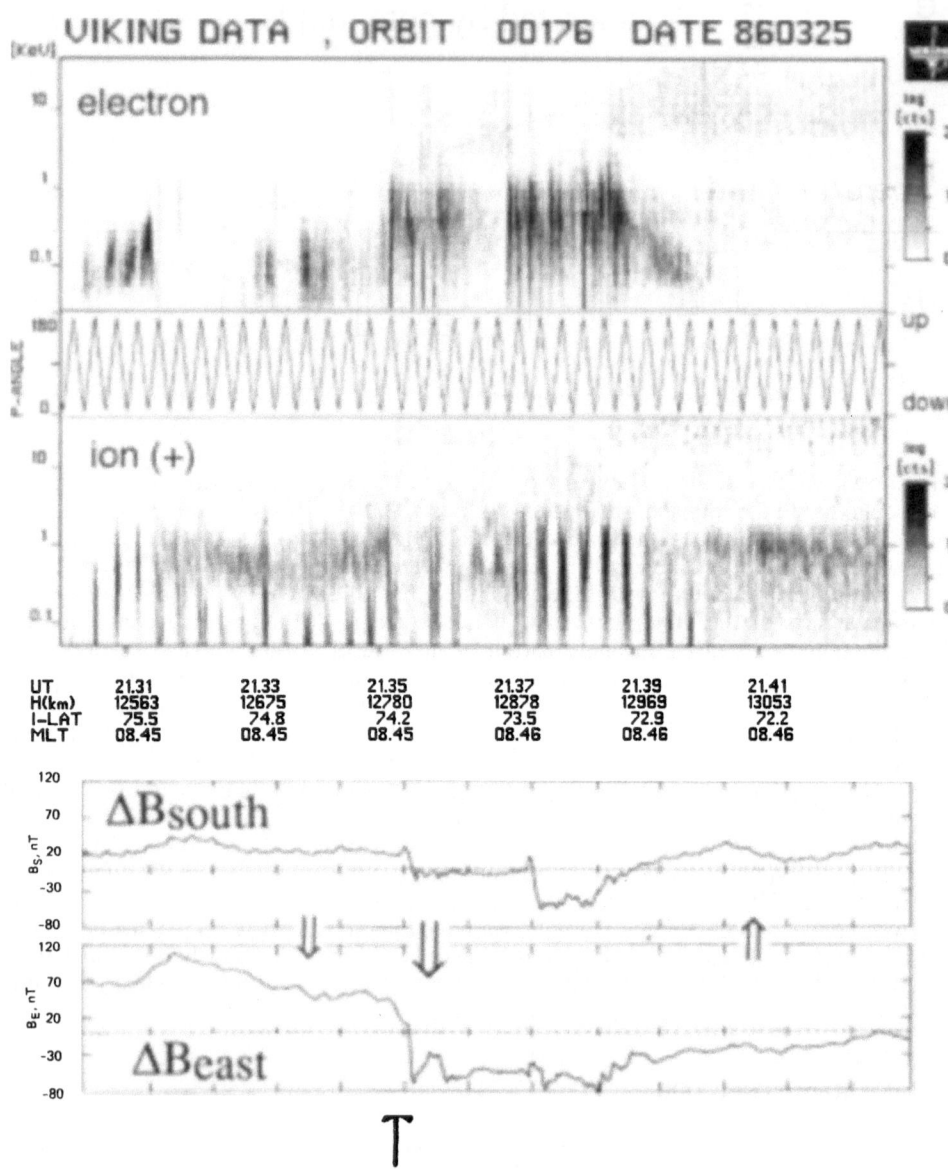

Figure 2. Particle and magnetic field data from Viking orbit 176 (25 March, 1986). Positive (negative) slopes of B_E mean upward (downward) FACs. Region-1 FAC is found at around 74°-75° Inv., and Region-2 FAC is found at around 72°-73° Inv. (after [43]).

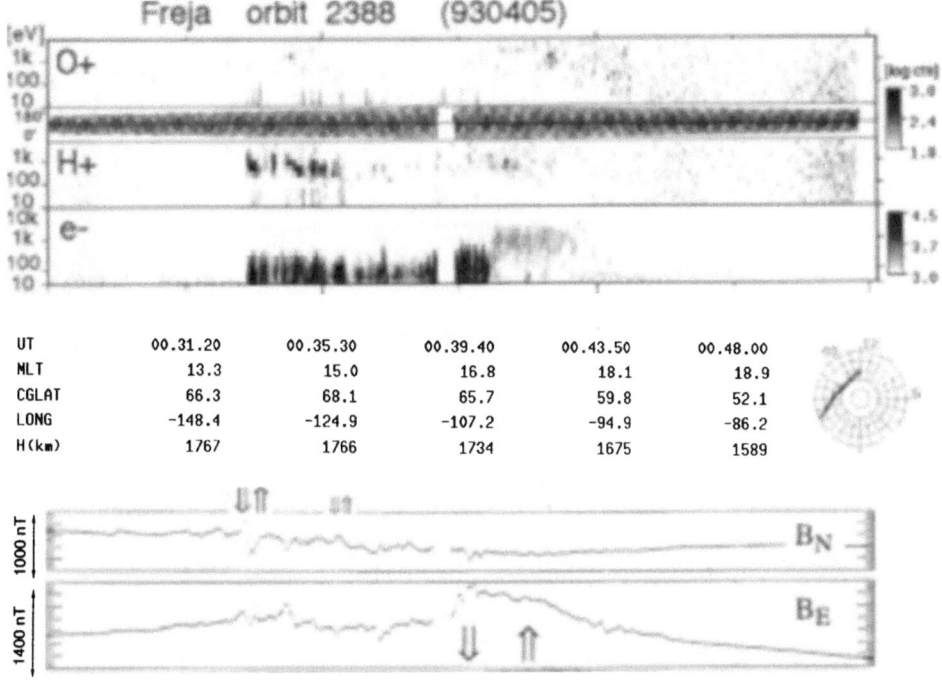

Figure 3. Particle and magnetic field data from Freja orbit 2388 (5 April, 1993). Positive (negative) slopes of B_E mean upward (downward) FACs at 17 MLT, and positive (negative) slopes of B_N mean upward (downward) FACs at 13 MLT. Corresponding to cusp, BPS, and CPS, we find cusp region FACs, upward Region-1 FAC, and Region-2 FAC, respectively.

period). For example, the ion conics which is seen all the way from 21:31 UT to 21:36 UT indicates waves activities and possible electron burst in this region. Therefore, we can conclude only that some portion of downward Region 1 FAC is composed of upward electron burst associated with the electrostatic shocks (and cavitons), but contributions from the thermal electrons are still unknown. None of the past studies give sufficient answer to this.

All Region-2 FACs are found in the region of CPS particles. Except for its poleward boundary where CPS electrons and BPS electrons coexist, we do not find any electron bursts below 73° Inv. The large-scale FAC in this part is not the sum of the small-scale paired FACs, but is carried by the precipitating CPS electrons accelerated by the large-scale steady electrostatic potential structure. This is consistent with previous results [40, 41]. However, its poleward part (21:38-21:39 UT) is different. It is apparently composed of many paired FACs which agrees with the small-scale structures shown in Figure 1 or the meso-scale magnetic fluctuations associated with Pc 4-5 pulsations [16] and/or transient magnetosheath plasma injections (PTE) [46] inside CPS. According to the total magnetic deviation, this part contributes about 50% of the total upward Region-2 FAC. The question is if the FAC in this part is contributed by such small- (or meso-)scale FACs. The majority of FAC in this region could still be carried by thermal electrons if the small- (meso-)scale FACs adds only minor spikes (up and down) without net FACs. Since the FAC regions are determined simply by its sense (upward or downward), one may not draw a general conclusion on the co-existing population without thorough statistics. Yet we see some cases of Region-2 FACs

178

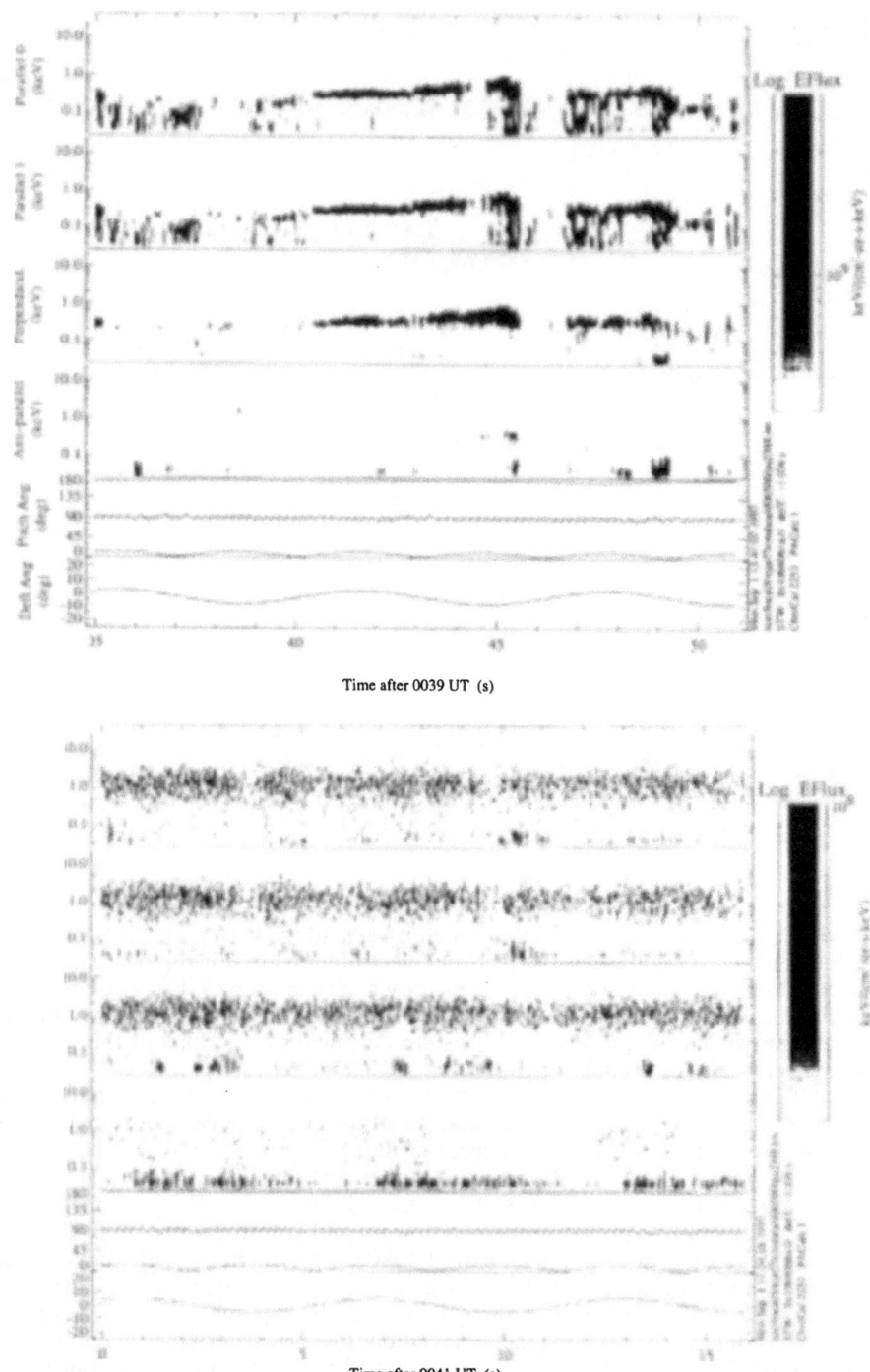

Figure 4. High-resolution electron data of Figure 3: (a) BPS part (upward Region-1 FAC); (b) CPS part (downward Region-2 FAC). The upper two panels in each figures show downward field-aligned electrons, the third panel shows trapped electrons (90° pitch angle), and the last panel shows upward field-aligned electrons.

without any small-scale spikes related to the electrostatic shocks (e.g., Figure 1 of [38]). So, we may at least conclude that CPS-related upward FACs are carried by these CPS electrons and are not the sum of small-scale FACs. Contribution from the small-scale FACs to the dawnside Region-2 FAC is a future problem.

2.2. AFTERNOON SECTOR

Let us move to the dusk sector. BPS electrons are now associated with the inverted-V structures, discrete aurora, and upward FACs. The relation is self-consistent. However, we have a puzzle concerning the CPS-related Region 2 FAC. It flows downward, i.e., the current carriers (electrons) must move upward whereas the plasma regime in this region (CPS) is in favour of upward FACs (cf. morning sector). So, we have to find a different type current carrier in this region.

Figure 3 shows the FREJA particle and magnetic field observations in the noon and dusk sectors. The satellite traversed the BPS region and CPS region at around 00:40 UT. Region-1 and Region-2 FACs are found exactly in the BPS and CPS regions, respectively. Enlarged plots of the electron data [62] in these regions are shown in Figure 4. Figure 4a (BPS = upward FAC) clearly shows intermittent inverted-V potential structures, indicating that the Region-1 FAC is subdivided into many small-scale FACs as is expected. However, the upward electron burst (expected in Figure 1) is barely seen. Certainly, the small-scale FAC structures are quite different between dawn and dusk. We instead see upward low-energy electrons (< 70 eV) continuously. The same type of upward low-energy electrons extends to the CPS region as is clear from Figure 4b. These electrons must be the carriers for the downward FAC, and hence Region-2 FAC may not be divided into small-scale FACs contrary to Region-1 FAC.

In summary, the relation between the large-scale and the small-scale FACs is similar in both sectors. The major differences between dawn and dusk are the roles of thermal electrons and electron burst. Contributions to the FACs are largest from the auroral precipitating electron (upward FAC), the second largest from the low-energy upward electrons (downward FAC), the third largest from the CPS-origin precipitating electrons (upward FACs), and barely from electron burst (downward FAC). The majority of upward FAC in the BPS is associated with intermittent inverted-V structures and the majority of the downward FAC in the CPS is carried by low-energy electrons .

2.3. NOON SECTOR

Obtaining the current carrier is more difficult because the cusp is normally filled with various wave activities [63], which makes it difficult to identify the electron population. Burch et al. [42] showed that at mid-altitude upward electron beams can explain the cusp Region-1 FAC in the prenoon sector, but the situation is not so simple. Electron data for the cusp part in Figure 3 is enlarged in Figure 5. There is no difference between the upward FAC region and the downward FAC region. Both regions are filled with supra-thermal electron bursts and the injecting electrons. Ion conics and beams [64] are another possible clue, because the ion beams usually indicate inverted-V potential structure below the satellite whereas ion conics usually indicate wave activity below the satellite (same source for the electron burst). However, we again find a lack of good correlation between the ion beams/conics and the senses (upward/downward) of FACs. Apparently the cusp large-scale FAC is not a simple extension of the dawnside/duskside Region-1 or Region-2 FACs [4, 27, 33, 36, 37]. This is one of the counter-evidence against the reconnection-related cusp models, e.g., by Cowley et al. [65; and references therein].

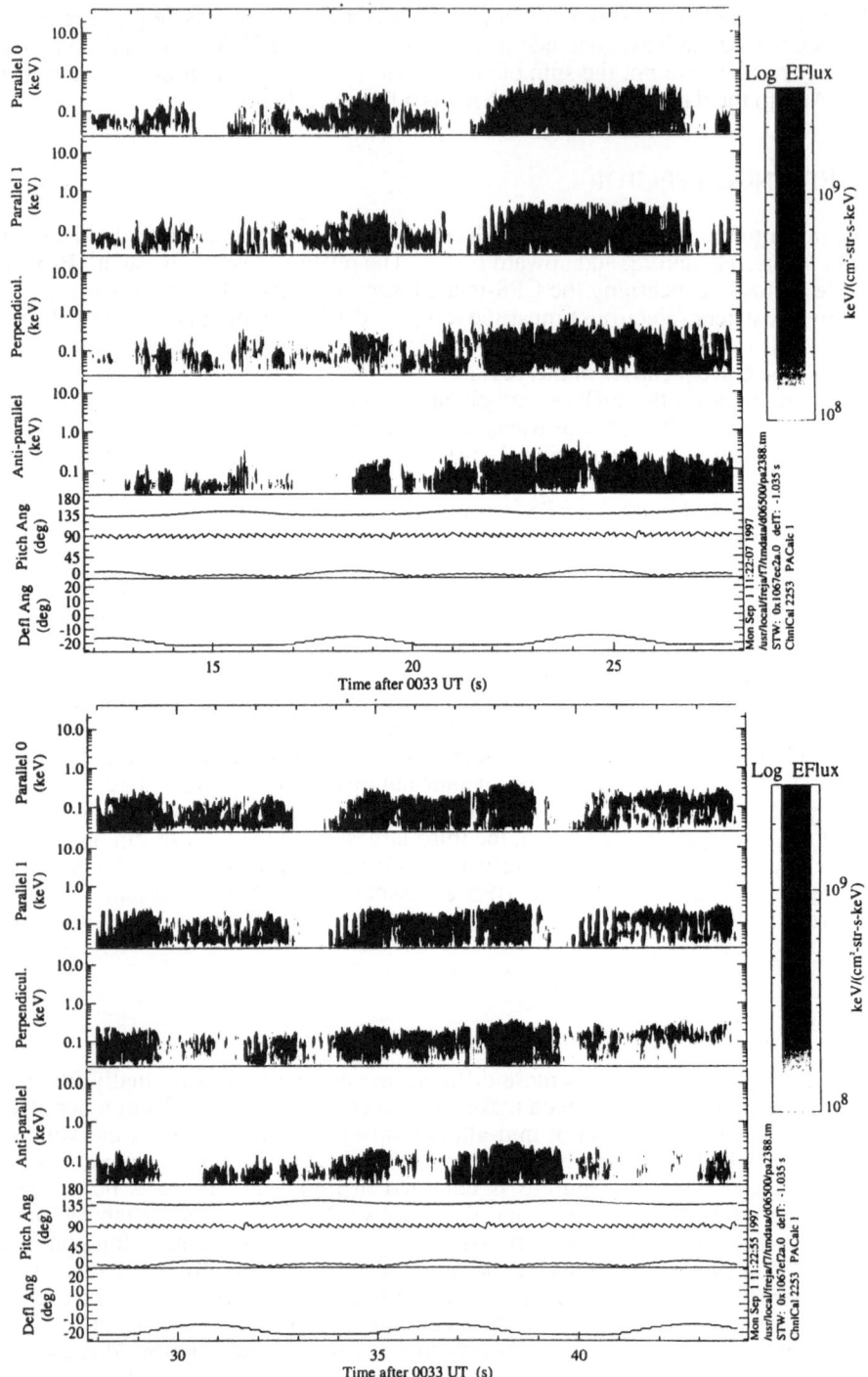

Figure 5. High-resolution electron data for the cusp part of Figure 3: (a) upward FAC region; (b) downward FAC region. The format is the same as Figure 4.

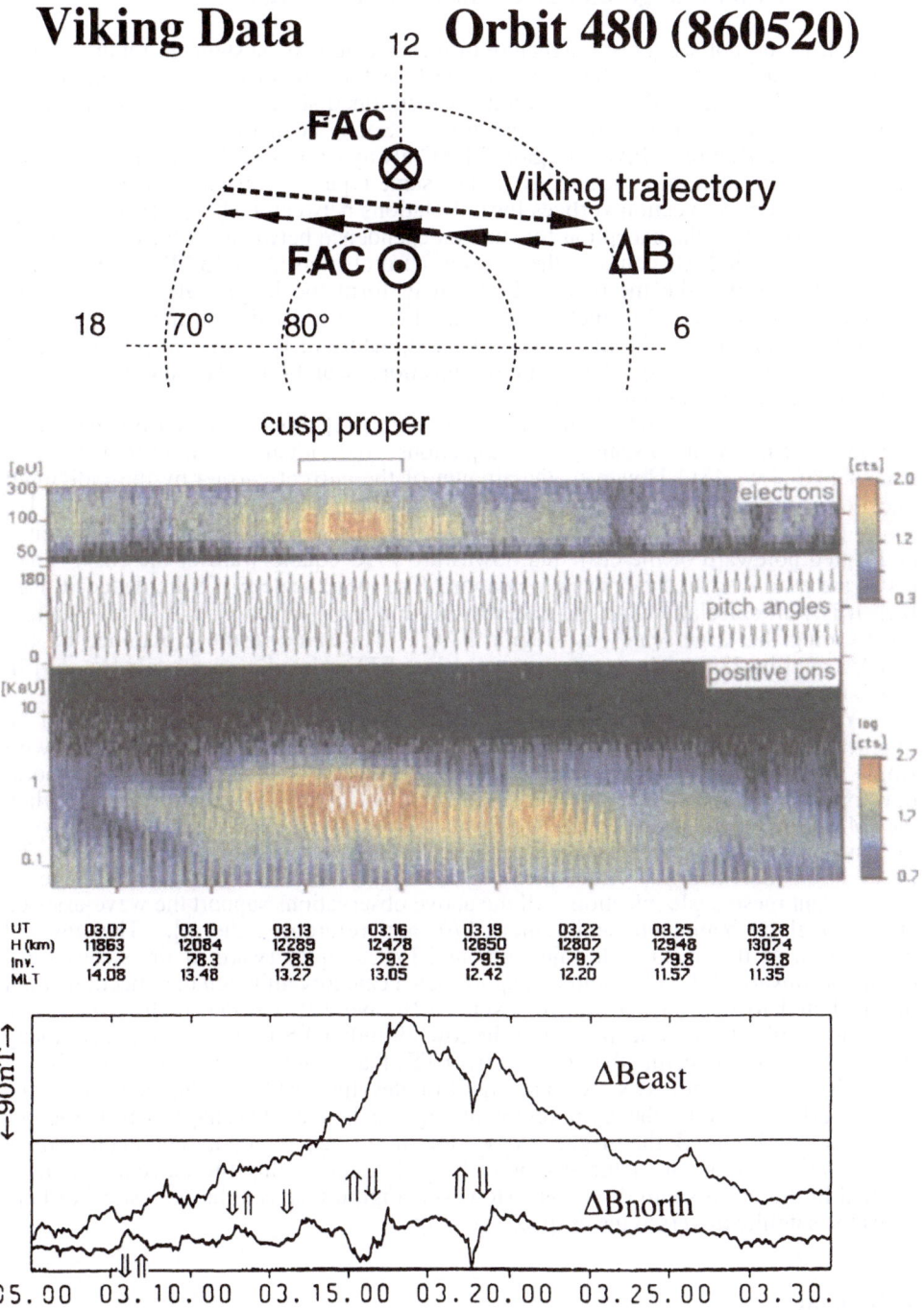

Figure 6. Particle and magnetic field data from Viking orbit 480 (20 May, 1986). Positive (negative) slopes of B_N mean downward (upward) FACs. Positive deviation B_N indicates upward FAC poleward of the satellite and/or downward FAC equatorward of the satellite [45].

3. Relation Between Large-scale FACs and Meso-scale FACs

Although we do not see a clear signature of meso-scale FAC systems in Figures 2 and 3, they are often observed in the dawn and dusk sectors together with transient plasma injections (PTE) and/or Pc 4-5 pulsations as mentioned above [16, 46]. These meso-scale FACs usually have bipolar signatures. So, it is difficult to estimate their contribution to Region-2 FAC. Region-2 FAC always exists in the CPS regardless of the existence of such meso-scale FACs. The same type of semi-independence is seen for the travelling convection vortices [66]. Relations between PTE and TCV are under investigation. Only the exception for such independence between the meso-scale FAC and the large-scale FAC could be the series of auroral spots at 14-15 MLT [13, 67, and references therein]: the meso-scale FACs may form the large-scale Region-1 and Region-2 FACs there. But, these meso-scale FACs are smaller in intensity than the cusp FAC. In Figure 3, the first large bipolar signature of B_N is the cusp FAC but all the other minor bipolar signatures and ion injections near 14-15 MLT are probably the LLBL-related afternoon auroral spots.

What about the cusp? Yamauchi and Lundin [52] showed that cusp is generally composed of many meso-scale plasma injections. Individual injection often carries a pair of FAC [39, 45]. However, the amount of the current carried by the individual injection is far below the large-scale cusp FACs. Figure 6 (after [45], their Figure 15) shows one example. IMF is strongly duskward (By = + 5 nT); i.e., upward FACs must be located poleward of the cusp and downward FAC equatorward of the cusp. Such FACs are easily verified by the B_E deviation. The meso-scale FACs are deduced from B_N. It is well correlated with the magnetosheath plasma injections but is nevertheless not forming the large-scale cusp-region FACs. The same result is obtained by the FREJA satellite [39]. The result argues against FTE-related cusp models [68, 69] in which the entire large-scale cusp region FAC is composed of meso-scale FACs.

Yet, meso-scale FACs carried by the meso-scale plasma injections could be important in the formation of the cusp FAC. Although such studies are difficult with single satellite observations because of the separation problem between the temporal development and the spatial structure, Yamauchi *et al.* [51] showed a clear case when the IMF turned from steady southward to steady northward. The cusp current flows in the cusp part, but some meso-scale FAC is also found in the equatorward injection. The observation indicates that either the cusp is moving back and forth, or we have an independent meso-scale injection. All the above observations support the wave-assisted cusp model by Yamauchi and Lundin [70; and references therin]. The unusual observation of the meso-scale injections and FACs equatorward of the steady cusp during northward IMF [71] could belong to such a category although the injection could be attributed to the plasma transfer event [72]. Note that some of the meso-scale magnetosheath plasma injections into the low-altitude CPS region do not accompany FACs at all [73], indicating that they are not FTEs that Southwood has modelled [74].

The cusp region FACs are somewhat troublesome. Although the plasma regime (ion) clearly determines the entire region of large-scale FAC, the shape of FAC regions is unclear. In Figure 3, the magnetic deviation in the cusp region is seen mainly in the north-south component of the magnetic field, indicating that the current system is aligned in the north-south direction. However, Figure 6 shows the opposite. In both cases the satellite traversed the cusp azimuthally.

4. Summary

It is yet too early for far-reaching conclusions without thorough statistics. Here we just list some summaries of several case studies .

The Region-1 and 2 FACs are well located in BPS/LLBL and CPS, respectively, in both the dawn and dusk sectors. This fact naturally causes some asymmetry of the current carrier between the dawn and dusk. The dawn Region-2 FAC is mostly carried by CPS electrons precipitating into the ionosphere whereas the duskside Region-2 FAC is carried by ionospheric low-energy (mostly thermal) electrons. The current carriers in the Region-1 FAC could be symmetric if it is mostly composed of many small-scale FACs as shown in Figure 1. However, its fine structure is different between dawn and dusk because the upward small-scale FACs inside an inverted-V potential structure is wide-spread compared to the downward small-scale FACs which are carried by the bursty electrons next to the inverted-V structure. Furthermore, the upward electron burst does not necessarily exist in the duskside Region-1 FAC. The roles of the thermal electron for the dawnside Region-1 FAC and of small- and meso-scale FACs for the dawnside Region-2 FAC must be investigated in the future.

The cusp FACs are not well understood. Meso-scale FACs are related with individual plasma injections, but unlike these injections, the meso-scale FACs do not constitute the large-scale FACs. They add minor bipolar signatures to the steady large-scale current system. The current carriers are quite unknown in the cusp region. We need thorough statistics, and any studies from a limited data set are not sufficient to solve the current carrier problem in the cusp region.

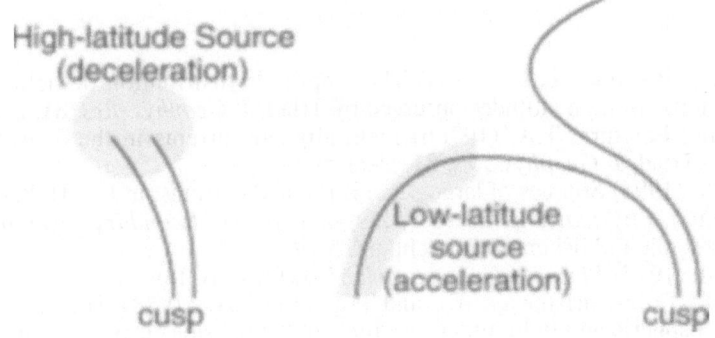

Figure 7. Proposed locations of the cusp (region-1) FAC. If the source is high-latitude, the decelerated solar wind directly drives a dynamo [76, 77]. If the source is low-latitude, the solar wind is accelerated and energy must come somewhere else. In this case, the cusp region-1 FAC must be dispersed and weak compared to cusp region-0 FAC, which is against the observation.

5. Restriction in Modelling

Before closing this rather diagnostic paper, let us mention some important notes for modelling. Since the FAC is a sink or a source of energy, any 2-D model needs an additional sink and source as the 3rd dimension effect [75-77]. For instance, the FAC transports energy from the source region to the ionosphere.

One may classify the source into two types: (1) constant dynamo (deceleration of convection or solar wind) and (2) constant total energy (conversion of "stored" electromagnetic energy). If the flow (e.g., solar wind) is steady, the former system predicts a long-term development of FAC toward a steady value whereas the latter predicts a decay of FAC system due to the ionospheric joule dissipation. According to

MAGSAT observations, the cusp FAC system is quite steady and never decays [23]. So, the FAC system must take energy from the solar wind (decelerate the solar wind somewhere) rather than from the "stored" magnetic field. Next question is the location of dynamo; i.e., poleward of the cusp (cusp region-1 FAC) or equatorward of the cusp (cusp region-0 FAC). During southward IMF, the equatorward FAC is always stronger than the poleward FAC, indicating that the dynamo is connected to the cusp region-1 FAC rather than the cusp region-0 FAC (mantle FAC). This gives us a restriction in mapping the FAC system, as shown in Figure 7 because the solar wind must be decelerated in the source region. Apparently, we have to map the cusp to the high-latitude dynamo region [70, 76] rather than the dayside magnetopause [78, 79]. The same conclusion is also obtained from the high-altitude observations of the cusp by HAWKEYE [80, 81], INTERBALL [82], and POLAR [83] spacecrafts. One may not simply map the low-altitude FAC beyond the high-altitude cusp [84, 85].

Acknowledgments. The Freja project is supported by Swedish National Space Board and German Space Agency. Freja and Viking magnetic field data are provided by L. J. Zanetti and T. A. Potemra at JHU/APL. Freja electron data (TESP) is provided by M. Boehm at Max-Planck-Institute for Extraterrestrial Physics (currently at JPL).

6. References

1. Iijima, T. and Potemra, T.A. (1976a) The amplitude distribution of field-aligned currents at northern high latitudes observed by Triad, *J. Geophys. Res.* **81**, 2165.
2. Iijima, T. and Potemra, T.A. (1976b) Field-aligned currents in the dayside cusp observed by Triad, *J. Geophys. Res.* **81**, 5971-5979.
3. Potemra, T.A. (1994) Sources of large-scale Birkeland currents, in J.A. Holtet and A. Egeland (eds.), *Physical signatures of magnetospheric boundary layer process*, Kluwer Academic Publishers, Dordrecht, pp. 3-27.
4. Akasofu, S.-I. (1977) *Physics of magnetospheric substorms*, Reidel.
5. Wilhjelm, J., Friis-Christensen, E., and Potemra, T.A. (1978) The relationship between ionospheric and field-aligned currents in the dayside cusp, *J. Geophys. Res.* **83**, 5586-5594.
6. Brekke, A., Doupnik, J.R., and Banks, P.M. (1974) Incoherent scatter measurements of E region conductivities and currents in the auroral zone, *J. Geophys. Res.* **79**, 3773-3790.
7. Levitin, A.E., Afonina, R.G., Belov, B.A., and Feldstein, Y.I. (1982) Geomagnetic variation and field-aligned currents at northern high-latitudes, and their relationship to the solar wind parameters, *Phil. Trans. R. Soc. Lond.* **A304**, 253-301.
8. Friis-Christensen, E., Kamide, Y., Richmond, A.D., and Matsushita, S. (1985) Interplanetary magnetic field control of high-latitude electric fields and currents determined from Greenland magnetometer data, *J. Geophys. Res.* **90**, 1325-1338.
9. Feldstein, Y.I. and Galperin, Yu.I. (1985) The auroral luminosity structure in the high-latitude upper atmosphere: its dynamics and relationship to the large-scale structure of the earth's magnetosphere, *Rev. Geophys.* **23**, 217.
10. Burch, J.L., Reiff, P.H., Menietti, J.D., Heelis, R.A., Hanson, W.B., Shawhan, S.D., Shelley, E.G., Sugiura, M., Weimer, D.R., and Winningham, J.D. (1985) IMF B_Y dependent plasma flow and birkeland currents in the dayside magnetosphere 1. dynamics explorer observations, *J. Geophys. Res.* **90**, 1577-1593.
11. Heppner, J.P. and Maynard, N.C. (1987) Empirical high-latitude electric field models, *J. Geophys. Res.* **92**, 4467-4489.

185

12. Blomberg, L.G. and Marklund, G.T. (1991) High-latitude convection patterns for various large-scale field-aligned current configurations, *Geophys. Res. Lett.* **18**, 717-720.
13. Elphinstone, R.D., Hearn, D., Murphree, J.S., Cogger, L.L., Johnson, M.L., and Vo, H.B. (1993) Some UV dayside auroral morphologies, in *Auroral Plasma Dynamics*, American Geophysical Union, Washington, 31-45.
14. Sugiura, M. (1984) A fundamental magnetosphere-ionosphere coupling mode involving field-aligned currents as deduced from DE-2 observations, *Geophys. Res. Lett.* **11**, 877-880.
15. Weimer, D.R., Goertz, C.K., Gurnnet, D.A., Maynard, N.C., and Burch, J.L. (1985) Auroral zone electric fields from DE 1 and 2 at magnetic conjunctions, *J. Geophys. Res.* **90**, 7479.
16. Potemra, T.A., Zanetti, L.J., Bythrow, P.F., Erlandson, R.E., Lundin, R., Marklund, G.T., Block, L.P. and Lindqvist, P.-A. (1988) Resonant geomagnetic field oscillations and Birkeland currents in the morning sector, *J. Geophys. Res.* **93**, 2661-2674.
17. Ishii, M., Sugiura, M., Iyemori, T., and Slavin, J.A. (1992) Correlation between magnetic and electric fields in the field-aligned current regions deduced from DE-2 observations, *J. Geophys. Res.* **97**, 13877.
18. Knudsen, D.J., Kelley, M.C., and Vickerey, J.F. (1992) Alfvén waves in the auroral ionosphere: A numerical model compared with measurement, *J. Geophys. Res.* **97**, 77.
19. Marklund, G. (1993) Viking investigations of auroral electrodynamical processes, J. Geophys. Res., 98, 1691-1704, 1993.
20. Peterson, W.K., Abe, T., André, M., Engebretson, M.J., Fukunishi, H., Hayakawa, H., Matsuoka, A., Mukai, T. Persoon, A.M., Retterer, J.M., Robinson, R.M., Sugiura, M., Tsuruda, K., Wallis, D.D., and Yau, A.W. (1993) Observations of a transverse magnetic field perturbation at two altitudes on the equatorward edge of the magnetospheric cusp, *J. Geophys. Res.* **98**, 21463-21470.
21. Lühr, H., Warnecke, J., Zanetti, L.J., Lindqvist, P.A., and Hughes, T.J. (1994) Fine structure of field-aligned current sheets deduced from spacecraft and ground-based observations: Initial Freja results, *Geophys. Res. Lett.* **21**, 1883-1886.
22. Fujii, R. and Iijima, T. (1987) The control of the ionospheric conductivities on large-scale birkeland current intensities under geomagnetic quiet conditions, *J. Geophys. Res.* **92**, 4505-4513.
23. Yamauchi, M. and Araki, T. (1989) The interplanetary magnetic field By-dependent field-aligned current in the dayside polar cap under quiet conditions, *J. Geophys. Res.* **94**, 2684-2690.
24. Lu, G., Richmond, A.D., Emery, B.A. Reiff,, P.H., de la Beaujardiere, O., Rich, F.J., Denig, W.F., Kroehl, H.W., Lyons, L.R., Ruohoniemi, J.M., Friis-Christensen, E., Opgenoorth, H., Persson, M.A.L., Lepping, R.P., Rodger, A.S., Hughes, T., McEwin, A., Dennis, S., Morris, R., Burns, G., and Tomlinson, L. (1994) Interhemispheric asymmetry of the high-latitude ionospheric convection pattern, *J. Geophys. Res.* **99**, 6491-6510.
25. Kamide, Y., Craven, J.D., Frank, L.A., Ahn, B.-H., and Akasofu, S.-I. (1986) Modeling substorm current systems using conductivity distributions inferred from DE auroral images, *J. Geophys. Res.* **91**, 11235-11256.
26. Iijima, T. and Shibaji, T. (1987) Global characteristics of northward IMF-associated (NBZ) field-aligned currents, *J. Geophys. Res.* **92**, 2408-2424.
27. Erlandson, R.E., Zanetti, L.J., Potemra, T.A., Bythrow, P.F., and Lundin, R. (1988) IMF By dependence of region 1 birkeland currents near noon, *J. Geophys. Res.* **93**, 9804-9814.

28. Hoffman, R.A., Sugiura, M., Maynard, N.C., Candey, R.M., Craven, J.D., and Frank, L.A. (1988) Electrodynamic patterns in the polar region during periods of extreme magnetic quiescence, *J. Geophys. Res.* **93**, 14515-14541.

29. Murphree, J.S., Johnson, M.L., Cogger, L.L., and Hearn, D.J. (1994) Freja UV imager observations of spatially periodic auroral distortions, *Geophys. Res. Lett.* **21**, 1887-1890.

30. Yamauchi, M., Lundin, R., and Aparicio, B. (1992) Viking observation of the substorm current wedge, *ESA SP* **335**, 495-497.

31. Marklund, G., Blomberg, L., Fälthammar, C.-G., and Lindqvist, P.-A. (1994) On the diverging electric fields associated with black aurora, *Geophys. Res. Lett.* **21**, 1859-1962.

32. Ohtani, S., Zanetti, L.J., Potemra, T.A., Baker, K.B., Ruohoniemi, J.M., and Lui, A.T.Y. (1994) Periodic longitudinal structure of field-aligned currents in the dawn sector: large-scale meandering of an auroral electrojet, *Geophys. Res. Lett.* **21**, 1879-1882.

33. Ohtani, S., Potemra, T.A., Newell, P.T., Zanetti, L.J., Iijima, T., Watanabe, M., Yamauchi, M., Elphinstone, R.D., de la Beaujardiere, O., and Blomberg, L.G. (1995) Simultaneous prenoon and postnoon observations of three field-aligned current systems from Viking and DMSP-F7, *J. Geophys. Res.* **100**, 119-136.

34. McDiarmid, I.B., Burrows, J.R., and Wilson, M.D. (1979) Large-scale magnetic field perturbations and particle measurements at 1400 km on the dayside, *J. Geophys. Res.* **84**, 1431-1441.

35. Bythrow, P.F., Potemra, T.A., and Hoffman, R.A. (1982) Observations of field-aligned currents, particles, and plasma drift in the polar cusps near solstice, *J. Geophys. Res.* **87**, 5131-5139.

36. Bythrow, P.F., Potemra, T.A., Erlandson, R.E., Zanetti, L.J., and M. Klumpar, D. (1988) Birkeland currents and charged particles in the high-latitude prenoon region: A new interpretation, *J. Geophys. Res.* **93**, 9791-9803.

37. Yamauchi, M., Lundin, R., and Woch, J. (1993a) The interplanetary magnetic field By effects on large-scale field-aligned currents near local noon: Contributions from cusp part and noncusp part, *J. Geophys. Res.* **98**, 5761-5767.

38. Woch, J., Yamauchi, M., Lundin, R., Potemra, T.A., and Zanetti, L.J. (1993) The low-latitude boundary layer at mid-altitudes: Relation to large-scale Birkeland currents, *Geophys. Res. Lett.* **20**, 2251-2254.

39. Andersson, L., Nilsson, H., and Wahlund, J.-E. (1997) Meso-scale structure in the cusp/cleft region, *Phys. Chem. Earth*, in press.

40. Anderson, H.R. and Vondrak, R.R. (1975) Observations of Birkeland currents at auroral latitudes, *Rev. Geophys. Space Phy.* **13**, 243.

41. Klupmar, D.M. (1979) Relationships between auroral particle distributions and magnetic field perturbations associated with field-aligned currents, *J. Geophys. Res.* **84**, 6524-6532.

42. Burch, J.L., Reiff, P.H., and Sugiura, M. (1983) Upward electron beams measured by DE-1: a primary source of dayside region-1 birkeland currents, *Geophys. Res. Lett.* **10**, 753-756.

43. Potemra, T.A., Zanetti, L.J., Erlandson, R.E., Bythrow, P.F., Gustafsson, G., Acuna, M.H., and Lundin, R. (1987) Observations of large-scale Birkeland currents with Viking, *Goephys. Res. Lett.* **14**, 419-422.

44. Ergun, R.E., Carlson, C.W., McFadden, J.P., Chaston, C., Delory, G.T., Peria, W., Mozer, F.S., Temerin, M., Elphic, R., Strangeway, R., Klumpar, D.M., Shelley, E.G., Peterson, W.K., Moebius, E., Kistler, L., Cattell, C., and Pfaff, R. (1997) Initial results from the FAST satellite: Solitary waves and AKR, *Ann. Geophys.* **15** *Sup.*, 683.

45. Lundin, R., Woch, J., and Yamauchi, M. (1991) The present understanding of the cusp, *ESA SP* 330 83-95.

187

46. Woch, J. and Lundin, R. (1992) Signature of transient boundary layer processes observed with Viking, *J. Geophys. Res.* **97**, 1431-1447.
47. Ohtani, S., Blomberg, L.G., Newell, P.T., Yamauchi, M., Potemra, T.A., and Zanetti, L.J. (1996) Altitudinal comparison of dayside field-aligned current signatures by Viking and DMSP-F7: Intermediate-scale FAC systems, *J. Geophys. Res.* **101**, 15297-15310.
48. Primdahl, F. and Spangslev, F. (1983) Does IMF By induce the cusp field-aligned current? *Planet. Space Sci.* **31**, 363-367.
49. Clauer, C.R. and Banks, P.M. (1986) Relationship of the interplanetary electric field to the high-latitude ionospheric electric field and currents: Observations and model simulation, *J. Geophys. Res.* **91**, 6959-6971.
50. Knipp, D.J., Richmond, A.D., Emery, B., Crooker, N.U., de la Beaujardiere, O., Evans, D., and Kroehl, H. (1991) Ionospheric convection response to changing IMF direction, *Geophys. Res. Lett.* **18**, 721-724.
51. Yamauchi M., Lundin, R., and Potemra, T. (1995) Dynamic response of the cusp morphology to the IMF changes: An example observed by Viking, *J. Geophys. Res.* **100**, 7661-7670.
52. Yamauchi, M. and Lundin, R. (1994) Classification of large-scale and meso-scale ion dispersion patterns observed by Viking over the cusp-mantle region, in J.A. Holtet and A. Egeland (eds.), *Physical signatures of magnetospheric boundary layer process*, Kluwer Academic Publishers, Dordrecht, pp. 99-109.
53. Goertz, C.K. (1984) Kinetic Alfvén waves on auroral filed lines, *Planet. Space Sci.* **32**, 1387-1392.
54. Lysak, R.L. (1997) Propagation of Alfvén waves through the ionosphere, *Phys. Chem. Earth,* in press.
55. Lundin, R., Eliasson, L., Haerendel, G., Boehm, M., and Holback, B. (1994) Large-scale auroral plasma density cavities observed by Freja, *Geophys. Res. Lett.* **21**, 1903-1906.
56. Winningham, D.J., Yasuhara, F., Akasofu, S.-I., and Heikkila, W.J. (1975) The latitudinal morphology of 10 eV to 10 keV electron fluxes during quiet and disturbed times in the 2100-0300 MLT sector, *J. Geophys. Res.* **80**, 3148.
57. Kremser, G. and Lundin, R. (1990) Average spatial distributions of energetic particles in the midaltitude cusp/cleft region observed by Viking, *J. Geophys. Res.* **95**, 5753-5766.
58. Newell, P.T. and Meng, C.-I. (1992) Mapping the dayside ionosphere to the magnetosphere according to particle precipitation characteristics, *Geophys. Res. Lett.* **19**, 609-612.
59. Woch, J. and Lundin, R. (1993) The low-latitude boundary layer at mid-altitudes: Identification based on Viking hot plasma data, *Geophys. Res. Lett.* **20**, 979-982.
60. Alfvén, H. and Fälthammar, C.G. (1963) *Cosmical Electrodynamics*, Fundamental Principles, Clarendon, Oxford.
61. Fridman, M. and Lemaire, J. (1980) Relationship between auroral electron fluxes and field-aligned electric potential difference, *J. Geophys. Res.* **85**, 664.
62. Boehm, M., Paschmann, G., Clemmons, J., Höfner, H., Fremzel, R., Ertl, M., Haerendel, G., Hill, P., Lauche, H., Eliasson, L., and Lundin, R. (1994) The TESP electron spectrometer and correlator (F7) on Freja, *Space Sci. Rev.* **70**, 509-540.
63. Pottelette, R., Malingre, M., Dubouloz, N., Aparicio, B., Lundin, R., Holmgren, G., and Marklund, G. (1990) High-frequency waves in the cusp/cleft regions, *J. Geophys. Res.* **95**, 5957-5971.
64. Thelin, B. and Lundin, R. (1990) Upflowing ionospheric ions and electrons in the cusp-cleft region, *J. Geomag. Geoelectr.* **42**, 753-761.
65. Cowley, S.W.H., Morelli, J.P., and Lockwood, M. (1991) Dependence of convective flows and particle precipitation in the high-latitude dayside ionosphere

on the X and Y components of the interplanetary magnetic field, *J. Geophys. Res.* **96**, 5557-5564.

66. Friis-Christensen, E., McHenry, M.A., Clauer, C.R., and Vennerstrom, S. (1988) Ionospheric traveling convection vortices observed near the polar cleft: a triggered response to sudden changes in the solar wind, *Geophys. Res. Lett.* **15**, 253-256.

67. Lundin, R., Yamauchi, M., Woch, J., and Marklund, G. (1995) Boundary layer polarization and voltage in the 14 MLT region, *J. Geophys. Res.* **100**, 7587-7597.

68. Smith, M.F. and Lockwood, M. (1990) The pulsating cusp, *Geophys. Res. Lett.* **17**, 1069-1072.

69. Lockwood, M. (1994) Ionospheric signatures of pulsed magnetopause reconnection, in J.A. Holtet and A. Egeland (eds.), *Physical signatures of magnetospheric boundary layer process*, Kluwer Academic Publishers, Dordrecht, pp. 229-243.

70. Yamauchi, M. and Lundin, R. (1997) The wave-assisted cusp model: comparison to low-altitude observations, *Phys. Chem. Earth*, in press.

71. Potemra, T.A., Erlandson, R.E., Zanetti, L.J., Arnordy, R.L., Woch, J., and Friis-Christensen, E. (1992) The dynamic cusp, *J. Geophys. Res.* **97**, 2835-2844.

72. Lemaire, J. (1977) Impulsive penetration of filamentary plasma elements into the magnetospheres of the Earth and Jupiter, *Planet. Space Sci.* **25**, 887-890.

73 Yamauchi, M., Woch, J., Lundin, R., Shapshak, M., and Elphinstone, R. (1993b) A new type of ion injection event observed by Viking, *Geophys. Res. Lett.* **20**, 795-798.

74. Southwood, D.J. (1987) The ionospheric signature of flux transfer event, *J. Geophys. Res.* **92**, 3207.

75. Sato, T. and Iijima, T. (1979) Primary sources of large-scale Birkeland current, *Space Sci. Rev.* **24**, 347-366.

76. Yamauchi, M., Lundin, R., and Lui, A.T.Y. (1993c) Vorticity equation for MHD fast waves in geospace environment, *J. Geophys. Res.* **98**, 13523-13528.

77. Yamauchi, M. (1994) Numerical simulation of large-scale field-aligned current generation from finite-amplitude magnetosonic waves, *Geophys. Res. Lett.* **21**, 851-854.

78. Lee, L.C., Kan, J.R., and Akasofu, S.-I. (1985) On the origin of the cusp field-aligned currents, *J. Geophys.* **57**, 217-221.

79. Onsager, T.G. (1994) A quantitative model of magnetosheath plasma in the low latitude boundary layer, cusp and mantle, in J.A. Holtet and A. Egeland (eds.), *Physical signatures of magnetospheric boundary layer process*, Kluwer Academic Publishers, Dordrecht, pp. 385-400.

80. Zhou, X.-W. and Russell, C.T. (1997) The location of the high-latitude polar cusp and the shape of the surrounding magnetopause, *J. Geophys. Res.* **102**, 105-110.

81. Fung, S.F., Eastman, T.E., Boardsen, S.A., and Chen, S.-H. (1997) High-altitude cusp positions samples by the Hawkewe satellite, *Phys. Chem. Earth*, in press.

82. Sandahl, I., Lundin, R., Yamauchi, M., Eklund, U., Safrankova, J., Nemecek, Z., Kudela, K., Lepping, R.P., Lin, R.P., Lutsenko, V.N., and Sauvaud, J.-A. (1997) Cusp and boundary layer observation by INTERBALL, *Adv. Spcae Res.*, in press.

83. Zhou, X.-W., Russell, C.T., Le, G., and Tsyganenko, N. (1997) Comparison of observed and model magnetic fields at high altitudes above the polar cap, *Geophys. Res. Lett.* **24**, 1451-1454.

84. Yamauchi, M. and Blomberg, L. (1997) Problems on mappings of the convection and on the fluid concept, *Phys. Chem. Earth*, in press.

85. Yamauchi, M. (1997) Discussion 3: On the convection and velocity filter, *Phys. Chem. Earth*, in press.

THE DAYSIDE AURORA AND ITS REGULATION BY THE INTERPLANETARY MAGNETIC FIELD

P. E. SANDHOLT

Department of Physics, University of Oslo, Oslo, Norway

C. J. FARRUGIA

University of New Hampshire, Durham, New Hampshire, USA

J. MOEN

University Courses on Svalbard, Longyearbyen, Svalbard, Norway

AND

B. LYBEKK

Department of Physics, University of Oslo, Oslo, Norway

1. Introduction

The processes which couple the solar wind mass, energy and momentum to the magnetosphere are known to be strongly dependent on the orientation of the interplanetary magnetic field (IMF; e.g. [1, 2] , and many others). When the IMF points south the predominant coupling process is believed to be reconnection between the interplanetary and terrestrial magnetic fields in the manner first proposed by Dungey [3]. Conversely, when the IMF points strongly north, other processes become important, maybe of a "viscous type" nature, such as Kelvin - Helmholtz instability, diffusion etc., but there is also the possibility of reconnection at neutral points poleward of the cusp, where the local rotation of the field across the magnetopause - the magnetic shear - is large.

Recent work (e.g. [4, 5, 6]) have shown further that the "dividing line" between these magnetospheric states does not quite coincide with the the transition from negative to positive north - south component of the IMF, B_z. If we use the IMF clock angle, θ (where $\theta = \arctan(\text{abs}(B_y)/B_z)$ for $B_z > 0$ and $= 180 - \arctan(\text{abs}(B_y/B_z))$ for $B_z < 0$), as the ordering parameter, these studies show that the major change in the state of the magnetosphere occurs at $\theta < 90°$. B_y is the east - west component of the IMF. According to Phan et al. [7] the transition between the regimes of low -

189

J. Moen et al. (eds.), Polar Cap Boundary Phenomena, 189–208.

shear(no local reconnection) and high - shear(reconnection is possible) at the subsolar magnetopause occurs at $\theta \sim 45°$.

The electromagnetic coupling between the outer magnetosphere and the underlying ionosphere leads us to expect that the dependence on θ at the magnetopause will be also reflected in the ionosphere, in particular in the morphology and dynamics of dayside auroral forms. The aim of this paper is to investigate if this is the case by examining a variety of dayside auroral configurations and activities in relation to the IMF clock angle. We find, indeed, that θ orders the dayside aurora rather well. This leads us to put forward a classification of dayside auroral forms and transitions between them based on three clock angle regimes (called hereafter CAR 1, CAR 2, and CAR 3). CAR 1 corresponds to the IMF $B_z >> 0$ orientation, and is defined for our purposes by $0° < \theta < 45°$. The clock angles in CAR 2 lie within $45-90°$, and are termed the intermediate regime. In CAR 3, corresponding to southward IMF ($B_z < 0$), θ lies within 90 - 180°. Some essential features of the auroral configurations seen in the respective clock angle regimes, as emerging from the study of the individual cases presented below, are summarized in the overview in Figure 1.

The upper two panels of the Figure show the auroral configurations for CAR 1, the left hand panel referring to positive B_y polarity and vice versa. This is a polar plot of MLAT($>\sim 70°$) and MLT for the dayside (06 - 18 MLT). In the same format the lower two panels show the auroral situation for clock angle regime 3 ($B_z < 0$) and both polarities of B_y.

First we describe the aurora for CAR 1. In the pre - noon sector(~ 06 - 09 MLT) three major types of auroral forms may be identified. Moving from lower to higher latitudes the different forms are called types 3, 4, and 2'. Type 3 is the diffuse green line aurora caused by the precipitation of electrons from the dayside extension of the plasma sheet. This aurora can in principle extend all around to dusk, although in practice its local time extent varies from case to case. Then between typical latitudes ~ 73 - 78° MLAT the pre - noon aurora consists of multiple, discrete forms which are inhomogeneous in the east - west dimension. This type 4 aurora is marked by dashed lines in the figure. Poleward of this form we often observe a more homogeneous form which generally contains higher red line intensities. The latter aurora appear as a longitudinal extension (also in the post - noon sector) of the midday gap aurora, which we call type 2.

In the midday sector (09 - 15 MLT), besides forms 2 and 3, we observe a further discrete form, whose green line intensity is enhanced ($\sim 1kR$) with respect to type 2. This aurora is typically observed at the poleward or equatorward boundary of the type 2 (midday gap) aurora, depending on IMF By polarity.

In the dusk sector (15 - 18 MLT) the most prominent auroral feature(for

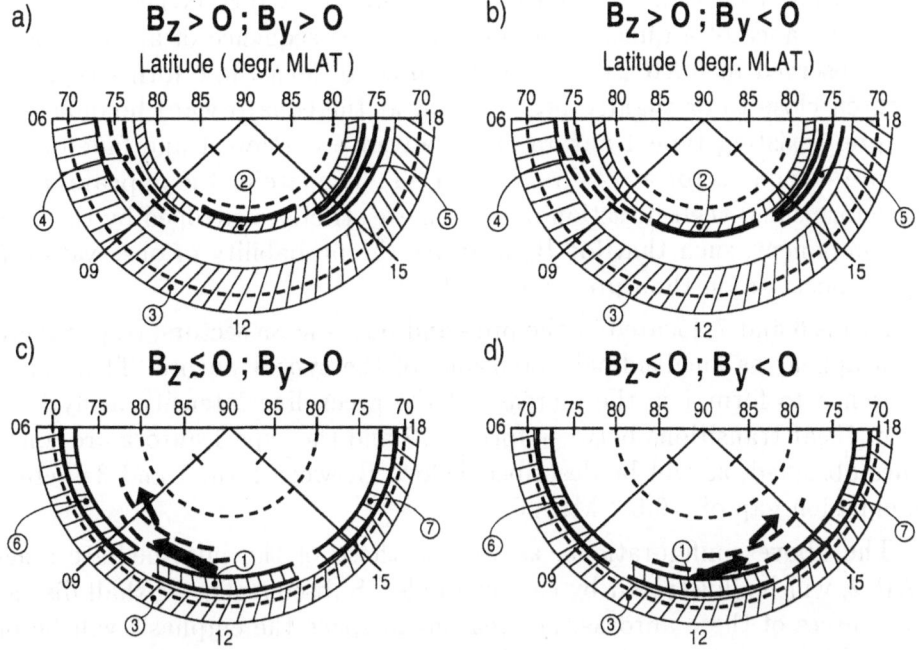

Figure 1. Schematic overview of dayside auroral forms for different regimes of IMF orientation. Forms 1 and 2 are due to particles of magnetosheath origin. Form 3 is the diffuse aurora due to particles from the central plasma sheet. Forms 4/6 and 5/7 are related to precipitation from the outer plasma sheet(BPS)/ low - latitude boundary layer in the morning and afternoon sectors, corresponding to northward and southward IMF - orientation, respectively.

CAR's 1 and 2) is the multiple, discrete arcs located between latitudes ~ 73 - 78°MLAT, called type 5. This aurora contains relatively strong emissions(~1 - 10kR) in both the green and red lines.

In the $B_y<0$ situation (top right panel) a similar auroral configuration is observed. The main difference is the auroral structure in the midday sector, where the discrete forms at the poleward edge of the gap aurora (type 2) is absent. Instead a discrete form at the equatorward boundary appears at times. The poleward and equatorward forms may be associated with different plasma convection and field - aligned current configurations in the cusp region.

Turning now to the large clock angle regime (CAR 3) and positive B_y polarity (bottom panel, left) we describe the midday sector first. Form 3 is still present and has been discussed above. In this case form 2 is absent. It is replaced by form type 1 (rayed bands), which is also dominated by the red line emission, typically much stronger than the type 2 aurora and located at much lower latitudes, within ~72 - 74°MLAT. A further new feature is the appearance of a quasi - periodic (T~5min) sequence of auroral forms (also observed in CAR 2). During $B_y>0$ conditions these forms typically brighten close to or slightly post - noon, at the equatorward boundary of the pre - existing type 1 aurora. They advance northwest and fade out in the pre - noon sector at ~ 75 - 77° MLAT. The site of the appearance of these forms and their direction of motion are strictly controlled by the IMF B_y component, such that for B_y negative the probability of observation is higher post - noon than pre - noon.

Forms 6 and 7, located in the pre - and post - noon sectors, respectively, often appear as longitudinal extensions of the type 1 aurora. Their main difference to form 1 is the increase of the green line intensity away from noon. Clear transitions between forms 6/7 and the type 1 aurora are sometimes observed, as will be described below. Between form 1 and 3 there is an emission gap of ~0.5 ° MLAT.

The above configurations, as well as those of the intermediate stage CAR 2, will be illustrated by case examples. Subsequently we shall discuss the sources of these auroras. For reasons of space the emphasis will be on the midday sector.

2. Observations

2.1. CASE 1: JANUARY 12, 1997

Figure 2 shows solar wind proton and magnetic field data from the Solar Wind Experiment, SWE [8] and Magnetic Field Investigation, MFI [9] on the Global Geospace Mission Spacecraft WIND for the time interval 00 - 14UT on January 12, 1997. The data are about 1.5 min. averages. From top to bottom the panels show the density (cm^{-3}), the bulk speed (km/s), the temperature(K), the GSM components of the IMF(nT), and the clock angle(°). The WIND spacecraft was at (102, -55, -6)Re and (105, -54, -6)Re at 04 and 14UT, respectively. During this interval the clock angle varies over its whole range of definition (bottom panel). B_y is positive in this interval except for four negative excursions at 0745, 0800, 0905, and from 0915 to 1000UT. The average variations of the IMF components and the clock angle are indicated by heavy traces in the figure. The proton data show an average speed solar wind (~530 km/s) of temperature $2*10^5$ K and density around 7 cm^{-3}.

We further subdivide the interval from 0410 to 1400UT into 7 subintervals, separated by vertical guidelines and labelled A - G. Horizontal lines delineate 3 clock angle(θ) ranges (CARs): i) $\theta < 45°$(CAR 1), ii)45 $< \theta < 90°$(CAR 2), and iii) $90< \theta < 180°$(CAR 3).

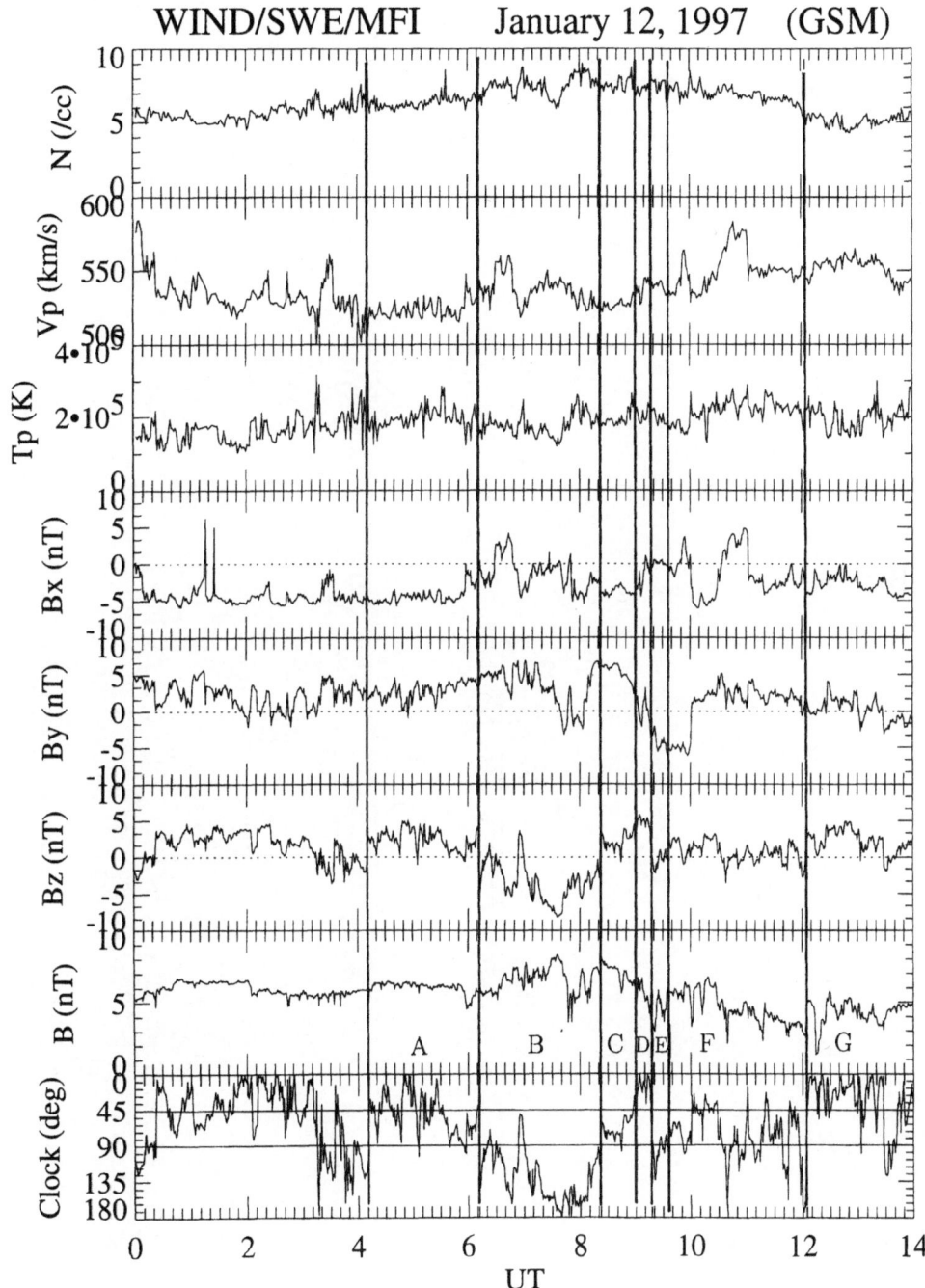

Figure 2. Solar wind proton and IMF data from WIND for the interval 00 - 14UT on January 12, 1997. See text for details.

194

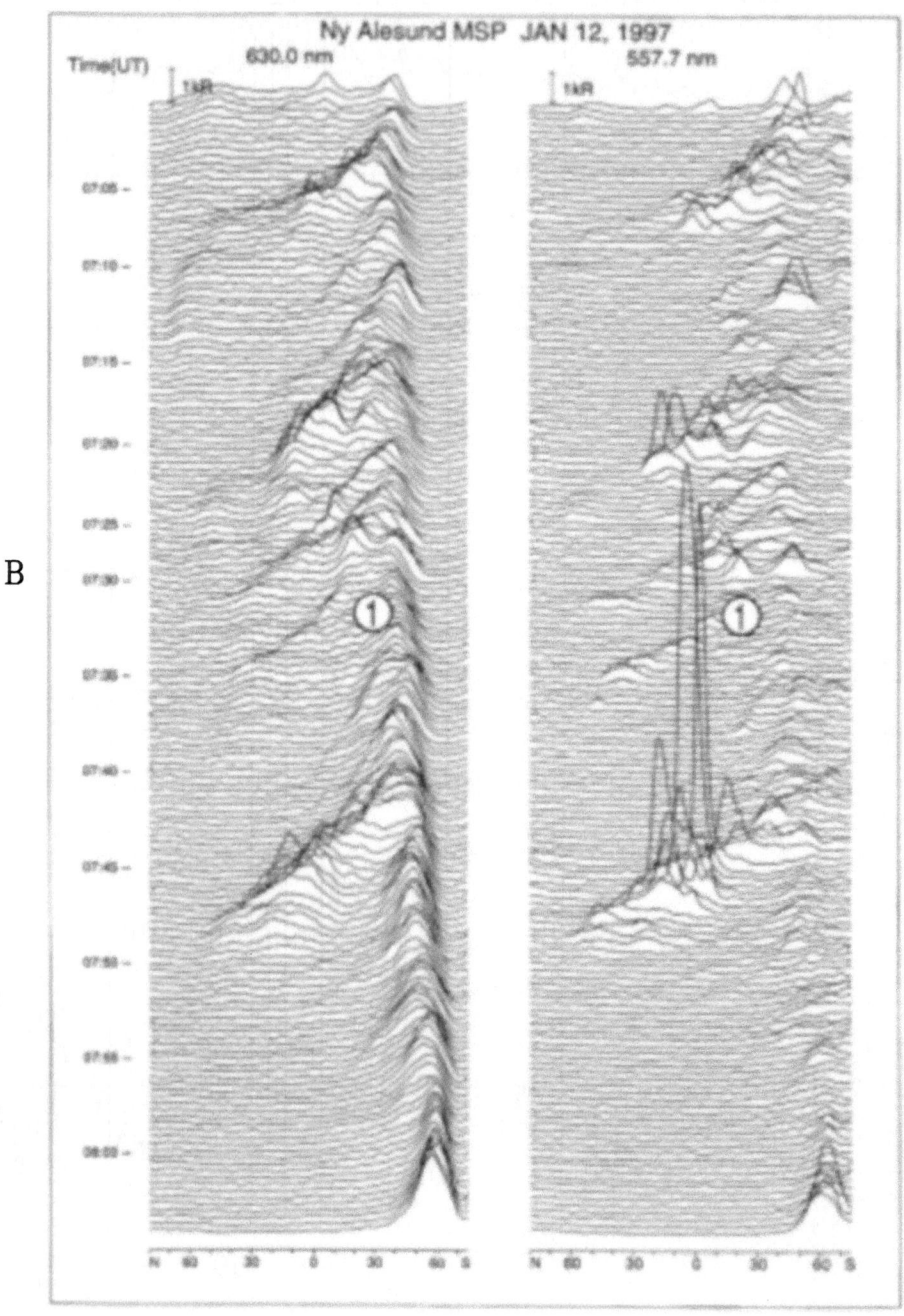

Figure 3. Meridan scanning photometer(MSP) observations from Ny Ålesund for the
0700 - 0800UT interval on January 12, 1997.

We now proceed to investigate the auroras corresponding to these clock angle regimes and transitions between them. Meridian scanning photometer (MSP) data for interval 0600 - 0700 UT (not shown) illustrates auroral changes corresponding to IMF transition from CAR 2 to CAR 3, characterized by the disappearance of the type 2 aurora at high latitudes ($\sim$78 - 79° MLAT) and the appearance of the type 1 aurora at lower latitude, taking place at $\sim$ 0615 UT. Figure 3 shows MSP data for interval B. It shows a sequence of poleward - moving forms (PMAFs), i.e. type 1 aurora. The peak auroral intensity migrated southward from $\sim$ 38 to 60° zenith angle in the south, which corresponds to a latitudinal displacement of $\sim$250km, in this period (0700 - 0800 UT) of large clock angle (CAR 3).

All - sky images of the PMAF at 0740 - 0748 UT (not shown) indicate that the initial brightening of this form has apparently occurred before its appearance at the eastern boundary of the field - of - view at 0740 UT. The subsequent images show the poleward and westward expansions of this form. At 0745 UT this form has detached itself from the background type 1 aurora located on its southern side. The fading phase of this event, occurring beyond 30° north of zenith, is seen at 0747 - 0750 UT. This case example illustrates the northwestward expansion of PMAFs as illustrated in figure 1.

We will next describe auroral observations during subintervals C to F. Figure 2 shows that interval C starts with a northward rotation of the IMF, followed by about $\sim$45 min. of clock angle regime CAR 2. An expanded view of the clock angle variation between 0800 and 1000 UT is shown in Figure 4. The auroral observations in Figure 5 show the transition that occurred in response to the northward rotation of the IMF observed by WIND at 0822 UT shown in the previous Figure. The type 1 aurora is still present and advancing slowly northward towards zenith. A type of bifurcation takes place at the interface between intervals B and C (at 0845 UT). A type 2 aurora appears poleward of the type 1 and subsequently moves northward becoming well separated from type 1. The poleward advance of type 2 continues up to near the end of clock angle regime CAR 2 at which time its separation from type 1 is maximum (not shown). In CAR 2, we again observe a sequence of type 1 auroral forms which move poleward, similar to those observed in interval B. When, then, the clock angle is CAR 1 (in interval D), form 2 is stable in position and weaker compared to interval C (Figure 6). Moreover, poleward - moving forms are no longer seen. A sharp southward rotation of the IMF terminates interval D. Interval E consists of CAR 3 and negative IMF B_y polarity. In this interval (Figure 6) form 2 persists for some time ($\sim$5 min.) and then disappears. Type 1 forms reappear, initially slightly south of zenith, and the equatorward boundary recedes to the south. A sequence of poleward - moving forms is present

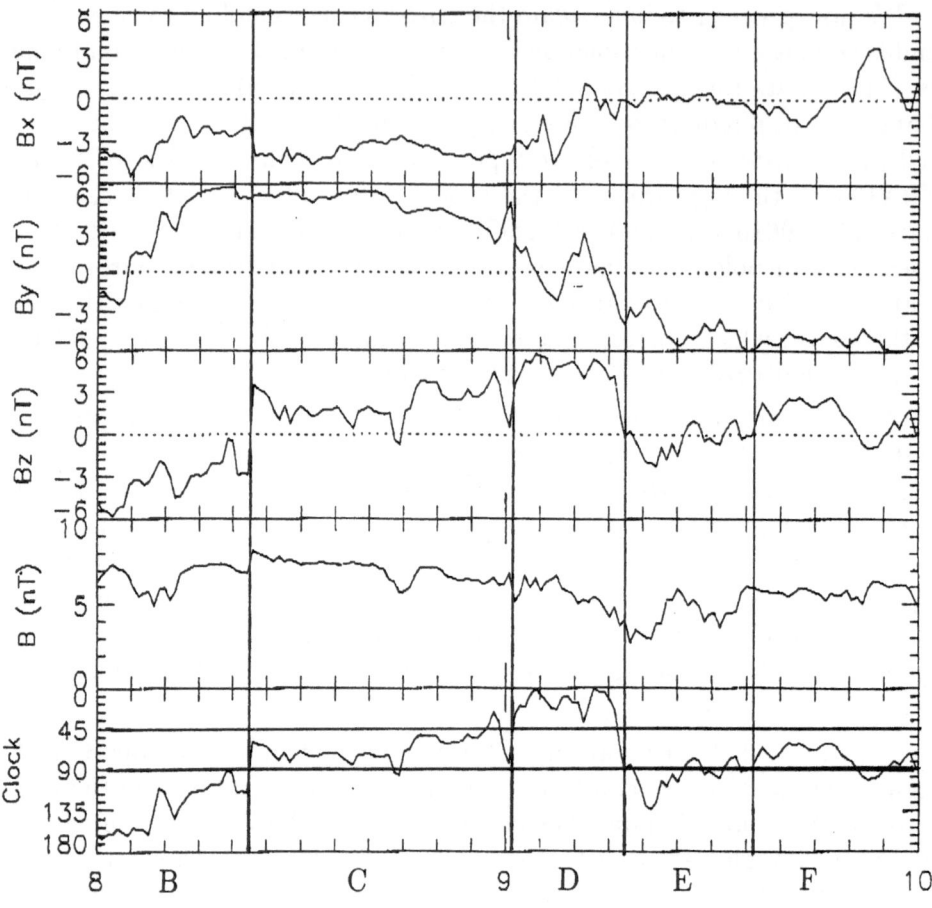

Figure 4. IMF observations from WIND for the 0800 - 1000 UT interval. Subintervals B, C, D, E, and F have been marked in the figure.

and continues until ~1030, i.e., into subinterval F, which is clock angle regime CAR 2. The aurora in interval F, shown in Figure 6, consists of the coexistence of forms 1 and 2, located on either side of zenith. The location of the different forms and the motion pattern of the type 1 forms (sequence of events with a mean repetition period of 7.5 min.) are shown in Figure 7 , which is based on images taken by the an auroral imager in Longyearbyen, Svalbard(74° MLAT).

2.2. CASE 2: JANUARY 4, 1995

Interplanetary data for January 4, 1995 is shown in Figure 8 in the same format as in Figure 2. At 09UT the the position vector of the WIND space-

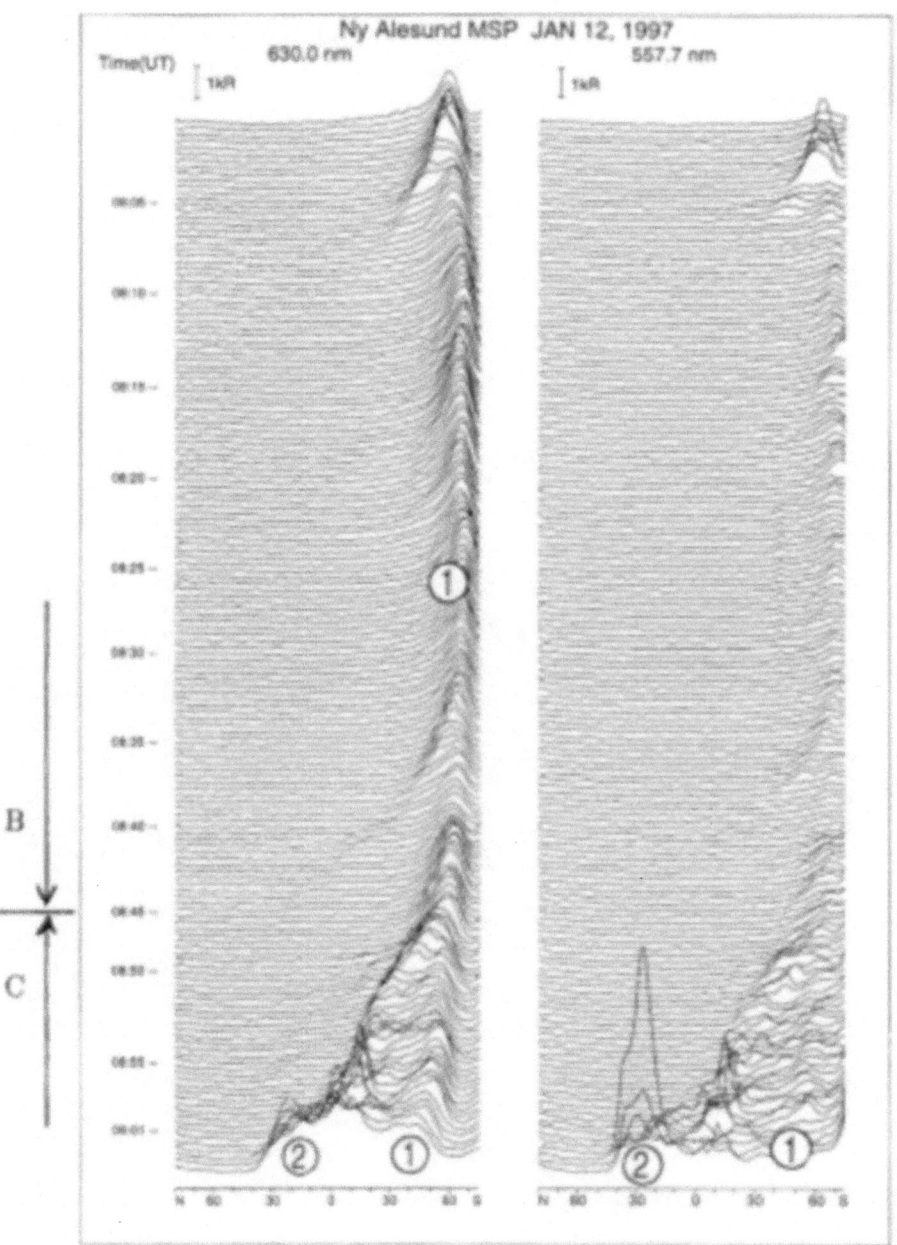

Figure 5. MSP observations for the 0800 - 0900 UT interval. Subintervals B and C have been marked.

198

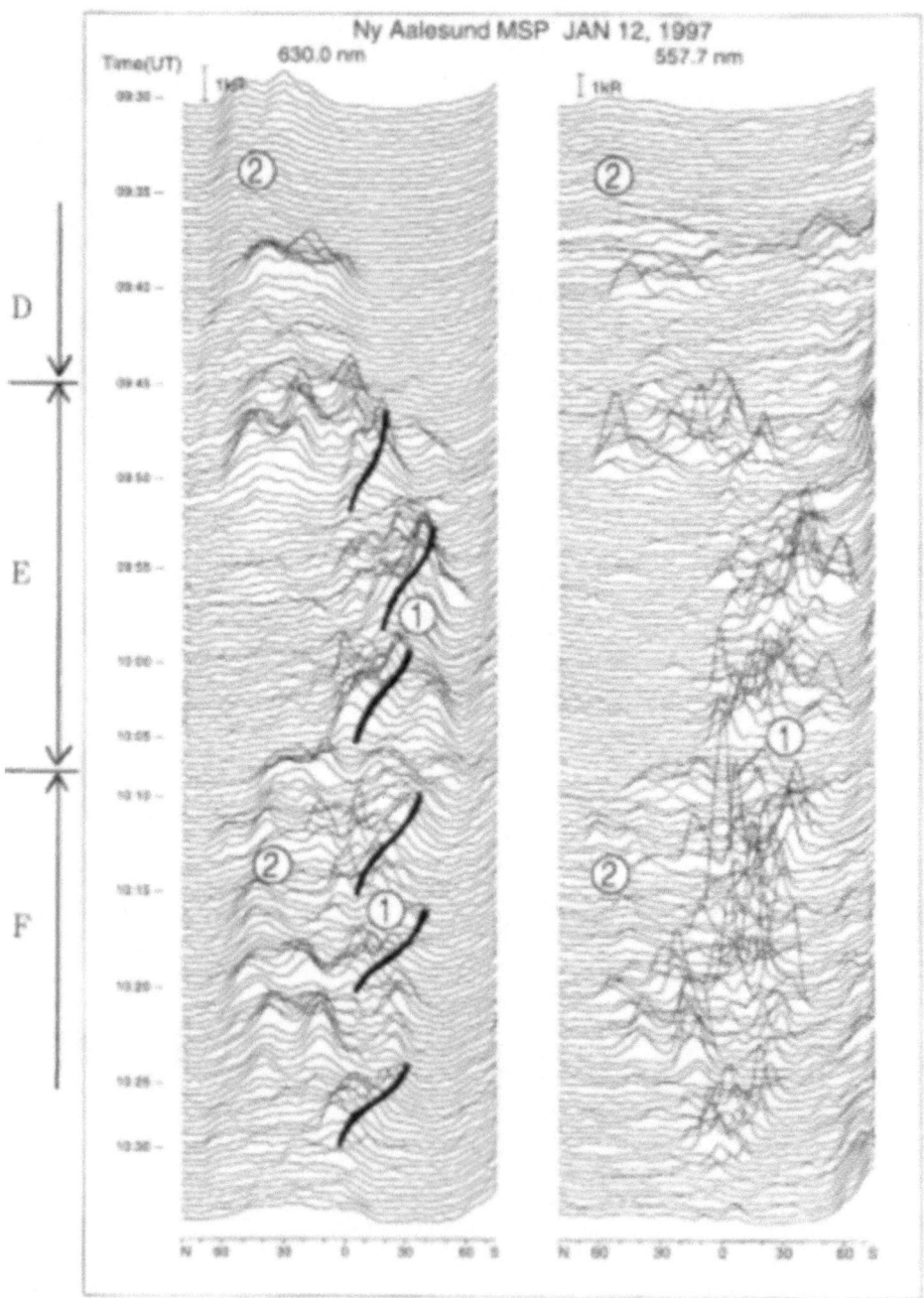

Figure 6.   MSP observations for the 0930 - 1030 UT interval. Subintervals D, E, and F and sequence of north - eastward moving forms have been marked.

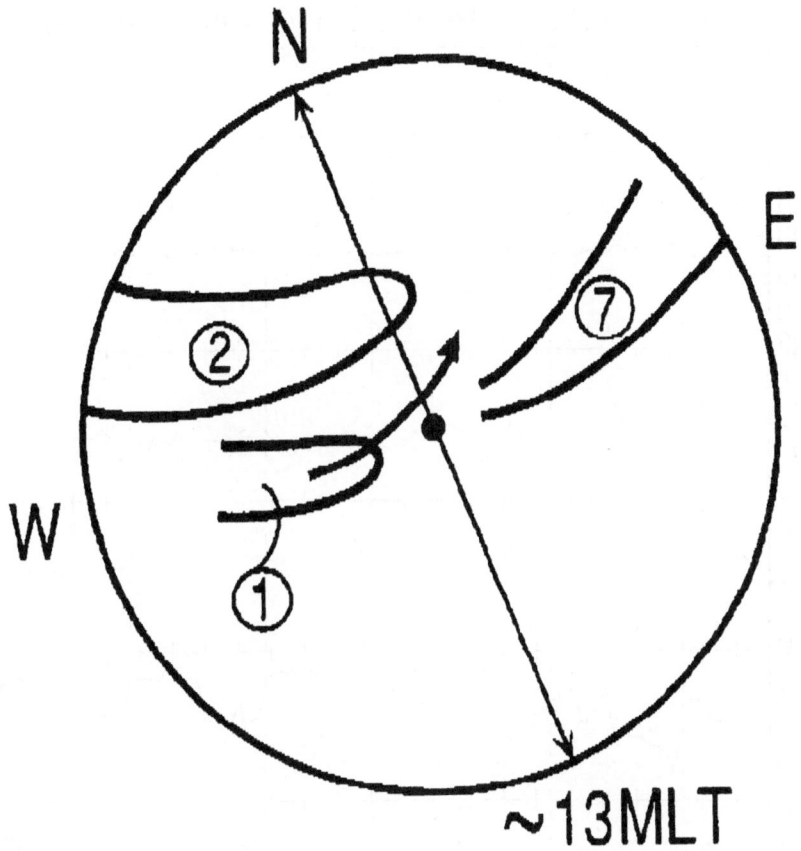

Figure 7. Schematic illustration of auroral forms observed at the transition between intervals D and E on January 12, 1997. Auroral forms of type 1, 2, and 7 have been marked. The motion pattern of the type 1 event sequence is indicated by arrow.

craft was (98, -75, -7)Re. Assuming that the relevant signal propagation time to the ground is the convection speed divided by the distance along the X - axis, we obtain a delay from WIND to the ground of ~15 min. The top three panels indicate a very steady and fast solar wind with little variation about the average values: proton density 4 cm^{-3}, bulk flow ~650 km/s and a temperature about $2*10^5$ K. The dynamic pressure is about 2.5 nPa. The bottom four panels show the magnetic field. Major southward rotations are marked by vertical guidelines. IMF B_y is of negative polarity throughout. All the three above defined clock angle regimes are present in this interval.CAR 1 starts at 0800 UT and continues up to the southward turning at 0813 UT. The corresponding MSP observations at two wavelengths are shown in Figure 9. As in the previous example the first CAR 1

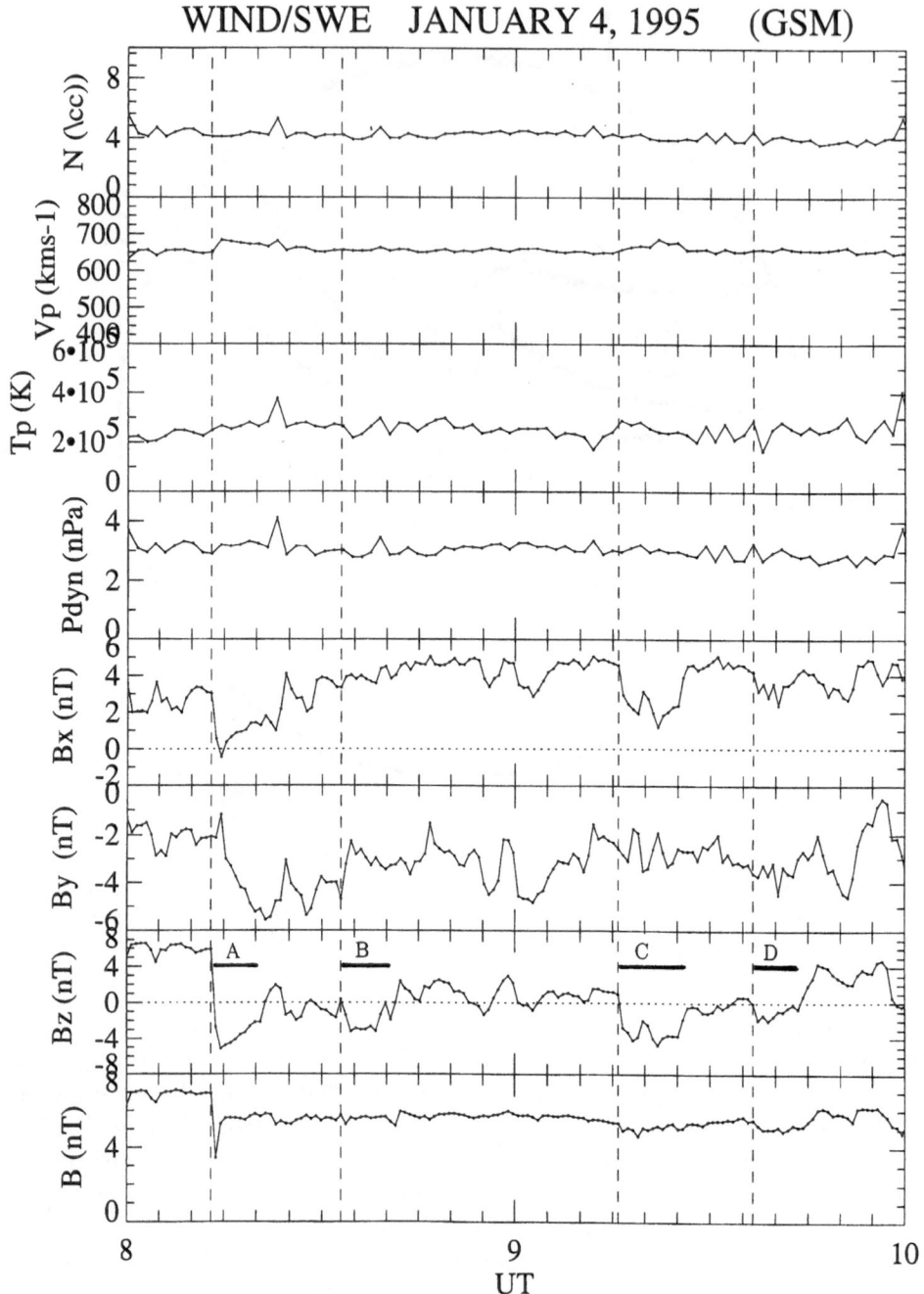

Figure 8. Solar wind proton and IMF observations from WIND for the 0800 - 1000 UT interval on January 4, 1995. Four IMF southward turnings (A, B, C, and D) have been marked.

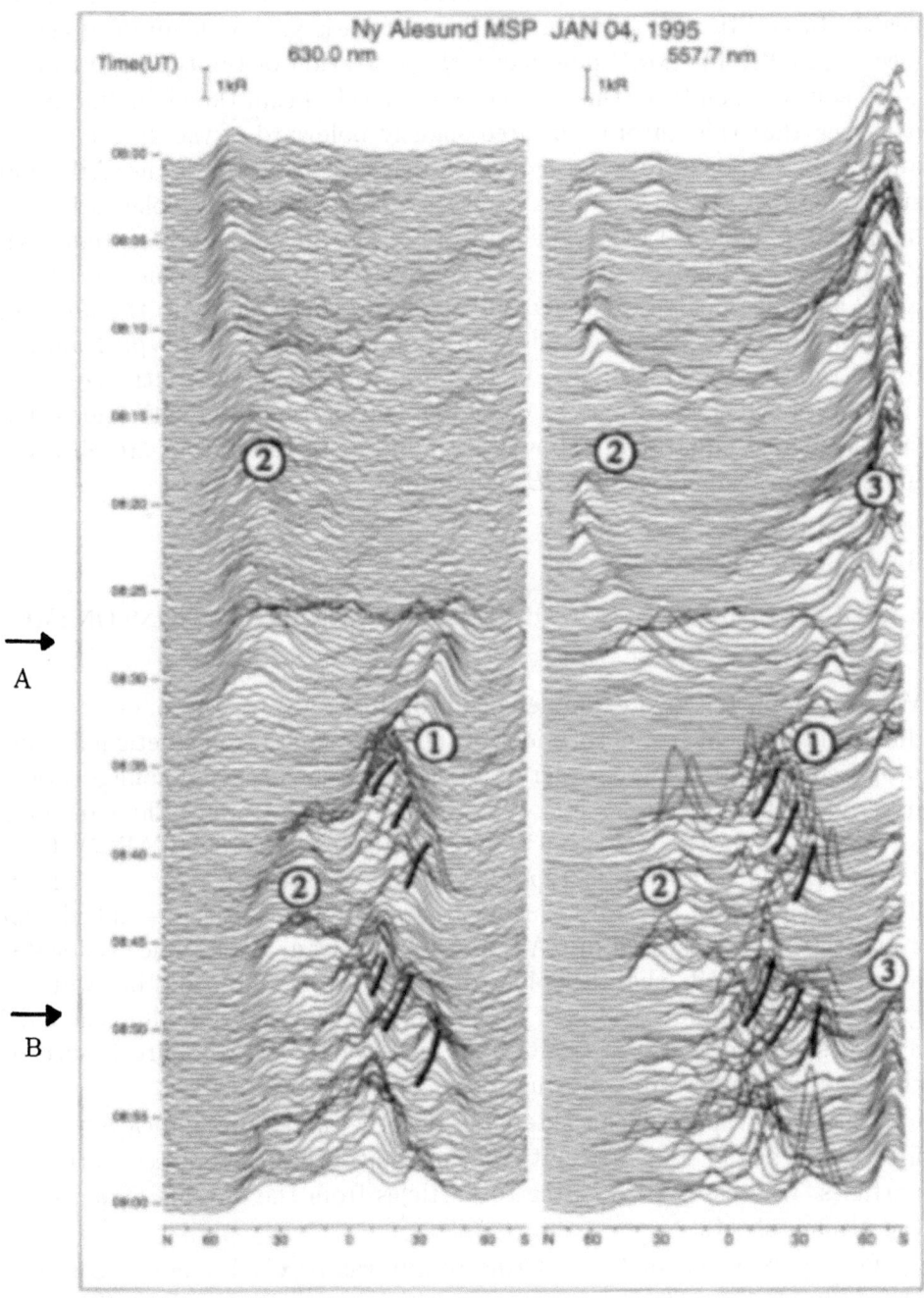

Figure 9. MSP observations for the 0800 - 0900 UT interval on January 4, 1995. Auroral forms 1, 2 and 3 and responses to the IMF southward turnings A, B, and C have been marked.

shows the presence of the type 2 aurora (rayed band) located far north of zenith. In addition, a diffuse green line aurora is seen well south of zenith (see Figure 9). The MSP plot shows the activation of the type 1 aurora at ~45° south of zenith in the red line at 0827 UT. From the green line panel it is seen that this aurora appeared slightly poleward of the diffuse type 3 aurora. In response to a subsequent northward rotation of the IMF (but still being in CAR 3) the type 1 aurora advanced slightly poleward. At a later time (~0838 UT) the type 2 aurora reappears as the IMF enters CAR 2 (at ~0822 UT). In CAR 2, we again observe the coexistence of type 1 and 2 auroras, located on either side of zenith, consistent with the previous example. In this example we also notice new activations of type 1 auroral forms (marked by thick lines) in response to subsequent southward rotations into CAR 3, i.e. events B, C, and D. During the latter activations (not shown) strong equatorward boundary motions of the aurora are observed under the prevailing negative B_y conditions.

3. Discussion

3.1. ON THE DEPENDENCE OF AURORAL CONFIGURATIONS ON IMF CLOCK ANGLE

We have reported two examples of auroral displays in the midday sector during intervals of strong rotations of the interplanetary magnetic field, and proposed an approximate ordering of the different observed configurations in terms of the IMF clock angle. We subdivided the latter into three regimes: a) 0 - 45°(CAR 1); b) 45 - 90°(CAR 2); and c) 90 - 180°(CAR 3). CAR 1 corresponds to aurora of type 2, typically located at ~78 - 79°MLAT. The coexistence of auroral forms located at different latitudes, called type 1 and type 2, is observed in CAR 2. The type 1 aurora is typically located at ~72 - 74°MLAT, slightly poleward of a third auroral form, i.e. a diffuse green line - dominated aurora which we call type 3. At midday in CAR 3 (and sometimes also in CAR 2) we observe the type 1 aurora with its characteristic poleward - moving forms.

The optical spectral characteristics of these three types indicate that the first two types (1 and 2) result from precipitation of magnetosheath type particles, whereas type 3 is due to particles from the dayside extension of the plasma sheet.

Our motivation for the ordering of auroras by clock angle is based on the study of plasma transfer processes at the magnetopause. Broadly speaking, the dominant process at low latitudes is believed to be reconnection when the magnetic shear is high (see Introduction). When it is low, it is believed to be of "viscous type". At high latitudes there are also reconnection - associated processes initiated at the neutral points poleward of the

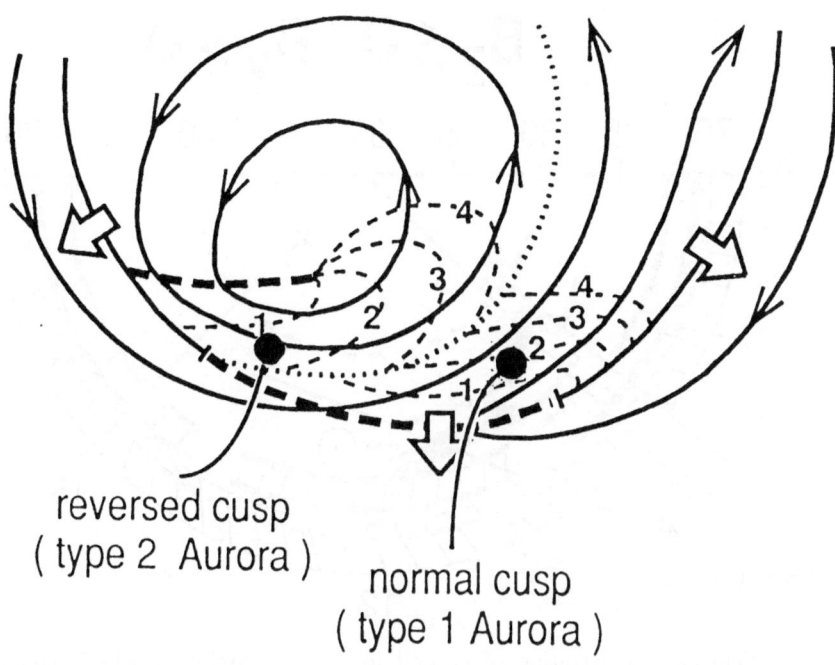

reversed cusp
(type 2 Aurora)

normal cusp
(type 1 Aurora)

Figure 10. Schematic illustration of spatial relationship between type 1 (normal) and 2 (reversed) cusp auroras and merging and lobe convection cells for the intermediate state defined by clock angle regime CAR 2 and IMF $B_y<0$. Ionospheric footprints of reconnection lines at low and high magnetopause latitudes are marked by heavy dashed lines. Numbered lines mark field line sequence, i.e. loci of equal times since reconnection [14].

cusp [10]. But in this case additional conditions come into play, such as seasonal effects (dipole - tilt) and the sunward or earthward tilt of the IMF (component B_x [11]).

Auroral signatures of the reconnection - processes are expected to occur, as confirmed in the present study. Thus, any process which is not dependent on clock angle (or the local magnetic shear), will not fit into the scheme proposed.

We ascribe the simultaneous presence of type 1 and 2 auroral forms, corresponding, respectively, to the normal and reversed cusp ion precipitation to the simultaneous occurrence of reconnection at low and high magnetopause latitudes. This dual mode of solar wind - magnetosphere coupling is supported by the simultaneous presence of the so - called merging and lobe convection cells [13]. A possible relationship between the two cusp au-

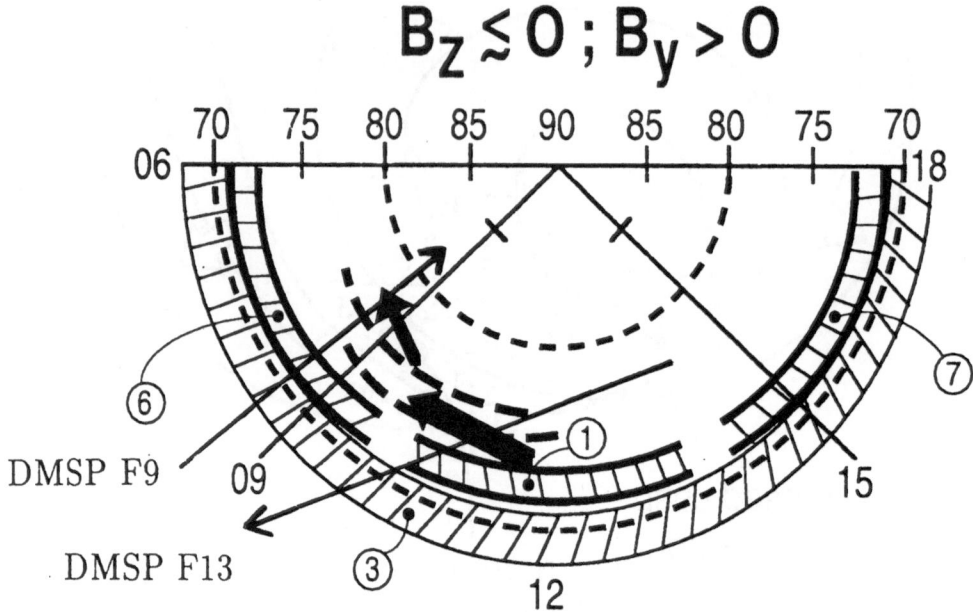

$$B_z \lesssim 0 \; ; \; B_y > 0$$

Figure 11. Schematic illustration of auroral forms and satellite tracks intercepting different phases of poleward - moving auroral forms, i.e. initial and fading phases. DMSP F 13 observed a stepped cusp ion dispersion signature, corresponding to a sequence of three PMAFs.

roral forms and the convection cells in the intermediate state (CAR 2, IMF $B_y < 0$) is indicated in Figure 10. Further details on this are given in [14]. Initial observations of the correspondence between type 1 and 2 auroral forms and the merging and lobe convection cells, respectively, have been reported by [15, 16, 17].

3.2. POLEWARD - MOVING AURORAL FORMS

The type 1 aurora is associated with a sequence of short - lived auroral forms(rayed bands) whose location and motion are regulated by the east - west component of the IMF. They are most frequently observed prenoon for positive B_y and vice versa [18]. In the present case example (IMF $B_y > 0$) the initial brightening was often observed on the postnoon side (at the boundary of the field - of - view) followed by a northwestward expansion, as illustrated in Figure 1. As documented in coordinated ground and satellite studies the fading phase is observed to occur in the regime of mantle precipitation ~ 5 - 10 min. after the initial brightening [19]. Two satellite passes which intercepted such forms are marked in Figure 11. Whereas the F9 pass (Jan. 12, 1991) crossed one event at ~ 09 MLT as it faded, DMSP F13 overflew three events at different phases of their evolution [20]. The

205

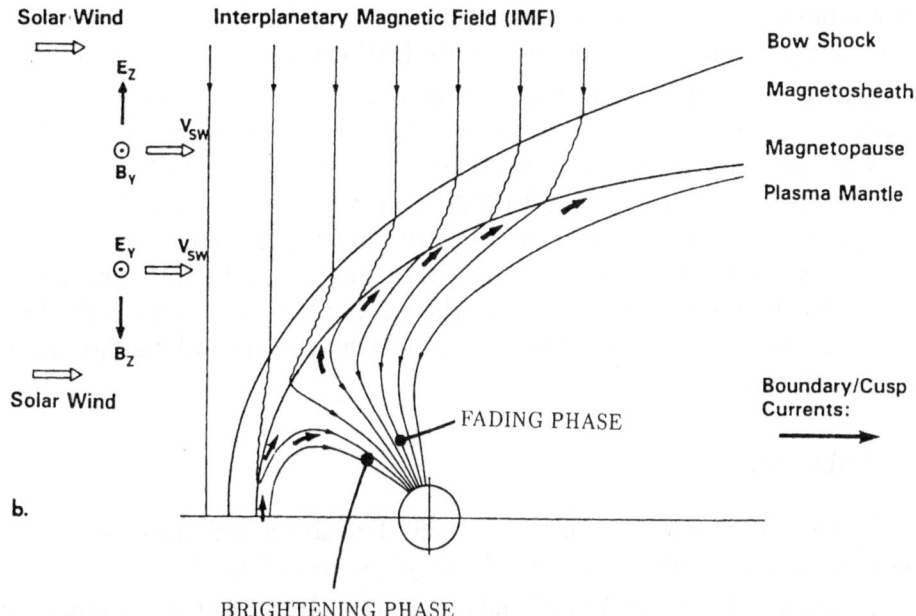

Solar Wind   Interplanetary Magnetic Field (IMF)   Bow Shock

Magnetosheath

Magnetopause

Plasma Mantle

Boundary/Cusp
Currents:

FADING PHASE

b.

BRIGHTENING PHASE

Figure 12. Schematic illustration of the open magnetosphere in the noon - midnight meridian plane with precipitations corresponding to brightening(along open LLBL field lines) and fading(mantle field lines) phases of PMAFs (adapted after [12]).

cusp ion precipitation was of the type commonly known as "staircase" or "stepped" cusp [21, 22, 23]. Corresponding to each form we observed a low - energy cutoff which was lower for higher latitude. The staircase type ion dispersion has been interpreted in terms of pulsed reconnection [24, 25]. Earlier, the PMAFs had also been interpreted in terms of pulsed reconnection [26, 27]. Thus, as argued by Farrugia et al. [20] the association of the two independent signatures strenghtens the reconnection - interpretation first proposed by Sandholt et al. [26, 28]. The combined ground and satellite studies show that these events contain a level of detailed structure in the electron precipitation and associated field - aligned currents. The brightening on newly - opened field lines and the fading phase along mantle field lines for the B_y positive case are marked in Figure 12.

For negative IMF B_y polarity, with the forms now preferentially expanding north - east, the associated field - aligned currents are now reversed [29]. In this case the events are most frequently observed on the post - noon side. The event sequence gives rise to a stepwise recession of the equatorward boundary of the aurora towards the south, as illustrated in Figure 9 (see also [30]). We have now documented that such events are activated every time the IMF rotates sharply southward. In the January 4, 1995 case the auroral boundary motion amounted to ~100 - 200 km in 5 - 10 min. events, i.e. more pronounced than in the steady IMF cases.

Acknowledgments. The auroral observation program in Svalbard is supported by the Norwegian Research Council(Norges forskningsråd) and the Norwegian Polar Research Institute. This work is partially supported by NASA grant NAG5-2834 and UNIS grant 9/963. We thank Torleif Sten for preparation of the optical instruments and the principal investigators on the SWE investigation, Dr. K. Ogilvie, and the MFI investigation on WIND, Dr. R. P. Lepping, for use of the data. Thanks are due to Dr. Stan Cowley for discussions on convection in relation to auroral configurations in the cusp region.

4. References

1. Fairfield, D. H. and Cahill, L. J. (1966) Transition region magnetic field and polar magnetic disturbances, *J. Geophys. Res.*, **71**, 155.

2. Arnoldy, R. L. (1971) Signature in the interplanetary medium for substorms, *J. Geophys. Res.*,**76**, 5189.

3. Dungey, J. W.(1961) Interplanetary magnetic field and the auroral zones, *Phys. Rev. Lett.*, **6**, 47.

4. Freeman, M. P., Farrugia, C. J., Burlaga, L. F., Hairston, M. R., Greenspan, M. E., Ruohoniemi, J. M., and Lepping, R. P. (1993) The interaction of a magnetic cloud with the Earth: Ionospheric convection in the northern and southern hemispheres for a wide range of quasi - steady interplanetary magnetic field conditions, *J. Geophys. Res.*, **98**, 7633.

5. Knipp, D. J., et al. (1993) Ionospheric convection response to a magnetic cloud: A case study for 14 January 1988, *J. Geophys. Res.*, **101**, 19273.

6. Greenwald, R. A., Bristow, W. A., Sofko, G. J., Senior, C., Cerisier, J.-C., and Szabo, A., (1995) Super Dual Auroral Radar Network radar imaging of dayside high - latitude convection under northward interplanetary magnetic field: Toward resolving the distorted two - cell versus multicell controversy, *J. Geophys. Res.*, **100**, 19661.

7. Phan, T. - D., Paschmann, G., and Sonnerup, B. U. O. (1996) Low - latitude dayside magnetopause and boundary layer for high magnetic shear, 2. Occurrence of magnetic reconnection, *J. Geophys. Res.*, **101**, 7817.

8. Ogilvie, K. W., Chornay, D. J., Fitzenreiter, R. J., Hunsaker, F., Keller, J. Lobell, J., Miller, G., Scudder, J. D., Sittler, E. C., Torbert, R. B., Bodet, D., Needell, G., Lazarus, A. J., Steinberg, J. T., Tappan, T. H., Mavretic, A., and Gergin, E. (1995) SWE, A comprehensive plasma instrument for the WIND spacecraft, *Space Science Rev.*, **71**, 55.

9. Lepping, R. L., Acuna, M. H., Burlaga, L. F., Farrell, W. M., Slavin, J. A., Schatten, K. H., Mariani, F., Ness, N. F., Neugebauer, F. M., Whang, Y. C., Byrnes, J. B., Kennon R. S., Panetta, P. V., Scheifele, J. and Worley, E. M. (1995) The WIND magnetic field investigation, *Space Science Rev.*,**71**, 207.

10. Gosling, J. T., Thomson, M. F., Bame, S. J., Elphic, R. C., and Russell, C. T. (1991) Observations of reconnection of interplanetary and lobe magnetic field lines at the high - latitude magnetopause, *J. Geophys. Res.*, **96**, 14097.

11. Crooker, N. U. (1992) Reverse convection, *J. Geophys. Res.*, **97**, 19363.

12. Stauning, P., Friis-Christensen, E., Rasmussen, O. and Vennerstrøm, S. (1994) Progressing polar convection disturbances: Signature of an open magnetosphere, *J. Geophys. Res.*, **99**, 11303.

13. Reiff, P. H., and Burch, J. L. (1985) By - dependent dayside plasma flow and Birkeland currents in the dayside magnetosphere, 2, a global model for northward and southward IMF, *J. Geophys. Res.*, **90**, 1595.

14. Sandholt, P. E., Farrugia, C. J., Moen, J., Cowley, S. W. H. (1997) Dayside auroral configurations: Responses to southward and northward rotations of the interplanetary magnetic field *J. Geophys. Res.*, submitted Sept. 1997.

15. Sandholt, P. E., Farrugia, C. J., Øieroset, M., Stauning, P., and Cowley, S. W. H. (1996) Auroral signature of lobe reconnection, *Geophys. Res. Lett.*, **23**, 1725.

16. Sandholt, P. E., Farrugia, C. J., Stauning, P., Cowley, S. W. H., and Hansen, T. (1996) Cusp/cleft auroral forms and activities in relation to ionospheric convection: Responses to specific changes in solar wind and interplanetary magnetic field conditions, *J. Geophys. Res.*, **101**, 5003.

17. Øieroset, M., Sandholt, P. E., Denig, W. F., and Cowley, S. W. H. (1997) Northward interplanetary magnetic field cusp aurora and high - latitude magnetopause reconnection, *J. Geophys. Res.*, **102**, 11349.

18. Karlson, K., Øieroset, M., Moen, J., and Sandholt, P. E. (1996) A statistical study of flux transfer event signatures in the dayside aurora: The IMF B_y - related prenoon - postnoon asymmetry, *J. Geophys Res.*, **101**, 59.

19. Sandholt, P. E., Moen, J., Rudland, A., Opsvik, D., Denig, W. F., and Hansen, T. (1993) Auroral event sequences at the dayside polar cap boundary for positive and negative interplanetary magnetic field By, *J. Geophys. Res.*, **98**, 7737.

20. Farrugia, C. J., Sandholt. P. E., Denig, W. F., and Torbert, R. B. (1997) Observation of a correspondence between poleward - moving auroral forms and stepped cusp ion precipitation, J. Geophys. Res., in press.

21. Newell, P. T., and Meng, C.-I. (1991) Ion acceleration at the equatorward edge of the cusp: Low altitude observations of patchy merging, *Geophys. Res. Lett.*, **18**, 1829.

22. Cowley, S. W. H., Freeman, M. P., Lockwood, M., and Smith, M. F. (1991) Ionospheric signature of flux transfer events, in C. I. Barron(ed.), *Cluster Dayside Polar Cusp*, Eur. Space Agency Spac. Publ., ESA SP - 330, pp. 105 - 112.

23. Escoubet, C. P., Smith, M. F., Fung, M. F., Anderson, P. C., Hoffmann, R. A., Basinska, E. M., and Bosqued, J. M. (1992) Staircase ion signature in the polar cusp: A case study, *Geophys. Res. Lett.*, **19**, 1735.

24. Lockwood, M. and Smith, M. F.(1992) The variation of reconnection rate at the dayside magnetopause and cusp ion precipitation, *J. Geophys. Res.*, **97**, 14841.

25. Lockwood, M. and Smith, M. F.(1994) Low - altitude signatures of the cusp and flux transfer events, *J. Geophys. Res.*, **99**, 8531.

26. Sandholt, P. E., Farrugia, C. J., Øieroset, M., Stauning, P., and Cowley, S. W. H. (1986) Signatures in the dayside aurora of plasma transfer from the magnetosheath, *J. Geophys. Res.*, **91**, 10063.

27. Sandholt, P. E., M. Lockwood, T. Oguti, Cowley, S. W. H. , Freeman, K. S. C., Lybekk, B., Egeland, A., and Willis, M. (1990) Midday auroral breakup events and related energy and momentum transfer from the magnetosheath, *J. Geophys. Res.*, **95**, 1039.

28. Sandholt, P. E., Egeland, A., Holtet, J. A. , Lybekk, B., Svenes, K., Åsheim, S., and Deehr, C. S. (1985) Large - and small - scale dynamics of the polar cusp, *J. Geophys. Res.*, **90**, 4407.

29. Øieroset, M., Luhr, H., Moen, J. , Moretto, T., and Sandholt, P. E. (1996) Dynamical auroral morphology in relation to ionospheric plasma convection and geomagnetic activity: Signatures of magnetopause X - line dynamics and flux transfer events, *J. Geophys. Res.*, **101**, 13275.

30. Sandholt, P. E., Farrugia, C. J., Øieroset, M., Stauning, P., and Denig, W. F. (1997) Auroral activity associated with unsteady magnetospheric erosion: Observations on December 18, 1990, *J. Geophys. Res.* , in press.

AURORAL AND GEOMAGNETIC SIGNATURES OF FLUX TRANSFER EVENTS AND ASSOCIATED CURRENT SYSTEMS FOR POSITIVE AND NEGATIVE IMF B_Y

M. ØIEROSET

International Space Science Institute, Berne, Switzerland

P. E. SANDHOLT

Department of Physics, University of Oslo, Oslo, Norway

Abstract. We interpret certain characteristic sequences of ionospheric poleward moving events near the cusp, seen simultaneously in dayside aurora and ground magnetometer observations, as flux transfer event (FTE) signatures and propose associated FTE current systems. This current system consists of a central filament or sheet-like field-aligned current, located on newly opened field lines with oppositely directed return currents on the poleward and equatorward sides. The propagation speed is found to be 1-2 km/s, and the size of the current system 2-300 km in the north-south direction. The FTE current system has opposite polarities corresponding to positive and negative IMF B_y component. The events are typically observed in the prenoon sector for positive B_y and postnoon during negative B_y conditions. In this study we present two case examples, one of each category.

1. Introduction

Russell and Elphic [1] interpreted transient magnetic field signatures observed near the magnetopause in terms of pulsed magnetic reconnection between the interplanetary magnetic field (IMF) and the Earth's field at the low-latitude magnetopause. The observed magnetic data signature was named flux transfer event (FTE) and this has later also become the term used for the candidate physical process, that is, pulsed magnetic reconnection.

FTEs are expected to be accompanied by bursts of particles transferred from the magnetosheath to the magnetosphere with a few minutes repetition period [2]. These particle bursts can then be transferred down along discrete flux tubes gener-

J. Moen et al. (eds.), Polar Cap Boundary Phenomena, 209–218.

ated by the pulsed reconnection process to the cusp ionosphere [3]. The expected ionospheric signature is thus a burst of particle precipitation in the cusp region, moving northward as the reconnected field lines are convecting antisunward [4, 5].

Sandholt *et al.* [6, 7, 8] have interpreted sequences of poleward moving auroral events at cusp latitudes as evidence for particle precipitation in FTEs. These events are mainly seen when IMF B_z is negative, they have a recurrence rate of 5-8 min, and are observed at the cusp equatorward edge, moving poleward. Later it has been shown that these poleward moving auroral events have an east-west motion and a prenoon-postnoon asymmetry related to the IMF B_y component [9, 10]. This has further supported the interpretation of these events in terms of FTE signatures, because the events were shown to expand westward into the prenoon sector for positive IMF B_y and eastward into the postnoon sector for negative B_y, consistent with predictions associated with the asymmetry of the reconnection geometry [11, 12]. The ionospheric prenoon-postnoon asymmetry is also consistent with observations by Watanabe *et al.* [13].

FTEs should be accompanied by field-aligned currents which close by Hall currents in the ionosphere [14, 15, 16]. These Hall currents will induce geomagnetic disturbances. From such current systems McHenry and Clauer [17] predicted ground magnetic signatures of the order of 100 to 150 nT and concluded that ground magnetometers can be used to study FTEs if there is a chain of magnetometers under the events. However, ground magnetic signatures of FTEs are in general not as easily identified as the optical footprints, and the combination of optical and ion drift observations with the geomagnetic observations has been shown to be useful (e.g., Sandholt *et al.* [7, 8], Lockwood *et al.* [18, 4]). Øieroset *et al.* [19, 20] used this combination of data sets to infer the detailed dynamical structure of the current system for the poleward moving events. Both prenoon and postnoon cases were presented and it was found that the inferred field-aligned current systems were qualitatively similar but had opposite polarity in the two sectors, corresponding to different IMF B_y polarity, as expected from theory (e. g., Lee [21]).

In this paper we will summarize the results from Øieroset *et al.* [19, 20] and discuss the two data sets in a larger perspective. These data sets consist of auroral and geomagnetic observations of poleward-moving events. The first set is from the prenoon sector during positive IMF B_y conditions, and the second set from the postnoon sector for negative B_y conditions. The inferred current system is then presented and discussed for both cases.

2. Instrumentation

We here present dayside auroral observations from Ny Ålesund (NAL), Svalbard (78.9°N, 11.9°E, 75.9° MLAT). The optical instrument used was a meridian scanning photometer (MSP). The MSP scans along the magnetic north-south meridian and measures the intensity for two different wavelengths of auroral emissions, the red line (630.0 nm) and the green line (557.7 nm). At 630.0 nm the MSP covers the latitude range from approximately 70° to 80° MLAT.

The ground magnetometer data presented are from the International Monitor for Auroral Geomagnetic Effects (IMAGE) magnetometer network in Svalbard and Scandinavia. In addition we use Greenland magnetometer data to investigate the east-west motion of the observed events and to estimate the large-scale ionospheric convection pattern.

3. Case Study I: Prenoon Events and Positive IMF B_y Conditions

Our prenoon case is from January 3, 1995, 0730-0755 UT, near 11 MLT at our station at Ny Ålesund. Wind IMF data show that this time interval corresponds to a period of negative B_z and positive B_y [20]. Data from the DMSP F11 satellite, crossing the Ny Ålesund meridian near 0735 UT, show that the ionospheric convection was 1-2 km/s and strongly antisunward, and that the poleward moving events were initiated at the equatorward boundary of cusp-type precipitation. The equivalent convection patterns indicate northwestward convection at the Svalbard stations, consistent with the DMSP ion drift measurements [20].

The Ny Ålesund MSP optical data are shown in Figure 1. The red line (630.0 nm) is shown in the left panel and the green line (557.7 nm) in the right. The MSP zenith angle is given at the bottom of each panel, with north to the left, south to the right, and zero zenith angle in the middle. The universal time (UT) is given on the vertical axis.

Before ∼0730 UT there was a broad belt of auroral emission extending several degrees MLAT to the north and south of Ny Ålesund. Then between 0730 and 0740 UT the auroral morphology changed with the disappearance of the high-latitude aurora and the brightening of a latitudinally narrow form south of Ny Ålesund, near 71° MLAT. A multiple structure of auroral emission, denoted 1a and 1b appeared at 0730 and 0732 UT and moved poleward. Three other poleward moving events (2, 3, and 4) followed, initiated at 0736, 0743, and 0748 UT. If we interpret event 1a and 1b as a multiple structure within one event, we get a recurrence period of 5-7 min for the poleward moving events. The average poleward propagation velocity was ∼1 km/s. It is noted that the initial brightening of the optical events may occur outside the field-of-view of the scanning photometer.

Figure 2 shows ground magnetic data (H and D component) from selected stations in the IMAGE magnetometer network. The six stations between SOR (67.1°MLAT) and NAL (75.9°MLAT) are included. These stations are situated near the magnetic meridian through Ny Ålesund. Near 0728 UT there was a major change in the level of magnetic perturbation, marked by positive H component deflections at all stations from BJN to NAL. The 0730-0748 UT interval was characterized by three large-amplitude bipolar (+/-) H component deflections (marked with the lines 1, 2, and 3) from BJN to NAL. These events are poleward moving,

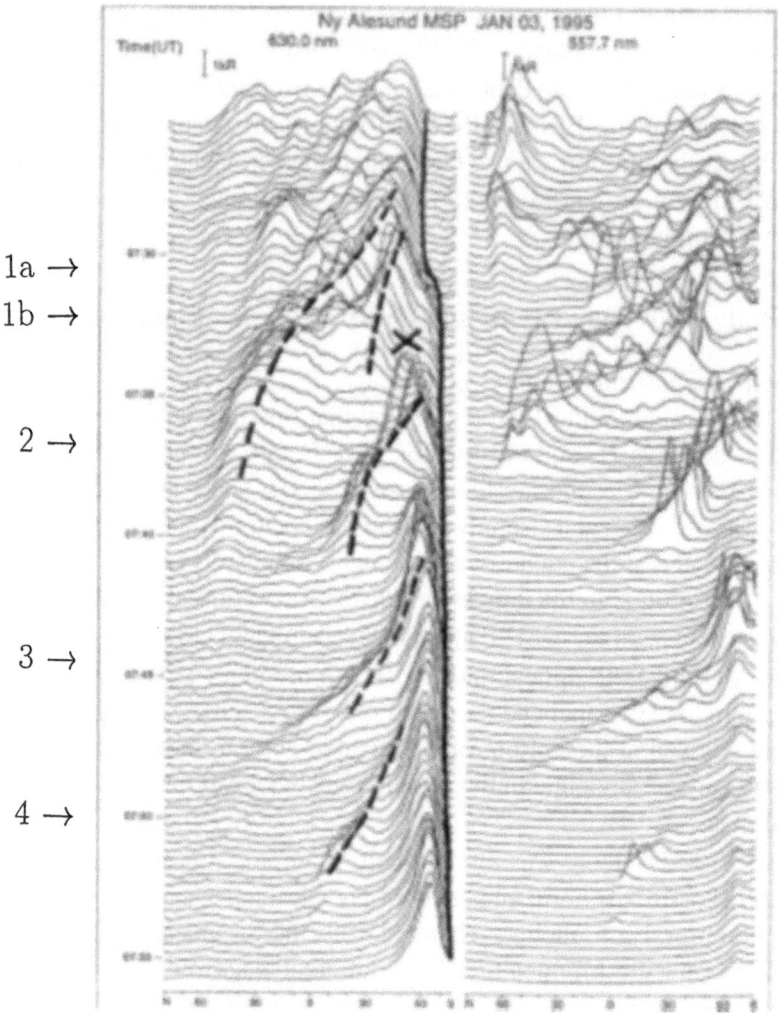

Figure 1. January 3, 1995 Ny Ålesund MSP data from the prenoon sector. The onset of the poleward moving events are marked by arrows. The equatorial boundary of the aurora and the optical events are given by the lines. From Øieroset *et al.* [20].

as indicated by the slope of the lines marking the events. The events are initiated near BJN/HOP, exhibiting positive H component peaks at 0729, 0735, and 0744 UT, almost simultaneously with the first three optical events. The magnetic events move poleward with a velocity of 1-2 km/s, comparable with the poleward propagation speed of the optical events. A precise timing of the magnetic and the optical events shows that the emission maxima of the optical events appeared at a station at the time of the steep negative slope of the H component deflection

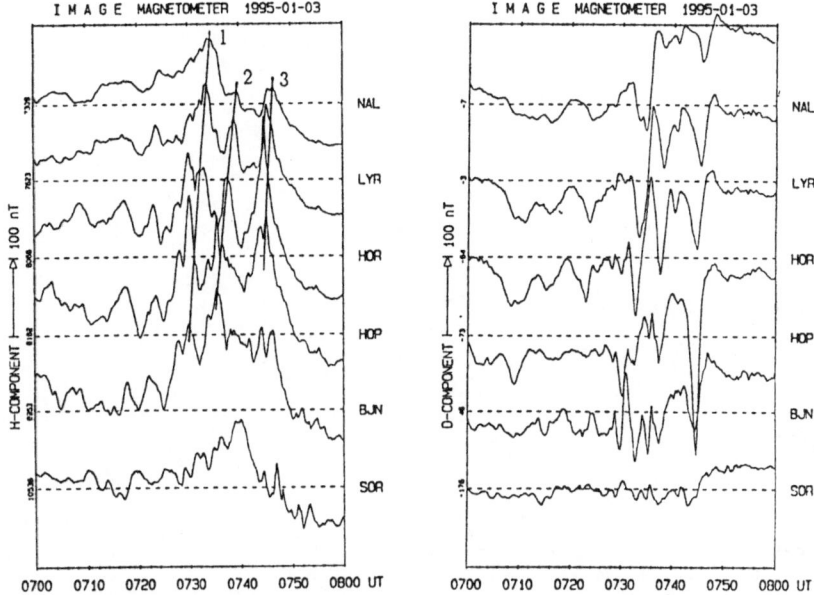

Figure 2. January 3, 1995 IMAGE data, H and D components. The poleward moving events are marked. From Øieroset *et al.* [20].

following the initial positive deflection. This indicates that the auroral event was located between the maxima of the eastward and westward Hall currents during the poleward propagation.

Events 1 and 3 were also identified in geomagnetic data from the Greenland east coast (not shown), 1-2 hours MLT earlier, where they appeared 1-2 min later than at NAL. Thus the events moved north-westward into the prenoon sector with a velocity 1-2 km/s.

4. Case Study II: Postnoon Events and Negative IMF B_y Conditions

The postnoon case is from January 14, 1994, and the event sequence occurred between 1050 and 1120 UT. This corresponds to approximately 14 MLT. Unfortunately, no solar wind data are available, but the IMF conditions can be estimated from ground-based observations of equivalent ionospheric convection. The equivalent convection plots show a distorted two-cell convection pattern with a strong zonal component in the cusp region [19]. This is a typical situation when IMF B_z is negative. The large prenoon cell and the smaller postnoon cell indicate negative B_y conditions [22].

MSP data for the postnoon case are shown in Figure 3. The course of events are similar to the prenoon case shown in Figure 1. A rapid southward shift of the aurora occurred at 1050 UT. Five poleward-moving auroral events, each event initiated at the equatorward boundary of the preexisting luminosity, were observed

214

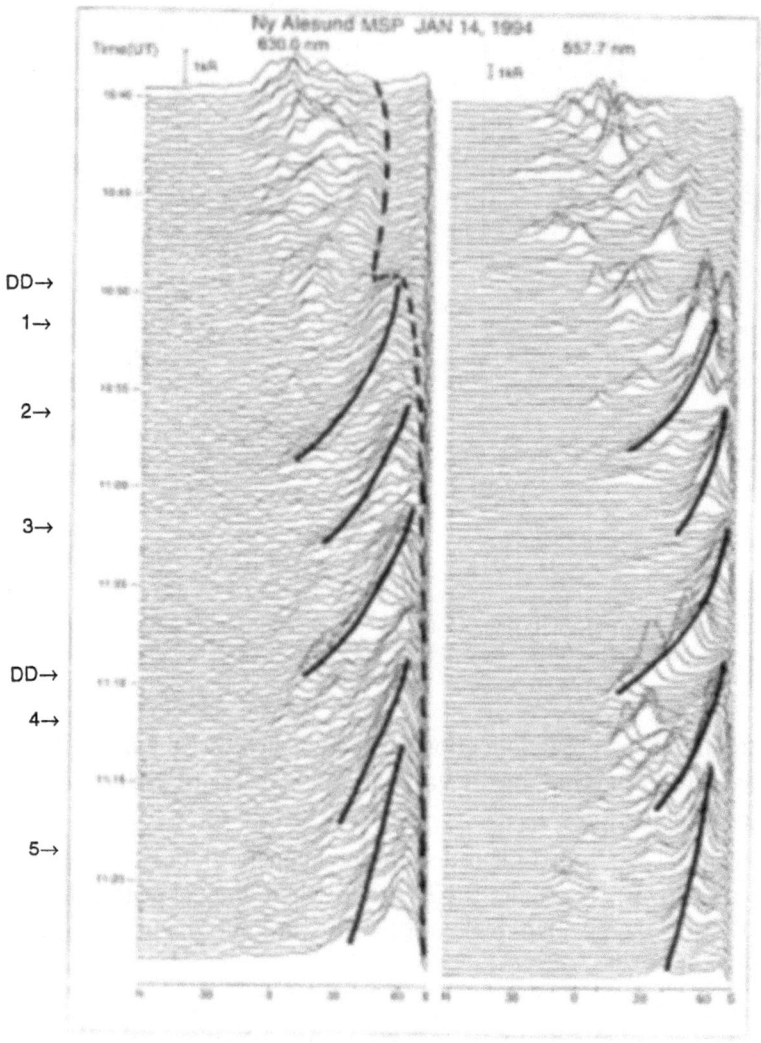

Figure 3. January 14, 1994 Ny Ålesund MSP data from the postnoon sector. The onset of the poleward moving events are marked by arrows. The poleward motion of the events are marked by the solid lines, and the equatorward boundary of the aurora with the dotted line. From Øieroset *et al.* [19].

at 1052, 1056, 1102, 1112, and 1118 UT. The average repetition period is 5 min and the poleward propagation velocity ~2 km/s.

The corresponding IMAGE magnetometer data in Figure 4 show only small variations in the H and D components until 1047 UT, when a significant negative H component deflection occurred at all stations from HOP to LYR. Five bipolar (-/+) H component deflections (marked with the lines 1 to 5) were observed, initiated

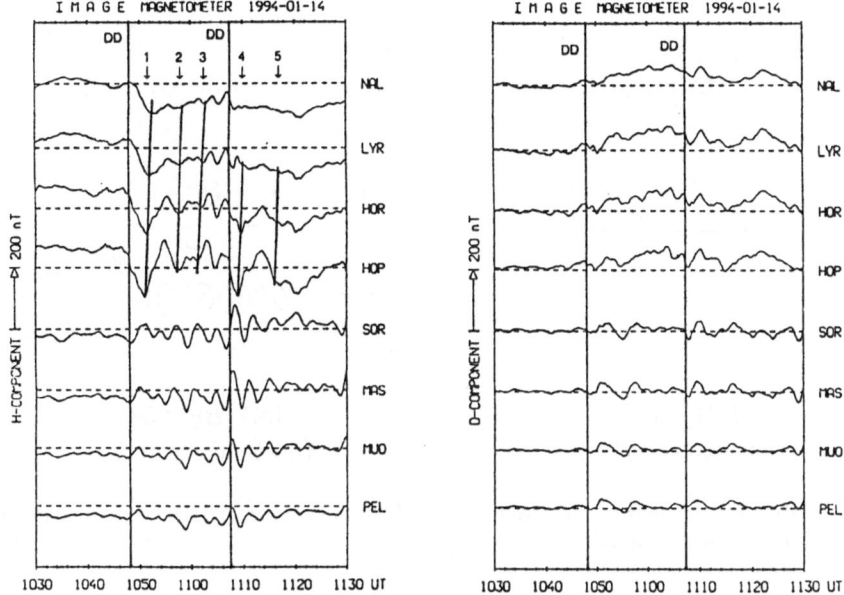

Figure 4. January 14, 1994 IMAGE data, H and D components. The poleward moving events are marked. From Øieroset *et al.* [19].

at 1051, 1057, 1101, 1109, and 1116 UT. These events moved poleward with a velocity of ∼2 km/s.

The magnetic signature observed on the low latitude IMAGE stations (PEL-SOR) differ significantly from those at Svalbard. At these latitudes Pc5 pulsations start at 1047 UT. These pulsations have a mean period of 250 s (4 mHz). There is a clear phase relation between the variations observed on the high-latitude stations and the pulsations in the south. Major negative deflections in the north are accompanied by positive peaks in the south.

5. Discussion

The poleward moving auroral events described above are interpreted in terms of FTE signatures. This interpretation is based on earlier studies of similar events [6, 7, 8, 9, 10]. Such events are known to be initiated at the cusp equatorward edge, move poleward, and also eastward or westward depending on the IMF B_y polarity. The motion pattern and repetition period of the events described here are consistent with the previous observations and interpretation.

For both the prenoon and postnoon cases presented above, the poleward moving auroral and geomagnetic events are initiated simultaneously at nearly the same latitude and move poleward with the same velocity. This coincidence strongly suggests a common source for the auroral and magnetic observations. We conclude that the poleward moving geomagnetic deflections are induced by the FTE cur-

216

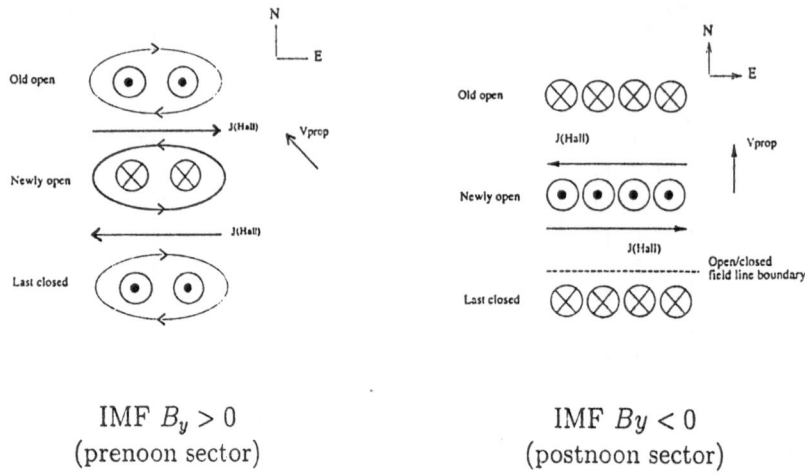

IMF $B_y > 0$
(prenoon sector)

IMF $By < 0$
(postnoon sector)

Figure 5. Sketch of the ionospheric current system associated with an FTE for a positive (left) and a negative (right) IMF B_y situation.

rent system moving poleward over the magnetic stations. Thus the geomagnetic deflections are geomagnetic FTE signatures.

Having established the ground magnetic FTE signatures we will now discuss the associated current system of these events. The prenoon (IMF $B_y > 0$) signature presented above is a bipolar (+/-) poleward moving signature in the H component. This is caused by an eastward and then a westward Hall current moving over the station. The postnoon (IMF $B_y < 0$) signature is also bipolar, but with opposite phase (-/+) compared to the prenoon signature. Thus the postnoon signature must have been caused by first a westward, then an eastward Hall current, moving poleward over the station. By plotting the changes in the magnetic field as horizontal perturbation vectors (see Øieroset *et al.* [19], their Figure 4) the location of field-aligned currents can be estimated from the divergence of the magnetic field vectors.

The resulting current systems for both events are shown in Figure 5. These FTE current systems may be obtained by a combination of the Southwood model [16] and the Saunders *et al.*-Lee model [14, 15]. For the prenoon ($B_y > 0$) case the downward field-aligned currents are assumed to be associated with newly opened geomagnetic field lines, and the auroral signature is located in this area. For the postnoon case the region of aurora and newly opened field lines are associated with the upward FAC. Poleward and equatorward of the intense Hall currents are field-aligned return currents. The northernmost FAC are on old open field lines *convecting* tailwards. The equatorward return current may correspond to the last closed field line. The whole current/convection system moves poleward and expand east- or westward dependent on the IMF B_y polarity.

The distance between the maxima of the two Hall currents are estimated to be 2-300 km. For the prenoon case, the large D component deflections indicate a current system that are limited in the east-west direction, i.e., a current filament. However, according to the smaller D component deflections in the Greenland data, the filament may have changed into a more sheet-like structure as it moved north-westward [20]. The geomagnetic data from the postnoon case indicate that these signatures were caused by a current sheet, probably with an east-west extent of more than 1000 km [19].

6. Acknowledgments

We greatly appreciate discussions with H. Lühr, GeoResearch center, Potsdam, Germany, and T. Moretto, Danish Meteorological Institute, Copenhagen, Denmark. IMAGE magnetometer data was provided by H. Lühr, and Greenland magnetometer data and equivalent ionospheric convection plots by T. Moretto. The coastal Greenland magnetometer stations are operated by the Danish Meteorological Institute (DMI) and the central Greenland stations are operated jointly by DMI and the University of Michigan. The IMAGE magnetometer network is operated jointly be the Finnish Meteorological Institute, the Sodankylä Geophysical Observatory, the University of Tromsø, the Technical University of Braunschweig, the GeoResearch Center Potsdam, and the Polish Academy of Sciences. We thank W. F. Denig, Phillips Laboratory, Hanscom Air Force Base, Massachusetts for DMSP F11 particle and convection data. The optical campaign in Ny Ålesund benefited from economical and technical support from the Norwegian Polar Research Institute and the Research Council of Norway. We are grateful to R. Lepping and B. Ignacio, Goddard Space Flight Center, Greenbelt, Maryland for providing Wind IMF data. This work originally started as a case study of the GEM (Geospace Environment Modeling) Working Group on Current Systems and Mapping, and it has benefited greatly from contributions and discussions within this community.

7. References

1. Russell, C. T. and Elphic, R. C. (1978) Initial ISEE Magnetometer Results: Magnetopause Observations, *Space Sci. Rev.*, **Vol. no. 22**, p. 681.
2. Lockwood, M. and Wild, M. N. (1993) On the Quasi-Periodic Nature of Magnetopause Flux Transfer Events, *J. Geophys. Res.*, **Vol. no. 98**, p. 5935.
3. Lockwood, M. and Smith, M. F. (1994) Low and Middle Altitude Cusp Particle Signatures for General Magnetopause Reconnection Rate Variations, 1, Theory, *J. Geophys. Res.*, **Vol. no. 99**, p. 8531.
4. Lockwood, M., Cowley, S. W. H., and Sandholt, P. E. (1990) Transient Reconnection - the Search for Ionospheric Signatures, *EOS, Trans.*, Am. Geophys. Union, **Vol. no. 71(20)**, p. 709.
5. Cowley, S. W. H., Freeman, M. P., Lockwood, M., and Smith, M. F. (1991) The Ionospheric Signature of Flux Transfer Events, *Rep. ESA SP-330*, p. 105, Eur. Space Agency, Neuilly, France.

218

6. Sandholt, P. E., Deehr, C. S., Egeland, A., Lybekk, B., Viereck, R., and Romick, G. J. (1986) Signatures in the Dayside Aurora of Plasma Transfer from the Magnetosheath, *J. Geophys. Res.*, **Vol. no. 91**, p. 10,063.
7. Sandholt, P. E., Lybekk, B., Egeland, A., Nakamura, R., and Oguti, T. (1989) Midday Auroral Breakups, *J. Geomagn. Geoelectr.*, **Vol. no. 41**, p. 371.
8. Sandholt, P. E., Lockwood, M., Oguti, T., Cowley, S. W. H., Freeman, K. S. C., Lybekk, B., Egeland, A., and Willis, D. M. (1990) Midday Auroral Breakup Events and Related Energy and Momentum Transfer from the Magnetosheath, *J. Geophys. Res.*, **Vol. no. 95**, p. 1039.
9. Sandholt, P. E., Moen, J., Rudland, A., Opsvik, D., Denig, W. F., and Hansen, T. (1993) Auroral Event Sequences at the Dayside Polar Cap Boundary for Positive and Negative Interplanetary Magnetic Field B_y, *J. Geophys. Res.*, **Vol. no. 98**, p. 7737.
10. Karlson, K. A., Øieroset, M., Moen, J., and Sandholt, P. E. (1996) A Statistical Study of Flux Transfer Event Signatures in the Dayside Aurora: The IMF B_y-Related Prenoon-Postnoon Asymmetry, *J. Geophys. Res.*, **Vol. no. 101**, p. 59.
11. Jørgensen, T. S., Friis-Christensen, E., and Wilhjelm, J. (1972) Interplanetary Magnetic-Field Direction and High-Latitude Ionospheric Currents, *J. Geophys. Res.*, **Vol. no. 77**, p. 1976.
12. Gosling, J. T., Thomsen, M. F., Bame, S. J., Elphic, R. C., and Russell, C. T. (1990) Plasma Flow Reversals at the Dayside Magnetopause and the Origin of Asymmetric Polar Cap Convection, *J. Geophys. Res.*, **Vol. no. 95**, p. 8073.
13. Watanabe, M., Iijima, T., and Rich, F. (1996) Synthesis Models of Dayside Field-Aligned Currents for Strong Interplanetary Magnetic Field B_y, *J. Geophys. Res.*, **Vol. no. 101**, p. 13,303.
14. Saunders, M. A., Russell, C. T., and Sckopke, N. (1984) Flux Transfer Events: Scale Size and Interior Structure, *Geophys. Res. Lett.*, **Vol. no. 11**, p. 131.
15. Lee, L. C. (1986) Magnetic Flux Transfer at the Earth's Magnetopause, *Solar Wind-Magnetosphere Coupling*, edited by Y. Kamide and J. Slavin, p. 297, Terra, Tokyo.
16. Southwood, D. J. (1987) The Ionospheric Signature of Flux Transfer Events, *J. Geophys. Res.*, **Vol. no. 92**, p. 3207.
17. McHenry, M. A. and Clauer, C. R. (1987) Modeled Ground Magnetic Signatures of Flux Transfer Events, *J. Geophys. Res.*, **Vol. no. 92**, p. 11,231.
18. Lockwood, M., Sandholt, P. E., Cowley, S. W. H., and Oguti, T. (1989) Interplanetary Magnetic Field Control of Dayside Auroral Activity and the Transfer of Momentum Across the Dayside Magnetopause, *Planet. Space Sci.*, **Vol. no. 37**, p. 1347.
19. Øieroset, M., Lühr, H., Moen, J., Moretto, T., and Sandholt, P. E. (1996) Dynamical Auroral Morphology in Relation to Ionospheric Plasma Convection and Geomagnetic Activity: Signatures of Magnetopause X-line Dynamics and Flux Transfer Events, *J. Geophys. Res.*, **Vol. no. 101**, p. 13,275.
20. Øieroset, M., Sandholt, P. E., Lühr, H., Denig, W. F., and Moretto, T. (1997) Auroral and Geomagnetic Events at Cusp/Mantle Latitudes in the Prenoon Sector During Positive IMF B_y Conditions: Signatures of Pulsed Magnetopause Reconnection, *J. Geophys. Res.*, **Vol. no. 102**, p. 7191.
21. Lee, L. C., Kan, J. R., and Akasofu, S.-I. (1985) On the Origin of the Cusp Field-Aligned Currents, *J. Geophys.*, **Vol. no. 57**, p. 217.
22. Heppner, J. P., and Maynard, N. C. (1987) Empirical High-Latitude Electric Field Models, *J. Geophys. Res.*, **Vol. no. 92**, p. 4467.

COHERENT-SCATTER RADAR STUDIES OF THE DAYSIDE CUSP

M. LESTER
Department of Physics & Astronomy, University of Leicester
University Road, Leicester LE1 7RH, United Kingdom

1. Introduction

It is now commonly accepted that in the cusp regions the entry of magnetosheath particles into the magnetosphere, and ultimately, the ionosphere is most direct [1]. Furthermore, magnetic reconnection at the magnetopause between the interplanetary magnetic field (IMF) and Earth's magnetic field, as proposed by Dungey [2], facilitates mass, momentum and energy transfer from the solar wind into the magnetosphere and ionosphere. This coupling process at the magnetopause leads to large-scale convective flows in the magnetosphere-ionosphere system. The time variation of these flows is most clearly observed in the cusp region and hence study of the ionospheric footprint of the cusp gives direct information on the dynamics of the magnetopause. Many recent in situ, ground based and theoretical studies of the cusp have been reviewed by Smith and Lockwood [3]. It is not the purpose of this paper to revisit that material, rather to consider observations of the ionospheric footprint of the cusp that can be made by coherent scatter radars such as those that form the network of HF radars termed SuperDARN [4]. It is pertinent, however, to ask what coherent radars can provide which are important for studies of the cusp. The key element is the spatial coverage obtained by the radars which can be achieved at relatively high spatial and temporal resolution, typically 45 km and 2 minutes respectively, although these can be improved. The extended coverage of such radars can then provide measurements over a large part of the polar cap, especially when combined in the manner of the SuperDARN network.

Coherent radars, such as those that form the SuperDARN network, receive scattered signals from irregularities on the E and F regions of the ionosphere. From the returned signal a number of parameters are routinely measured, including the power of the backscatter return, the Doppler velocity and the so called spectral width. Many of the radars can also measure the altitude at which the scatter occurred. Apart from the large fields of view, another advantage of these radars is the paired or bistatic nature. The large fields of view means that comparisons with spacecraft overflights or images can be made more easily than with many other ground based instruments. Furthermore, in imaging the cusp from the ground we are potentially able to track it for a considerable period of time with one pair of radars. With a network of such radars, and given the right IMF conditions, the cusp could in principle be followed for between 12 and 15 hours. The paired nature of the radars allows the Doppler, or line of sight velocities, to be merged to produce vector velocities, and hence estimates of convection flow boundaries can be made and also the voltages produced at reconnection sites can be estimated.

The remainder of the paper is divided into a discussion of what parameters can be used to identify specifically the cusp and then a discussion of the flows associated with

J. Moen et al. (eds.), Polar Cap Boundary Phenomena, 219–232.
© 1998 *Kluwer Academic Publishers.*

the cusp. This is then followed by an assessment of what may be learnt in the near future given the range of instrumentation available currently.

2. Identification of the Cusp by Coherent Radars

It has been well know since the early 1970s that the electron precipitation at energies below 300 eV which is associated with the cusp and cleft is also associated with significantly enhanced and intense spread-F in the topside ionosphere [5]. It is no surprise therefore that these regions are also associated with HF radar backscatter. Papers by Baker and colleagues in 1990 and 1995 [6, 7] have investigated this scatter with the Goose Bay and Halley radars in association with coincident particle measurements by one of the DMSP spacecraft. These two radars have similar fields of view in latitude and local time coverage but in different hemispheres, Goose Bay in the north and Halley in the south. By comparing the location of the cusp as identified in the particle measurements on a single pass through the Halley Bay field of view, Baker et al. [6] demonstrated that the spectral width in the region of the cusp was typically larger than 150 m s^{-1}, with values often exceeding 750 m s^{-1}, whereas elsewhere the widths were more like 100 m s^{-1} or less. Furthermore, these authors demonstrated that the ion velocity measurements made by DMSP in the region of the cusp on this pass were highly variable, with rapid and large velocity variations compared to elsewhere on the spacecraft track.

A more detailed piece of work investigated a number of passes through the Goose Bay and Halley fields of view by DMSP spacecraft [7]. In this work, there was an attempt not only to identify spectra associated with the cusp, but also to separate these spectra from those associated with the low-latitude boundary layer (LLBL). Some typical spectra from a pass which demonstrates the characteristic behaviour of spectra apparently in the cusp region are shown in Figure 1. These two spectra are taken from consecutive range gates, along one beam direction.

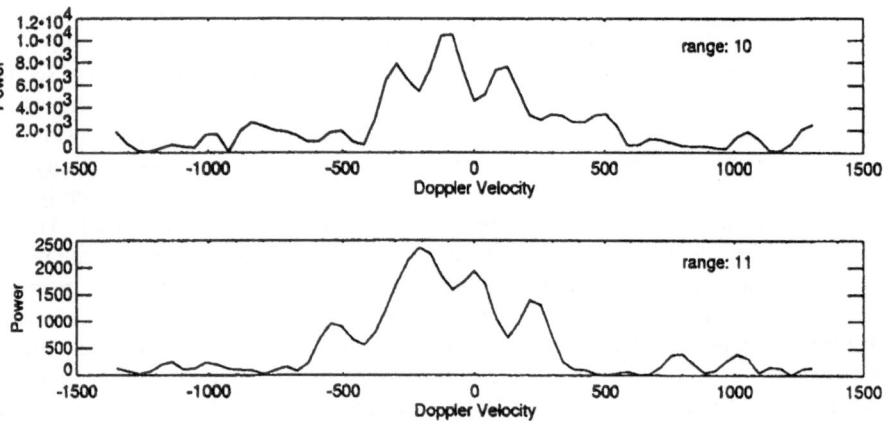

Figure 1. Radar Doppler power spectra observed in a region associated with the ionospheric cusp identified in DMSP particle data (adapted from [7]).

The spectra are multi component and vary considerably from range gate to range gate. We note that the multiple components each have a "width", i.e. full width at half maximum, which is similar to that of a "normal", single peaked spectrum taken from the same range but a different local time. Clearly processes in the region where the multi component spectra occur are very different from those which create the scatter at similar latitudes but a different local time region. Comparing these spectra with spectra which occur in regions identified as the LLBL by the particle data from the DMSP spacecraft (Figure 2), we note some major differences. The spectra apparently associated with the LLBL are relatively narrow and have a single component. The spectral width here is typically between 100 and 150 m s^{-1} in these examples. However, it is not always the case that the spectral width of scatter which maps to particle populations characteristic of closed magnetic field lines, or LLBL, is always single peaked and narrow. Multi component spectra which vary significantly from range gate to range gate and look more like so-called cusp scatter can also occur on field lines which map to regions identified as LLBL by the DMSP particle measurements.

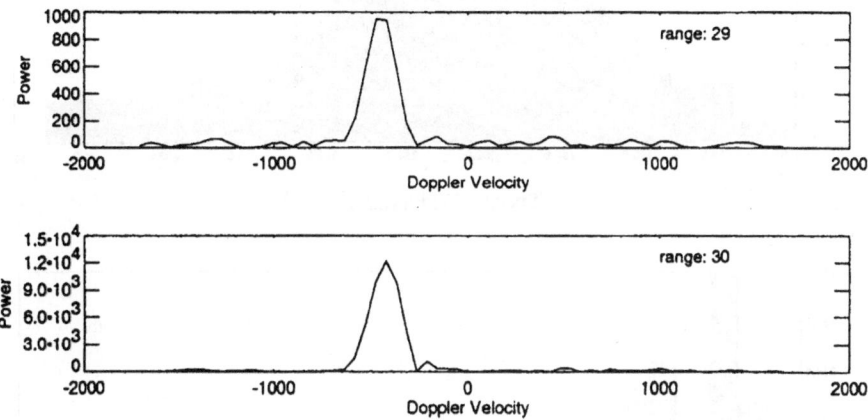

Figure 2. Radar Doppler power spectra observed in a region associated with the LLBL based upon DMSP particle data (adapted from [7]).

The potential for confusion of spectra in the cusp region with those in the LLBL is further demonstrated by an analysis of the scatter during 8 DMSP passes through the cusp region and 8 passes through the closed field line regions [7]. The occurrence distributions of spectral width for the scatter associated with the two different particle populations are given in Figure 3. The top panel is for scatter associated with cusp particles and the lower panel for the particles on closed magnetic field lines. There are significant differences between the two distributions. The distribution for the scatter associated with the cusp is broadly Gaussian with a maximum at ~220 m s^{-1}, whilst the scatter associated with the LLBL has a more exponential distribution with the maximum around 50 m s^{-1}. We note, however, that whilst 30% of the LLBL distribution has spectral widths which are less than 100 m s^{-1} and only 5% of the cusp scatter occur in that range, a larger percentage of LLBL scatter, 20%, has widths greater than 300 m s^{-1} compared with the cusp scatter, 13%. It may be that these larger widths in the LLBL are associated with temporal variability in the radar scatter. Further problems with spectral width as a unique identifier of the cusp occur at latitudes poleward of the cusp, where particle measurements indicate mantle type precipitation. The spectral width, however, can often be as large in this region as in the cusp.

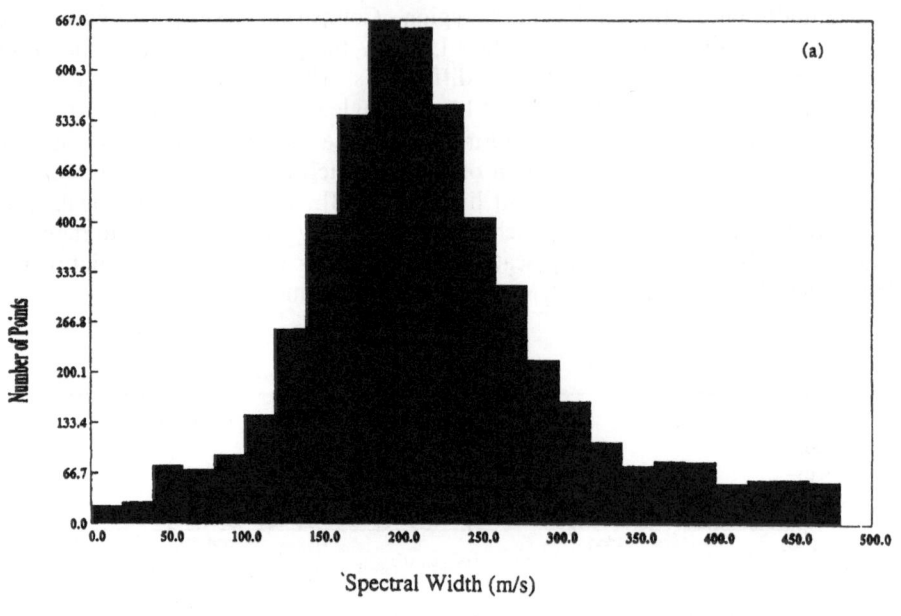

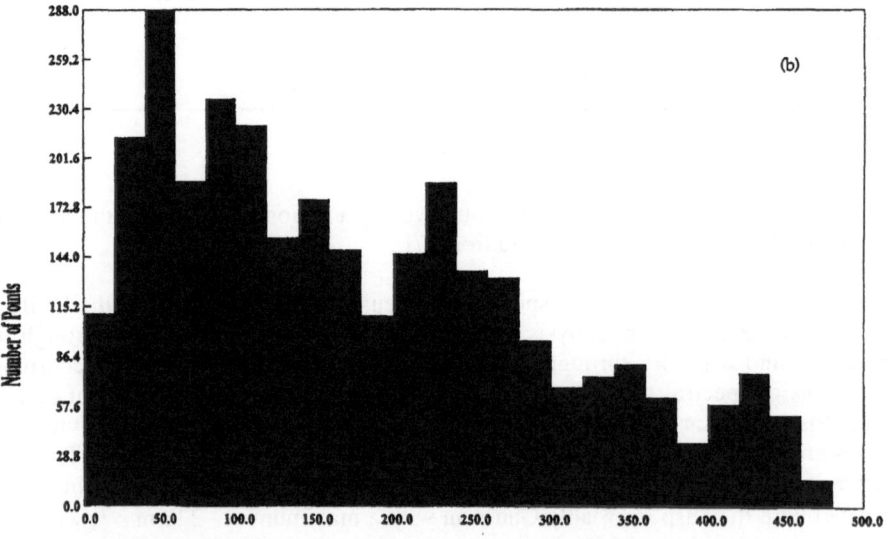

Figure 3. Distribution of spectral widths for 8 different passes associated with the cusp (a) and 8 different passes associated with the LLBL (b) (adapted from [7]).

One final study of the radar backscatter associated with the cusp has been undertaken by Yeoman et al. [8], who not only compared scatter from the CUTLASS Finland radar with DMSP particle data but also simultaneous red line (630.0 nm) and green line

(557.7 nm) emissions from the same region observed by meridian scanning photometers (MSPs). The red line emission in particular has been utilised in the past for studies of the cusp region since the intensity of this emission is associated with precipitating soft electrons (energy ~100 eV), which deposit their energy at F-region altitudes [9]. The MSP data for the interval 0710 to 0750 UT are shown in Figure 4. The MSPs are situated at Ny Ålesund, which has geomagnetic coordinates of 76.0° N, 112.8° E (these are Altitude Adjusted Corrected Geomagnetic (AAGCM) co-ordinates based on the PACE co-ordinates [10]), and scans vertically in a plane 35° west of geographic north. Although position in this data format is displayed as angle of the scan from vertical to north and south, it is possible to convert these pointing directions into a geomagnetic position by assuming an altitude of dominant emission. Here we assume that this altitude is 250 km. The DMSP spacecraft passed through the CUTLASS Finland radar field of view between 0737 and 0742 UT and this interval is highlighted on the MSP data. Concentrating on this interval, the intensity of the red line emission peaks at an elevation of 30° S, equivalent to 75.2° N, whilst the emission is present from 74.0° to 75.6° N. Furthermore, the red line emission intensity during this interval is much higher than the green line emission intensity, which is mainly from the E-region.

The equatorward boundary of the red line emission, 74.0° N, is coincident with the equatorward boundary of the HF backscatter at 0738 UT. Furthermore the latitude is close to that of the equatorward limit of the cusp-like (0.1 - 2 keV) ion precipitation detected by the DMSP spacecraft at 73.7° some 4 degrees longitude to the west. The peak of the red line emission at 75.2° coincides with an enhancement of softer cusp precipitation at <1 keV measured by DMSP.

A detailed comparison of the ionisation rates expected by the particle fluxes measured by DMSP indicates that the F region irregularities, which cause the HF backscatter, are more likely controlled by the ion energy flux rather than the precipitating electrons. This is unexpected as previous work [11] had demonstrated that on average the precipitating electron flux is an order of magnitude larger than that of the ions, and dominates even in the cusp region. Here, it does seem that for short intervals, however, the ion energy flux may dominate. This work clearly needs to be extended beyond this single case study as the relationship between the optical aurora, the radar backscatter and particle precipitation needs clarifying as does the cause of the radar backscatter itself. Furthermore, there needs to be work which looks at the spectral width associated with these irregularities. In general, throughout the interval described above, the spectral width of the scatter associated with the red line emission was less than 200 m s^{-1}, which does not indicate the cusp like spectra of earlier work. The red line emission, however, has been associated with the cusp [9].

In summary, we have some evidence that the spectral width may be a potential diagnostic of the region of the cusp. However, it is clear that so called wide spectra do not only occur on field lines which are open, and therefore connected to the magnetosheath. Furthermore, there is some evidence for spectra which apparently occur on open magnetic field lines and which are relatively narrow. Future studies involving the CUTLASS Finland radar, the MSPs at Ny Ålesund and Longyearbyen, and recently installed All Sky Cameras operating in the same wavelengths as the MSPs at Ny Ålesund should play a key role in determining a quantitative relationship between the red line emission, indicative of magnetic field lines connected to the magnetosheath, and the radar spectral widths.

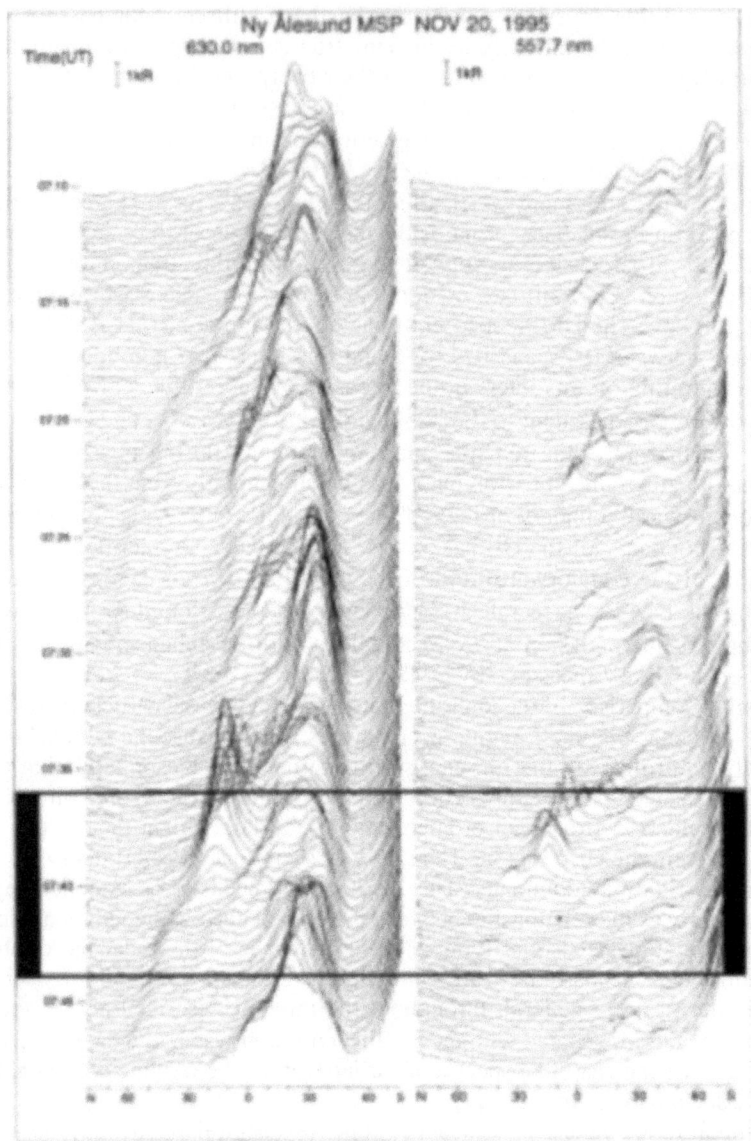

Figure 4. Meridian Scanning Photometer Data for the interval 0710 - -0750 UT on November 20 1995. The red line emission is on the left and the green line emission on the right. The interval of interest is highlighted by the box.

3. Ionospheric flows in the Cusp Region

Moving now to the ionospheric flows associated with the low-altitude cusp, one of the first observations of transient flow signatures in the vicinity of the cusp was made with the STARE radar [12]. Other early studies of convection associated with the cusp were also made with the VHF radars, STARE and SABRE [13, 14, 15]. At VHF frequencies there is little or no refraction of the ray due to the ionospheric electron density and thus in order to match the required orthogonality condition between radar wave vector and the magnetic field, coherent backscatter at auroral latitudes is likely to occur only from E-region altitudes. Furthermore, it is difficult to reach orthogonality poleward of auroral latitudes and this implies that the observations by STARE and SABRE were made when the cusp was at a lower latitude than its normal location, which requires exceptional solar wind conditions. HF radars not only provide larger fields of view but also the refraction of the wave means that investigation of the cusp is possible under normal solar wind conditions.

The control of ionospheric convection by the orientation of the IMF is no longer beyond doubt. Average statistical patterns for various orientations have been presented by a number of authors [e.g. 16]. However, it has become clear that the ionospheric convection pattern cannot be adequately described in this time averaged manner and the dynamics of ionospheric convection, either as magnetic flux is added to the dayside through magnetic reconnection, or through destruction of magnetic flux in the magnetotail during magnetospheric substorms, or through other processes, remain poorly understood. Indeed this is the central challenge for the next few years: to provide a comprehensive view of the dynamics of ionospheric convection as solar wind conditions change.

Simultaneous observations of ionospheric convection in northern and southern hemispheres during a period of changing IMF B_y in the vicinity of the cusp were reported by Greenwald et al. [17]. In this investigation, data from the Goose Bay and Halley HF radars measured the response of the convection pattern in the cusp region to changes in IMF B_y measured by IMP-8. A time delay of order 3 - 5 minutes from the time the change in IMF was incident at the magnetopause to the onset of the reconfiguration of the ionospheric convection pattern was found. This time delay is similar to those estimated from EISCAT measurements [18]. Furthermore it took a further 6 - 8 minutes for a new convection pattern to be established within the two radar fields of view, which covered some 2.5 hours of local time. During this interval the IMF B_z component remained negative. The authors were also able to identify a region within the cusp ionosphere as being the footprint of magnetic reconnection at the magnetopause.

Further investigation of the data set used in [17] demonstrated that following an enhancement in the southward component of the IMF, this region moved equatorward for a period of 20 minutes [19]. During this time, there were two bursts of poleward directed flow which reached 2 km s^{-1} in the southern hemisphere and which occurred close to the ionospheric footprint of this reconnection site. A weaker response was observed in the northern hemisphere, the poleward flow was of order 1 km s^{-1}. These two events were separated by 10 minutes, which is similar to the repetition rate reported for flux transfer events (FTEs) at the magnetopause [20] and for dayside auroral transients and flow bursts [21]. The asymmetry in the flow response in the two hemispheres to the same change in IMF conditions is due in this case to the IMF B_y component which favoured reconnection in the southern hemisphere.

An investigation of a similar flow enhancement observed by the Halley radar was responsible for the idea of an enhanced convection channel in the region of the cusp [22]. The radar observations were supported by ion velocity measurements by DMSP F13 which measured a peak ion drift transverse to the spacecraft track of 2.7 km s^{-1}. In the radar data an isolated patch of high flow was seen to evolve over a period of 4 minutes moving poleward and westward before disappearing from the radar field of view. By simulating a variety of possible models of ion velocity flow, the authors concluded that the best fit was associated with a narrow channel of enhanced flow superposed upon the weaker, about 1 km s^{-1}, background ionospheric flow. There was also evidence for weaker return flow outside the convection channel. The conclusion that the flow channel is a signature of enhanced magnetic merging at the magnetopause seems well supported by the data, but it seems that it arises from a more continuous period of reconnection, of several minutes, rather than short-lived pulses.

Bearing this last conclusion in mind it is interesting to move to two recent studies of ionospheric flows associated with the cusp which have utilised a different scanning mode from that normally run [23, 24]. The scanning mode is slightly different in each case, but the main idea of a high time resolution beam is the same. In these modes the radar does not simply complete a full scan of the 16 beam directions, but after each beam direction returns to a special beam, often referred to as the camping beam, such that the scan is 0, n, 1, n, 2, n,...., 15, n where n is the special beam and has a temporal resolution which is twice the beam dwell time. The full scan for such a mode takes twice the normal time to complete. In the first of these studies [23] a series of poleward moving transient features in the line of sight velocity measured along the special beam were observed. The start location of these transients was up to 3 degrees equatorward of the region of scatter with high spectral width, previously associated with the cusp, and moving into this region. These velocity transients occurred more frequently as the region of high spectral width moved equatorward. A total of 17 transients were seen during an interval of 2 hours, a mean repetition rate of one every 7 minutes which is similar to the repetition rate for FTEs [20]. A DMSP pass occurred during the interval and the measurements of ion flux by this spacecraft show an example of a stepped ion cusp which is consistent with theoretical predictions of a dynamic, time varying cusp [25]. Investigating the velocity transients in the full scan data demonstrated the existence of similar convection channels to that discussed above [22]. Thus, since the velocity features occurred immediately adjacent to the cusp, and subsequently propagated into the cusp region, it would seem that these velocity transients were ionospheric signatures of FTEs. If this is so, and if the high spectral width is associated with the cusp, then the FTEs appear to originate in the LLBL, although this is discussed further below. Furthermore, the authors also comment that there is evidence for velocity transients which actually occur more frequently than the 7 minute repetition rate mentioned above. This latter point is important as evidence for such a higher repetition rate has been found in another data set involving the CUTLASS Finland radar and the Ny Ålesund MSP, which is shown in Figure 5. The second panel shows the line of sight ion velocity for the interval 0923 - 0947 UT as a function of geographic latitude. The MSP red line data are in the third panel and the green line in the bottom panel.

The MSP data have been transformed from the zenith angle coordinate system to a geographic latitude coordinate system by assuming the altitude of the emissions to be 250 km for the red line and 110 km for the green line. The red line data show the existence of a number of poleward moving auroral transients which have a repetition rate of about 1 every 2 minutes. These poleward moving transients have been identified in the figure by the black and white dashed lines. Looking at the ion velocity data, there is also evidence for coincident poleward moving forms. Thus there is clear evidence for the poleward moving auroral transients to occur more frequently than the commonly accepted frequency of flux transfer events. Note also that the poleward moving transients in both data sets start from approximately the same latitude, close to the equatorward boundary of the red line emission.

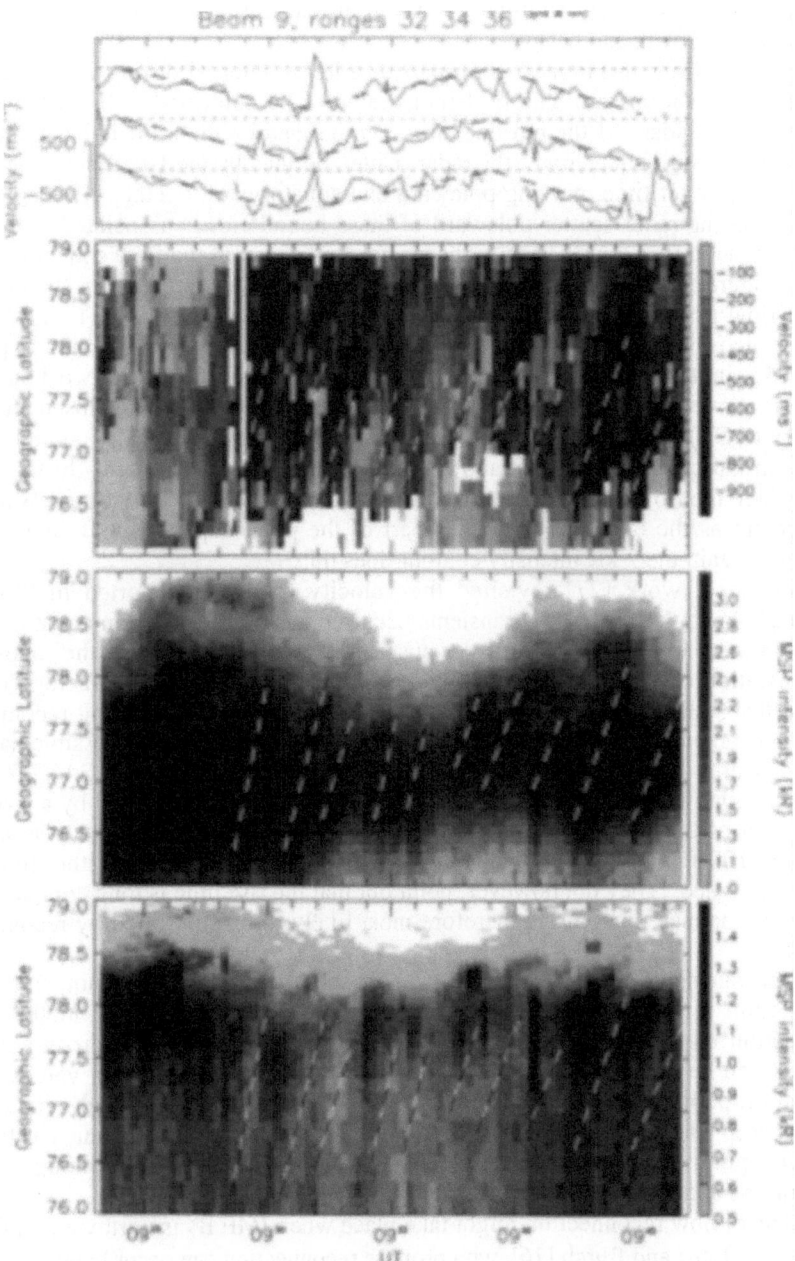

Figure 5. Radar and MSP observations during 0923 - 0947 UT 17 December 1995.

The top panel of Figure 5 illustrates three time series of the line of sight velocity for the interval for 3 range gates, 32, 34 and 36. The velocities decrease (i.e. become more poleward) and increase (become less poleward) in a relatively smooth manner between 0924 and 0939 UT before again decreasing until 0945 UT. The time scale for growth and decay is approximately 15 minutes. The transients discussed above are superposed on this smooth growth and decay.

The observations in Figure 5 form part of a larger interval during which the MSP and radar observations are clearly demonstrated to be colocated. In particular, the equatorward boundary of the red line emission, defined by the 2 kR intensity contour, and the equatorward boundary of the radar scatter, defined by the 10 dB power level, are closely related over time, moving poleward and equatorward together. This colocation suggests that the radar scatter poleward of these boundaries is associated with newly opened magnetic flux. Furthermore, the growth and decay of the flow shown in the top panel of Figure 5 is consistent with predictions of a two stage process for the evolution of cusp flow associated with FTEs [26]. Flow is initiated by the FTEs as a single vortex, which then evolves into a twin cell flow and then decays, the whole process taking 15 minutes. This is further supported by the motion of the two boundaries discussed above. In the first phase of the flow evolution, the open/closed field line boundary (OCB) is expected to move equatorward. The 2 kR intensity contour can be used as an indicator of the OCB and moves equatorward as expected. In the second phase, where the flow is predicted to evolve into a twin cell pattern with poleward flow at the centre as the open flux becomes part of the polar cap, the OCB should move poleward. Again the 2 kR intensity contour does this.

More recent work [27] revisited the velocity transients reported in [23]. As mentioned above these velocity transients occurred some 100 - 200 km equatorward of the region to which magnetosheath, or cusp, precipitation maps to in the ionosphere. This separation can be considered to be due to a combination of two factors, one the distance that will be travelled poleward by an open flux tube following reconnection while the magnetosheath ions travel along it, and secondly the incompressible nature of the ionosphere. The combined distances from these two elements is 115 - 225 km, which is similar to the observations. Thus the flow transients are caused by a process at the location of the OCB. It is also interesting to note that the transient rise and fall times are ~100 s and ~ 10 s respectively. The authors do not explain either time scale nor the difference. They do, however, point out that the rise time is long compared with the electron bounce time and is therefore most likely to be controlled by reconnection processes.

By measuring the flow across the proposed location of the footprint of the reconnection site at the dayside magnetopause, or separatrix, estimates of the reconnection electric field, and hence the contribution to the total cross polar cap potential, can be made [28]. In a case study of data from Goose Bay, values for the potential drop along the separatrix of up to 60 kV were found, although typically the potential drop was between 20 and 40 kV. These values were often as much as the total polar cap potential, estimated from the AMIE procedure [29], although mainly contributions of 50% or more were estimated.

Models of how reconnection might take place when IMF B_z is northward have been discussed by Reiff and Burch [16], who propose reconnection can occur between the IMF and the tail lobe magnetic field, poleward of the cusp. In these situations there is no addition of open magnetic flux to the polar cap. For weakly positive B_z, a single lobe cell forms whose direction is clockwise in the northern hemisphere for positive B_y.

Alternatively, when the clock angle of the IMF is greater than 70°, reconnection between the IMF and closed dayside magnetic field lines can occur [30], when a twin cell pattern is observed. It should be noted that reconnection between the IMF and the lobe cell is not precluded under these circumstances, but in the case study reported in [30], the lobe cell was observed to decay as the clock angle exceeded 70°. In one study of convection in the vicinity of the cusp during IMF northward conditions [31], a well defined boundary in the flow was seen to move poleward at a rate of 4 degrees of latitude in one hour. The flows on the poleward side of the boundary were largest, typically larger than 800 m s^{-1} toward the radar, while the flows on the equatorward side were typically 200 m s^{-1} away from the radar. There were however several bursts of larger flow away from the radar on the equatorward side of the boundary, indicating either that the orientation of the pattern was changing, or that the cause of the flows was varying with time. The poleward motion of this boundary was accompanied by a poleward motion of the overall region of scatter, which could not be accounted for by simple rotation of the field of view underneath the auroral oval. Finally, the spectral widths on either side of the boundary were large. The evidence from the spatial plots would suggest that the flows may have been driven by lobe reconnection.

To summarise the observations of ionospheric flows observed by coherent radars in the vicinity of the cusp, we note that the work has produced several new and important results. The existence of flow transients is of crucial importance to our understanding of the dynamics of the cusp. Furthermore, work involving the evolution of the flow after the initial opening of magnetic flux in the FTE and the following addition of the open flux to the polar cap has crucially confirmed theoretical predictions. The spatial and temporal nature of the observations have been crucial in developing these pieces of work.

4. Summary

The observations reported here only form part of the body of work of coherent radars on the cusp and the initiation of ionospheric flow in that region. A key element for these types of study is the identification of the cusp, which has typically been by the use of the spectral width parameter. Although there remains a degree of uncertainty in the use of this parameter, there is growing evidence that in certain cases it can be used successfully to identify the open field line region. Flow transients have often been seen, sometimes in association with high speed convection channels, at other times in association with the evolution of the flow as magnetic flux is added to the polar cap. It is clear from the measurements, however, that there are limitations in the experiments performed so far. These stem mainly from the lack of sufficient temporal resolution. This is particularly highlighted in the case of the transient features seen at a rate of one every 2 minutes. This is barely the time for a complete scan normally. Even the transient features which occur at 7 minutes or so, cannot be properly imaged spatially. It is clear therefore that improved temporal resolution will be of considerable benefit in future experiments.

The observations reviewed above have mainly concentrated on single radar observations. The move towards the bistatic nature will provide the ability to locate the convection flows with respect to the cusp, either by comparison with optical images or by using the radar scatter. This will be crucial. At the moment the spatial nature of the flows has been inferred from the line of sight velocity data. Furthermore, a number of theoretical predictions involving the dynamic nature of the cusp can be tested in association with radar, optical and spacecraft data.

Acknowledgements. I would like to thank Steve Milan, Stan Cowley, Alan Rodger and Jøran Moen for many illuminating discussions. Thanks also to Tim Yeoman, Jackie Davies and John Taylor for help with the figures. I am grateful to Per Even Sandholt for providing the data for Figure 4.

5. References

1. Newell, P.T. and Meng, C.-I. (1988) The Cusp and the Cleft/LLBL: Low-Altitude Identification and Statistical Local Time Variation, *Journal of Geophysical Research* **93**, 14549 - 14556.
2. Dungey, J.W. (1961) Interplanetary Magnetic Field and the Auroral Zones, *Physics Review Letters* **6**, 47 - 49.
3. Smith, M. and Lockwood, M. (1996) Earth's Magnetospheric Cusps, *Reviews of Geophysics* **34**, 233 - 260.
4. Greenwald, R.A. et al. (1995) Darn/Superdarn: A global view of the dynamics of high-latitude convection, *Space Science Reviews* **71**, 761 - 796.
5. Dyson, P.L. and Winningham, J.D. (1974) Top Side Ionospheric Spread F and Particle Precipitation in the Day Side Magnetospheric Clefts, *Journal of Geophysical Research* **79**, 5219 - 5230.
6. Baker, K.B., Greenwald, R.A., Ruohoniemi, J.M., Dudeney, J.R., Pinnock, M., Newell, P.T., Greenspan, M.E. and Meng, C.-I. (1990) Simultaneous HF-Radar and DMSP Observations of the Cusp, *Geophysics Research Letters* **17**, 1869 - 1873.
7. Baker, K.B., Dudeney, J.R., Greenwald, R.A., Pinnock, M., Newell, P.T., Rodger, A.S., Mattin, N, and Meng, C.-I. (1995) HF Radar Signatures of the Cusp and Low-Latitude Boundary Layer, *Journal of Geophysical Research* **100**, 7671 - 7695.
8. Yeoman, T.K., Lester, M., Cowley, S.W.H., Milan, S.E., Moen, J. and Sandholt, P.E. (1997) Simultaneous Observations of the Cusp in Optical, DMSP and HF Radar Data, *Geophysics Research Letters* In Press.
9. Rodger, A.S.,Mende, S.B., Rosenberg, T.J. and Baker, K.B. (1995) Simultaneous Optical and HF Radar Observations of the Ionospheric Cusp, *Geophysics Research Letters* **22**, 2045 - 2048.
10. Baker, K.B. and Wing, S. (1989) A New Magnetic Coordinate System for Conjugate Studies at High Latitudes *Journal of Geophysical Research* **94**, 9139 - 9143.
11. Hardy, D.A., Gussenhoven, M.S and Brautignan, D. (1989) A statistical model of auroral ion precipitation, *Journal of Geophysical Research* **94**, 370 - 392.
12. Goertz, C.K., Nielsen, E., Korth, A., Glassmeier, K.H., Haldoupis, C., Hoeg, P. and Hayward, D. (1985) Observations of a Possible Ground Signature of Flux Transfer Events, *Journal of Geophysical Research* **90**, 4069 - 4078.
13. Sofko, G.J., Greenwald, R.A., Korth, A. and Kremser, G. (1979) STARE Ionospheric Electron Flows during the August 28/78 GEOS-2 Magnetopause Crossing, Proceedings of Magnetospheric Boundary Layers Conference, ESA SP-148, 183 - 185.
14. Waldock, J.A., Jones, T.B. and Nielsen, E. (1984) SABRE Observations of the Morning Sector Convection Reversal, *Planetary and Space Science* **32**, 837 - 843.

15. Freeman, M.P. and Southwood, D.J. (1988) The Effect of Magnetospheric Erosion on Mid- and High-Latitude Ionospheric Flows, *Planetary and Space Science* **36**, 509 - 522.
16. Reiff, P.H. and Burch, J.L. (1985) IMF By-Dependent Plasma Flow and Birkeland Currents in the Dayside Magnetopause. 2. A Global Model for Northward and Southward IMF, *Journal of Geophysical Research* **90**, 1595.
17. Greenwald, R.A., Baker, K.B., Ruohoniemi, J.M., Dudeney, J.R., Pinnock, M., Mattin, N., Leonard, J.M. and Lepping, R.P. (1990) Simultaneous Conjugate Observations of Dynamic Variations in High-Latitude Dayside Convection Due to Changes in IMF By, *Journal of Geophysical Research* **95**, 8057 - 8072.
18. Etemadi, A., Cowley, S.W.H., Lockwood, M., Bromage, B.J.I., Willis, D.M. and Lühr, H. (1988) The Dependence of High-Latitude Dayside Ionospheric Flows on the North-South Component of the IMF: A High Time Resolution Correlation Analysis Using EISCAT "Polar" and AMPTE UKS and IRM Data, *Planetary and Space Science* **36**, 471.
19. Pinnock, M., Rodger, A.S., Dudeney, J.R., Greenwald, R.A., Baker, K.B. and Ruohoniemi, J.M. (1991) An Ionospheric Signature of Possible Enhanced Magnetic Field Merging on the Dayside Magnetopause, *Journal of Atmospheric and Terrestrial Physics* **53**, 201 - 212.
20. Rijnbeek, R.P., Cowley, S.W.H., Southwood, D.J. and Russell, C.T. (1984) A survey of Dayside Flux Transfer Events Observed by ISEE 1 and 2 Magnetometers, *Journal of Geophysical Research* **89**, 786 -.800
21. Lockwood, M., Sandholt, P.E., Cowley, S.W.H. and Oguti, T. (1989) Interplanetary Magnetic Field Control of Dayside Auroral Activity and the Transfer of Momentum Across the Dayside Magnetopause, *Planetary and Space Science* , 1347.
22. Pinnock, M., Rodger, A.S., Dudeney, J.R., Baker, K.B., Newell, P.T., Greenwald, R.A. and Greenspan, M.E. (1993) Observations of Enhanced Convection Channel in the Cusp Ionosphere, *Journal of Geophysical Research* **98**, 3767 - 3776.
23. Pinnock, M., Rodger, A.S., Dudeney, J.R., Rich, F. and Baker, K.B. (1995) High Spatial and Temporal resolution Observations of the Ionospheric Cusp, *Annales Geophysicae* **13**, 919 - 925.
24. Milan, S.E., Lester, M., Cowley, S.W.H., Moen, J., Sandholt, P.E. and Owen, C.J. (1997) Meridian-Scanning Photometer, Coherent HF Radar and Magnetometer Observations of the Cusp: A Case Study, *Annales Geophysicae*, Submitted.
25. Lockwood, M. and Smith, M.F. (1994) Low- and Middle-Altitude Cusp Particle Signatures for General Magnetopause Reconnection Rate Variations: 1. Theory, *Journal of Geophysical Research* **99**, 8531.
26. Cowley, S.W.H. and Lockwood, M. (1992) Excitation and Decay of Solar Wind-Driven Flows in the Magnetosphere-Ionosphere System, *Annales Geophysicae*, **10**, 103.
27. Rodger, A.S. and Pinnock, M. (1997) The Ionospheric Response to Flux Transfer Events: The First Few Minutes, *Annales Geophysicae* **15**, 685 - 691.
28. Baker K.B., Rodger, A.S. and Lu, G. (1997) HF-Radar Observations of the Dayside Magnetic Merging Rate: A Geospace Environment Modeling Boundary Layer Campaign Study, *Journal of Geophysical Research* **102**, 9603 - 9617.
29. Richmond, A.D. and Kamide, Y. (1988) Mapping Electrodynamic Features of the High-Latitude Ionosphere from Localized Observations: Combined Incoherent

Scatter Radar and Magnetometer Measurements for January 18 - 19, 1984, *Journal of Geophysical Research*, **93**, 5760.

30. Freeman, M.P., Farrugia, C.J., Burlaga, L.F., Hairston, M.R., Greenspan, M.E., Ruohoniemi. J.M. and R.P. Lepping (1993) The Interaction of a Magnetic Cloud with the Earth' Ionospheric Convection in the Northern and Southern Hemispheres for a Wide Range of Quasi-Steady Interplanetary Magnetic Field Conditions, *Journal of Geophysical Research* **98**, 7633 - 7655.

31. Lester, M., Thomas, E.C., Pinnock, M., Senior, C. and Owen, C.J. (1997) CUTLASS Observations of Convection in the Cusp Region, *Advances in Space Research* In Press.

IONOSPHERIC RADIOWAVE ABSORPTION PROCESSES IN THE DAYSIDE POLAR CAP BOUNDARY REGIONS

PETER STAUNING
Solar-Terrestrial Physics Division
Danish Meteorological Institute
DK-2100 Copenhagen, Denmark
(e-mail: pst@dmi.dk)

Abstract. Radio wave absorption processes in the upper, partly ionized atmosphere, the ionosphere, hold a unique source of information on the physics of these regions. The detection of absorption processes has been greatly augmented by the advance of the imaging riometer technique and the deployment of a number of imaging riometer installations in the polar regions. The present report discusses the theoretical foundation for the use of absorption observations for the diagnostics of the ionospheric plasma conditions and for the examination of the precipitation of high-energy charged particles into the upper atmosphere. A survey of absorption event types of relevance for the polar cap boundary regions is given. Example cases of some of these absorption event types are presented to illustrate the physics involved in such events and to describe the observing systems, the data handling techniques, and the analysis and display tools available.

1. Introduction

For ionospheric radiowave absorption processes the detection techniques comprise four main types, which could be classified as:

  (i)   Attenuation of reflected continuous radiowave (CW) signals
  (ii) Attenuation of reflected pulsed radiowave signals
  (iii) Attenuation of transmitted signals
  (iv) Cosmic noise absorption (riometer)

The first three methods rely on the operation of transmitters to generate the probing signals and calibrated receivers to detect the amplitude of the received signal after it has passed through the ionospheric regions. The first two methods use ground-based transmitter-receiver systems. A transmitter sends signals upward to be reflected at some ionospheric level and turned downward to be detected by the receiver. Both these methods, however, are severely affected by the variable ionospheric reflection conditions. The third method needs the transmitter to be mounted on a rocket or a satellite while the receiver at ground detects the signals attenuated by absorption processes along the connecting path. All three methods are heavily influenced by effects like focusing, scintillations, and scattering other than absorption processes along the signal path and by

233

J. Moen et al. (eds.), Polar Cap Boundary Phenomena, 233–254.

effects from the variable match of polarization, phase and frequency between the transmitting and receiving system.

Compared to the first three techniques the cosmic noise absorption method (iv) is fairly simple and quite reliable. It uses the rather uniform distribution of galactic radio sources in the sky to provide the probing signals and a calibrated receiver to detect the level of the signals after their passage through the ionosphere (e.g., Little [1]). The so-called RIOmeter (Relative Ionospheric Opacity meter) [2] uses a servo feed-back loop to adjust a calibrated noise generator to supply the same signal power as that received from the sky via some suitable antenna system. The noise generator level then provides a measure independent of receiver sensitivity of the received cosmic noise power. Since the beam of a fixed antenna sweeps over the same region of the sky each sidereal day (4 min shorter than the UT day) the signal should repeat itself regularly if not attenuated by the ionospheric absorption. For relatively high observing frequencies (f > ~30 MHz) and during undisturbed conditions the attenuation is very small during local night. The recorded levels at local nighttime (which will slide through the day in sidereal time) during quiet conditions may thus serve to identify the reference "Quiet-Day" (QD) signal level. The ionospheric absorption in relative units (for instance dB) may now be scaled as the deviation of the recorded values from the QD level.

It appears that only observations of types (ii) and (iv) are actually conducted in the polar cap regions. Such observations use ionosondes (digisondes) and riometers, respectively. A body of observations of f_{min} by ionosondes (digisondes) are available from around ten instruments in the northern and southern polar cap boundary regions. This parameter is the minimum frequency (usually a few hundred kHz) for which a return signal can be detected. It is a measure of the ionospheric absorption in the D-region up to the usual reflection height of around 100 km. The parameter is very equipment-sensitive. Furthermore, f_{min} is usually derived once every quarter of the hour at most.

Polar riometer observations are made with a number (around 40) of instruments of the traditional type which has a single, relatively broad antenna beam (normal-beam), and with some (around 10) of the recently developed imaging type of riometer instrument which employs an array of multiple narrow beams to provide a local image of absorption intensities over the sky. Figure 1 displays a map over a part of the northern polar cap region where the location of digisonde, riometer and imaging riometer stations operated by the Danish Meteorological Institute (DMI) are indicated. The map also presents the location of further northern imaging riometer sites and indicates the magnetically conjugate location to imaging riometer sites in the southern polar cap region. Coordinates and other parameters for imaging riometer installations are listed in Stauning [3].

The imaging riometer for ionospheric studies (IRIS) technique and data handling procedures are explained in Detrick and Rosenberg [4]. The first instrument of this kind was deployed at South Pole station in 1988. The upper part of Figure 2 (after Detrick and Rosenberg [4]) depicts the contours of the intersections between the 3 dB beam cones and an ionospheric reference altitude level for the absorption processes of 90 km. In the figure the dotted circle indicates the beam contours for a typical normal-beam riometer antenna system. The dashed square illustrates the typical dimensions, 240 x 240 km, of the absorption images derived by image processing techniques from observations made with the array of imaging riometer beams.

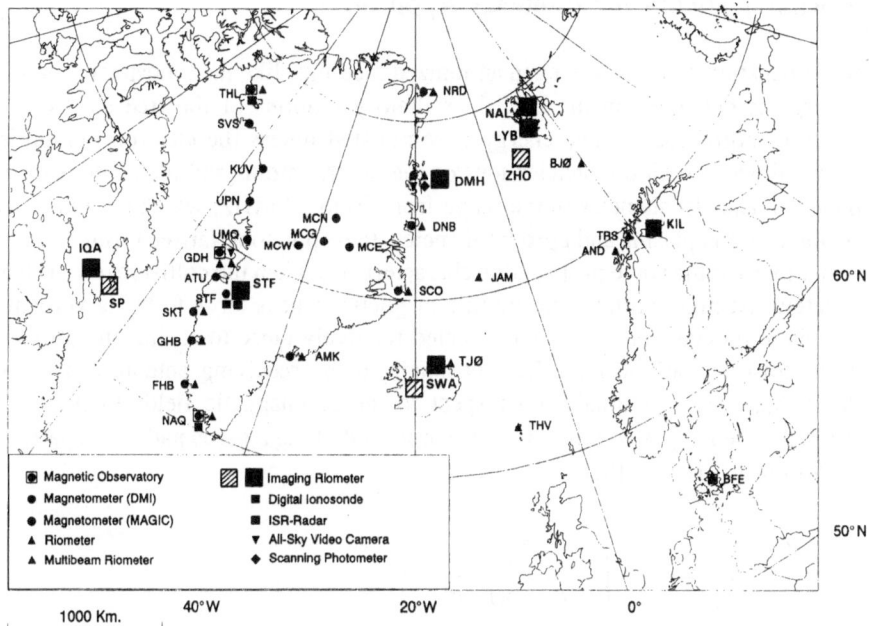

Figure 1. Map of DMI geophysical observing stations in the north-atlantic polar regions. Markings of the imaging riometers in Canada, Iceland, Finland, Ny-Ålesund and the conjugate locations for South Pole, Syowa and Zhongshan have been added.

Sequences of absorption images are well suited to illustrate the dynamical developments in the temporal and spatial structures of absorption events. The lower part of Figure 2 illustrates the construction of a "keogram-type" diagram of absorption intensities versus time along the horizontal axis and latitude (could also be longitude) along the vertical axis. The stacked vertical "lines" of absorption intensities, for instance in colour coding, represent cross sections through each of a sequence of absorption images. They are placed along the horizontal axis at positions which corresponds to the actual time of the observations. Such displays are well suited to illustrate the combined temporal and spatial developments in cases that have dominant one-dimensional features like simple drift motions.

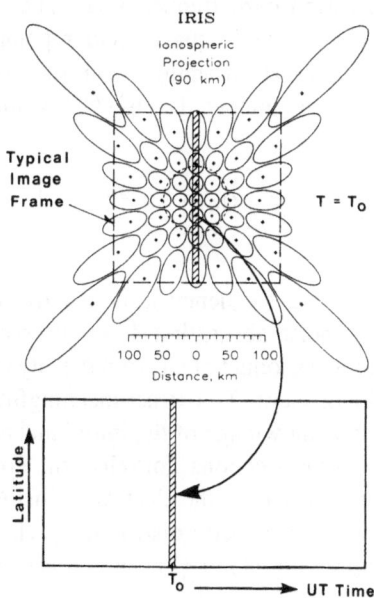

Figure 2. (Upper) Projection of IRIS beams to 90 km altitude. (Lower) Construction of keogram-type intensity contour plots.

2. Absorption Theory for Riometer Applications

During propagation of radio waves in an ionized medium, like the ionosphere, part of the wave energy is deposited in the forced oscillatory motions of the free electrons. This intermittently stored radio wave energy is re-radiated unless the electron motion is disturbed by collisions with the heavier ions or the neutral atoms and molecules of the gas. At typical riometer frequencies in the range from 20 to 50 MHz, which are much greater than the ionospheric plasma and gyro frequencies, the Appleton-Hartree formulas are good approximations for the wave propagation characteristics when the collision term is handled appropriately. At high latitudes the Earth's magnetic field is close to being vertical. For systems where the antenna beams are directed relatively close to zenith, the propagation of cosmic noise signals through the ionosphere to the receiving antennas may thus be considered 'quasi-longitudinal' with respect to the geomagnetic field. In this case the expression for the integrated radio wave absorption A along a given path has a particularly simple form (e.g., Rawer [5]):

$$A = \frac{q^2}{2\epsilon_o mc} \int_S \frac{N_e \nu}{(\omega \pm \omega_L)^2 + \nu^2} ds \tag{1}$$

where ϵ_o and c are the permittivity and speed of light in free space respectively; q and m are the electron charge and mass. N_e and ν are the local electron density and the (effective) collision frequency values while $\omega_L = \omega_c \cos\theta$ is the longitudinal component of the electron gyro frequency ω_c. The $+$ and $-$ signs are used for one or the other of the two characteristic quasi-circular polarizations. ω is the angular frequency of the wave propagating at an angle θ with the geomagnetic field along the path S of which ds is a differential element. Using MKSA units (N_e in [m^{-3}], ds in [m]) gives:

$$A[dB] = 4.58 \; 10^{-5} \int_S \frac{N_e \nu_{eff}}{(\omega \pm \omega_L)^2 + \nu_{eff}^2} ds \tag{2}$$

The wave frequency ω is usually the dominant term in the denominator. Hence the cosmic noise absorption A, as observed by the riometer method, is largely related to the integrated product of electron density and collision frequency along the signal path through the ionosphere. To extract meaningful information from this parameter we need the best possible knowledge of the individual profiles of electron density and collision frequency.

The cross-sections for electron collisions with other species are strongly velocity dependent. Hence the electron collision frequency must be found through integration over the electron velocity distribution. The 'effective' electron collision frequency to be used in expression (1) or (2) can be derived from:

$$\nu_{eff} = \sum_j N_j \int_{v=0}^{\infty} \sigma_m^j(v) \, v \, \mathcal{F}(v) dv \bigg/ \int_{v=0}^{\infty} \mathcal{F}(v) dv \tag{3}$$

where N_j is the density of ions or neutrals of type j, j=1 to J, while σ^j_m is the velocity-dependent cross-section for momentum transfer in electron collisions with species of type j. v is the electron velocity and $\mathscr{F}(v)$ is the electron velocity distribution function (e.g. Maxwellian).

The ionosphere is a complex mixture of neutral molecules and atoms, and various singly or multiply charged ion types. Different electron collision types dominates in the different regions of the ionosphere. As a coarse guide line the electron-neutral collisions (ν_{e-n}) dominate in the D-region (~ 60-90 km) and in the lower E-region (~ 90-120 km). In the upper E-region (~ 120-150 km) the electron-ion collisions (ν_{e-i}) are also important and in the F-region ($> \sim 150$ km) they are dominant.

The velocity-dependent cross-sections for various ion types and atomic and molecular species and the resulting specific collision frequencies are given, for instance, in Mantas [6] and Aggarwal [7]. For electron-neutral interactions the specific collision frequencies can be expressed as simple polynomials in $T^{\prime2}$, where T is the electron temperature assuming a Maxwellian distribution. Hence we have:

$$\nu_{e-n,eff} = \sum_j N_j (A_j T^{1/2} + B_j T + C_j T^{3/2} + D_j T^2)$$

(4)

where A_j, B_j, C_j, and D_j are given in [6] as tabulated coefficients for the major constituents of the upper atmosphere. For relevant cases the electron-neutral collision frequency increases a little more steeply than linearly with increasing electron temperature. Further, the collision frequency is proportional to the number density of neutrals. Hence the electron-neutral collision frequency is approximately proportional to the atmospheric pressure provided the electrons are in thermal equilibrium with the ambient neutral atmosphere. It is now seen from (1) or (2) that the resulting radiowave absorption intensities are high when ionization is created at low altitudes (D-region) where the neutral densities are high. Such low-altitude ionization processes need high-energy particle precipitation or strong EUV- or X-ray radiation.

An exception to this general situation is the occurrence of E-region electron heating by unstable plasma waves in strong electric fields (e.g., Schlegel and St.-Maurice [8]). In such cases the temperature dependence of the electron-neutral collisions can result in very high electron-neutral collision frequency values (comparable to those found in the D-region) which in turn can cause appreciable radiowave absorption enhancements for signals passing through the E-region [9, 10].

Turning now to electron-ion collisions through the elastic scattering of electrons in the Coulomb field of ions of charge Z_i, density N_i, and temperature T_i, which may differ from the electron temperature T_e, the effective electron-ion collision frequency is given in Majumdar [11] (in cgs units):

$$\nu_{e-i,eff} = \sum_i N_i \left(\frac{8\pi}{3} \frac{Z_i^2 e^4}{(2m\pi)^{1/2} (kT_e)^{3/2}} \ln\Lambda \right)$$

(5)

where:

$$\Lambda = \frac{k^{3/2}}{1.78 Z_i e^3} \left(\frac{T_e^3}{\pi N_e} \right)^{1/2} \left(\frac{T_i}{Z_i T_e + T_i} \right)^{1/2}$$

(6)

and where k is the Boltzmann constant while m and N_e are the electron mass and density.

For the upper F-region where singly charged O^+ ions dominate, we can approximate $Z_i \approx 1$ and $\Sigma N_i \approx N_e$. Using cgs units the expression for $\nu_{e\text{-}i}$ [s^{-1}] may be reduced to a simple numerical expression in N_e [cm^{-3}] and T_e, T_i [°K]:

$$\nu_{e\text{-}i,\text{eff}} = [29.4 + 3.64 \ln T_e (\frac{2T_e T_i}{N_e(T_e + T_i)})^{1/2}] N_e T_e^{-3/2} \tag{7}$$

It is important to note, that the value of $\nu_{e\text{-}i,\text{eff}}$ (disregarding the slow variations in the logarithmic term) is proportional to the electron density in contrast to $\nu_{e\text{-}n,\text{eff}}$ which is independent of electron density (eq. 4). Hence the radiowave absorption, as given by (1) or (2), is quadratic in N_e when electron-ion collisions are dominant (F-region). Another important feature to note from (7) is the inverse temperature dependence compared to (4). The electron-ion collision frequencies decrease with increasing temperature in contrast to the electron-neutral collision frequencies that increase with temperature. This implies that the combination of high electron densities and low electron temperatures can result in high values of $\nu_{e\text{-}i}$ and hence high radiowave absorption intensities.

At auroral and polar latitudes the ionospheric electron densities that control the radiowave absorption intensities are often quite variable. There is some steady ionization in the upper atmosphere which is largely created by solar UV and EUV radiation. Additional ionization is produced by precipitation of energetic particles of magnetospheric, solar or cosmic origin and by bursts of solar EUV and X-ray radiation. The electron densities are further affected by horizontal or vertical drifts.

The time varying electron densities $N_e(h)$ at height h can, in principle, be derived from the continuity equation

$$dN_e/dt = Q^{EUV} + Q^* - K - N_e \nabla v - v \cdot \nabla N_e \tag{8}$$

where Q^{EUV} is the ionization rate due to solar EUV radiation and Q^* the corresponding ionization rate due to precipitation of energetic particles, while K is the recombination rate and v the electron drift velocity at h. Calculations of the ionization rate Q^* associated with particle precipitation can be based on the ionization functions for energetic electrons and protons developed by Rees [12, 13, 14]. In numerical calculations the recombination rate K is often expressed in terms of a variable effective recombination coefficient defined through:

$$K = \alpha_{\text{eff}} N_e^2 \tag{9}$$

In the upper regions of the ionosphere the use of empirically determined recombination coefficients adjusted for varying composition and temperatures works well. The effective lifetime of the ionization is quite long (minutes to hours). Hence the drift contributions, that are expressed through the last two terms of (8), are very important. Horizontal drift processes carry substantial amounts of ionization around in the polar cap regions, for instance, from the solar-illuminated high-density dayside ionosphere to the low-density nightside ionosphere. Neutral winds push the ionized constituents up and down the field lines. For the calculation of the electron densities in the upper ionosphere several statistical or semi-empirical models are available like the International Reference Ionosphere (IRI) [15], the Time-Dependent Ionospheric Model (TDIM) [16], the Thermosphere-Ionosphere General Circulation Model (TIGCM) [17], and the Thermosphere-Ionosphere Nested Grid Model (TING) [18].

In the polar regions, with the absorption calculations using eqs. (2) and (7), the upper ionospheric (F-region) electron densities and temperatures predicted by the models will hardly ever give absorption enhancements beyond 0.1-0.2 dB at typical riometer frequencies ($\sim$ 30 MHz) for conditions ranging from quiet to very disturbed. From calculations of absorption intensities in realistic cases it was concluded in Wang et al. [19] that the F-region plasma frequencies need to approach or exceed 10 Mhz to produce substantial absorption which could be unambiguously identified in riometer recordings. It cannot be ruled out, however, that special conditions that could enhance the electron densities like the intense, soft precipitation found in the cusp regions or the soft "polar rain" electron fluxes in the polar cap can result in conditions favourable for F-region absorption events. Further possibilities for discernible absorption features may exist at transpolar drifting high-density F-region electron patches where the low electron temperatures at the nightside polar cap or the effects of gradient drift plasma instabilities at the patch edges may enhance the effective electron collision frequencies.

For the lower ionosphere the recombination processes are complicated functions of ion chemistry and electron density. Chemical models have been developed, for instance, by Torkar and Friedrich [20] and Smirnova et al. [21] to assist calculations of K during various geophysical conditions. In the lower ionosphere (D- and E- regions) where the ionization lifetime is short (msec to sec) the two drift terms are less important and often neglected. Eq. (8) then transforms into the simpler expression:

$$dN_e/dt = Q^{EUV} + Q^* - \alpha_{eff}N_e^2 \tag{10}$$

hence in steady state where $dN_e/dt=0$ we have:

$$N_e = [(Q^{EUV} + Q^*)/\alpha_{eff}]^{1/2} \tag{11}$$

In the auroral zone and the polar cap boundary regions events of intense precipitation of high-energy particles are frequent occurrences. To model the absorption levels associated with such events one needs to calculate the resultant electron densities in the lower ionosphere ($\sim$ 60-140 km) from eq. (11). Numerical values for the composition and temperatures of the neutral atmosphere are, for instance, provided by the MSIS-90 model (a further development of the MSIS-86 model, Hedin [22, 23]) using as input to the model program the coordinates of the actual location, the time and date, the solar 10.7 cm flux, and geomagnetic activity parameters. The atmospheric composition and temperatures are needed for calculations of the two ion production terms Q and for defining the ion chemistry that controls the resulting recombination coefficient α_{eff}.

The atmospheric parameters are also needed to calculate the effective collision frequency ν_{eff} entering eq. (1) or (2). There is the added problem that the electron and ion temperatures, which enter the collision frequency formulas eqs. (4) and (7), may differ from the temperature of the neutral atmosphere. Usually, the main part of the difference is related to the Joule heating of the ionized constituents in the horizontal ionospheric electric fields. In strong electric fields the Farley-Buneman ion-acoustic plasma instabilities could be excited as the electron drift velocities relative to the ions exceed the phase velocity of ion-acoustic waves. Under the combined influence of strong electric fields and the unstable plasma waves the electron temperatures could be increased by up to an order of magnitude (e.g., Schlegel and St.-Maurice [8]). Such cases may only occur in the E-region where the ions are slowed down by collisions with the neutrals while the electrons move almost freely in the electric fields. Further up in the rarefied atmosphere the ions and electrons move together at the same drift speed. Further down the drift of the electrons is impeded by collisions with neutrals in the denser atmosphere to never reach the ion-acoustic sound speed. Some additional, usually small, ion and electron heating effects are related to the energetic charged particle precipitation.

Combining the knowledge of atmospheric composition and motion, ionospheric modelling, electrodynamic conditions, and energetic charged particle precipitation enables the modelling of anticipated absorption intensities. Comparisons with actually observed absorption intensities during varying geophysical conditions then hold a great potential for verifying the various sectors of the modelling of physical processes.

3. Absorption event types occurring in polar cap boundary regions

The identification of ionospheric radiowave absorption events from their appearance and giving them names is a useful step in order to enable sorting and categorization of events. Table 1 gives a summary of radiowave absorption processes in the polar cap boundary regions and lists some of their characteristics. These types are commented below.

- Contrary to *solar proton PCA events* the *solar electron PCA* type of event is not well documented since only few cases have been reported. One example is presented in Barcus *et al.* [24], where the absorption event lasted around 6 hours and reached a peak intensity of 2 dB in the central polar cap (Thule).

- The three types of absorption events named "Slowly varying", "Pulsating" and "Flaming" are all caused by precipitation from high-energy magnetospheric electron populations generated during auroral substorms. They differ by the nature of the precipitation mechanism. The *slowly varying* type of absorption event is related to the steady "drizzle" precipitation caused by pitch-angle scattering due to the fluctuating wave fields in the outer trapped radiation zones (e.g., Baker *et al.* [25, 26]). The slow variations in absorption amplitudes reflect the substorm source mechanism rather than the actual precipitation processes. The *pulsating* type of event is characterized by large-amplitude absorption pulsations with periods of the order of a few minutes associated with pulsations in the local geomagnetic field of corresponding periodicities (e.g., Kikuchi *et al.* [27]). These two types occur on closed field lines holding a quasi-trapped high-energy electron population. The *"flaming"* absorption type is not officially named yet. The *concept* describes the precipitation in the noon region at the border between the "closed" magnetic fields of the outer radiation belts and the "open" polar cap fields. As the sunward flow of plasma and embedded geomagnetic fields within the auroral oval turns into an antisunward flow across the polar cap in the typical two-cell convection patterns, the field lines are either opened up to merge with the interplanetary fields or just stretched and extended into the magnetospheric tail region. In either case they are no longer capable of holding a trapped high-energy electron population so the remaining electrons are dumped off. In colour-coded latitude-time keogram-type absorption plots (see Figure 2) such events will appear as "flames" extending with decreasing intensities from the auroral-latitude regions into the polar regions. These events are observed equatorward of the region of soft cusp ion and electron precipitation, but their appearance will be much like that of the cusp events since the plasma convection patterns that control both types are continuous across the border between open and closed fields (e.g., Nishino *et al.* [28]).

TABLE 1. Summary of characteristics of ionospheric radiowave absorption events as observed by a ~30 MHz riometer located in the polar cap boundary region (detection limit ~0.1 dB).

Event name	Location/ local time	Occurrence frequency	Typical duration	Typical intensities	Direct cause of absorption
Solar flare proton event	polar cap	few a year	few days	1-10 dB	~1-100 MeV protons
Solar flare electron event	central cap	few a cycle	few hours	1-3 dB	~0.1-10 MeV electrons
Auroral absorption	night sector	daily	10-100 min	1-3 dB	~10-100 keV electrons
Slowly varying absorption	morning	daily	1-2 hours	1-3 dB	~30-300 keV electrons
Pulsating absorption	morning/day	few a week	1-2 hours	.5-1 dB	~30-300 keV electrons
"Flaming" noon absorption	noon sector	few a week	½-1 hours	.5-1 dB	~30-300 keV electrons
Magnetic impulse absorption	dayside	daily	1-10 min	.1-.5 dB	not known
Daytime absorption spikes	dayside	few a week	.5-2 min	.1-.5 dB	not known
Electron heating events	dayside	few a month	½-2 hours	.1-.5 dB	E-field > ~50 mV/m
F-region patches	dayside	few a month	½-1 hour	.1-.2 dB	high F-region density

- The absorption events associated with *magnetic impulses* have not yet been explained properly. They occur at high latitudes but within the closed magnetospheric region. The absorption features are most likely associated with enhanced precipitation of substorm-related high-energy electrons due to the disturbances causing the magnetic impulses.

- The corresponding situation exists for *dayside absorption spike events*. They, typically, have a duration of 1-2 min. Contrary to nightside substorm absorption spike events which are often very intense (several dB) the dayside spike events typically have peak absorption values of 0.2-0.3 dB (at 38 MHz). They have a preferred occurrence at and slightly equatorward of the afternoon polar ionospheric convection reversal boundary (e.g., Stauning and Rosenberg [29]).

- *Electron heating events* are related to the enhanced electron collision frequencies caused by electron heating in the E-region through the interaction of plasma waves and strong electric fields. The electric field must exceed approximately 25 mV/m to excite Farley-Buneman ion-acoustic plasma waves. To accomplish the strong electron temperature enhancements required to enable the detection by riometer observations, the electric fields need to be at least of the order of 50 mV/m (e.g., Stauning [9]; Stauning and Olesen [10]).

- The occurrence of absorption events associated with *F-region patches* has been inferred by Rosenberg *et al.* [30] from the simultaneous observations of drifting F-region patches seen by HF backscatter radars and drifting absorption enhancements recorded by an imaging riometer. The drift speeds of the absorption features were found to match those of the F-region density enhancements for an assumed effective altitude for the absorption processes of ~300 km corresponding to the height of the F-region electron density peak.

In the following a few examples of the above absorption event types shall be analyzed to illustrate the observing systems, the data handling techniques, and the available analysis and display tools.

4. Examples of absorption events

4.1 POLAR CAP ABSORPTION EVENTS

The absorption effects of solar flare protons, Polar Cap Absorption (PCA), are usually extended rather uniformly over the entire polar cap including the boundary regions down to latitudes of 60-65° inv.lat. or even lower during the larger events. The PCA events are caused by precipitating high-energy solar protons which arrive to the Earth's environment within a quarter of an hour to a few hours after a major flare in a solar region facing the Earth (e.g., Reid [31]; Van Allen *et al.* [32]). The protons stay trapped in the heliospheric magnetic fields for some days or a few weeks and are precipitated, among others, into the planetary atmospheres. The softer protons are constrained to follow the "open" geomagnetic field lines (polar field lines merged with the interplanetary magnetic field) to be precipitated into the polar cap atmosphere. The more energetic ones have cut-off latitudes well equatorward of the auroral oval and may thus penetrate deep into the "closed" magnetospheric regions. The proton fluxes may be observed directly by satellites in the solar wind, like for instance IMP8, and by polar orbiting satellite during their polar passes. Protons at the high-energy end of the spectrum may also be observed from the geostationary orbit.

When appropriately located the imaging riometer observations allow the precise detection of the transition between the polar cap uniform proton fluxes and the fluxes reduced by increasing energy cutoff at lower latitudes. These effects include the socalled midday recovery, a minimum in absorption intensity sometimes observed around local noon.

The solar proton energy spectra are usually rather hard. Hence the protons ionize the neutral atmosphere down to quite low altitudes. The PCA absorption processes are usually weighted such that the main effects are seen at altitudes of 50-70 km. This is substantially lower than the corresponding ranges for other types of events like auroral absorption, where the dominant height range is usually 80-100 km. At the low altitudes of PCA absorption events the chemistry of the ionized constituents is markedly different from that further up in the ionosphere. Instead of having a plasma of electrons and a few types of positive ions we have at these altitudes (50-70 km), in addition to the electrons, a variety of positive and negative ions, of which some are simple ionized species of the main atmospheric constituents (N_2, O_2, NO, O, etc.), while other are complicated cluster types. The formation of negative ions is greatly influenced by the solar illumination. During daytime the negative ions are dissociated such that the electrons are released, while during night the free electrons are attached to neutral atoms, molecules or clusters to form negative ions (e.g., Hultqvist [33]).

Ions have little effect on radio wave propagation and attenuation at the frequencies in question, hence the absorption is minimum during night when a large fraction of the electrons are removed through the formation of negative ions. During daytime the abundance of electrons is thus significantly larger than in the night for equal fluxes of precipitating high-energy solar protons. Consequently, at daytime the absorption of radiowaves is stronger than at nighttime. A day/night ratio in absorption intensities of around 4-8 is often observed during PCA events. Other absorption types have much smaller ratios (1-2) between day and night intensities for equal ionizing fluxes.

The day-night effect in PCA absorption intensities is seen clearly around sunrise and sunsets. The variations in solar proton fluxes are usually rather slowly compared to the rate of change in ion chemistry associated with the zenith angle variations at these times. One should thus expect a steep change in absorption intensities as the absorbing region enters or leaves the sunlight. One of the relatively few published reports referring to riometer data is the study by Reagan and Watt [34] where they find a large asymmetry between sunrise and sunset with ionization lag times of the order of 45 min from sunset ($\chi=90°$) at 70 km altitude and lead times of the order of 10 min from sunrise.

Their results were based on normal-beam riometer observations. The imaging riometer technique has improved the quality of the absorption observations in two ways. The narrowness of the beams enables a precise definition of the solar zenith angle at the ionospheric intersects. Further, the diversity of beams allows the simultaneous observation of sunrise/sunset effects at different solar zenith angles whereby effects related to variations in the solar proton fluxes can be minimized.

We shall use the PCA event in late October, 1992, as an example. The event is discussed in Stauning [35] on the basis of IRIS data from Sdr. Strømfjord (STF). Here we present imaging riometer data from Danmarkshavn (DMH) which is located 10° poleward in geographic latitude and 32° eastward compared to STF. Hence the sunrise and sunset times and the slope of zenith angle variations will take different values.

The solar flare activity responsible for the PCA event occurred on 29 and 30 October. The PCA event developed very slowly in the afternoon hours of 30 October. Figure 3 presents the recorded data for 0000 to 2400 UT on 31 October 1992. The upper panel hold traces for 1-min averages of the cosmic noise intensities recorded in a central N-S column of beams. The intensities have been normalized such that the undisturbed recordings would be horizontal traces. 100% deflection corresponds to 4 of the divisions along the vertical axis. The beams have been arranged in order from south at the bottom to north at the top. The traces in the bottom panel presents the horizontal northward geomagnetic component (H), the eastward component (E), and the downward component (Z) recorded by the co-located magnetometer.

244

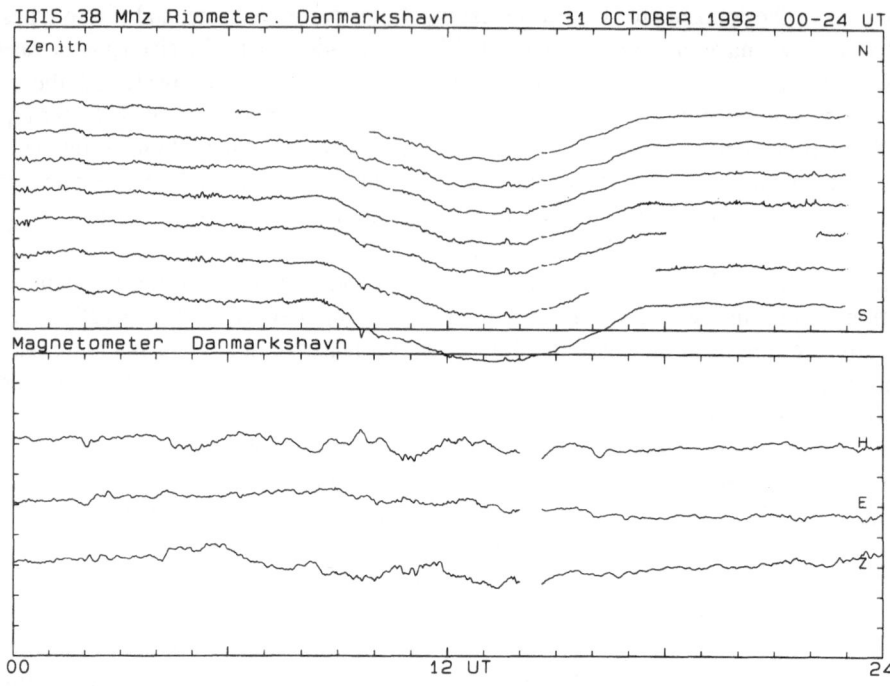

IRIS 38 Mhz Riometer. Danmarkshavn   31 OCTOBER 1992   00-24 UT

Figure 3. Summary plot of Danmarkshavn imaging riometer and magnetometer data for PCA event on 31 October 1992. Absorption intensity (positive downward) in the middle of the day is 3 dB.

In the early morning of 31 October (before sunrise) the absorption intensity was around 0.5 dB at the 38 MHz riometer in DMH (and in STF). At 0700 UT, shortly before twilight time in DMH the absorption intensities had increased slowly to 0.7 dB. Starting a little after 0700 UT the absorption intensities increased more rapidly to reach a maximum of around 3 dB at noon at around 1300 UT. In the afternoon the absorption intensities decreased to reach a steady level of around 0.9 dB in the evening. During the steady rise before noon there were some different variations at 0930-1000 UT in the southernmost beams coming from substorm activity. These variations are also reflected in the magnetic recordings in the bottom panel which otherwise show fairly weak magnetic activity.

The variations of the absorption intensities with solar illumination are presented in Figure 4. The "dots" (prenoon) and "crosses" (postnoon) depict absorption intensity vs. solar zenith angle χ for samples averaged over 5 min of data for the 4 central beams. The "knees" at sunrise and sunset are evident from the diagram. The sunrise knee occurs at a solar zenith angle of 97.6° and that of sunset occurs at a zenith angle of 98.6°. These values are very close to those found from the imaging riometer recordings from STF for this PCA event and for another PCA event [35]. For sunrise the corresponding STF values were 97.5° and 97.0°, while for sunset the STF solar zenith angles were 98.5° and 98.0° respectively. Across the imaging riometer field-of-view the UT time of the knee will shift by 20-25 min due to the varying location of the ionospheric intersect.

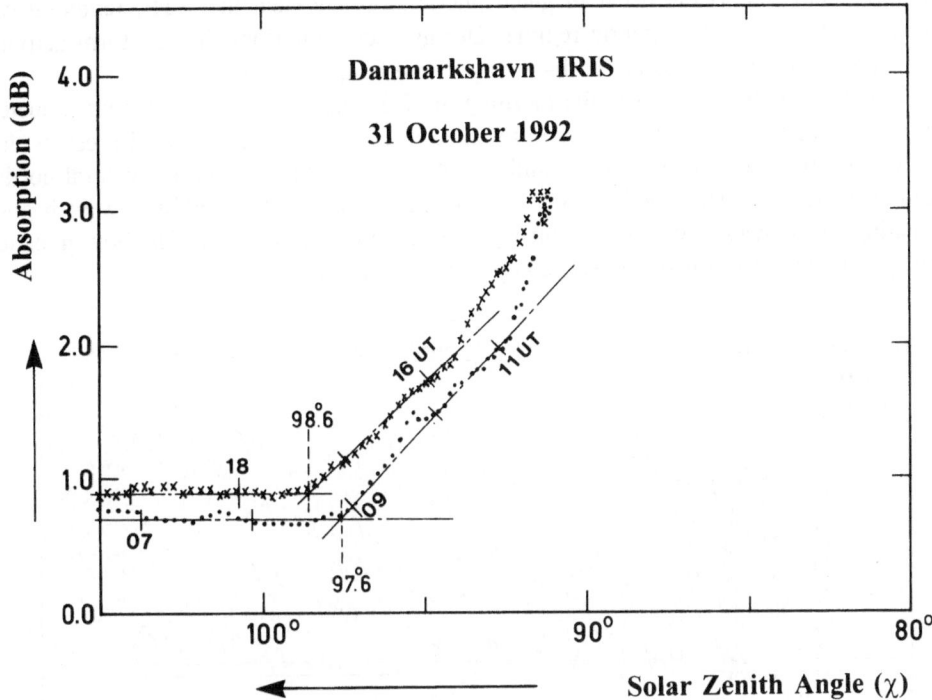

Figure 4. Display of PCA absorption intensities against solar zenith angle at sunrise (dots) and sunset (crosses) for PCA event on 31 October 1992. Solar zenith angles for the "knees" in absorption are marked.

It is seen in Figure 4, that there is a difference in sunrise and sunset transitions corresponding to 1° in zenith angle value for the knee, but the asymmetry is not nearly the amount suggested in Reagan and Watt [34]. Figure 4, furthermore, illustrates the large day/night ratio in absorption intensities. In this case for DMH, contrary to STF, the variation in the absorption against zenith angle did not reach a "plateau" level since the zenith angle remained above 90°. Hence the true daytime level of absorption intensities is never attained. For DMH the maximum absorption values are around 3 dB corresponding to a day/night ratio of around 4. For STF the daytime absorption intensity reached 4 dB which corresponds to a day/night ratio of around 5.

4.2 PULSATING MORNING ABSORPTION EVENTS

Sdr. Strømfjord (STF) at an invariant latitude of 73.7° is usually located in the polar cap region void of high-energy electrons during most local times except for a few hours around local magnetic noon at ~1400 UT. During very quiet conditions, when the auroral oval retracts to a small diameter, STF could be located in the closed magnetospheric region exposed to the precipitating debris of high-energy electrons drifting eastward from the source region of substorm activity in the night sector. But then the substorm activity is usually very weak and so are the derived events. Alternatively, the solar wind

246

conditions may distort the usual magnetospheric structure such that STF, temporarily, enters the closed magnetospheric regime. During such conditions the substorm activity augmented by the magnetospheric distortions could be quite strong.

One such example occurred in the morning of 12 January 1997. The imaging riometer and magnetometer data from STF are presented in Figure 5. The top panel presents the recorded (normalized) traces for the middle column of beams providing a central north-south cross section. The middle panel presents the traces for the middle row of beams providing an east-west cross section through the antenna field of view. The bottom panel displays the three local geomagnetic components, *H*, *E*, and *Z*.

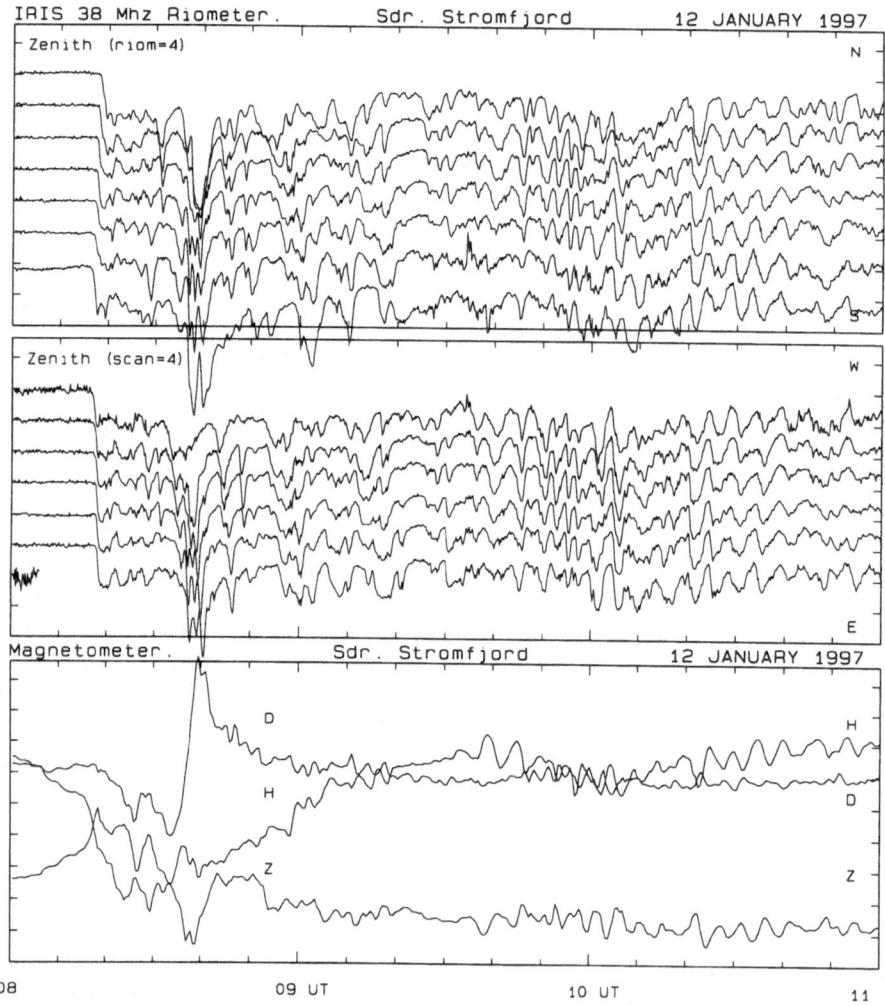

Figure 5. Pulsating absorption event at 0817-1100 UT on 12 January 1997. The upper panels display signal strength in the central north-south column and east-west row of beams. Absorption is positive downward. The bottom panel displays local geomagnetic components.

Prior to 0816 UT, all imaging riometer traces (two top panels) in Figure 5 display very quiet conditions typical of the polar cap region. The geomagnetic components indicate the build-up of stresses. The *H*-component decreases (the *H*-trace in the figure dips below the *E* and *Z* traces at 0810 UT), and the *Z*-component increases. Then, at 0816 UT a large disturbance hits the south-westernmost beams. There is a clear north-east motion at a speed of approximately 1 km/s of the "bulge" of disturbances as it sweeps across the field-of view from southwest to northeast in a couple of minutes. After this onset STF remains in the closed-field region for a while and is now continually exposed to high-energy precipitation much like usual auroral zone conditions at invariant latitudes of ~60-65°.

At 0832 UT there is a further strong enhancement. In this case there is a marked delay from west to east and only minor delays from south to north. So this "bulge" is north-south aligned and is drifting eastward at a speed somewhat lower than that of the first onset. At 0845 UT and 0850 UT one may again observe eastward drifting north-south aligned features. At 0900 UT there is a further enhancement drifting south-eastward at a speed of ~0.5 km/s.

Between 0900 and 0930 UT we just see random-like motions or enhancements. Then, from around 0930 and lasting a couple of hours, there are strong pulsation activities in the absorption as well as in the magnetic patterns. At 0930-0950 we may note a few large pulsations which have periods of ~5 min. At around 0950-1000 UT we observe regular absorption pulsations at periods of around 2 min while at 1030 to 1050 UT we see strong and regular pulsations at periods of ~5 min. There are no clear drift motions associated with the pulsations occurring after 0930 UT.

The above disturbances are indicative of several features. Firstly, the magnetospheric configuration is greatly distorted, possibly compressed, to enable STF (at inv.lat. ~73.7°) to be located in the trapped radiation region at this local time (early morning). Secondly, STF observes the eastward drifting high-energy electron debris from substorm activity further into the night sector. Thirdly, as the substorm injection ceases (or moves further to the nightside) we are left with an intense high-energy electron population extending rather uniformly in latitude and longitude. The electrons of this population are pitch-angle scattered by what appears to be resonant hydromagnetic wave activity. The wave activity is probably formed through resonances between longitudinal waves bouncing between the hemispheres and the transverse azimuthal or radial cavity-mode waves at the L-shell of STF. The resonance beats at either of two characteristic periods, ~2 min or ~5 min.

The precipitation of high-energy electrons into the atmosphere above STF is strong throughout the pulsation interval. This is indicative of a continuous refill of the loss cone to maintain an isotropic electron distribution at the connecting field lines. In other reported cases the occurrence of isotropic high-energy particle distributions have been related to the effects of the magnetic perturbations introduced by the tail current sheet at the equatorial crossing of the field lines. The perturbations affect the particle motion along the field lines such that pitch-angle scattering results. The tail current sheet scattering mechanism has been discussed, for instance, in Sergeev and Malkov [36]; Sergeev *et al.*, [37]; Sergeev [38]. The occurrences in our case of large-amplitude magnetic field oscillations in the equatorial regions of the field lines extending from STF, most likely, have the corresponding pitch-angle scattering effects since the anticipated amplitudes of the magnetic perturbations have the same order of magnitude in the two cases. The results here are the wave-like magnetic perturbations, and the periodical modulation of the precipitated high-energy electron fluxes which causes corresponding variation in the ionospheric absorption of radiowaves. The absorption enhancements and the magnetic variations need not have the same phase since the two features propagate at different speeds from the equatorial source regions to the ionosphere (e.g., Stauning et al. [39]).

4.3 POLEWARD PROGRESSING ABSORPTION EVENTS

The interplanetary magnetic field (IMF) has a profound influence on the polar plasma convection. The north-south component, IMF B_Z, in coarse terms, controls both the size of the polar cap by affecting the front side merging processes, and the transpolar plasma drift by affecting the cross-polar potential. The azimuthal component, IMF B_Y, controls the dayside east-west oriented ionospheric DPY currents by affecting the north-south oriented electric fields in the noon sector of the polar cap boundary region. During cases of positive (northward) or numerically small IMF B_Z, it appears that the IMF B_Y effects, such as the ionospheric plasma flow related to the DPY currents, are constrained to the cusp region at around 75° inv.lat. Contrary, during cases of large negative IMF B_Z, these DPY disturbances move from the cusp region, where they are initiated, across a considerable fraction of the dayside polar cap. Hence they form the class of events termed "poleward progressing convection disturbance events" (e.g., Clauer et al. [40]; Stauning et al. [41, 42]; Stauning [43,44]).

A typical example observed during 1130-1530 UT on 2 August, 1991, is presented in Figure 6a from Stauning et al. [42]. The upper traces in the diagram display the horizontal geomagnetic component, H, recorded by several stations of the north-south chain of magnetometer stations operated by DMI at the west coast of Greenland (see Figure 1). The stations are located at or poleward of cusp latitudes. Local magnetic noon is around 1400 UT for these stations. The bottom trace depicts the IMF B_Y component measured in front of the magnetosphere by the IMP8 satellite. The interplanetary data have been exposed to an averaging procedure equivalent to the (natural) averaging in the magnetometer response to ionospheric current systems. The IMF B_Z component was consistently negative holding values around -10 nT during this interval.

One may readily observe the reproduction of IMF B_Y variations in the H-component traces. There is a delay of around 10-15 minutes accounted for in the propagation of the IMF variations with the solar wind from the IMP8 position to the front magnetosphere and the propagation of the induced disturbances along the field lines from the outer magnetosphere to the ionospheric foot point. Furthermore, the varying patterns are observed further and further delayed at observatories further and further north (poleward). The latter shift is the progression which occurs at speeds of around 0.5-1.0 km/s.

The physical motions of the ionosphere producing the displayed magnetic perturbations are east-west oriented flows. These plasma motions are driven by horizontal ionospheric electric fields which for the larger excursions may take values of up to 50-150 mV/m. Electric fields this large generate E-region ion-acoustic plasma waves that, in turn, enable the electric field to selectively heat the E-region plasma electrons to temperatures up to an order of magnitude above the temperature of the ambient neutrals. The ion species are heated to moderately enhanced temperatures through ordinary Joule heating processes.

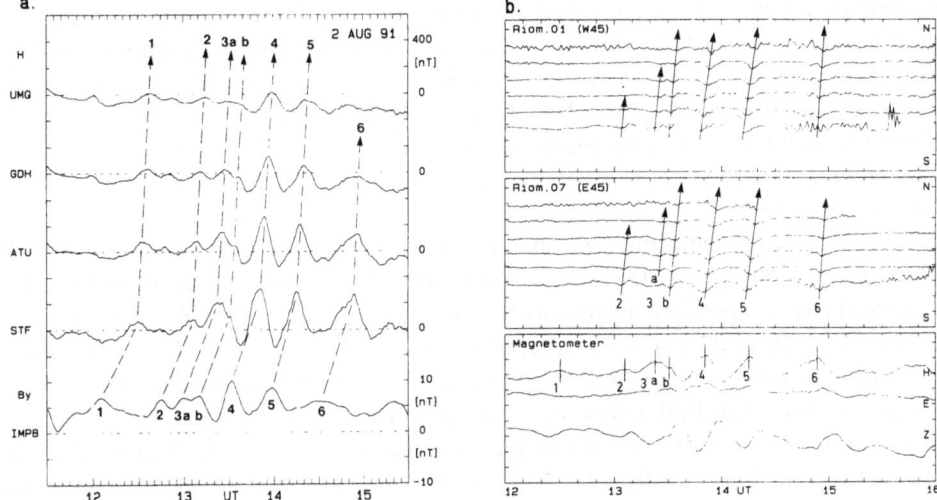

Figure 6. (a) Poleward progressing convection disturbances on 2 August 1991. Upper traces are H-component data from a chain of magnetometers. Bottom trace displays averaged IMF B_Y values. (b) Display of poleward progressing absorption event in Sdr. Strømfjord around local magnetic noon on 2 August 1991. The top panels show signal strength in N-S columns of imaging riometer beams.

The response in the imaging riometer observations to the progressing convection disturbances encountered at STF around local noon on 2 August, 1991, are displayed in Figure 6b. The two upper panels displays the signal intensities in the westerly (W45°) and easterly (E45°) north-south cross-sections of imaging riometer beams. Absorption is positive downward. One may note the northward (poleward) progressing absorption enhancements tracking the variations in the magnetic H-component at STF. When an altitude of 120 km (E-region) is used for the calculation of absorption progression then velocities derived from the monostatic absorption observations closely match the progression velocities derived from the multipoint magnetic data displayed in Figure 6a.

The electron heating process was first detected in incoherent scatter radar observations (e.g., Schlegel and St-Maurice [8]). The absorption processes associated with electron heating events have been described in Stauning [9]; Stauning and Olesen [10]; Stauning [44]. Theoretical calculations of the electron heating and the expected absorption intensities using the actually observed electric fields and electron density profiles match the observed absorption values (Stauning [44]). The mutual consistency supports the above interpretation of this class of absorption events.

4.4 DAYSIDE ABSORPTION SPIKE EVENTS

The imaging riometer observations of cosmic noise absorption intensities with high temporal and spatial resolutions have revealed a new type: the high-latitude daytime absorption spike event (Stauning and Rosenberg [29]). These events, typically, have

250

durations of 1-2 min. They are rather stationary and are limited in spatial extent to a region of typically 50-100 km in dimensions; they are rather weak, typically 0.2-0.3 dB at the IRIS observing frequency of 38 MHz. The accompanying signatures in the geomagnetic recordings are weak. They have no obvious relations to planetary magnetic activity nor to interplanetary magnetic field (IMF) variations. The bulk of these events occur at or a little equatorward of the afternoon flow reversal in the polar plasma convection patterns.

This class of events constitutes one among several types of small-scale impulsive high-latitude dayside ionospheric disturbance events. They have occurrence characteristics like the isolated electron precipitation events reported from satellite X-ray observations (e.g., Imhof *et al.* [45]), or the dayside auroral bright spots in Lui *et al.* [46]. Hence they could be part of or otherwise related to these processes. They are not associated with magnetic impulse events or travelling convection vortices since both these types, which do have absorption signatures, have dominant magnetic signatures, much larger spatial scales, and different dynamic characteristics. The high-latitude daytime absorption spike events appear to form a population well separated from substorms at the nightside. However, the generation mechanism may have some similarities with the processes which, acting in a different plasma environment in the midnight sector, are responsible for the generation of large-amplitude absorption spike events observed during substorms (e.g., Nielsen and Axford [47]; Hargreaves *et al.* [48]; Nielsen [49]).

A strong, but otherwise typical spike event, that occurred at 1805 UT on 21 July, 1991, is presented in Figure 7. This figure has a field for each of the 49 beams of the IRIS instrument in Sdr. Strømfjord (STF). Geomagnetic north (compas direction) is upward.

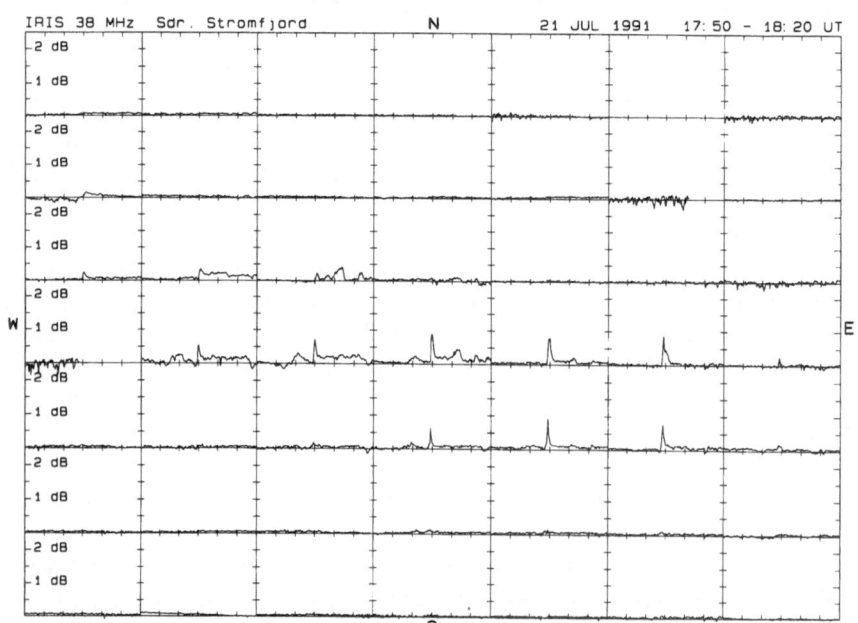

Figure 7. Distributed frame diagram of absorption intensities for a spike event at 1805 on 21 July 1991. The frames present imaging riometer beams positioned according to their direction.

The fields in Figure 7 have been placed according to the beam direction with the field for the northwesternmost beam placed in the upper left corner and so on. Each field has a time scale corresponding to the time interval: 1750-1820 UT, noted at the top of the diagram. The spacing between the marks on the horizontal axes is 5 min. The vertical scales for absorption have marks every 0.5 dB as indicated by the annotation to the left in the diagram. In some fields the traces are missing. They have been removed from the plot due to signal scintillations caused by one of the strong, discrete radio sources, Cassiopeia A or Cygnus A, entering the antenna beam. One trace (second upper-left) has been partly removed only. The remaining section displays the increasing scintillation level.

It is seen from the figure, that the absorption intensities are very low ($< \sim 0.1$ dB) in most of the different fields (beams). However, in the middle beams one observes fluctuating absorption intensities at a level of around 0.2-0.4 dB, and then, a little before 1805 UT, the onset of a sharp absorption spike extending to an amplitude of almost 1 dB. The duration of the spike is around 1 min. The latitudinal width corresponds to two fields, that is, ~ 40 km, while the longitudinal extent corresponds to 3-4 fields, that is, ~ 80 km.

These daytime absorption spike events have a preferred occurrence at or just equatorward of the afternoon convection reversal. Most of the events are associated with shear reversals. The explanation offered in Stauning and Rosenberg [29] assumes upward field-aligned currents, carried by downward precipitating electrons, at or in the vicinity of the convection reversal. If the magnetospheric electron densities are too small to sustain the currents flowing against the converging magnetic field then large field-aligned potentials could be formed. In that case the magnetospheric electrons are accelerated and precipitated into the ionosphere whereby its conductivity is increased, and the large-scale horizontal electric fields reduced. This, in turn, has the effect that the field-aligned potentials are further increased in a positive feedback process driving the event into saturation. The feedback effect operates equally well in the reverse process whereby the conditions may quickly revert to the undisturbed level.

Since the saturation of the field-aligned currents is an element in the process then, in agreement with observations, one should not expect any significant magnetic signature of the event. An unclear point is whether the absorption effects are directly associated with the accelerated electrons involved in the field-aligned currents or whether they originate from the enhanced precipitation, caused by these disturbances, of an existing high-energy electron population.

5. Summary

The use of radiowave absorption observations provided by imaging riometer systems offers useful features for geophysical investigations. Such observations can be used for, among others:
- Characterization of the general geophysical conditions, identification of the actual type of disturbance event, and grading the level of activity.
- Detection of the fine-structure in the spatial distributions of high-energy particle precipitation such as arc-like or spike-like features and identification of dynamics features such as transient, periodical or fluctuating variations, and drift motions.

- The absorption observations can be related to studies of wave-particle interactions and to studies of ionospheric plasma instability processes
- Analysis of observed absorption intensities can provide quantitative verification of the relations between high-energy particle precipitation, ionization profiles, chemical reaction schemes and neutral atmospheric composition. Such work may assist to develop models for the ionization balance in the lower ionosphere (D-region).
- The detection of energetic particle radiation can be used in tracing of magnetospheric morphology, detection of trapping boundaries, for the definition of magnetosphere-ionosphere conjugacy and arctic-antarctic conjugacy used for verification of magneto-spheric models, and for studies of polar cap boundary processes related to solar wind-magnetosphere interactions.
- The absorption observations can be used to supplement incoherent scatter radar observations to compensate for the limited spatial and temporal resolution available with this instrument.
- The observations are useful as supplements to satellite experiments to help separating spatial and temporal variations in the one-dimensional satellite observations.

In recognition of the useful features, efforts are presently made to explore the possibility for making processed data available (e.g., in the form of absorption images) from several of the observing imaging riometer systems by using modern telescience tools.

6. Acknowledgments

We gratefully acknowledge the use of IMP 8 interplanetary data provided by R. Lepping at NASA Goddard Space Flight Center. Also, we gratefully acknowledge the use of the Greenland magnetometer data made available by O. Rasmussen, Danish Meteorological Institute (DMI). The careful operation of the DMI magnetometers at the extended net of observatories in Greenland and the operation of the incoherent scatter radar and the imaging riometer at the station in Sdr. Strømfjord, managed for the National Science Foundation and DMI by SRI International, is most gratefully acknowledged. The processing of imaging riometer data and the assistance in the reproduction efforts provided by Søren Henriksen, DMI, is greatly appreciated. The operation of imaging riometers at Sdr. Strømfjord, Danmarkshavn and Longyearbyen has been supported by the Danish Research Council for Natural Sciences.

7. References

1. Little, C. G. (1954) High latitude ionospheric observations using extraterrestrial radio waves, *Proc. Inst. Radio Engrs.*, *42*, 1700-1701.
2. Little, C. G. and H. Leinbach (1959) The Riometer - a device for continuous measurement of iono-spheric absorption, *Proc. IRE*, *47*, 315-320.
3. Stauning, P. (1996) Investigations of ionospheric radio wave absorption processes using imaging riometer techniques. *J. Atmos. Terr. Phys.*, *58*, 753-764.
4. Detrick, D. L. and T. J. Rosenberg (1990) A phased-array radiowave imager for studies of cosmic noise absorption, *Radio Science*, *25*, 325-338.

5. Rawer, K. (ed.) (1976) Manual on ionospheric absorption measurements, World Data Center A for Solar-Terrestrial Physics, *Report UAG-57*, Boulder.

6. Mantas, G. P. (1974) Electron collision frequencies and energy transfer rates, *J. Atm. Terr. Phys.*, *36*, 1587-1600.

7. Aggarwal, K. M., N. Nath and C. S. G. K. Setty (1979) Collision frequency and transport properties of electrons in the ionosphere, *Planet. Space Sci.*, *27*, 753-768.

8. Schlegel, K. and J. P. St.-Maurice (1981) Anomalous heating of the polar E-region by unstable plasma waves, *J. Geophys. Res.*, *86*, 1447-1452.

9. Stauning, P. (1984) Absorption of cosmic noise in the E-region during electron heating events. A new class of riometer absorption events, *Geophys. Res. Lett.*, *11*, 1184-1187.

10. Stauning, P. and J. K. Olesen (1989) Observations of the unstable plasma in the disturbed polar E-region, *Physica Scripta*, *40*, 325-332.

11. Majumdar, R. C. (1972) Generalization of Appleton-Hartree equation of theories of collision frequency, *Indian J. Radio Space Phys.*, *1*, 31.

12. Rees, M. H. (1963) Auroral ionization and excitation by incident energetic electrons, *Planet. Space Sci.*, *11*, 1209-1218.

13. Rees, M. H. (1969) Auroral electrons, *Space Sci. Rev.*, *10*, 413-441.

14. Rees, M. H. (1982) On the interaction of auroral protons with the Earth's atmosphere, *Planet. Space Sci.*, *30*, 463-472.

15. Rawer, K. and D. Bilitza (1989) International reference ionosphere, *J. Atmos. Terr. Phys.*, *51*, 781-790.

16. Schunk, R. W. (1988) A mathematical model of the middle and high latitude ionosphere, *Pure Appl. Geophys.*, *127*, 255-303.

17. Roble, R. G., E. C. Ridley, A. D. Richmond, and R. E. Dickinson (1988) A coupled thermosphere/ionosphere general circulation model, *Geophys. Res. Lett.*, *15*, 1325-1328.

18. Wang, W., T.L. Killeen, A. G. Burns, and R. G. Roble (1997) A thermosphere-ionosphere nested grid (TING) high-resolution, three-dimensional, time-dependent model, to appear in *J. Atm. Solar-Terr. Phys.*

19. Wang, Z., T. J. Rosenberg, P. Stauning, Su. Basu, G. Crowley (1994) Calculations of riometer absorption associated with F-region plasma structures based on Sondre Stromfjord incoherent scatter radar observations, *Radio Science*, *29*, 209-215.

20. Torkar, K. M. and M. Friedrich (1983) Tests of an ion-chemical model in the D- and lower E-region, *J. Atm. Terr. Phys.*, *45*, 369-385.

21. Smirnova, N., O. Ogloblina, and V. A. Vlaskov (1988) Modelling of the lower ionosphere, *Pure and Appl. Geophys.*, *127*, 353-379.

22. Hedin, A. E. (1987) MSIS-86 Atmospheric model, *J. Geophys. Res.*, *92*, 4649-4662.

23. Hedin, A. E. (1991) Extension of the MSIS thermospheric model into the middle and lower atmosphere, *J. Geophys. Res.*, *96*, 1159-1172.

24. Barcus J.R., K.D. Hudnut, *P. Stauning*, I.B. Iversen, and J. Phillips (1986) Observations of Bremsstrahlung from Solar electron in the Earth's atmosphere, *J. Atmos. Terr. Phys.* 48, 375-384.

25. Baker, D. N., P. Stauning, E. W. Hones,jr., P. R. Higbie, and R. D. Belian (1979) Strong electron pitch angle diffusion observed at geostationary orbit, *Geophys. Res. Lett.*, *6*, 205-208.

26. Baker, D. N., P. Stauning, E. W. Hones jr., P. R. Higbie, and R. D. Belian (1981) Near equatorial, high-resolution measurements of electron precipitation at L=6.6, *J. Geophys. Res.*, *86*, 2295-2313.

27. Kikuchi, T., H. Yamagishi and N. Sato (1988) Eastward propagation of Pc 4-5 range pulsations in the morning sector observed with scanning narrow beam riometer at L=6.1, *Geophys. Res. Lett.* *15*, 168-171.

28. Nishino, M., H. Yamagishi, P. Stauning, T. J. Rosenberg, and J. A. Holtet (1997) Location, spatial scale and motion of radio wave absorption in the cusp-latitude ionosphere observed by imaging riometers. *J. Atmos. Solar-Terr. Phys.*, *95*, 903-924.

29. Stauning, P. and T. J. Rosenberg (1996) High-latitude daytime absorption spike events, *J. Geophys. Res.*, *101*, 2377-2396.

30. Rosenberg, T. J., Z. Wang, A. S. Rodger, J. R. Dudney, K. B. Baker (1993) Imaging riometer and HF radar measurements of drifting F-region electron density structures in the polar cap, *J. Geophys. Res.*, *98*, 7757-7764.

31. Reid, G. C. (1961) A study of the enhanced ionization produced by solar protons during a polar cap absorption event, *J. Geophys. Res.*, *66*, 4071-4085.

32. Van Allen, J. A., W. C. Lin, and H. Leinbach (1964) On the relation between absolute Solar cosmic ray intensity and riometer absorption, *J. Geophys. Res.*, *69*, 4481-4491.

33. Hultqvist, B. (1963) On the height distribution of the ratio of negative ion and electron densities in the lowest ionosphere, *J. Atm. Terr. Phys.*, *25*, 225-240.

34. Reagan, J. B. and T. M. Watt (1976), Simultaneous satellite and radar studies of the D-region ionosphere during the intense solar particle event of August 1972, *J. Geophys. Res.*, *82*, 4579-4596.

35. Stauning, P. (1996) High latitude D- and E-region investigations using imaging riometer observations. *J. Atmos. Terr. Phys.*, *58*, 765-783.

36. Sergeev, V. A., and M. V. Malkov (1988) Diagnostics of the magnetic configuration of the plasma layer from measurements of energetic electrons above the ionosphere, *Geomagn. Aeron.*, *28*, 549.

37. Sergeev, V. A., M. V. Malkov, K. Mursula (1993) Testing the isotropic boundary algoritm method to evaluate the magnetic field configuration in the tail, *J. Geophys. Res.*, *98*, 7609-7620.

38. Sergeev, V. A. (1996) Energetic particles as tracers of magnetospheric configuration, *Adv. Space Res.*, *18*, 161-170.

39. Stauning, P., H. Yamagishi, M.Nishino, and T.J. Rosenberg (1995b) Dynamics of Cusp-latitude absorption events observed by imaging riometers. *J. Geomag. Geoelectr.*, *47*, 823-845.

40. Clauer, C. R., P. Stauning, T. J. Rosenberg, E. Friis-Christensen, P. M. Miller, and R. J. Sitar (1995) Observations of a solar wind driven modulation of the dayside ionospheric DPY current system, *J. Geophys. Res.*, *100*, 7697-7713.

41. Stauning, P., E. Friis-Christensen, O. Rasmussen, and S. Vennerstrøm (1994a) Progressing polar convection disturbances: Signature of an open magnetosphere, *J. Geophys. Res.*, *99*, 11,303-11,317.

42. Stauning, P., C. R. Clauer, T. J Rosenberg, E. Friis-Christensen, and R. Sitar (1995a) Observations of solar-wind-driven progression of interplanetary magnetic field B_Y-related dayside ionospheric disturbances, *J. Geophys. Res.*, *100*, 7567-7585.

43. Stauning, P. (1994) Coupling of IMF B_Y variations into the polar ionospheres through interplanetary field-aligned currents, *J. Geophys. Res.*, *99*, 17,309-17,322.

44. Stauning, P. (1995) Progressing IMF B_Y-related polar ionospheric convection disturbances, *J. Geomag. Geoelectr.*, *47*, 735-757.

45. Imhof, W. L., H. D. Voss, D. W. Datlowe; and J. Mobilia (1988) Isolated electron precipitation regions at high latitudes, *J. Geophys. Res.*, *93*, 2649-2660.

46. Lui, A. T. Y., D. Ventekatesan, and J. S. Murphree (1989) Auroral bright spots on the dayside oval, *J. Geophys. Res.*, *94*, 5515-5522.

47. Nielsen, E., and W. I. Axford (1977) Small scale auroral absorption events associated with substorms, *Nature*, *267*, 502-504.

48. Hargreaves, J. K., H. J. A. Chivers and E. Nielsen (1979) Properties of spike events in auroral absorption, *J. Geophys. Res.*, *84*, 4245-4250.

49. Nielsen, E. (1980) Dynamics and spatial scale of auroral absorption spikes associated with the substorm expansion phase, *J. Geophys. Res.*, *85*, 2092-2098.

RESPONSE OF THE POLAR CAP IONOSPHERE TO CHANGES IN (SOLAR WIND) IMF

HERBERT C. CARLSON, JR
AFRL/Chief Scientist: Geophysics, HAFB, Bedford MA, 01731, USA

Abstract: The character of the polar cap ionosphere is dominated by the component of the solar wind IMF (Interplanetary Magnetic Field) that is parallel/anti-parallel to earth's magnetic field. Changes in the polar cap ionosphere properties, driven by changes in the IMF, necessarily drive changes in both the polar thermosphere and in growth rates of plasma instabilities and polar plasma structuring over a range of scale sizes exceeding 10^4. The physical processes determining this character are mutually interactive. Selected interactive physical processes are reviewed, to help clarify what is known, and suggest where important new findings may be found, particularly at Svalbard. The ionosphere/ thermosphere transient response to IMF reversals should clarify magnetospheric topology for northward IMF. The cusp thermosphere should exhibit transient upwelling and molecular enrichments. Certain cusp aurora should be excited by thermal electrons; their greater altitude may impact understanding of polar ionospheric convection. Polar ionospheric patch research requires more rigor in specific areas. A number of clear signatures are suggested for polar cap arcs, cusp reconnection events, and related phenomena. The thermosphere, as the rest frame for plasma response to electric fields (currents, heating, chemical loss, instabilities), can have transpolar winds order a km/s, which require measurement.

1. Introduction

Recall that the solar atmosphere, too hot for the Sun's gravitational field to contain, is continually escaping the sun. This escaping (solar wind) atmosphere, too hot for electrons to remain bound to nuclei, is a fully ionized plasma carrying it's own magnetic field. Earth's magnetic field represents an obstacle to this "solar wind". At that sunward distance from the earth where the Earth's magnetic field pressure approximates the opposing plasma pressure of the solar wind (~10 Earth radii towards the sun), the solar wind flows around the earth's magnetosphere (~50 Earth radii diameter cylinder, extending downstream beyond the distance to the Moon). Mechanical (kinetic) energy of the solar wind plasma is converted to electrical energy by the interaction, carried downward along magnetic fields into the ionosphere in the

J. Moen et al. (eds.), Polar Cap Boundary Phenomena, 255–270.
© 1998 *Kluwer Academic Publishers.*

form of energetic electrons and ions (current carriers) and Poynting flux. A magnetic reconnection interpretation of the interaction at the magnetopause best predicts observed behavior of ionospheric circulation and a great many other properties of earth's space environment, and is thus employed here.

This electrical energy is conveyed into the high latitude ionosphere and upper atmosphere. The Poynting flux manifests itself there as joule or ohmic heating. The particle flux leads directly to: production of secondary electrons; thermal plasma ionization which can be thought of as chemically stored energy; heat into the thermal electron gas; and excited states relaxing via optical emissions, which escape optically thin but redeposit energy in optically thick regions of the upper atmosphere. Ultimately the energy is carried away in the form of photons, Alfven waves, electromagnetic emissions and plasma instabilities, "polar wind" light and heavy ions, currents driven by polar thermosphere "flywheel" effects, but mostly downward heat conduction to the lower thermosphere heat sink.

Above 110 km in the thermosphere, near auroral latitudes the rate of heat deposition tracing to solar wind driven processes at times well exceeds the global average solar UV/EUV heating rate of ~0.5 erg cm^{-2} sec^{-1}. This has global impact on the ionosphere and thermosphere. For example midday thermospheric winds, which under magnetically quiet conditions blow poleward from the sub-solar atmospheric hot spot towards the anti-solar cold region, will abate (reverse and blow equatorward) for moderate (high) heating rates at auroral latitudes [1]. More complex processes redistribute global thermosphere composition, pressure, and density, which thereby significantly effects ionospheric composition, transport, and density.

The effectiveness with which solar wind energy couples to Earth's magnetosphere-ionosphere-thermosphere depends on the IMF (interplanetary magnetic field), as the magnetic reconnection model would predict [2, 3].

The strong southward IMF coupling-effects on polar ionospheric and thermospheric circulation is relatively well understood. The weak northward IMF coupling case is much more poorly understood. Critical and fundamental questions still remain for both cases however [4]. These will be discussed under Plasma Properties.

The global thermospheric effects have been comprehensively studied [5, 1]. Although much has been learned, here again fundamental questions remain outstanding [e.g. 6]. These global effects are beyond the scope of this paper. However variability of both the local cusp and polar thermosphere must be discussed as necessary context for understanding Plasma properties. This is discussed in the Thermosphere Properties section.

Plasma structuring in the high latitude, polar, and cusp ionosphere, is also of special interest here. The forcing plasma instability processes involve key scientific issues uniquely subject to study from the location of Svalbard, and depend intimately on thermosphere-ionosphere interactions. These processes, and consequent structuring of polar ionospheric plasma, particularly as amenable to study from the cusp region, is discussed in the Instability Properties section.

Processes discussed in each of these three sections are interdependent, as noted in the final section.

2. Plasma Properties

Polar cap ionospheric convection depends dramatically on whether the solar-wind magnetic field is more nearly parallel or antiparallel to the sub-solar magnetic field of the earth. When the fields are antiparallel (the IMF is southward), there is a well ordered two cell polar ionospheric convection pattern, responding to strong large scale ordered magnetospheric convection driven by solar wind coupling. When the fields are antiparallel, although solar wind coupling still drives convection, the convection becomes very anisotropic [7]. Convection flows are ordered over scale sizes of thousands of km in the earth-sun direction, but tens of km in the dawn-dusk direction. These strong differences between how the solar wind couples to plasma circulation (convection) in the magnetosphere and polar ionosphere for parallel vs. antiparallel IMF orientation lead to dramatic differences in the character of the polar ionosphere. One can say it is in one of two states. The magnetic reconnection view has the best success at predicting many of the contrasting properties of these two states [3]. Contrasts between these two states is sufficiently dramatic, that ground based observations typically can identify the sign of the B_z (north/south) component of the IMF vector if stable for 10-15 minutes [8]. Often this is also true of the B_y (dawn/dusk) component of IMF, if stable for at least 5 minutes.

The character of the Southward and Northward IMF state of the polar ionosphere has been well observed by satellites [9, 10, 11] and ground based instrumentation [12, 13]. The Southward IMF state is relatively well defined and understood [9, 14], while the Northward IMF state is not [15, 4]. Here we will focus on problems especially ripe for attention at Svalbard, for both states. Strongest emphasis addresses how the solar wind forces the magnetosphere, ionosphere, and thermosphere.

Understanding of some aspects of the coupling between the polar ionosphere and solar wind is illuminated by studying the transient response of the polar ionosphere to sharp changes in IMF [16].

Table 1 summarizes the time delays between reversals of the sign of the IMF B_z and B_y component, and response of observed properties in or near the polar cap.

2.1. POLAR CAP ARCS

Polar cap arcs are the optical emission signatures of steep electric field gradients in the dawn-dusk direction within the polar cap. Given their simple arc electrodynamics [17], a straight forward first principles approach to analyzing a sun aligned arc reveals many properties it must have, with consequent signatures it must have [4]. Briefly, solar wind mechanical energy is transformed to electrical energy at the magnetopause interface. This is conveyed down magnetic field lines as Poynting flux to the

Table 1: POLAR CAP RESPONSE TIME TO IMF REVERSAL

Worker	Time Delay (minutes)	IMF*	Observation
Bargatze	Dayside Merging / Tail Substorm	1	AL Index 20 min, 60 min
Hairston & Heelis	Southward Turning / Northward Turning	1,2	Ion Driftmeter 15-25 min, 29-44 min
Lockwood	As soon as	1	Magn. Merging Effects (ISR) 10 min (or more)
Trochichev	1 / 2	1,2	PC Arcs: Disappear, Appear (4) 20 min, 55 min
Valladeres et al	Disappear before	1	PC Arcs: Disappear less than 30 min
Rodriguez et al		1	PC Arcs: Antisunward decay 16-31 min
Carlson	By reversal	3	PC Arcs: Transpolar Reorientation 5 min
Clauer & Friis - Christiansen	Initial / Polar Reconfiguration	1	Magnetometer and ISR 3 min, 22 min

(Time Delay axis marked at 20, 40, 60)

*1. Northward to Southward turning
2. Southward to Northward turning
3. Note this is By Reversal
4. Low sensitivity delays arc detection until >1kR

ionosphere where it drags high speed curtains of plasma antisunward, dissipating the electrical energy as joule heating of the ions at the feet of the field lines (ion-neutral collisions for particle view). The resulting dawn/dusk plasma velocity gradient has high velocity on the dawn side and low antisunward or sunward velocity around the dusk side of the velocity gradient (converging electric field gradient). The sense of the velocity difference across the boundary determines whether it tends to drive a horizontal convergence or divergence, and thus whether a field aligned upward or downward current is required to maintain a divergence free state. Here the horizontal current gradient requires a field aligned current up and out of the arc, carried by (soft and/or hard) precipitating suprathermal electrons (electrons being far more mobile than ions). The sheet of precipitating electrons ($\sim$ or > 0.1 keV) produces secondary electrons, ambient electron gas heating, ionization which thereby modifies the conductivity across the arc, and by electron impact the optical signature of the arc. A self-consistent 3D current system is set up, adjusted to the modified conductivity, and with a return current sheet adjacent to the arc and carried by thermal electrons. As an arc grows stronger, the precipitating electrons grow harder (faster electrons meaning more current per unit particle flux), and multiple arcs form.

Thus *signatures* of these arcs must include:

1. Curtain (sun aligned) of rapid antisunward moving plasma on dawn edge of arc
2. Slowing of plasma velocity towards dusk and often reversal towards sunward around dusk edge
3. Upward current sheet within this converging electric field, carried by a precipitating sheet of soft and/or hard electrons (producing arc optical signature at their feet)
4. Core of enhanced E (hard electrons) and lower F (soft electrons) region ionization under upward current sheet, (and separately, modified ratio of molecular to atomic ions, as per modified production and recombination rates)
5. Adjacent sheet of downward current carried by upgoing thermal electrons
6. Core of high ion temperature in the E region on the dawn edge of the arc (joule heating under Poynting flux)
7. Sheet of hot electrons over arc (heated by precipitating suprathermal electron flux) (for thermal balance, both heating by precipitating electrons and cooling to enhanced ionization vary across the arc).

Furthermore, these signatures must be also present in any velocity-shear ohms-law arc. They are observed in the quiet dawn auroral oval, flow channels in the southward IMF cusp, the poleward edge of the dusk auroral oval, post midnight "hook" arcs, and others. For non earth-sun aligned features, e.g. hook arcs, the dawn/dusk side of the arc instead becomes the fast antisunward plasma flow side (vice dawn), and slow or sunward flow side (vice dusk).

Not all of these signatures have been necessarily appreciated in cusp velocity shear (reconnection) events.

In the future, there are several internal consistency matters that are in need of testing models against observations. These include understanding the three dimensional currents in a system of sun aligned arcs, the distinction between ionospheric flow shears and magnetospheric boundaries, the transient aspects of arc growth and decay, and arc effects on other phenomena.

The key outstanding issue related to sun-aligned arcs however, is what can we learn about northward-IMF solar wind coupling to the magnetosphere, from their properties. This is closely tied to the topological problem of how they project to and through the magnetosphere.

2.2. POLAR PATCHES

For southward IMF, polar ionospheric convection [15] transports plasma antisunward across the cusp and polar cap, then returns it sunward, equatorward of the polar cap. When the noon cusp is sunlit, one might expect (Knudsen) a continuous tongue of enhanced ionization to traverse the polar cap along these convection flow lines. This has been well modeled [18]. Large midday electron density, produced by solar EUV, enters the polar cap through the cusp and traverses the surrounding lower density plasma, as a large density tongue or peninsula.

In sharp contrast, what is typically observed is a series of patches, i.e. islands of high density ionospheric plasma, fully surrounded by lower density polar plasma. These islands move antisunward when observed sufficiently close to the dayside auroral region [19, 20, 21]. To understand the origin of polar cap patches, as these islands of enhanced plasma density have come to be called, remains an active area of research. Crowley [22] has presented an excellent review of work to date in this area.

Progress in this research area would be greatly facilitated however if the community could agree on three matters: 1) reaffirm definition of the problem to be solved; 2) determine what data is/isn't relevant; 3) review goals.

1) *First*, the problem to be solved clearly is not how to get enhanced daytime plasma densities into the dark polar cap. Tongues or peninsulas of ionization were discussed and modeled a decade ago. To the contrary, the problem is how to chop peninsulas into islands. Some recent work has muddied this distinction.

Therefor we need to agree on a common definition of a patch, such as:

"An island of high density plasma, approaching daytime values, surrounded by clearly lower density plasma within the polar cap."

2) *Second*, data must address the problem. All patches are density enhancements in the polar cap. Not all density enhancements in the polar cap are patches (that is islands). An antisunward moving peninsula of enhanced plasma density, following a meandering path as the IMF B_y varies, can look discontinuous if observed only in a single plane, or only in an overhead line of sight vs. time. Ground based data of this nature (line of sight altitude profiles or single parameter vs. time without imager or other data to establish whether the enhancement is an island or a line segment) can do more damage than good in guiding our understanding! Satellite data

along an orbit cutting patches (as identified by other means) can be quite valuable, as can ground based data complimented by some island discriminating capability, e.g. imager data with airglow intensity capability (tens of R vs. hundreds to kR sensitivity).

Therefor we must define the data used, and how we know it is data for a patch (island).

Anyone wanting to study snake-like peninsulas should identify them as such, or at the very least acknowledge that it is not known which the data apply to.

3) *Third*, we must recognize that even chopping a continuous "peninsula" into a series of islands can be done in several ways. In addition, there are other ways to produce enhanced polar plasma density structures and islands without "chopping" a continuous peninsula into segments, and these too must be considered.

The easiest way to chop a continuous tongue of ionization entering the polar cap through the cusp, is with a strong transient velocity flow channel. This idea is a familiar from trough generation research, variously called chemical recombination rate dependence on plasma velocity in neutral rest frame, velocity dependent recombination, or electric field enhanced dissociation of O^+ with neutral molecular nitrogen to form NO^+, which very rapidly recombines. What is attractive about this is that such velocity transients are known to occur.

Transient reversals or changes in the IMF B_y and B_z components, as well as solar wind transient pressure enhancements, also often occur. In principle, a rapid change in B_y would just make a peninsula in the form of islands connected by an "umbilical cord" of enhanced ionization (as beautifully modeled by Anderson and co-workers). However, in the real world other processes likely exist which for sufficiently narrow connections in effect "cut the cord". Likewise, patches from reversals in B_z require realistic description of the transient response.

Still another process, that applies even for steady IMF is production of patches by transient magnetopause reconnection [23]. This mechanism produces patches with realistic sausage shapes, as typically observed (Buchau, private comm., 1992), with no problem of connectivity between adjacent patches.

Therefor the time has come to stop asking how can one produce a patch, and focus on the question of which patch generating mechanism(s) is (are) dominant. The models (e.g. Anderson, Schunk, and co-workers) are ready whenever the data are.

The Svalbard ISR is in a unique position to address these questions too. It will be difficult though to do so with a standard patrol observing-program. An observing mode tailored to this scientific question would serve best.

2.3. THERMAL BALANCE

Magnetic field aligned currents, carried by 0.1 to several keV electrons in aurora, polar cap arcs, and the cusp will heat the ambient thermal electron gas. For the latter two cases the precipitating electrons are soft. For a given energy flux they produce relatively more secondary electrons at F region altitudes and greater F region electron gas heating rates. The story doesn't stop with heating rates however. For the polar cap arc and cusp soft electron situation it is essential to keep in mind that the temperature to which the ambient electrons are heated depends on a simple thermal balance between well defined heating and cooling rates.

A common mistake when studying temperature enhancements is to focus too strongly on heating rates. Often it is the cooling rate of the electron gas in the ionosphere that is the primary variable determining the quasi steady state electron gas temperature. The electron gas cools by: downward heat conduction through the lower altitude electron gas to the thermosphere heat sink at much lower altitudes (below 200 km) where the electron-neutral collisions become frequent; and by collisions locally with the ions. The latter cooling rate is proportional to the number of electrons times the number or ions with which to collide. The ions quickly cool by collisions with the local neutral particles. By charge neutrality, the ion and electron number densities are equal. The cooling rate of the electron gas is thus proportional to the square of the electron density. For the relevant collision cross sections, when the electron density is much below 3×10^5 cm^{-3}, the electron gas becomes in poor thermal contact with the local ion gas, and can only cool by downward heat conduction, the latter being directly dependent on the altitude gradient of the electron temperature. The electron temperature increases until its altitude gradient so large that cooling by downward heat conduction can carry away all the locally deposited heat. For electron densities much above 3×10^5 cm^{-3}, the electron gas is in good thermal contact with the ion gas. It then cools directly to the local ion gas, and thermal balance is achieved for much lower electron temperatures. (The ion gas is clamped closely to the neutral particle gas temperature by ion-neutral collisions for altitudes below 450 km.) This introduces an approximation to a two-position switch, driven by the electron density, between high or low electron gas temperature.

The actual magnitude of the "high" temperature of course depends on the heating rate. For sufficiently high heating rates that the thermal balance is achieved with electron temperatures above 3000 K, a third electron gas cooling term comes in-excitation of the lowest energy excited state of atomic oxygen O(^{1}D). The very strongly temperature-dependent excitation of 630 nm optical emission by thermal electrons becomes appreciable.

The idea that the cooling term in the thermal balance can be the critical variable controlling electron temperatures (and 630 nm emissions) has been of considerable value recently. It led to explanation and modeling of 630 nm airglow observations by thermal electron excitation in ionospheric HF heating experiments [24, 25], and SAR arc intensities vs. solar cycle and by event [26, 27].

In the cusp there is also an important application of these ideas. We have found that as F region electron densities switch from much above to much below 3×10^5 cm^{-3}, electron temperatures switch from low to high. As upper F region electron temperatures switch from below to above 3000 K, thermally excited 630 nm emissions go from negligible to significant. Cusp electron fluxes and consequent heating rates can be large enough to produce thermal excitation of 630 nm emissions if the cooling rate is low (electron gas insulated from the local ions and thermosphere). Otherwise the dominant 630 nm excitation is from direct suprathermal electron impact excitation rather than by electrons in the tail of the thermal electron gas distribution function.

This can be of importance beyond the simple question of the correct 630 nm excitation mechanism. The 630 nm emission from thermal excitation is from significantly higher altitude than direct suprathermal electron impact excitation. If the cusp 630 nm comes from thermal vs. direct suprathermal electron impact, the role of transient magnetopause reconnection in driving global ionospheric convection is much greater than presently realized [28].

A straightforward triangulation measurement of the height of 630 nm emission over Longyearbyn/Ny Alesund, with coincident ISR electron gas temperature and electron density measurements would directly address this question. During mid winter it should confirm the expectation of thermally excited 630 nm in the cusp for low F region plasma densities and not uncommon cusp precipitation.

3. Thermosphere Properties

The global average energy flux to the thermosphere above 120 km, by solar UV/EUV radiation, is ~ 0.5 erg cm^{-2} sec^{-1} = $\sim 0.5 \times 10^{-3}$ W m^{-2}. Any sustained additional heating rate comparable to or greater than this can thus materially alter the nearby thermosphere. During major magnetic storms the energy flux into the auroral region can be ten times this amount. This is sufficient to reverse midlatitude daytime F region thermosphere winds. Normally forced poleward by sub-solar UV/EUV heating, they then become driven equatorward by auroral heating. Also changed are the global thermal, neutral composition, and density structure of the upper atmosphere.

Global thermosphere modulation of this nature has been well studied [29], although important problems in this area remain to be solved [6].

Here global impact is discussed only for context. Here we will focus instead on the polar cap and/or the cusp, which is much less well studied or understood. It too can have large scale impact [30].

3.1. DYNAMICS

The main effect of auroral heating on the local thermosphere is the upwelling of the overhead atmosphere. The particle and electromagnetic (joule heating) energy goes into heat or internal energy in the gas, producing adiabatic expansion. The heated parcel of gas raises in geopotential altitude until its adiabatic cooling matches the heating. This moves nitrogen rich gas to altitudes higher than otherwise, and leads to increased molecular to atomic gas ratios, temperature, pressure, and density at a fixed overhead altitude. For representative auroral energy deposition rates of ~ 5 erg cm^{-2} sec^{-1}, the instantaneous auroral energy flux into the thermosphere above about 110 km (into the finite width auroral region) during a magnetic storm is ten times that from the

264

solar UV/EUV over the earth. It is reasonable to expect this to have global impact on the thermosphere. In fact it does [5, 1].

To better quantify and validate global impact, Crowley *et al* [31] presented a well documented model simulation, based on an unusually complete set of input data to maximize the degree of reality of the model output and degree of validation possible [32]. Good overall agreement was found. The vertical winds exceed 1 ms^{-1}, the F region enhancement of molecular nitrogen to atomic oxygen more than doubles, and the number density of molecular nitrogen near 200 km is near twice its quiet time value. A theoretical study has also been done with this same model, focusing on the global response of the thermosphere to an impulsive (order hour vs. day) geomagnetic storm auroral heating event [36]. The global response of the thermosphere to even this transient auroral heating was significant, but the contribution to the global disturbance from the cusp (vs. nighttime auroral heating) was negligible. The net energy, duration, and horizontal extent of particle and joule heating in the cusp is substantially less than in the auroral region for a large magnetic storm, so it is not surprising that the impact on the global thermosphere is substantially less.

However, let's ask a closely related but importantly different question. How does the local overhead thermosphere response compare for the cusp vs. auroral region? A few simple principles go a long way towards providing a clear expectation. Breaking this question down into its component elements, three effects all conspire to amplify the overhead cusp thermosphere response, relative to overhead response per unit energy flux into night aurora.

How does the response to a given energy flux compare? In the auroral zone the expansion of the heated gas cell is upward, because the horizontal width of the heated band well exceeds the vertical neutral particle scale height. The same is true of the cusp-heated band. Thus, although the total effect on the global thermosphere will be far smaller, the local overhead effect would still be as great for a given energy flux. Energy flux enters in the form of particles and joule heating, which actually will be different.

How do the particle fluxes compare in the cusp vs. non-dayside aurora? Cusp particle energy fluxes are typically ~0.5 to 5 erg cm^{-2} sec^{-1}, less than auroral particle energy fluxes, but not by a great amount. However the particle energy distribution is significantly softer. This particle energy difference is highly significant, and enhances the cusp thermosphere density and composition response in two ways.

First, the lower energy cusp electrons deposit their energy at much higher altitudes. A 0.2 keV representative cusp electron has unity optical depth near 200 km, vs. a 2 keV representative auroral electron with unity optical depth near 110 km. Since particles deposit most of their energy within a neutral scale height of their altitude of unity optical depth, the thermospheric density about the altitude of the cusp electron heating is two orders of magnitude smaller than that about the altitude of the auroral electron heating. The higher altitude more tenuous atmosphere responds far more dramatically to a given energy deposition. The thermosphere responds to heat deposition primarily by adiabatic expansion, the energy going into raising the atmosphere against gravity.

The cusp case vertical winds will correspondingly also be about two orders of magnitude greater (order meters per second near 200 km altitude) than the auroral. Recall that upwelling is the primary way to change thermospheric composition (vertical gradients of composition far exceed horizontal gradients).

Second, joule heating, which often exceeds particle heating, is favored in the cusp. To see this, recall that the downward Poynting flux from the magnetosphere can be dissipated in the ionosphere-thermosphere and do mechanical work against **JxB** forces there. The Poynting flux, which originates as mechanical energy in the solar-wind generator, and is conveyed down the geomagnetic field lines into the ionosphere, is thus dissipated as joule heating. Joule heating in the ionosphere is $\mathbf{J} \cdot \mathbf{E}$ in the ionosphere rest frame. For an applied electric field, the energy dissipated is by the current component parallel to **E** (the Pedersen current). Hall current is nondissipative. This leads to a dependence of joule heating on the electron characteristic energy. Higher energy (auroral) particles penetrate more deeply into the atmosphere, where the Hall conductivity exceeds the Pedersen, and currents tend to be nondissipative. Lower energy (cusp) particles produce ionization at altitudes where Pedersen conductivities exceed Hall, so joule heating is relatively more important.

Third, the cusp heating effect on its overhead thermosphere is further amplified, relative to the auroral case, because of the mechanism driving the joule heating. The Pedersen current density is the Pedersen conductivity times the electric field **E**. Thus the height-integrated joule heating is proportional to the square of **E**. In the cusp for southward IMF, after reconnection, transient plasma flow velocities are about double that in the polar cap. Because of the square law dependence on **E**, double the plasma velocity means four times the joule heating.

Thus one should expect to see transient increases in: vertical neutral particle and ion velocities; neutral density; and molecular to atomic ratios for neutral particles and ions. How should one look? Experimental confirmation by satellite would be very challenging for most satellite techniques, and unrealistic for rockets. The incoherent scatter radar (ISR) technique, including its time continuous capability, is ideally suited to verifying this expectation. The sporadic overhead atmospheric-upwelling would show in enhancements of the molecular ion concentrations, and ion collision frequencies, both of which narrow the ISR spectral return as they change its shape. Thermospheric upwelling and subsequent subsidence should be manifest in ISR spectral Doppler shifts, as the ions respond to the neutral drag. The transient reconnection high velocity flow channels, and thus velocity shear electrodynamics, should lead to an associated current sheet (heating the overhead F region electron gas) that adjoins the joule heating channel (heating the E region ion gas). In the plasma properties section, see the discussion of polar cap arc electrodynamics and signatures. One can look for this current effect as an ISR transient spectral asymmetry.

3.2. REST FRAME FOR ELECTRODYNAMICS

It is the rest frame of the neutral gas that determines currents and effective electric field strengths. This is partly explained by recalling that it is collisions between the charged and neutral gas particles, and 'in particular the different collision frequencies (mobilities) of the ions vs the electrons with the neutral gas particles, which produce a finite conductivity and currents.

Transpolar thermospheric winds can vary enormously (km/s). Let's consider how this happens and why it is so important. The collision frequency of ions with neutral particles in the upper thermosphere (F region altitudes) is determined be the charge exchange rate between the dominant neutral and ion species, respectively atomic oxygen O, and atomic oxygen ions O^+. When an electron leaps from an O to an O^+, the neutral/ion particle energies are of course likewise exchanged (each becomes the other). Near 300 km the neutral density (predominantly O) is order 10^9 cm^{-3}, while the ion density (predominantly O^+) is order 10^5 to 10^6 cm^{-3}. The average O^+ ion collides with an O atom about once per second. If a large electric field is applied to the ion, giving it ordered energy in the direction of **E**, the charge exchange collision leaves the resultant ion going in a randomly different direction, thereby changing the ordered energy into disordered energy or heat. This is simply a particle view of F region joule heating. However this particle view clarifies many aspects of momentum transfer and heating rates. For instance, if an electric field is applied to ions, the once per second collision frequency means the entire ion gas will heat up in a few seconds. The change in velocity between the initial O and the product O^+ after the electron exchange, clearly depends on the difference between the mean ion and neutral particle velocity. Thus the ion heating rate must vary with the square of the velocity difference. This in turn means the square of the electric field as seen in the rest frame of the neutral gas. Quantitatively, for midlatitude ion velocities of order 100 m/s the ion heating rate is negligible, for high latitude velocities exceeding 1 km/s the ions heating exceeds 1000 K.

If the electric field persists long enough, the mean ion drift velocity will drag the neutrals up to their velocity. How long will this take? For 10^6 ions cm^{-3} ions and 10^9 cm^{-3} atomic oxygen atoms, each atom will on average experience one collision per 1000 ion collisions. Thus if the transpolar ionospheric velocity is 1 km/s, the thermosphere will come up to a transpolar velocity of 1 km/s in a few times 1000 seconds, or about half an hour. For an ionospheric density of 10^5 cm^{-3}, this would take about a quarter of a day, by which time the earth has rotated so much that the thermosphere is unaffected. Polar ionospheric plasma densities repeatedly alternate between these two values as patches pass across the polar cap for southward IMF, and also alternate over time scales for which the IMF alternates between northward and southward.

Thus the neutral rest frame can vary by 1 km/s or more! Ion/neutral gas velocity differences can vary by 2 km/s including flywheel effects. This can have enormous impact on plasma and thermospheric properties at high latitudes, and errors in theory or model calculations if thermospheric velocities are assumed or taken from climatological models vs. measured.

For electric field dependent ion chemical reaction rates, ion heating rates, currents in the ionosphere, plasma instability growth rates, i.e. anything dependent on the difference between ion and neutral gas velocities, the effective electric field is defined in the neutral (thermosphere) rest frame. The (thermosphere) rest frame velocity needs to be measured, not assumed.

3.3. ENERGY BUDGET

Models of the global thermosphere show a missing polar heating process, based on matching modeled to observed thermospheric/ionospheric densities and composition (D. Rees, R. G. Roble, private communication), and persistent equatorward winds at upper midlatitudes [6]. This continues to be an unresolved polar and global issue. The degree to which joule heating by polar cap arcs [33] and transpolar thermospheric processing by phenomena in the cusp (Killeen, private communication) and thermospheric agitation by patches, is a matter that the Svalbard ISR is in a special position to address.

4. Instability Properties

High latitude ionospheric plasma is subject to electromechanical forces that can structure it. Thus phrased, this could even capture the large end of the spectrum of ionospheric structure scale sizes already discussed as polar cap patches, which typically range from order 100 to 1000 km. But these are presently viewed as due to large scale transport processes, not plasma instability processes. We will confine discussion here to structuring by plasma instabilities.

Plasma instability processes commonly occur, and produce plasma structure in the nominal range of 10s of km to meters. These structures, and the plasma instability processes producing them, are a subject of intense research from both the perspective of understanding the associated plasma physics, and the perspective of their major effects on radio wave propagation for communications, navigation, and satellite imaging [34].

One major class of plasma instabilities in the ionosphere is the gradient drift instability. At the equator, the mechanical force of gravity acts across the horizontal altitude gradient of electron density, to produce a growth rate of plasma structuring that exceeds the damping rate of the structures, whence an instability and structuring. This produces the common phenomenon of spread F and scintillation in the equatorial ionosphere. Another example of this class of instability is at high latitudes, where a strong auroral or polar cap electric field acts perpendicular to a vertical gradient in electron density, again producing structuring growth rates that exceed the structuring decay rates and produce strong auroral clutter and scintillation. When strong currents flow, this modifies the geometry of the instability, but retains its basic gradient drift character. This class has been known and studied for decades.

A very different plasma structuring instability is also found at high latitudes- a shear driven instability [34, 35]. Theoretical studies had initially shown that the growth rate for this Kelvin-Helmholtz type instability was not sufficient great in the high latitude ionosphere to exceed the damping rate. This instability was deemed ineffective there, because model calculations concluded that the plasma shear carried away the free energy needed to feed the instability too rapidly for it to develop. When this instability was experimentally discovered [34], and validated as the shear instability, the theory was revisited. It emerged that collisions of ionospheric plasma with the neutral atmosphere had been omitted in the modeling. When added, this provided the needed free energy to drive the instability growth rate above the damping rate. Further refinement of the theory and modeling led to realization of the need to include capacitive coupling between the ionosphere and the magnetosphere [36].

In the cusp, strong transient velocity shears during reconnection events would be expected to provide strong initial growth rates for the shear driven instability. If this is a dominant patch generating mechanism [23], the patches as they continue across the polar cap will develop gradient drift instabilities. While the latter should favor instability growth on its edge, if the patch generation process has initially seeded it with shear instability structure, the spatial distribution of irregularity structure within the patch can be quite different. (For a given growth rate, growth from pre-structured plasma becomes strongly structured sooner than growth from smooth plasma.). This may explain why severe structured is observed throughout a patch, vs. predominantly on its edge as predicted by gradient drift instability theory.

The reconnection process, and strong ion friction moving through the thermosphere, provides the free energy to drive the growth of both of these instabilities. The instability decay rate depends strongly on the electrical conductivities at the feet of the ionospheric irregularity structures. These conductivities vary not only with particle precipitation, but predictably with solar zenith angle for a cusp that can experience both darkness and sunlight at local noon (i.e. Svalbard in mid winter vs. other times of year). The capacitive coupling of the cusp ionosphere to the magnetosphere varies significantly over the reconnection time cycle, influencing the balance of growth/decay rates. The Svalbard cusp provides a valuable "ionospheric laboratory" for study of the basic plasma physics, geophysics, and practical effects of these structuring phenomena [37]. Such studies are integral to understanding of the interaction of plasma, thermosphere, and instability processes in a realistic environment. Also important are instability growth rates relative to polar cap convection reconfiguration time scales [38], the latter being about five to tens of minutes for onset to completion (see table 1 here).

5. References

1. Roble, R. G. (1977) The Thermosphere: *The Upper Atmosphere and Magnetosphere*, National Research Council Monograph, Nat. Acad. Sci.

2.   Reiff, P. H. and Burch, J. L. (1985) IMF By-dependent plasma flow and Birkeland currents in the dayside magnetosphere, 2, A global model for northward and southward IMF, *J. Geophys. Res.*, **90**, 1595-1609.

3.   Cowley, S. W. H. and Lockwood, M. (1992) Excitation and decay of solar-wind driven flows in the magnetosphere-ionosphere system, *Ann. Geophys.*, **10**, 103.

4.   Carlson, H. C. (1994) The dark polar ionosphere: Progress and future challenges, *Radio Sci.*, **29**, 157-165.

5.   Richmond, A. D., and Roble, R. G. (1987) Electrodynamic effects of thermospheric winds from the NCAR thermospheric general circulation model, *J. Geophys. Res.*, **92**, 12365.

6.   Carlson, H. C., and Crowley, G. (1989) The equinox transition study: an overview, *J. Geophys. Res.*, **94**, 16861-16868.

7.   Carlson, H. C., Heelis, R. A., Weber, E. J., Sharber, J. R. (1988) Coherent mesoscale patterns during northward IMF, *J. Geophys. Res.*, **93**, 14,501-14514.

8.   Carlson, H. C. (1996) Incoherent scatter radar mapping of polar electrodynamics, *J. Atmos. Terr. Phys.*, **58**, 37-56.

9.   Heelis, R. A., Lowell, J. K., and Spiro, R. W. (1982) A model of the high-latitude ionospheric convection pattern, *J. Geophys. Res.*, **87**, 6339-6345.

10. Ismail, S., Wallis, D. D., and Cogger, L. L. (1977) Characteristics of polar cap sun-aligned arcs, *J. Geophys. Res.*, **82**, 4741.

11. Burke, W. J., Gussenhoven, M. S., Kelley, M. C., Hardy, D. A., and Rich, F. J. (1982) Electric and magnetic field characteristics of discrete arcs in the polar cap, *J. Geophys. Res.*, **87**, 2431-2443.

12. Weber, E. W., Klobuchar, J. A., Buchau, J., Carlson, H. C., Livingston, R. C., de la Beaujaudiere, O., McCready, M., Moore, J. G., Bishop, G. J. (1986) Polar cap F layer patches: Structure and dynamics, *J. Geophys. Res.*, **91**, 12,121-12,129.

13. Weber, E. W. and Buchau, J. (1981) Polar cap F layer auroras, *Geophys. Res. Lett.*, **8**, 125-128.

14. Heppner, J. P. and Maynard, N. C. (1987) Empirical high-latitude electric field models, *J. Geophys. Res.*, **92**, 4467.

15. Heelis, R. A., (1984) The effects of interplanetary magnetic field orientation on dayside high-latitude convection, *J. Geophys. Res.*, **89**, 2873.

16. Rodriguez, J. et al (in press, 1996) Decay of polar cap arcs after a southward turning of IMF, *J. Geophys. Res.*

17. Lyons, L. R. (1980) Generation of large scale regions of auroral currents, electric potentials, and precipitation by the divergence of convection electric field, *J. Geophys. Res.*, **85**, 17-24.

18. Schunk, R. W. and Sojka, J. J. (1987) A theoretical study of the lifetime and transport of large ionospheric density structures, *J. Geophys. Res.*, **92**, 12343.

19. Weber, E. J., Buchau, J., Moore, J. G., Sharber, J. R., Livingstone, R. C., Winningham J. D., and Reinisch, B. W. (1984) F-layer ionization patches in the polar cap, *J. Geophys. Res.*, **89**, 1683-1694.

20. Buchau, J., Weber, E. J., Anderson, D. N., Carlson, H. C., and Moore, J. G. (1985) Ionospheric structures in the polar cap: their origin and relation to 250 MHz scintillation, *Radio Sci.*, **20**, 325-338.

21. Anderson, D. N., Buchau, J., and Heelis, R. A. (1988) Origin of density enhancements in the winter polar cap ionosphere, *Radio Sci.*, **23**, 513-519.

22. Crowley, G. (1996) Critical review of ionospheric patches and blobs, Chapter 27, *Review of Radio Science 1993-1996*, URSI, Oxford Sci. Publ., Editor W. R. Stone, 619-648.

23. Lockwood, M. and Carlson, H. C. (1992) The production of polar cap electron density patches by transient magnetopause reconnection, *Geophys. Res. Lett*, **19**, 1731-1734.

24. Carlson, H. C. (1993) High power HF modification: Geophysics span of EM effects, and energy budget, *Adv. Space Res.*, **13**, 1015-1024.

25. Mantas, G. P. and Carlson, H. C. (1996) Reinterpretation of the 6300-Å airglow enhancements observed in the ionosphere heating experiments based on analysis of Platteville, Colorado, data, *J. Geophys. Res.*, **101**, 195-209.

26. Kozyra, J. U., Valladares, C. E., Carlson, H. C., Buonsanto, M. J., and Slater, D. W. (1990) A theoretical study of the seasonal and solar cycle variations of stable auroral red arcs, *J. Geophys. Res.*, **95**, 12219-12234.

270

27. Kozyra, J. U., Chandler, M. O., Hamilton, D. C., Peterson, W. K., Klumpar, D. M., Slater, D. W., Buonsanto, M. J., and Carlson, H. C. (1993) The role of ring current nose events in producing stable auroral red arc intensifications during the main phase: observations during the September 19-24 1984 storm, *J. Geophys. Res.*, **98**, 9267-9283.
28. Lockwood, M., Carlson, H. C., and Sandholt, P. E. (1993) Implications of the altitude of transient 630-nm dayside auroral emissions, *J. Geophys. Res.*, **98**, 15571-15587.
29. Burns, A. G., T. L. Killeen, G. Crowley, B. A. Emery, and R. G. Roble, (1989) On the mechanisms responsible for high-latitude thermospheric composition variations during the recovery phase of a geomagnetic storm, *J. Geophys. Res.*, **94**, 16961-16968.
30. Burns, A. G., Killeen, T. L., and Roble, R. G. (1991) A theoretical study of thermospheric composition perturbations during an impulsive geomagnetic storm, *J. Geophys. Res.*, **96**, 14153-14167.
31. Crowley, G., Emery, B. A., Roble, R. G., Carlson, H. C., and Knipp, D. J. (1989a) Thermospheric dynamics during September 18-19, 1984, 1, Model simulations, *J. Geophys. Res.*, 94, 16925-16944.
32. Crowley, G., Emery, B. A., Roble, R. G., Carlson, H. C., Salah, J. E., Wickwar, V. B., Miller, K. L., Oliver, W. L., Burnside, R. G., and Marcos, F. A. (1989b) Thermospheric dynamics during September 18-19, 1984, 2, Validation of the NCAR thermospheric general circulation model, *J. Geophys. Res.*, **94**, 16945-16959.
33. Valladares, C. E. and Carlson, H. C. (1991) The electrodynamics, thermal, and energetic character of intense sun-aligned arcs in the polar cap, *J. Geophys. Res.*, **96**, 1379-1400.
34. Basu, Su., Basu, Sa, MacKenzie, E., Coley, W. R., Sharber, J. R., and Hoegy, W. R. (1990) Plasma structuring by the gradient drift instability at high latitudes and comparison with velocity shear driven processes, *J. Geophys. Res.*, **95**, 7799.
35. Basu, Su., Basu, Sa., MacKenzie, E., Fougere, P. F., Coley, W. R., Maynard, N., Winningham, J. D., Sugiura, M., Hanson, W. B., and Hogey, W. R. (1988), Simultaneous density and electric field fluctuation spectra associated with velocity shears in the auroral oval, *J. Geophys. Res.*, **93**, 115.
36. Keskinen, M. J., and Huba, J. D. (1990) Nonlinear evolution of high latitude ionospheric interchange instabilities with scale-size dependent magnetospheric coupling, *J. Geophys. Res.*, **95**, 15157.
37. Basu, Sa., Basu, Su., Chaturvedi, P. K., and Bryant, C. M. (1994) Irregularity structures in the cusp-cleft and polar cap regions, *Radio Sci.*, **29**, 195-208.
38. Clauer, R. C. and Friis-Christensen, E. (1988) High-latitude dayside electric fields and currents during strong northward IMF: observations and model simulation, *J. Geophys. Res.*, **93**, 2749-2757.

POLAR CAP PHENOMENA AND THEIR RELATION TO BOUNDARY LAYERS AND THE IMF

D.J. McEWEN
Department of Physics and Engineering Physics
116 Science Place, University of Saskatchewan
Saskatoon, Saskatchewan, Canada S7N 5E2

1. Introduction

1.1 THE POLAR CAP

The polar cap is the area around the geomagnetic pole bordered by the poleward edge of the auroral oval. It has a nominal radius of about 15° but it contracts under the influence of the Interplanetary Magnetic Field (IMF) with increasing positive B_z [1-4]. This contraction is most evident in the dusk and dawn sectors. From a study of extensive DMSP particle data, Meng and Makita [2] have plotted the latitudes of the poleward boundaries of electron precipitation vs B_z(nT) and obtained the following empirical expressions showing polar cap contraction with increasing IMF B_z:

Evening sector $\quad$ $(78 + 0.8\ B_z)$ °corrected geomagnetic latitude (CGM)
Morning sector $\quad$ $(80 + 0.7\ B_z)$ °CGM

The quiet time polar cap is often seen to display a teardrop shaped void of emissions near the pole [1]. Since the low latitude boundary layer (LLBL) widens for northward B_z, the dayside poleward boundary connected to the LLBL must also expand (by as much as 2° from 83° to 85° CGM) according to studies by Makita et al. [3].

During major auroral substorms, with the IMF B_z southward, auroral expansion can extend to well above 80° CGM such as the November 4, 1993 substorm around 2330 UT [5]. During unusually major (global) events, the whole polar cap can fill with auroral emissions such as on November 8-9, 1991, February 21, 1994 and January 10-11, 1997 [6,7].

This whole picture of the effect of the IMF on the size and characteristics of the polar cap has been illustrated by Akasofu [8]. He shows a contracted polar cap with B_z positive, followed by an auroral substorm and poleward expansion subsequent to B_z going negative. Then a subvisual glow starts to fill the polar cap as the substorm subsides and there is an intrusion of arcs in the dusk or dawn sectors, depending on whether B_y is positive or negative. This illustrates the characteristics of emissions from the polar cap for a 10 hour period encompassing an auroral substorm.

J. Moen et al. (eds.), Polar Cap Boundary Phenomena, 271–280.

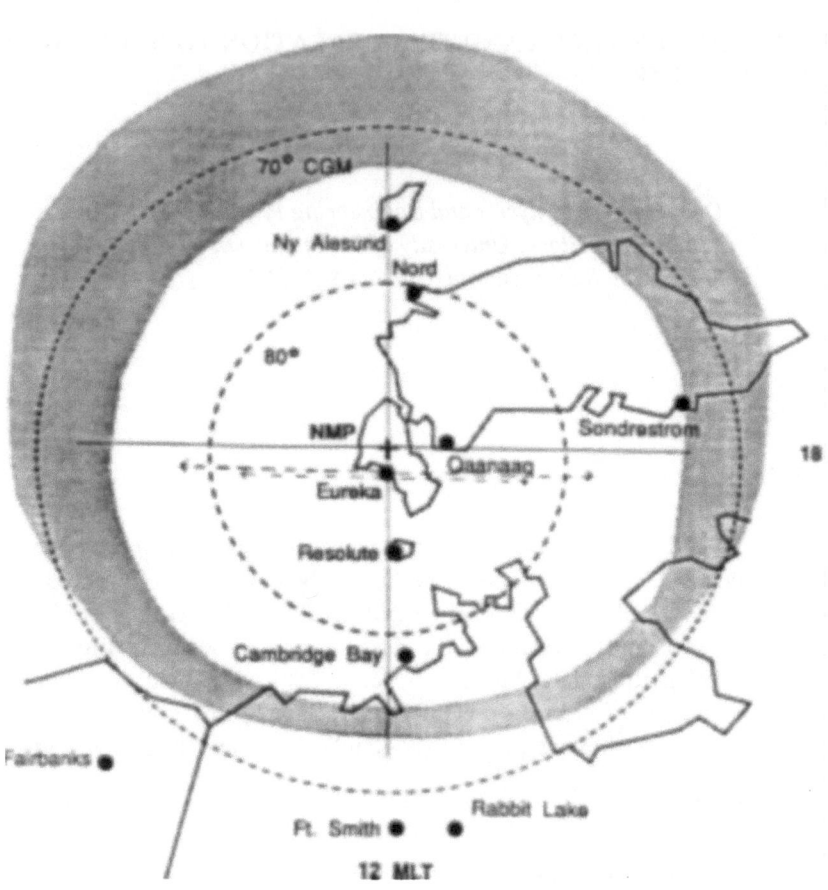

Figure 1. A polar map in magnetic coordinates of latitude and local time, with midnight at the top. The map is centered on the north magnetic pole (NMP). The location of Eureka (89° CGM) is shown relative to some other polar stations. The auroral oval at 1800 UT at Eureka for a K_p = 4 level is shown in grey.

1.2 EUREKA POLAR OBSERVATORY

The opening of a polar observatory at Eureka, Nunavut Territory, Canada (80.05°N, 86.42°W) in 1990 has permitted optical measurements of emissions in the central polar cap throughout each winter since due to its location at 89° CGM, near the north magnetic pole. Its location relative to other polar stations is shown in Figure 1. The auroral oval at 1800 UT at Eureka for a K_p=4 level is shown in grey.

The instruments there have included a 3λ all-sky camera (ASC) and a 6λ meridian scanning photometer (MSP). These instruments have provided a quasi-continuous monitor of optical activity in the polar cap each winter, yielding airglow intensities and detailed information on any enhancements superimposed on the airglow, such as polar auroral arcs or F-layer patches [9,10].

The MSP has a 1° field of view and steps through the meridian, horizon-to-hor-

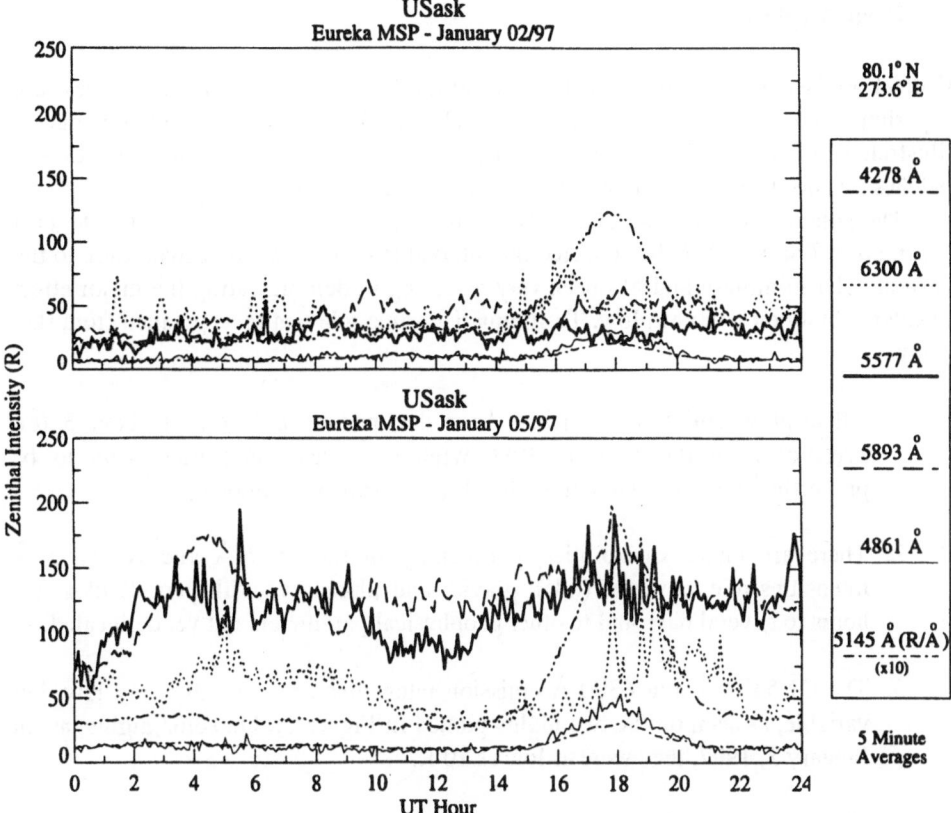

Figure 2. Zenithal intensities of emissions over Eureka on January 2 and 5, 1997. The six emissions are identified in the legend on the right. An enhancement is seen in some emissions around local noon (18 UT) due to twilight.

izon in 1° steps each 40s. The emissions at 4278, 6300, 5577, 5893 and 4861 Å plus a background wavelength at 5145 Å were monitored with sensitivities of the order of 1R/count. The zenithal intensities of these emissions are shown as summary plots in Figure 2 for two typical quiet days in early January, 1997. It has been advantageous to mount this instrument on a turntable which rotates to keep it scanning along a meridian fixed with respect to the sun-earth line. Since 1993 this direction has usually been along the dusk-dawn meridian. The fields of view in this meridian are shown with dashed lines in Figure 1 for the two OI 5577 Å and 6300 Å airglow emissions at assumed heights of 100 and 300 km respectively. The optical coverage of the MSP is thus from the magnetic pole down to approximately 83° and 76° CGM respectively, for these two oxygen emissions. Since the polar cap boundary in one or more time sectors is often above these latitudes, as discussed above, the ASC and MSP together provide optical coverage of most of the polar cap, particularly in the dawn-dusk direction.

2. Polar Airglow

Optical emissions associated with polar auroral arcs and F-region patches are superimposed on an airglow background which is itself variable and often dynamic as illustrated in Figure 2. And for extracting information on the sun-aligned arcs and patches one needs to be aware of this variable background.

The polar airglow levels of OI 5577 and 6300 Å have been monitored for the past 6 winters at Eureka. This has covered the interval from the last solar maximum to the current solar minimum and has been very revealing in demonstrating the major effect of solar activity on polar airglow. To summarize some relevant facts gleaned from that study:

1. All airglow emission intensities have declined by a factor of about 3 (on average) from the 1990 to 1996 winters. These intensities seem to be proportional to the solar activity levels (solar sunspot numbers).

2. There are major variabilities, particularly in the 5577 Å and Na 5893 Å intensities due to waves (tidal, gravity and planetary) with periods of a few hours to several days and to other geophysical parameters not yet delineated.

3. The OI 5577 Å and 6300 Å emission intensities are currently very low, but variable, as seen on the two nights plotted in Figure 2. On some nights, as on January 2, 1997, they were as low as 20 R.

4. These very low levels of airglow may make it more difficult to monitor ionospheric irregularities, such as F layer patches, but they give better opportunities for studying effects of electron precipitation.

3. Polar Auroral Arcs

Sun-aligned polar arcs do occur most of the time when the IMF B_z is positive (nearly half of the time) and can be seen if one has sufficient sensitivity to detect them. Zhang [9] found that from ASC records (with a threshold of ~300 R of 5577 Å) they appeared about 8% of the time, but by examining MSP records many more were observed (with a threshold of < 50 R). As polar auroras are excited by soft electrons (often < 1 keV characteristic energy) they can be detected most often by their 6300 Å emission, up to 40% of observing time.

A typical example of weak polar auroral arcs over the central polar cap on December 15, 1993 is shown in Figure 3. MSP dawn-dusk scans at 5577 Å and 6300 Å across the zenith meridian at Eureka from 1600 to 2100 UT show their evolution. Transverse cross-sections of several feeble arcs can be seen, some lasting for several hours. The strongest arc at around 1730 UT reaches 700 R in both the 5577 Å and 6300 Å emission intensities.   Some of the other arcs have much greater red line emission than green line, indicative of excitation by very soft electrons. The IMF B_z

MSP at Eureka, Dec 15, 1993

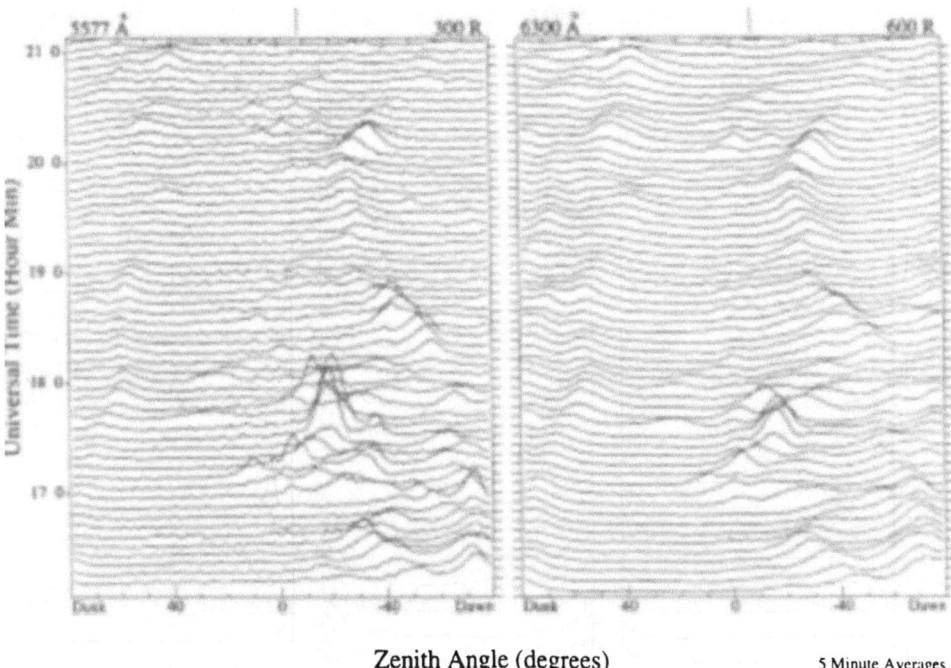

Zenith Angle (degrees)   5 Minute Averages

Figure 3. Meridian scans at 5577 Å (left) and 6300 Å (right) across several sun-aligned arcs viewed at Eureka from 1600 to 2100 UT December 15, 1993. Each plotted scan is averaged over 5 minutes. Vertical lines above the plots representing 300 R and 600 R respectively for the 5577 Å and 6300 Å plots give the intensity scales. The strongest arc, at around 1730 UT, reaches 700 R in both 5577 and 6300 Å and is detectable for about 1 hour. Softer and weaker arcs are present in both the dusk and dawn sectors through the 5 hour period.

was positive throughout this period.

It is generally believed that the IMF B_y influences the entry of arcs into the polar cap in the dawn sector when B_y is negative and in the dusk sector when B_y is positive. It follows then that the polar arcs should drift in a dawn-dusk direction during or following changes in B_y. We do see this _sometimes_ but it is difficult to find a classic case illustrating it, and at times the direction of motion of an arc is contrary to the direction of B_y change. Even taking into account delay times of 1-2 hours between changes in B_y and reconfiguration of the polar cap convection patterns, there is seldom a clear simple picture of ionospheric response.

February 11, 1996 was a case illustrating this, when the IMF B_y remained negative but active through the 12 hours pre-noon, as recorded by the WIND satellite. B_z was northward from 03 to 07 hours UT and southward for a few hours both before

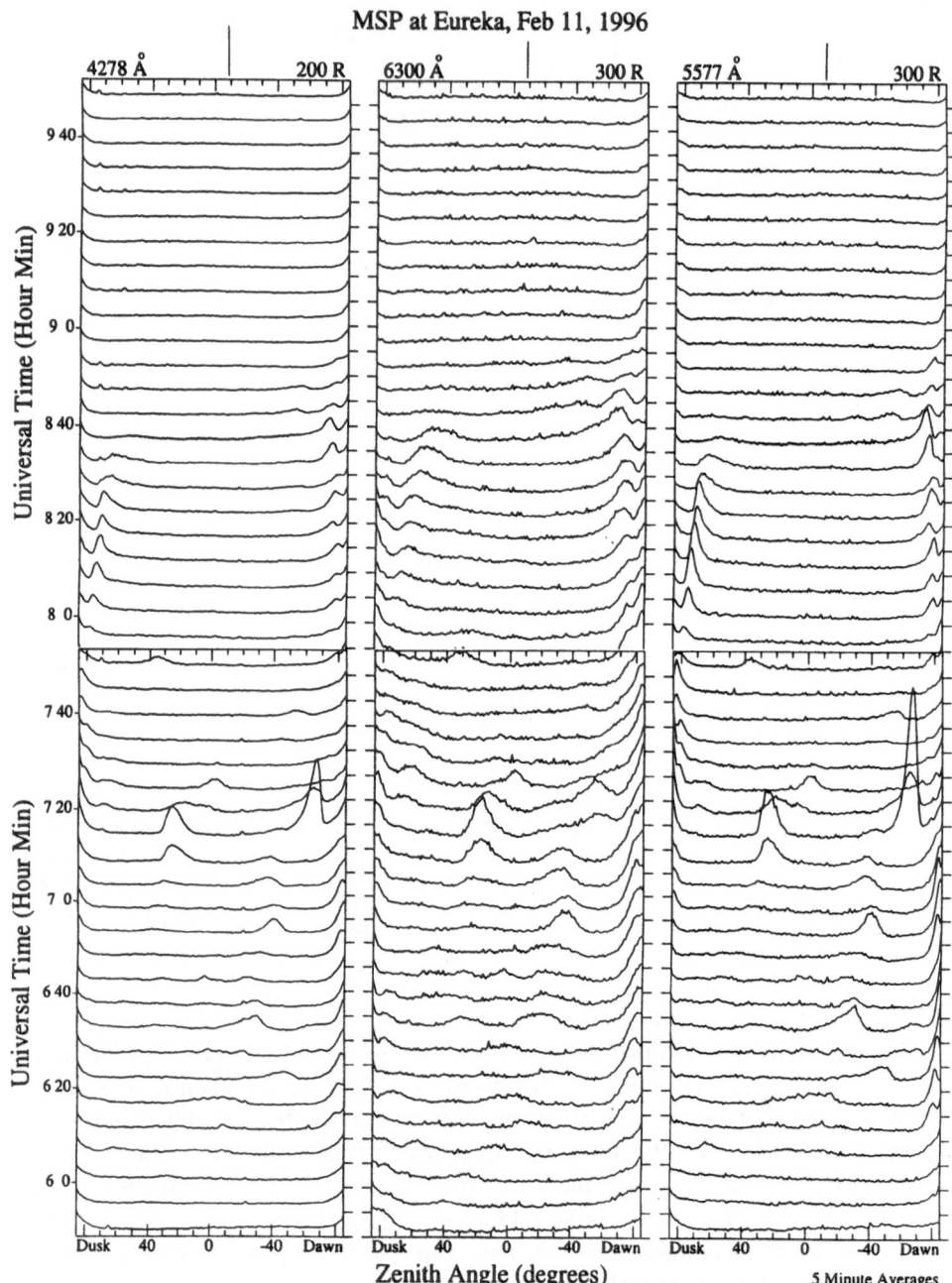

MSP at Eureka, Feb 11, 1996

Figure 4. A sequence of MSP scans at 4278, 6300 and 5577 Å recorded February 11, 1996 illustrating sun-aligned arcs in both the dusk and dawn sectors from 0600 to 0900 UT, with IMF B$_y$ negative. A broad F-layer patch appears after 0900 UT (seen as a 6300 Å enhancement) following B$_z$ turning southward. Each scan shown is averaged over 5 minutes; intensity scales for each emission are given above the 3 plots.

and after this period.

Figure 4 shows the intensities of the 4278, 6300 and 5577 Å emissions through the period 0550 to 0950 UT. Sun-aligned arcs appear in both the dusk and dawn sectors from 0600 to 0900 UT. In particular, around 0720 UT there are several arcs across the whole polar cap above 85° CGM while B_y has been continuously negative. From 03 to 05 UT (not shown) there were F-layer patches (to be discussed later) and following 0900 UT a weak patch can be seen as enhanced 6300 Å emission across the central polar cap. In summary , while arcs appear following B_z turning northward their positions across the central polar region do not seem to be controlled by B_y sign.

While the polar arcs are subvisual and transient in general, with lifetimes averaging about 8 minutes, some persist for an hour or more with one striking one (November 15, 1993) lasting for greater than 10 hours within the field of view of the Eureka instruments [9].

In general summary then, polar auroras occur commonly any time the IMF B_z is northward, during exceptionally large magnetic storms (when B_z may be southward), or following magnetospheric compressions and other such occasions when global auroras occur, regardless of the IMF B_z orientation. These illustrate that observations of polar auroras in general can be a useful indicator of the state of the magnetosphere and the solar wind.

4. F-Layer Patches

As is well established from several studies, F-layer irregularities occur over the polar region much of the time when B_z is southward. They were first reported by Buchau et al. [11], studied during the following years by Weber and co-workers [12] at the Phillips Lab from sites in Greenland, and more recently studied by McEwen and Harris [10] from the polar observatory at Eureka. First to summarize their optical and physical characteristics:

1. They appear as regions with plasma densities a few times higher than ambient and with 6300 Å emission significantly enhanced over background airglow levels. They can be observed either by radio or by optical means.

2. They drift antisunward over the magnetic pole at speeds of a few hundred m/s. They evidently are solar or particle produced plasmas that originate in the dayside ionosphere and are transported to the polar cap. They appear to survive transit across the polar cap into the midnight sector.

To illustrate their appearance and form, Figure 5 shows a photometer record from December 8, 1993 where patches drifted over Eureka every few minutes for most of the day. Each patch as it drifted through the zenith gave a 6300 Å enhancement of a few hundred R above the airglow background which was about 100 R. Some of these patches evidently originated over Svalbard as they were seen by the EISCAT radar some 1-2 hrs. earlier. B_z as recorded by IMP 8 was strongly negative at about -10 nT for most of the day, except for a 1-hour period following 2100 UT. There is a gap in

278

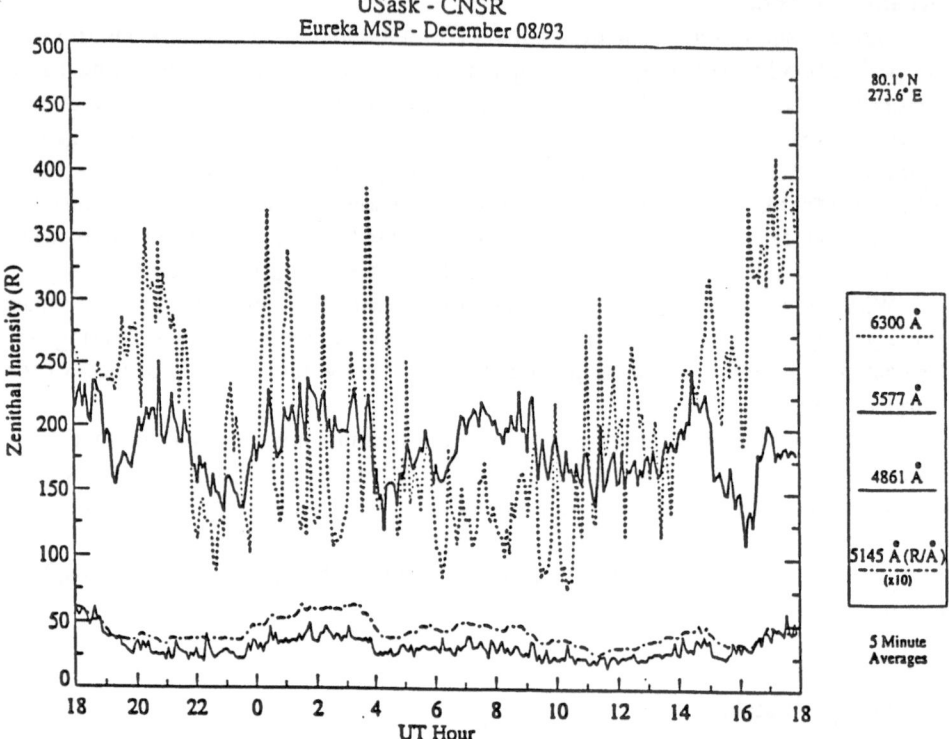

USask - CNSR
Eureka MSP - December 08/93

80.1° N
273.6° E

6300 Å

5577 Å

4861 Å

5145 Å (R/A)
(x10)

5 Minute
Averages

Figure 5. Zenithal airglow intensities measured by four channels of the meridian-scanning photometer (MSP) at Eureka for the 24-hour period centered on local midnight (0600 UT) on December 8, 1993. Enhancements in the 6300 Å intensity (dotted curve) due to F layer patches drifting antisunward through zenith can be seen through most of the period, with intensities up to 300 R above the background 6300 Å airglow.

the patches an hour after this time, illustrating IMF control on patch entry into the polar cap.

Patches evidently survive their passage across the whole polar cap from dayside to the midnight sector of the auroral oval, although the full transit of a patch or sequence has not been tracked. Modelling of patch lifetimes suggest that at 300 km they should survive about 10 hrs [13], and drift times across the polar cap are only about ½ of this, based on measured drift velocities. Eureka observes brighter patches on average (by 50 R) during the mid-day hours, when the station is closer (by 2°) to the dayside cusp, compared to the midnight hours [10].

An additional related observation on intensities is that the 6300/5577 R/G ratio is higher around mid-day compared to midnight hours (3.36±0.11 vs 2.72±0.18). The excitation of the two OI emissions is by a dissociative recombination with the observed branching ratio of 0.33 for excitation to the $O[^1S]$ state. This gives an average R/G ratio of 3.06.

The control on patch entry into the polar cap can be seen on January 15, 1994 [10] in a series of patches drifting over Eureka from 2000 to 2230 UT. The IMF B_z

switched northward at 2140 UT and the patches terminated by 2240 UT with 2 sun aligned arcs appearing.

Two of the theories regarding generation of these plasma patches need further investigations: magnetopause reconnection with flux transfer [14] or IMF B_y transients plus enhanced plasma flow (flow channel events, FCEs) [15]. Observations of patch recurrence patterns (frequencies of 20-30 min) give some support to the former. Transients in the IMF B_y are common during periods of patch observations, but a 1-1 correspondence has not been found. The occurrence of patches at all UT times is contrary to the prediction of formation in the early afternoon through FCEs. As shown in patch occurrence vs UT time through the day [10], for specific levels of magnetic activity (K_p = 1, 2 and 3 and > 4) there is no preferred time of day for patches to be seen over the magnetic pole.

Patches are much harder to detect optically now at solar minimum but under these quieter conditions it may be possible to better recognize their relationships with magnetospheric and solar wind states. Their connections with the dayside cusp will hopefully be clarified soon with the additional tools at Svalbard such as the EISCAT Svalbard radar.

5. Summary

A polar observatory operating at Eureka (89° CGM) since 1990 has permitted a study of the dynamics of the polar atmosphere and ionosphere. The 5577 and 6300 Å OI airglow emissions, themselves time variable, provide a particularly useful tool for these investigations. During much of the time either polar auroras or F-layer patches, depending on the state of the IMF, can be detected superimposed on the airglow background.

Polar auroras are of two main types - sun-aligned or substorm induced. The sun-aligned arcs are frequent but usually feeble and transient and often most easily viewed by their 6300 Å emission. The substorm induced auroras are infrequent but much more extensive and intense. Ionospheric F-layer patches seen as 6300 Å enhancements are detected drifting across the magnetic pole in an antisunward direction during much of the time when B_z is negative.

Continued observations of these phenomena in association with the data from the new Svalbard polar radar and the ISTP spacecraft data should soon provide further insights into their formation and direct linkages with solar wind parameters.

6. Acknowledgements

Funding for this research was provided by the Natural Sciences and Engineering Research Council of Canada. Facilities at Eureka and the ASTRO Lab were kindly provided by the Atmospheric Environment Service, Department of Environment. We acknowledge with thanks the IMP 8 and WIND IMF data kindly made available by R. Lepping, NASA.

7. References

1. Lassen, K. and Danielson, C. (1978) Quiet time patterns for auroral arcs for different directions of the interplanetary magnetic field in the Y-Z plane, *J. Geophys. Res.*, **83**, 5277-5284.

2. Meng, C.I. and Makita, K. (1986) Dynamic variations of the polar cap, *in Solar Wind - Magnetospheric Coupling*, Y. Kamide and J. Slavins, eds., Terra Scientific Publ. Co., Tokyo, Japan.

3. Makita, K. Meng, C.I. and Akasofu, S.I. (1988) Latitudinal electron precipitation patterns during large and small IMF magnitudes for northward IMF conditions, *J. Geophys. Res.*, **93**, 97-104.

4. Lundin, R., Eliasson, L. and Murphree, J.S. (1991) The quiet-time aurora, in Auroral Physics, C.I. Meng, M.J. Rycroft and L.A. Frank, eds., Cambridge Univ. Press, Cambridge, U.K., pp. 177-193.

5. McEwen, D.J. (1997) Ionospheric dynamics in the central polar cap, *Adv. Space Res.*, **17**, in press.

6. McEwen, D.J. and Huang, K. (1995) The polar onset and development of the November 8 and 9, 1991 global red aurora, *J. Geophys. Res.*, **100**, 19585-19594.

7. McEwen, D.J., Hammel, G. and Rolheiser, N. (1994) The magnetosphere-ionosphere perturbation on February 21, 1994, *EOS Trans. AGU*, **75**, 571.

8. Akasofu, S.-I. (1991) Auroral phenonema, in Auroral Physics, Cambridge University Press, Cambridge, U.K., pp. 3-12.

9. Zhang, Y. (1995) A study of polar auroral arcs at Eureka, NWT, Ph.D. Thesis, University of Saskatchewan.

10. McEwen, D.J. and Harris, D.P. (1996) Occurrence patterns of F layer patches over the north magnetic pole, *Radio Science*, **31**, 619-627.

11. Buchau, J.B.W., Reinisch, W., Weber, E.J. and Moore, J.G. (1983) Structure and dynamics of the winter polar cap F region, *Radio Sci.*, **18**, 995-1010,.

12. Weber, E.J., Klobuchar, J.A., Buchau, J., Carlson, H.C., Livingston, R.C. et al. (1986). Polar cap F patches: Structure and dynamics, *J. Geophys. Res.*, 91, 12121-12129.

13. Schunk, R.W. and Sojka, J.J. (1987) A theoretical study of the lifetime and transport of large ionospheric density structures, *J. Geophys. Res.*, **92**, 12343-12351.

14. Lockwood, M. and Carlson, H.C. (1992) The production of polar cap electron density patches by transient magnetopause reconnection, *Geophys. Res. Lett.*, **19**, 1731-1734.

15. Rodger, A.S., Pinnock, M., Dudeney, J.R., Baker, K.B. and Greenwald, R.A. (1994) A new mechanism for polar patch formation, *J. Geophys. Res.*, **99**, 6425-6436.

POLAR PATCHES - OUTSTANDING ISSUES

A.S. RODGER
British Antarctic Survey
Madingley Road, Cambridge, CB3 0ET, UK.

Abstract

Historically, the term polar patch was used to describe isolated regions of enhanced 630 nm emission observed in the polar cap. In recent years, signatures from other instruments, such as riometer, ionosonde and HF backscatter radar, have been used to identify related phenomena. In this paper, the relationship between these various signatures, ascribed the term of polar patches is briefly described. A summary of the formation processes of polar patches is presented, followed by a discussion of the sources of plasma that form polar patches. Some outstanding issues concerning polar patches are raised.

1. Introduction

Polar patch was originally the term selected to describe regions in the polar cap ionosphere where the intensity of 630 nm emissions is enhanced above some unspecified background. Normally there is little or no increase in the intensity of 557.7 nm emission at the same time. These patches are typically 100-1000 km in dimension, and drift anti-sunward across the polar cap e.g. [1]. In recent years, there has been a dramatic increase in the study of polar patches. Signatures of phenomena similar to those of polar patches have now been identified in many other types of ground- and space-based data. These observations have contributed significantly towards understanding the structure, dynamics and formation of polar patches. Crowley [2] provides a comprehensive summary of the literature on polar patches, and some of the related phenomena. See also the special section on high-latitude plasma structures in the May-June 1996 issue of Radio Science. This paper will focus on describing the relationship between the signatures of polar patches in different instrument data sets, and summarise the major issues associated with polar patches that still require further understanding.

281

J. Moen et al. (eds.), Polar Cap Boundary Phenomena, 281–288.

2. Polar patches, as observed by different methods

2.1 THE OPTICAL SIGNATURE OF POLAR PATCHES

All-sky camera data were used initially to determine the properties of polar patches. To date all measurements of optical polar patches have been somewhat subjective. There is no objective definition of a polar patch, i.e. the level of enhanced emission above the background intensity. This fact, together with some inherent limitations to the use of optical techniques, influences the observed properties. For example, all-sky cameras only provide data when in darkness. Also the height of the 630 nm emission is usually assumed to be near 250 km. With this assumption, the maximum size that can be measured reliably with an all-sky camera is <1500 km. Although there are now networks of all-sky camera in both hemispheres, coverage is not yet sufficient to allow tracking of polar patches from their formation near the ionospheric footprint of the cusp, right across the polar cap into the nightside auroral oval.

Polar patches are not the only cause of increased 630 nm intensity. The effects of energetic particle precipitation associated with polar rain or transpolar arcs are important. This is well illustrated by McEwan *et al.* [3] who show (their figure 4) several short-lived enhancements of 630 nm emission from observations made at Eureka, Canada, a station located deep in the polar cap. Only about half of the enhancements drift anti-sunward.

2.2 ELECTRON CONCENTRATION SIGNATURE OF POLAR PATCHES

Significant, short-lived enhancements of the maximum electron concentration in the F region of the ionosphere are considered to be the ionosonde and incoherent scatter radar (ISR) signatures of polar patches [4]. Whilst modern ionosondes and the ISR can both determine the plasma velocity, the ISR has the advantage of being able to determine plasma temperatures. Studies using ISR have shown that the electron temperature within polar patches is not elevated above the background level, indicating that energetic particle precipitation is not occurring within the patch.

There is no agreed limit on the enhancement of electron concentration necessary to be called a patch. Some authors have chosen an absolute increase in the maximum plasma frequency, taking no account of the background level of ionisation. This approach can lead to considerable bias in the diurnal, seasonal and solar cycle occurrence of polar patches. Furthermore there can be other reasons for short-lived enhancements of F-region concentration, e.g. the effects of gravity waves, transpolar arcs and polar rain. Differentiating these phenomena from the signature of a polar patch with the ionosonde or ISR data alone is very difficult.

Spatial variations in electron concentration, measured by polar orbiting satellite, are often interpreted as indicating the presence of polar patches. In this case also, an increase in electron concentration need not necessarily indicate the presence of a polar patch. The increase could be due to one of many processes such as energetic particle precipitation, increased plasma scale heights, Joule heating, etc.

There is not a precise relationship between an optical and electron concentration signature of a polar patch. Two illustrative examples are given. Sojka et al. [5] showed that the intensity of the emission of polar patches depended upon the height distribution of electron concentration. They find that raising the peak of an F-layer with a maximum concentration of 10^{12} m^{-3} from 300 to 360 km (approximately 1 scale height) reduces the emission intensity from about 500 to 200 Rayleighs. Secondly, Rosenberg et al. [6] showed observations of three patches of enhanced electron concentration over South Pole Station, but there was a single enhancement of the 630 nm emission.

The absorption of radio waves is determined mainly by the electron concentration, the electron-neutral collisions at D- and E-region altitudes, and the electron-ion collision frequency at F-region altitudes. Normally the electron-neutral collision frequency term dominates the observed absorption but occasionally in the polar cap, F-region effects are more important. Rosenberg et al. [6] and Wang et al. [7] showed some case studies where riometer absorption between 30 and 40 MHz is increased by up to about 0.5 dB as enhancements in electron concentration drift over the imaging and broad beam riometers. Therefore riometers can be another useful indicator of the presence of polar patches. Like all techniques it suffers from some limitations such as the requirement for measurements to be made during times of low D-region absorption, and the F-region electron concentration must exceed about 4.5×10^{11} m^{-3}. However patches can be observed both in daylight and in darkness.

2.3 THE IRREGULARITY SIGNATURE OF POLAR PATCHES

Observations by scintillation and HF backscatter radar techniques show that regions of intense irregularity activity are often associated with polar patches identified by one of the other polar patch signatures discussed above. Scintillation experiments are sensitive to irregularities in the order of 1-10 km, whereas the HF backscatter radars depend upon the presence of 10 m scale sized irregularities. The cause of the irregularities has not been determined but the spatial gradients in electron concentration associated with polar patches are likely to be unstable to the gradient drift instability [8]. The growth rate for this instability is directly related to the magnitude of the gradient, the electron temperature and the velocity of the plasma in the rest frame of the neutral atmosphere. It is also inversely proportional to the electron concentration. Observations deep in the polar cap show that irregularities are observed frequently on the leading and trailing edges of patches, and sometimes through the entire patch. Again the presence of irregularities is not a unique definition of a polar patch and may be caused by other processes (e.g., space and time variations of polar rain). However the irregularities associated with polar patches have been very valuable in determining the causative processes of polar patches and to demonstrate that patches **ExB** drift e.g. [9].

In summary, there is no physical characteristic that allows a unique definition of polar patches. Each experimental technique used to observe polar patches has strengths and weaknesses. The occurrence characteristics of polar patches are significantly influenced by the method of observation. There is general agreement however from all techniques that polar patches are regions of enhanced electron concentration and 630 nm emission of scale size 100-1,000 km. They usually occur when IMF Bz is negative and drift anti-sunward across the polar cap at 300-1,000 m s^{-1}. There is no precipitation occurring within polar patches when observed deep in the polar cap. Occasionally patches may form simultaneously in geomagnetically conjugate regions [10].

3. The tongue of ionisation and the polar patches

It has been established both by observation [11, 12] and by computer modeling [13], that there is often a large enhancement of electron concentration formed near the equatorward edge of the auroral oval when it is in sunlight. This ridge of ionisation arises because F-region plasma remains in this sunlit region for an extended interval owing to the counteracting corotation and convection electric fields. The ridge of ionisation is eventually extruded by the convection electric field, through the cusp and across the polar cap. The common parlance for this feature is the tongue of ionisation, a feature first identified in the ionosonde data from the International Geophysical Year 1957-58. The longitude extent of the enhanced ridge as it passes through the cusp region is about 1 h of MLT or about 500 km at 75° magnetic latitude [13].

In the morning sector, the corotation and convection electric fields act in the same sense. As a result F-region plasma is in sunlight for a much shorter time than in the afternoon sector, and the plasma levels entering the polar cap are well below those of the afternoon cell plasma. Therefore during periods when there is anti-sunward convection, the cusp region can be considered as a location where two nozzles of ionisation are located, one projecting high F-region electron concentrations across the polar cap, immediately adjacent to a second ejecting much lower concentration levels.

For stations deep in the polar cap, such as Thule in the North and Vostok in the South, the maximum F-region plasma concentration will depend critically upon whether afternoon or morning cell plasma passes over the station. The parameter which affects the trajectory of the anti-sunward convecting plasma to the greatest extent is the IMF By component (see [14] for example). Given that the IMF By component is rarely stable for more than a few tens of minutes, the tongue of ionisation will no longer form a simple cross polar cap feature but will become greatly distorted. This has been well illustrated by the modelling work of Sojka *et al.* [15].

4. The cause of polar patches

There is still not universal agreement on the causes of polar patches or indeed where they are formed. Part of this difficulty arises from a confusion in the use of the term polar patch. Some authors have applied the term polar patch to observations and model simulations of the distorted tongue of ionisation. These can be readily produced by varying the y-component of the IMF on a time scale of tens of minutes [15]. Other research scientists have been more specific and define a patch as a region of high F-region electron concentration completely surrounded by regions of lower concentration. In practice, it may be virtually impossible to separate these two possibilities from observations in the deep polar cap. However by observing the time evolution of polar patches, important insight into the formation processes is obtained.

To create a distinct and separate patch, it is necessary to cause a discontinuity in the plasma entering the polar cap. The most successful mechanism to date involves the effects of fast plasma jets of the type associated with the ionospheric signature of flux transfer events (FTEs) [9]. FTEs map to the ionospheric footprint of the cusp. The plasma jets associated with FTEs are fastest when there is a finite value of IMF By. The effects of the release of field line tension add to the electric field associated with reconnection at the magnetopause. The high plasma velocity in the ionosphere reduces markedly the lifetime of F-region plasma. For example, the lifetime falls from 1 h to 1 min when the plasma velocity in the rest frame of the neutral particles is about 2.5 km s^{-1}. By this mechanism, substantial (up to one order of magnitude) depletions of the F-region plasma concentration can occur, thus slicing the plasma entering the polar cap. Support for this hypothesis is provided by Valladares et al. [13],[16].

There may be other mechanisms to create discrete polar patches as a result of changes to the IMF. For example, Lockwood and Carlson [17] suggest that variations in the z-component of the IMF are important. In their mechanism, a rapid expansion of the high-latitude convection pattern entrains additional high concentration plasma at the equatorward edge of the auroral oval, bringing it over the polar cap, leading to enhancements in the existing electron concentration levels. Several limitations to this mechanisms have been identified [18]. Another way by which a discontinuity may be introduced into the plasma entering the polar cap is to take proper account of the known longitude motion of the convection throat, and the energetic cusp particle precipitation as IMF By changes (e.g. [19]). These variations are not included in the present modelling e.g. [15].

5. The source of plasma for polar patches

The source of plasma which forms polar patches is now almost universally accepted to be ionisation that is entrained in the two cell, anti-sunward convection pattern that has passed through the cusp region [4, 11, 12, 16]. The climatology of polar patches is very heavily influenced by the likelihood of the ridge of ionisation forming in the afternoon cell. The key factor is that the equatorward edge of the two cell anti-sunward convection pattern must be in sunlight. FTEs will cause 'holes' both in the morning and afternoon cell plasma, but the latter will be much deeper owing to the higher plasma concentration there. Considerable support for this hypothesis is provided from the predicted and observed climatology of polar patches in both hemispheres [20, 21, 22]. Perhaps the greatest uncertainty is the time required and the stability of the high-latitude convection electric field pattern necessary to form a large ridge of ionisation near the equatorward edge of the auroral oval.

A further unresolved uncertainty is the extent to which the cusp particle precipitation contributes to the plasma concentration levels entering the polar cap. Whilst there has been some modelling of this process, none to date has used realistic fluxes of energetic ions and electrons. Further the effect of an FTE in the cusp region is to heat the plasma, leading to considerable upflow into the topside ionosphere where it acts as a source of F-region plasma as it convects across the polar cap. These processes contribute significantly to the evolution of the patch as it convects towards the nightside ionosphere [23]. There is no doubt that polar patches and the tongue of ionisation are important sources of F-region structure in the nightside auroral oval [10, 24].

6. Conclusions

Considerable progress has been made in recent years in understanding the signatures of polar patches as observed by a variety of different instruments. As described above, the morphology of polar patches is heavily influenced by the observational method.

Key questions for future polar patch studies include:-

6.1.1   Definitions
Should there be a precise definition of a polar patch?

Should the signatures of polar patches as observed by different instruments be investigated further?

6.1.2   Sources of plasma
How long does it take to form and destroy the afternoon ridge of ionisation?

How important is cusp ion and electron precipitation in patch formation?

6.1.3   Transport of ionisation
Is it possible to differentiate between the tongue of ionisation and discrete polar patches deep in the polar cap?

Is it possible to model more realistically, the changes in convection pattern and particle precipitation arising from changes of IMF Bz and By to determine the relative importance of various mechanisms for polar patch formation?

How important is topside plasma resulting from FTEs in the evolution of polar patches as they cross the polar cap?

How important are polar patches in determining the structure and irregularity morphology in the nightside auroral oval?

To address these key questions, networks of observatories (digital ionosonde and all-sky camera) together with the imaging capabilities of SuperDARN radars and riometers, are essential. The new high-latitude ISRs at Svalbard and Resolute Bay will provide key plasma diagnostic observations also. Complementary computer modelling initiatives will be necessary.

7. Acknowledgements

I am most grateful to the organisers of this excellent meeting, and to NATO for its support. The work described in this paper has benefitted substantially from discussions with many scientists in recent years, particularly Dave Anderson, Helen Balmforth, Sunanda Basu, Herb Carlson, Dwight Decker, John Dudeney, Mike Lockwood, Don McEwan, Roy Moffett, Mike Pinnock and Ted Rosenberg.

8. References

1. Weber, E.J., Buchau, J., Moore, J.G., Sharber, J.R., Livingstone, R.C., Winningham, J.C., and Reinisch, B.W. (1984) F-layer ionisation patches in the polar cap, *J. Geophys. Res.*, **89**, 1683.
2. Crowley, G. (1996) Critical review of ionospheric patches and blobs, In *The review of Radio Science 1992-1996*, edited by W. Ross Stone, Oxford University Press, UK.
3. McEwan, D.J., Harris, D.P., MacDougal, J.W., and Grant, I.F. (1995) Drifting F-layer patches over the magnetic pole, *J. Geomag. Geoelectr.*, **47**, 527.
4. Anderson, D. N., Buchau, J., and Heelis, R.A. (1988) Origin of density enhancements in the winter polar cap ionosphere, *Radio Sci.*, **23**, 513.
5. Sojka, J.J., Schunk, R.W., Bowline, M.D., and Crain, D.J. (1997) Ambiguity in identification of polar cap F-region patches: contrasting radio and optical observation techniques. *J. Atmos. Solar-Terr. Phys.* **59**, 249.
6. Rosenberg, T.J., Wang, Z., Rodger, A.S., Dudeney, J.R., and Baker, K.B. (1993) Imaging riometer and HF radar density measurements of drifting F-region electron density structures in the polar cap, *J. Geophys Res.*, **98**, 7757.
7. Wang, Z., Rosenberg, T.J., Stauning, P., Basu, S., and Crowley, G. (1994) Calculations of riometer absorption associated with F-region plasma structures based on Sondre Stromfjord incoherent scatter radar observations, *Radio Sci.*, **29**, 209.
8. Tsunoda, R. T. (1988) High latitude F-region irregularities: A review and synthesis, *Rev. Geophys.*, **26**, 719.
9. Rodger, A.S., Pinnock, M., Dudeney, J.R., Baker, K.B., and Greenwald, R.A. (1994) A new mechanism for polar patch formation, *J. Geophys. Res.*, **99**, 6425.
10. Rodger, A.S., Pinnock, M., Dudeney, J.R., Watermann, J., Beaujardiere, de la, O., and Baker, K.B. (1994) Simultaneous two-hemisphere observations of the presence of polar patches in the night-side ionosphere, *Ann. Geophys.*, **12**, 642.
11. Foster, J.C. (1993) Storm time plasma transport at middle and high latitudes, *J. Geophys. Res.*, **98**, 1675.
12. Pinnock, M., Rodger, A.S., and Berkey, F.T. (1995) High-latitude F-region electron concentration measurements near noon - a case study, *J. Geophys. Res.*, **100**, 7723.
13. Valladares, C.E., Decker, D.T., Sheehan, R., and Anderson, D.N. (1996) Modeling the formation of polar patches using large plasma flows, *Radio Sci.*, **31**, 573.
14. Heppner, J.P. and Maynard, N.C. (1987) Empirical high-latitude electric field models, *J. Geophys. Res.* **92**, 4467.

288

15. Sojka, J.J., Bowline, M.D., Schunk, R.W., Decker, D.T., Valladares, C.E., Sheehan, R., Anderson, D.N., and Heelis, R.A. 1993 Modelling polar cap F-region patches using time varying convection, *Geophys. Res. Lett*, **20**, 1783.

16. Valladares, C.E., Basu, S., Buchau, J., and Friis-Christensen, E. (1994) Experimental evidence for the formation and entry of patches into the polar cap, *Radio Sci.*, **29**, 167.

17. Lockwood, M. and Carlson, Jr., H.C. (1992) The production of polar cap electron density patches by transient magnetopause reconnection, *Geophys. Res. Lett.*, **19**, 1731.

18. Rodger, A.S., Pinnock, M., Dudeney, J.R. (1994) Comments on the paper "Production of polar cap electron density patches by transient magnetopause reconnection" by M. Lockwood and H. C. Carlson, Jr., *Geophys. Res. Lett.*, **21**, 2335.

19. Cowley, S.W.H., Morelli, J.P., and Lockwood, M. (1991) Dependence of convective flows and particle precipitation in the high latitude day-side ionosphere on the X and Y components of the IMF, *J. Geophys. Res.*, **96**, 5557.

20. Sojka, J.J., Bowline, M.D., and Schunk, R.W. (1994) Patches in the polar ionosphere: UT and seasonal dependence, *J. Geophys Res.*, **99**, 14959.

21. Bowline, M.D., Sojka, J.J., and Schunk, R.W. (1996) Relationship of theoretical patch climatology to polar cap patch observations, *Radio Sci.*, **31**, 635.

22. Rodger, A.S., and Graham, A.C. (1996) Diurnal and seasonal occurrence of polar patches, *Ann. Geophys.*, **14**, 533.

23. Balmforth, H.F., Moffett, R.J., and Rodger, A.S. (in press) Modelling studies of the effects of cusp inputs on the polar ionosphere, *Adv. Space Res.*

24. Anderson, D. N., Decker, D.T., and Valladares, C.E. (1996) Modeling boundary blobs using time-varying convection, *Geophys. Res. Lett.*, **23**, 579.

CONJUGATE FEATURES OF AURORAS OBSERVED BY TV CAMERAS AND IMAGING RIOMETERS AT AURORAL ZONE AND POLAR CAP CONJUGATE-PAIR STATIONS

H. YAMAGISHI, Y. FUJITA, N. SATO
National Institute of Polar Research, Itabashi, Tokyo, Japan
P. STAUNING
Danish Meteorological Institute, Copenhagen, Denmark
M. NISHINO
Solar Terrestrial Environment Laboratory, Nagoya University,
Toyokawa, Japan
K. MAKITA
Takusyoku University, Hachiouji, Tokyo, Japan

1. Review of the Past Conjugate Aurora Studies

For the study of magnetospheric phenomena, geomagnetically conjugate point observations in northern and southern polar regions provide unique opportunities to study whether the symmetry of magnetospheric configuration is maintained, or breaks down at the time of auroral substorms. In addition, such studies can be used to study asymmetries between the sunlit and the dark polar ionosphere.

Conjugate observation of auroral phenomena was initiated in the IGY period by DeWitt [1], using Campbell Island-Farewell, Alska station pair at L=4. He confirmed the occurrence of similar forms and motions, and simultaneous break up of auroras at the conjugate points. Wescott [2] analyzed all-sky camera data from the Syowa Station (L=6.1) and Reykjavik (L=5.9) conjugate pair. He found that the conjugacy at this latitude was not so good as at L=4. Conjugacy was fair in some time intervals, but not so close in other intervals. It was followed by the famous experiments with conjugate aircraft observations by University of Alaska (1967-1971). Belon et al. [3] found that the conjugacy of auroras were well maintained from dipole latitude of $65° - 71°$ in the magnetically quiet period. Stenbaek-Nielsen et al. [4] reported the conjugacy during magnetically disturved periods. There were two arc systems, one at low latitudes (invariant lat. $63° - 65°$) and the other at high latitudes (invariant lat.$> 65°$). The former showed good conjugacy, while the latter did not. Stenbaek-Nielsen et al. [5] discussed differences of conjugate auroral intensities in the northern and southern hemisphere, I_N and I_S. For low latitude diffuse auroras, $I_N/I_S \sim 1.3$. The larger intensities were attributable to smaller magnetic field and lower mirror height in the northern hemisphere, which results in larger precipitating flux to this hemisphere. For high latitude discrete auroras,

J. Moen et al. (eds.), Polar Cap Boundary Phenomena, 289-300.

the ratio I_N/I_S was highly variable, and the northern hemisphere auroras tend to appear in higher latitude as compared to those appearing in southern hemisphere. Recently, Stenbaek-Nielsen et al. [6] attributed such features to the effect of the interplanetary magnetic field (IMF) B_y component as described in Section 2.3.

National Institute of Polar Research (NIPR) started Syowa Station-Iceland conjugate pair observations during the IMS period. The observations have been conducted on a continuous basis since 1983 with 3 Icelandic stations, Husafell, Tjornes and Isafjordul [7, 8]. Every year at equinox, optical auroral observations have been carried out at Syowa and Iceland when the sky is dark in both places. Some of the observational results related to auroral conjugacy are as follows. Sato and Saemundsson [9] reported that auroral break up occurred at almost exact conjugate latitudes, as calculated by the IGRF model. However, there were considerable differences in the poleward expanding speeds observed at the conjugate points. Fujii et al. [10] reported that small-scale features such as curls and ray structures grew independently in the both hemispheres, suggesting that the mechanism forming these small-scale features existed at lower altitudes rather than in the equatorial region. Minatoya et al. [11] used the synthesized view fields of two all-sky TV cameras of Syowa and Asuka Stations, and found a large longitudinal displacement of conjugate auroras with as much as 700 km. It became clear in the later analysis that the direction of the displacement was in accordance to the expected IMF B_y effect (westward displacement in southern hemisphere for $B_y<0$), but the amount of displacement was about twice the Tsyganenko 1996 model results.

Opportunities for conjugate auroral observations are very limited for optical methods because one hemisphere is usually sunlit. Even at eqinox when dark sky are present at both stations, weather condition impose the other limitation, the sky must be clear at both points. If one wants to know the conjugacy in summer, or in winter season, and wants to confirm whether the observed associations are statistically significant, or not, it is necessary to introduce other observational means that can work through all the year. One of such instruments is a radio wave imager. We have installed a pair of imaging riometers at Iceland in July 1990 and at Syowa Station in February, 1992. This system, Syowa-Iceland Radio Absorption Conjugacy Experiment (SIRACE) , provides auroral images at the conjugate points all through the year in the form of Cosmic Noise Absorption (CNA) images with 8x8 pixel resolution within the field of view of 200 km by 200 km .

2. Conjugate Studies Using Radiowave Imager in Auroral Zone

2.1 MODEL CALCULATION OF THE CONJUGATE POINT

According to fieldline tracing using the IGRF model, the geomagnetic conjugate point of Syowa Station is found in Iceland, and it moves toward northeast at a speed of 10 km/year [12] as shown in Fig. 1. Two squares in the figure indicate the view field of the imaging riometer at Tjornes (solid square) and that for Syowa (dashed square), respectively. The latter is a view field projected onto the northern hemisphere using IGRF model. If we include external fields caused by magnetospheric current systems, the calculated conjugate point of Syowa rotates daily around the conjugate point determined by the internal field alone as shown in Fig. 2. This calculation was made using Tsyganenko

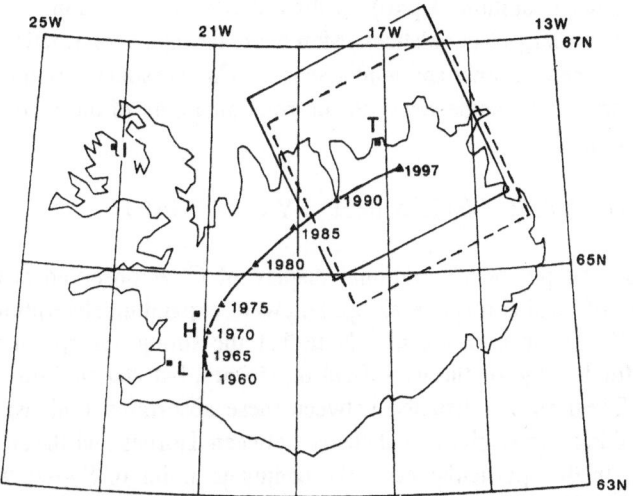

Fig. 1 Drift of geomagetic conjugate point of Syowa Station from 1960 to 1997 calculated from the IGRF model. The squares indicate field of view of the imaging riometers. Solid square is for the Tjornes riometer and the dashed is for the Syowa Station, mapped using the IGRF model. Modified from Ono [12].

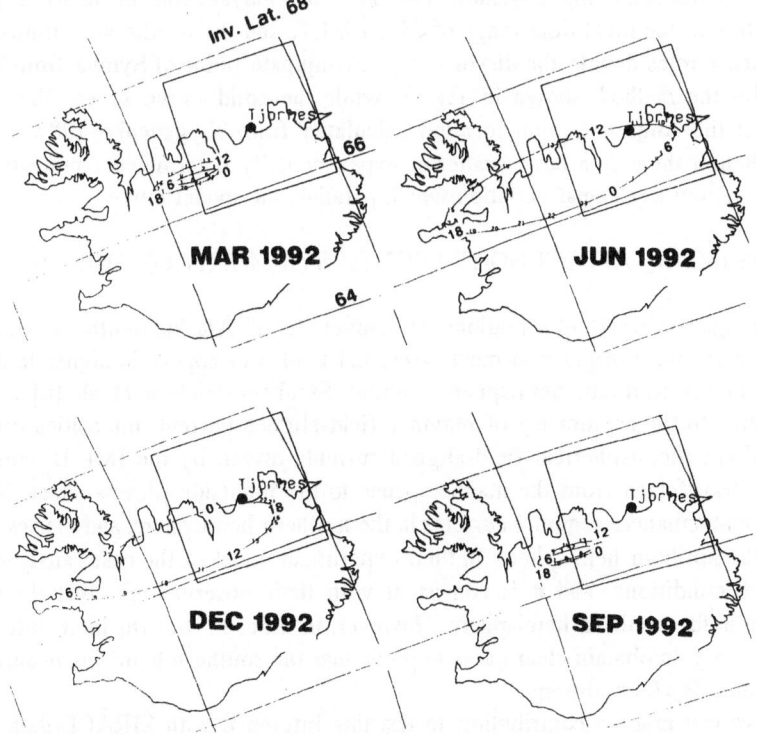

Fig. 2 Daily drift motion of the conjugate point of Syowa for equinoxes and solstices, calculated from Tsyganenko 1987 model for Kp=0.

1987 model for quiet condition (Kp=0). Radius of the daily rotation is the smallest in equinox time and the largest at solstices. Moreover, the phase of rotation is shifted by about 12 hours between summer and winter season. This seasonal variation is caused by the seasonal change of the angle between the rotational axis of the earth and the solar wind flow direction.

2.2 CONJUGATE POINT DETERMINED BY CNA IMAGES

Fig. 3 shows an example of a meridional display of CNA observed at the conjugate points at the time of local auroral break-up. Poleward expanding absorption region were recognized at 1959 UT at both stations. Note that the strong absorption region started from the low latitude edge of the view field of Tjornes, but started from the zenith for Syowa Station. The Relative distance between these absorption regions, shown in the bottom panel as thick curve, gives the distance between Tjornes and the actual conjugate point of Syowa. In this particular case, the conjugate point of Syowa was located at about 40 km south of Tjornes. It should be noted that the conjugate CNA regions can be well defined at the onsets of disturbances, but it becomes difficult to define in the later stages. This may be a consequence of the distorted magnetosphere configuration due to increased field-aligned current intensities. The relative distances of the CNA region are also calculated for later stages (thin curve in the bottom panel of Fig. 3), but not referenced in the following statistics. This kind of analysis was made for 2 years of SIRACE data in the local time range of 21-24 MLT, and the results are summarized in Fig. 4. Open circles denote the distance of the conjugate point of Syowa from Tjornes, obtained by the method shown in Fig. 3, while the solid curve gives the seasonal variation of the conjugate point location calculated from Tsyganenko 1987 model for $K_p=4$. Although there is a large scatter in experimentally determined conjugate points, they tend to show a seasonal variation which parallels the model curve.

2.3 DOES IMF-B_y AFFECT NORTH-SOUTH SYMMETRY OF AURORAS?

In the conjugate aircraft observations of University of Alaska, northern hemisphere discrete auroras were brighter in most cases, and tended to appear in higher latitudes as compared to the southern hemisphere auroras. Stenbaek-Nielsen et al. [6] attributed these features to the asymmetry of region 1 field-aligned current intensities caused by addition of interhemispherical field-aligned currents driven by the IMF B_y component which was transferred from the magnetopause to the nightside plasma sheet. Negative B_y component enhances region 1 currents in the northern hemisphere, and causes brighter aurora in the northern hemisphere. In their experiment, most of the cases analyzed were under $B_y<0$ conditions, and it is consistent with their observations that the brighter auroras are in the northern hemisphere. However, in order to confirm their inference, it is still necessary to obatain clear cases to prove that the southern hemisphere auroras are stronger under $B_y>0$ conditions.

Here, we can make a contribution to test this inference with SIRACE data. It is an advantage for SIRACE data, being possible to compare CNA intensity at the conjugate points all through the year. However, there is the question whether CNA intensity

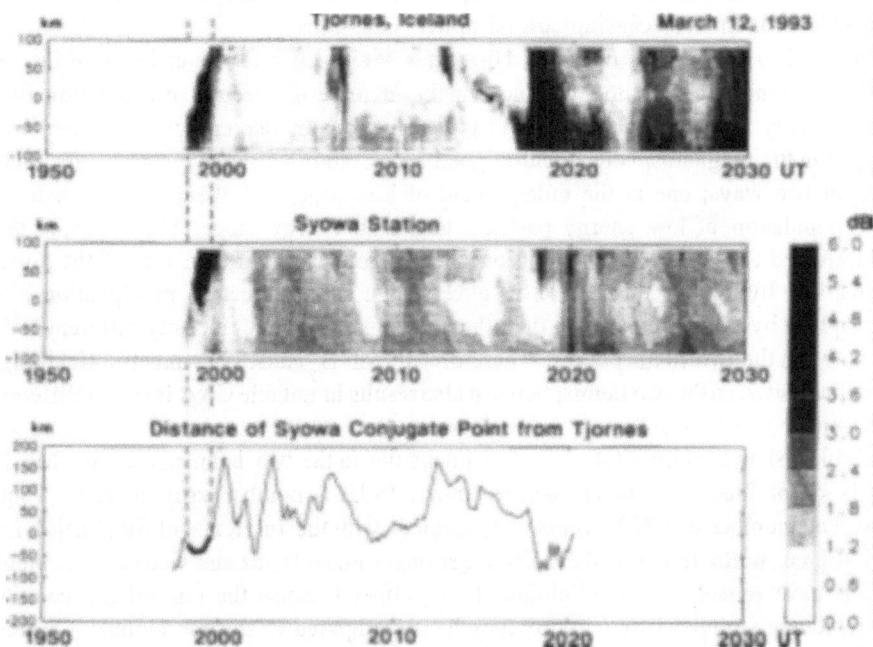

Fig. 3 Meridional displays of CNA at the conjugate points. Bottom panel shows conjugate point location determined from the relative distance of the two CNA regions.

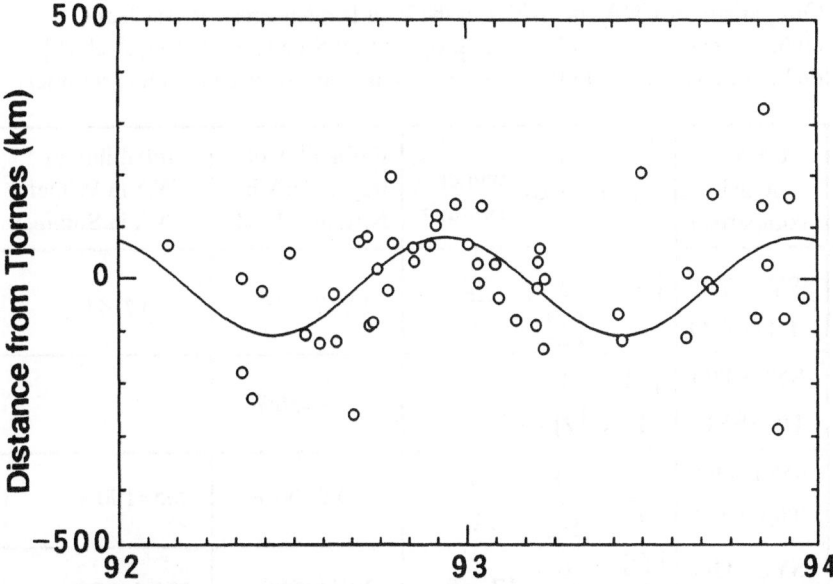

Fig. 4 Scatter plot of the conjugate point locations determined by the method shown in Fig. 3 for 1992 and 1993 in the local time range of 21-24 MLT. Solid curve denotes calculated conjugate point location.

294

differences between the conjugate points reflect the difference of optical auroral intensity, because CNA is caused by precipitating electrons of several tens of keV, about ten times higher than those causing auroral optical emissions.

In order to answer this question, Hirashima et al. [13] made an estimate of the effect of field-aligned acceleration voltages on the increase of electron precipitation flux in above energy range. Let us assume that the electrons in this energy range are already trapped in the magnetosphere. Field-aligned acceleration works to increase precipitating flux in two ways; one is the enlargement of loss cone, and the other is transfer of a large population of low energy particles to a high energy range. They showed that if field-aligned acceleration voltage in one hemisphere is larger than that of the opposite hemisphere by 10 kV, this results in excess high energy electron precipitation in one hemisphere by 50% compared to the other. This gives a CNA intensity difference of 1.8 dB between the two hemispheres. Therefore, if IMF B_y causes asymmetric field-aligned potentials between the two hemispheres, it also results in notable CNA intensity differences between the two hemispheres.

Table 1 shows comparisons of CNA intensities in the two hemispheres for IMF $B_y>0$ and $B_y<0$ for June, September and December, 1992. A number surrounded by a square represents number of CNA events in agreement with the inference of Stenbaek-Nielsen et al. [6], i.e., northern hemisphere CNA is stronger under $B_y<0$, and vice versa. September data is most reliable for highlighting the B_y effect because the ionospheric conditions are relatively symmetric in equinox months as compared to solstice months. Our result, however, shows that the number of cases in agreement with their inference is almost equal to those against their inference. That is, the B_y effect cannot be confirmed from our conjugate CNA observation.

Table 1 Comparison of CNA intensities in the both hemispheres for 3 seasons.
[   ] : Number of cases agree to B_y effect proposed by Stenbaek-Nielsen et al. [6]
___ : Number of cases agree to the seasonal assymmetry (CNA in winter>summer)

Month	CNA intensity comparison	By>0	By<0	winter $\gtrless$ summer	Probability of larger CNA in N.H. for By <0	Probability of CNA in Winter> CNA in Summer
June	SYO>TJO	[9]	3	12	12/17=71%	12/17=71%
	TJO>SYO	2	[3]	5		
Sep.	SYO>TJO	[3]	3		5/9=56%	
	TJO>SYO	1	[2]			
Dec.	SYO>TJO	[0]	0	0	3/5=60%	5/5=100%
	TJO>SYO	2	[3]	5		
Total	SYO>TJO	[12]	6	17 5	20/31=65%	17/22=77%
	TJO>SYO	5	[8]			

It has been reported that CNA intensities are larger in winter and smaller in summer at auroral zone stations [14, 15, 16]. Underlined numbers in Table 1 denote CNA events in agreement with this summer-winter asymmetry. The rate of CNA events showing this summer-winter asymmetry is 71% for June and 100% for December. Therefore, seasonal asymmetry is larger and clearer than any B_y related asymmetry. Seasonal asymmetry will be discussed in more detail in 2.4.

2.4 SUMMER-WINTER ASYMMETRY OF CNA INTENSITIES

What is the cause of summer-winter asymmetry of CNA? One candidate is the possibility that the field-aligned potentials for auroral electron acceleration are asymmetrically formed between summer and winter hemispheres. Newell et al. [17] analyzed a very large number of particle analyzer data of DMSP satellites, and found that the occurrence rate of accelerated electron precipitation events is larger in the dark hemisphere than in the sunlit hemisphere by 3 times. This fact suggests that field-aligned potentials can be higher in the winter hemisphere than in the summer hemisphere, and this difference causes excess energetic electron precipitation and enhanced CNA in the winter hemisphere. From our observations, summer-winter asymmetry of CNA intensity is typically ~1 dB. This can be explained by acceleration potential differences of several kV, according to the estimation by Hirashima et al. [13] discussed in 2.3.

Some fraction of the electrons injected in the nightside magnetosphere drift through morning toward dayside and cause daytime CNA events. Summer-winter asymmetry appears not only in nighttime, but also in daytime CNA events. Table 2 compares summer-winter asymmetry of CNA intensities for nighttime and daytime events. Underlined numbers denote CNA events in agreement with this summer-winter asymmetry, i.e., larger winter absorption. It is found that the asymmetry is more distinct for daytime events.

Table 2 Night and day comparison of the summer-winter assymmetry of CNA intensity

	CNA intensity comparison	June	Dec.	Probability of CNA in Winter> CNA in Summer
Nighttime	SYO>TJO	12	0	17/22=77%
	TJO>SYO	5	5	
Daytime	SYO>TJO	12	0	21/24=88%
	TJO>SYO	3	9	
Total	SYO>TJO	24	0	38/46=82%
	TJO>SYO	8	14	

According to Newell et al. [17], however, accelerated electron precipitation events were rarely found in daytime. Then, the summer-winter asymmetry of daytime CNA events are no longer explained by asymmetric formation of field-aligned acceleration potentials between summer and winter hemispheres. Another possible mechanism of daytime summer-winter asymmetry is the seasonal change of atmospheric structure in the D-region ionosphere. According to the MSIS 90 model, atmospheric temperature in the altitude range of 75-95 km is higher in winter (above 200K) and lower in summer (below 150K), just opposite to the seasonal variation in other altitude ranges [18]. It is possible that the larger CNA intensities are caused by increased electron-neutral collision frequencies in the warmer D-region at wintertime. This effect contributes to nighttime CNA events as well, in addition to the effect of asymmetric field-aligned acceleration between summer and winter hemispheres.

3. Conjugacy in Polar Cap Region

3.1 MODEL CALCULATION OF THE CONJUGATE POINT

The polar cap is a region of open field lines connected to the interplanetary magnetic field. From this definition, it seems difficult to define geomagnetic conjugate points in this region. However, if the invariant latitudes are not too high (less than ~77°), and the magnetic local times are away from the noon and midnight hours, geomagnetic conjugate points can be calculated using Tsyganenko 1996 model. As an example, Fig. 5 shows traces of the conjugate point of Zhongshan Station (69.37° S, 76.38° E) for vernal equinox, winter and summer solstices under IMF $B_z=B_y=0$ conditions. Daily drift of the conjugate point is recognized in similar manner to that for auroral zone conjugate point shown in Fig. 2. Radius of the daily drift is small (250 km) at equinox and large (1200 km) at solstices. There is a phase shift of a half day in the daily rotation of the conjugate point between summer and winter, i.e., the conjugate point of Zhongshan is located in the west and moves toward east in the morning hours in December, and just the opposite way in June. However, the radius of the drift motion is much larger (about 3 times) for polar cap conjugate points as compared to auroral zone conjugate points. Validity of the fieldline tracings in such high latitudes need evaluation in the future. The above results will give a good estimate on the drift of the conjugate regions when we analyze the conjugate data in the polar cap.

3.2 CONJUGATE CNA OBSERVATIONS IN THE POLAR CAP

In 1990s, a network of imaging riometers was installed in the polar cap region from east Greenland to Svalbard and in their conjugate region in Antarctica as shown in Table 3. Locations of the northern hemisphere stations are marked in Fig. 5 together with their field of view. This network is very convenient for conjugate studies because the conjugate point of Zhongshan Station moves between east Greenland and Svalbard, where northern hemisphere imaging riometers are located.

Conjugate observations using this network started from 1997, and only initial observational data in January-Febuary, 1997 are available at present. Fig. 6. shows a good example of a conjugate CNA event observed at Zhongshan and Longyearbyen on January 30, 1997. A very strong CNA event, reaching more than 6 dB, appeared at

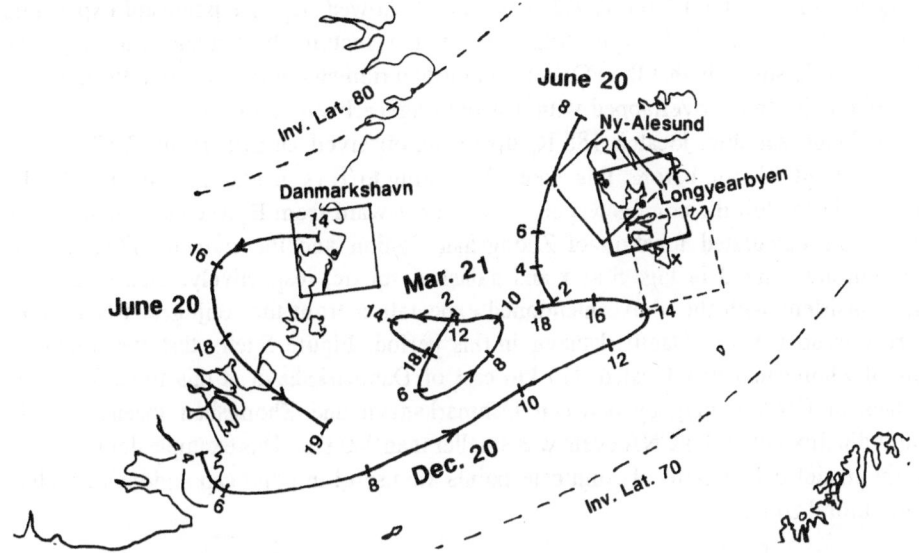

Fig. 5 Daily drift motion of the conjugate point of Zhongshan Station for vernal equinox and solstices. Solid squares denote FOV of imaging riometers.

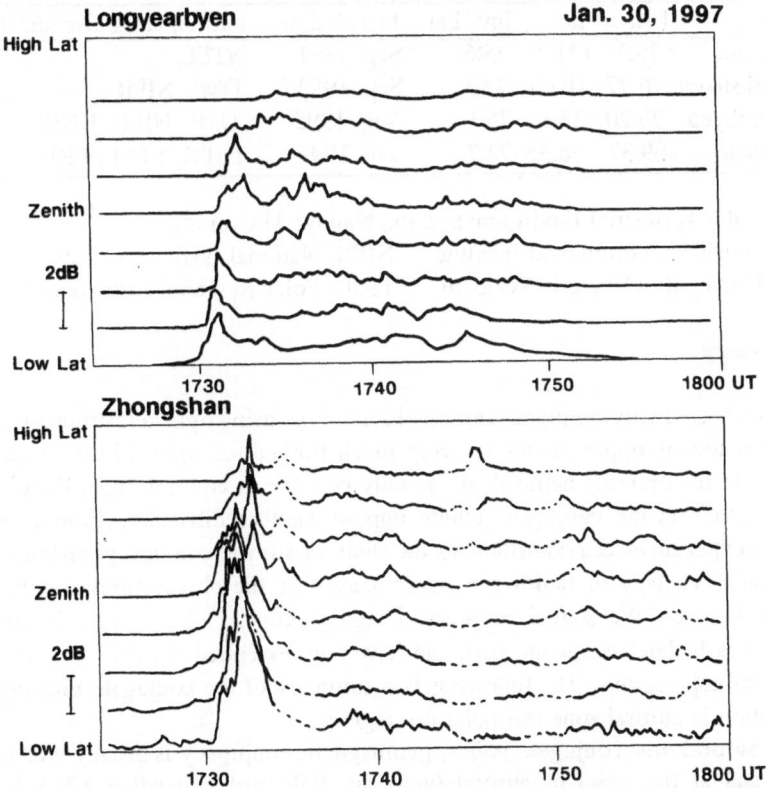

Fig. 6 Conjugate CNA events observed at Zhongshan Station and Longyearbyen.

Zhongshan a little after 1730 UT. This CNA event showed stepwise poleward expansion. At Longyearbyen, a CNA event with very similar feature, but smaller intensity was observed. This suggests that the FOV of Zhongshan riometer, projected onto the northern hemisphere, is almost overlapped with that of Longyearbyen riometer.

The Wind satellite, located $188\ R_E$ upstream, observed an IMF B_z of -2 nT and B_y of -4 nT about 40 min. before this event. According to Tsyganenko 1996 model, this B_y value works to shift the conjugate point 350 km eastward from $B_y=0$ condition shown in Fig. 5. The calculated location of Zhongshan Station and the riometer FOV in this condition are marked in Fig. 5 as x and a dashed square, respectively. This location is quite consistent with the above-mentioned expectation from the conjugate CNA event. There was no CNA at Danmarkshavn in this period. Figure 5 tells that the conjugate point of Zhongshan was located 940 km east of Danmarkshavn at this time. Therfore, the lack of CNA conjugacy between Danmarkshavn and Zhongshan means that the longitudinal extent of this CNA event was smaller than 940 km. This example demonstrates that the model calculations of conjugate points are useful in polar cap regions as well as in the auroral zone.

Table 3   Polar cap network of imaging riometers, extending from east Greenland to Svalbard, and their conjugate region in antarctica.

Station	Lat.	Lon.	Inv. Lat.	Installation	Participating Organization
Ny-Alesund	75.3	131.2	75.6	Sep. 1991	STEL
Danmarkshavn	76.77	-18.66	77.3	Sep. 1992	DMI, NIPR
Longyearbyen	78.20	15.80	75.1	Aug. 1995	DMI, NIPR, UNIS
Zhongshan	-69.37	76.38	74.7	Jan. 1997	NIPR, STEL, PRIC

STEL: Solar Terrestrial Environment Lab., Nagoya University
DMI: Danish Meteorological Institute   NIPR: National Institute of Polar Research
UNIS: University Course in Svalbard   PRIC: Polar Research Institute of China

4. Summary

There has been many conjugate auroral observations using optical instruments. However, the observational opportunities are very much limited for optical instruments, because dark sky in the opposite hemispheres is only possible in equinox time. Requirements of good weather at the conjugate points impose another difficulty. Conjugate imaging riometer experiment can contribute to the study of auroral conjugacy, making use of the particular advantage of radiowave instruments that the observations can be made all through the year. We started conjugate imaging riometer observations in auroral zone with Syowa-Iceland conjugate pair, and have now extended our conjugate observations into polar cap regions. The following is a summary of the conjugate imaging riometer observation in auroral zone and polar cap regions.

For auroral zone conjugate points, geomagnetic conjugacy is usually well maintained before and at the onset of auroral break-up. Poleward expanding CNA bands were observed at conjugate points. The latitude of the actual conjugate point of the station in the other hemisphere can be determined from the relative distance of the CNA bands which appeared in the common field of view of the two imaging riometers.

Conjugate point latitudes determined in this way showed seasonal variations by moving to higher latitude in winter and lower latitude in summer when we consider the conjugate point of a southern hemisphere station mapped into the northern hemisphere. This tendency qualitatively agrees with the seasonal drift of conjugate points calculated from Tsyganenko 1987 magnetosphere model.

Stenbaek-Nielsen et al. [6] proposed that IMF B_y component plays an important role in the formation of asymmetric field-aligned currents, and resultant intensity difference of discrete auroras between the two hemispheres. We have made a test of this effect using conjugate CNA data under the assumption that different field-aligned acceleration potentials between the two hemispheres can affect CNA intensity as well as the brightness of discrete auroras. The result was that the IMF B_y effect cannot be recognized from our CNA observations.

Many previous reports in literature indicate that CNA is stronger in wintertime than in summertime in the auroral zone. We confirmed this asymmetry as an interhemispheric difference between summer and winter hemisphere. We also found that this asymmetry was more pronounced in daytime CNA events as compared with nighttime ones.

Nighttime CNA asymmetry can be attributed to asymmetric formation of field-aligned potentials in the summer (sunlit) and winter (dark) polar ionosphere, suggested by Newell et al. [17]. For daytime CNA asymmetry, field-aligned potential mechanism does not work, and the cause can be seeked for in seasonal change of atmospheric structures in the D-region ionosphere. Atmospheric temperature in this region is higher in winter and lower in summer, just opposite to the seasonal variation of the atmospheric temperature in other altitude range [18]. This characteristics may cause larger CNA in wintertime through increased electron-neutral collision frequencies in the warmer D-region atmosphere.

It is generally difficult to define the conjugate points in the polar cap, which is the region of open field lines. However, under some limitations on latitudes and local times, the conjugate points can be calculated in the polar cap using Tsyganenko 1996 model. The calculated conjugate points show daily and seasonal drift, essentially the same as those calculated for auroral zone latitudes. However, the amplitude of the excursions are much larger (about 3 times) than in the auroral zone. Such calculations will give a good estimate on the drift of the conjugate regions when we analyze the conjugate data in the polar cap. A conjugate CNA event observed at Zhongshan Station and Longyearbyen supports the validity of conjugate point calculations in the polar cap regions.

5. Acknowledgments

Authors express their sincere thanks to Prof. Th. Saemundsson of University of Iceland for his long-term collaboration in Syowa-Iceland conjugare observations. Thanks are also due to Dr. Y. Sanoo for his laborious installation of imaging riometer at Zhongshan Station, and Prof. R. Liu of Polar Research Institute of China for collaboration at Zhongshan Station. We greatly appreciate the efforts of collaborating people operating imaging riometers in Danmarkshavn and Longyearbyen under harsh condition in the arctic. IMF data from WIND satellite were made available from ISTP key parameters provided by Dr. Lepping, NASA/GSFC.

6. References

1. DeWitt, R. N. (1962) The occurrence of aurora in geomagnetically conjugate areas, *J. Geophys. Res.* **68**,1347-1352.
2. Wescott, E. M. (1966) Magnetoconjugate phenomena, *Space Sci. Rev.* **5**, 507-561.
3. Belon, A. E., Maggs, J. E., Davis, T. N., Mather, K. B., Glass, N. W., and Hughes, G. F. (1969) Conjugacy of visual auroras during magnetically quiet periods, *J. Geophys. Res.* **74**,1-28.
4. Stenbaek-Nielsen, H. C., Davis, T. N. and Glass, N. W. (1972) Relative motion of auroral conjugate points during substorms, *J. Geophys. Res.* **77**, 1844-1858.
5. Stenbaek-Nielsen, H. C., Wescott, E. M., Davis, T. N. and Peterson, R. W. (1973) Differences in auroral intensity at conjugate points, *J. Geophys. Res.* **78**, 659-671.
6. Stenbaek-Nielsen, H. C. and Otto, A. (1997) Conjugate auroras and the interplanetary magnetic field, *J. Geophys. Res.* **102**, 2223-2232.
7. Sato, N., Fukunishi H. and Saemunsson, Th. (1984) Operation plan for the Iceland-Syowa conjugate campaign in 1983-1985, *Mem. Natl. Inst. Polar Res., Special Issue* **31**, 169-179.
8. Nagata. T. (1987) Research of geomagnetically conjugate phenomena in antarctica since the IGY, *Mem. Natl. Inst. Polar Res., Special Issue* **48**, 1-45.
9. Sato N. and Saemundsson, Th. (1987) Conjugacy of electron auroras, *Mem. Natl. Inst. Polar Res., Special Issue* **48**, 58-71.
10. Fujii, R., Sato, N., Ono, T., Fukunishi, H., Hirasawa, T., Kokubun, S., Araki, T. and Saemundsson, Th. (1987) Conjugacy of rapid motions and small-scale deformations of discrete auroras by all-sky TV observations, *Mem. Natl. Inst. Polar Res., Special Issue* **48**, 72-80.
11. Minatoya, H., Sato, N., Saemundsson, Th. and Yoshino T. (1996) Large displacements of conjugate auroras in the midnight sector, *J. Geomag. Geoelectr.* **48**, 967-975.
12. Ono, T. (1987) Temporal variation of geomagnetic conjugacy in Syowa-Iceland pair, *Mem. Natl. Inst. Polar Res., Special Issue* **48**, 46-57.
13. Hirashima, Y. (1991) Quantitative study of a localized Bremsstrahlung X-ray flux due to the field-aligned electric field, *J. Geomag. Geoelectr.* **43**, 539-547.
14. Basler, R. P. (1963) Radio wave absorption in the auroral ionospherte, *J. Geophys. Res.* **67**, 4665-4681.
15. Hartz, T. R., Montbriand, L. E. and Vogan, E. L. (1963) A study of auroral absorption at 30 Mc/s, *Can. J. Phys.*, **41**, 581-595.
16. Hargreaves, J. K. and Cowley, F. C. (1967) Studies of auroral radio absorption events at three magnetic latitudes-II, Differences between conjugate regions, *Planet. Space Sci.* **15**, 1585-1597.
17. Newell, P. T., Meng, C.-I. and Lyons K. M. (1996) Suppression of discrete aurorae by sunlight, *Nature*, **381.27**, 766-767.
18. Hedin, A. E. (1991) Extension of the MSIS thermosphere model into the middle and lower atmosphere, *J. Geophys. Res.* **96**, 1159-1172.

HF RADARS AS A TOOL FOR CONJUGATE STUDIES OF MAGNETOSPHERIC PHENOMENA

A. D. M. WALKER
Space Physics Research Institute, University of Natal, Durban, 4041 South Africa

Abstract. The SuperDARN chain of HF radars now covers a considerable portion of the auroral zone, cusp and polar cap regions in both hemispheres. The use of the radars to study magnetospheric convection in conjugate regions is reviewed. There is now a conjugate pair of bistatic radars, which allows conjugate convection velocity vectors to be determined. This leads to new opportunities for studying conjugate phenomena. Preliminary results from these radars are reviewed.

1. Introduction

Magnetically conjugate observations are important for understanding the physics of many polar cap phenomena. Recent extensions to the Super Dual Auroral Radar Network (SuperDARN) [1] in the Southern Hemisphere provide the capability of making conjugate observations of convection velocity vectors over a large area of the polar cap, cusp, and auroral zone. In this paper we review the current status of the SuperDARN (Super Dual Auroral Radar Network) radars, discuss the types of conjugate observations which are possible, and review conjugate radar observations of magnetospheric convection. We identify a number of problems which can be attacked using the radars.

2. The SuperDARN Network of HF Radars

In this section we briefly review the current status of the SuperDARN radars. The definitive review is by Greenwald *et al.* [1]. The radars operate in the frequency band 8–20 MHz. They observe scatter from field aligned

J. Moen et al. (eds.), Polar Cap Boundary Phenomena, 301–310.

irregularities in the E- and F-regions of the ionosphere. From the auto-correlation function of the returned multipulse pattern, the backscattered power, the Doppler velocity, and the spectral width of the irregularities can be found in sixteen directions at 45 different ranges. The field of view is typically 2000 km × 2000 km with ~50 km spatial resolution and 120 s time resolution during normal operation. It is possible to use other modes of operation during special periods. Strong backscatter only occurs for waves incident normally on the magnetic field. The radars exploit ionospheric refraction to get this correct aspect angle. The SuperDARN radars are located in positions suitable for observing the auroral zone, cusp and polar cap.

The velocity of the irregularities is the same as the background plasma $\mathbf{E} \times \mathbf{B}$ velocity. Two radars observing the velocity of the irregularities from different directions can thus find the drift velocity of plasma in the ionosphere at those locations in their common field of view. This provides a map of plasma drift, and hence electric field over a substantial area.

There are now ten SuperDARN radars in operation in both hemispheres covering a large fraction of polar cap and auroral zone over a wide range of local times. Six of these are in the northern hemisphere and four in the southern hemisphere. One is under construction in the Southern hemisphere, and three (two in the north and one in the south) are planned but not yet funded. Their fields of view are shown in Figure 1. Information about their location and status are given in Table 1. It can be seen that the fields of view of the northern hemisphere radars cover about 120° of longitude in the region of the northern polar cap, cusp and auroral zone. If the Alaska and British Columbia radars are funded this will be extended to almost 180°. There is currently no prospect, however, of any further extension into northern Siberia. The southern hemisphere radars are well placed to complement the northern hemisphere radars. With the commissioning of the SHARE radar in March 1997[1], there are now two pairs of radars which are conjugate — Goose Bay/Stokkseyri in the north and Halley/Sanae in the south. The Syowa East/Kerguelen pair will potentially extend the eastern longitude range. When the Tasmanian radar is complete it will provide coverage twelve hours of magnetic local time from Halley/Sanae and conjugate to the Siberian sector. In addition it will have a field of view extending to lower latitudes, covering more of the auroral zone and even the plasma-pause.

[1]The SHARE radar began operation on March 2 1997. After three and a half months of operation, during the preparation of this article, five of the antenna towers were severely damaged by an Antarctic storm. Operations continue with a severely degraded polar diagram. Due to the exigencies of Antarctic operations, full restoration can only occur during the austral summer 1997/8 at the earliest.

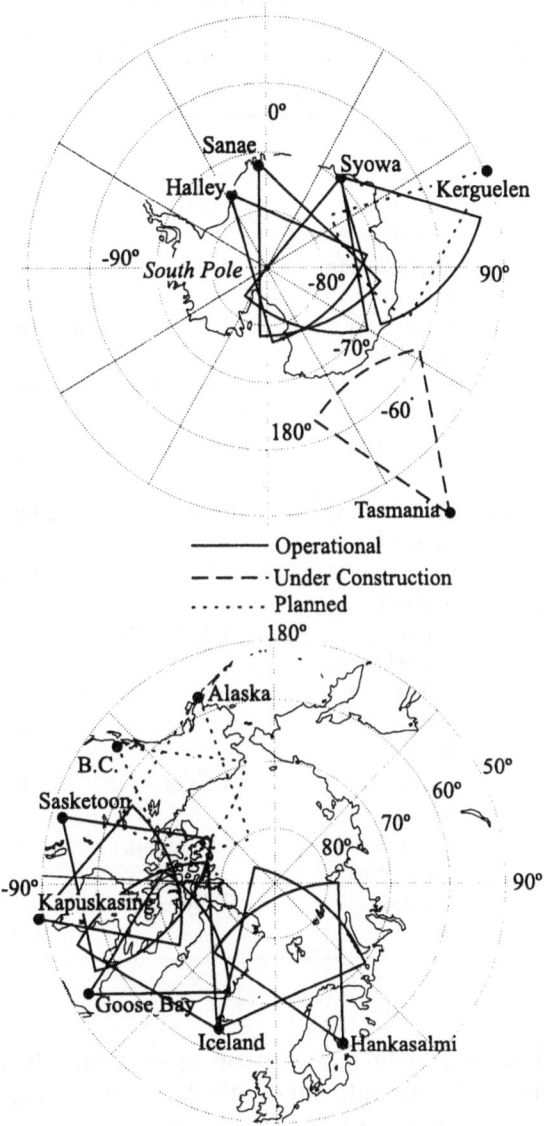

Figure 1. The SuperDARN chain.

3. Conjugacy

Magnetically conjugate phenomena occur on a single magnetic field line or flux tube. In this sense a low altitude satellite passing over a ground station can be magnetically conjugate to it. Phenomena in opposite polar caps with open field lines are not conjugate. The term "magnetic conjugacy" is, however, sometimes loosely used to describe phenomena in opposite

TABLE 1. SuperDARN radars

Radar	Location	Principal Investigator	Status
	Alaska	—	Proposed
	British Columbia	—	Proposed
Saskatoon	Saskatoon, Saskatchewan	G Sofko	Operational
Kapuskasing	Kapuskasing, Ontario	R A Greenwald	Operational
Goose Bay	Goose Bay, Labrador	R A Greenwald	Operational
Stokkseyri	Stokkseyri, Iceland	J-P Villain	Operational
CUTLASS-Iceland	Pykkvibær, Iceland	M Lester	Operational
CUTLASS-Finland	Hankasalmi, Finland	M Lester	Operational
PACE	Halley, Antarctica	J Dudeney	Operational
SHARE	Sanae 4, Vesleskarvet, Antarctica	A D M Walker	Operational
Syowa South	Syowa, Antarctica	N Sato	Operational
Syowa East	Syowa, Antarctica	N Sato	Operational
Kerguelen	Kerguelen I., Antarctica	J-P Villain	Proposed
TIGER	Tasmania	P Dyson	Under construction

hemispheres at locations which would be approximately conjugate if the Earth's main field were the only magnetic field. It is only worth studying conjugate phenomena if it adds to our understanding of the physics; if phenomena simply mapped from northern to southern hemispheres, there would be no need to venture into the extreme conditions in Antarctica when comparatively benign conditions are to be found in Arctic locations such as Svalbard.

We shall use a somewhat broader view of conjugacy. In opposite hemispheres the behaviour of formerly conjugate regions is influenced by their past history. A region on the ground which is strictly magnetically conjugate to a region in the opposite hemisphere may suddenly change to conjugacy with the solar wind magnetic field, as reconnection takes place on the

magnetopause. Phenomena occurring at the formerly conjugate locations
may, nevertheless, still continue to mirror each other, because of similar
behaviour in the parts of the solar wind which now connect to them. We
shall still study such behaviour under the label of conjugacy. It is possible
to distinguish four cases.

- The phenomenon is strictly conjugate. The behaviour at each end of
  the field line is the same. This might be anticipated, for example, if
  one was observing magnetospheric convection on closed field lines.
- The phenomenon is strictly conjugate. The behaviour at each end of
  the field line is different. An example would be the observation of a
  ULF pulsation at each end of a closed field line. The state of the
  polarisation would provide information about the harmonic which was
  being observed.
- Similar behaviour in each polar cap when the field lines are open.
  This is not a conjugate phenomenon but is affected by past conjugacy.
  An idealised exa~~ple~~ 2a. With a southward

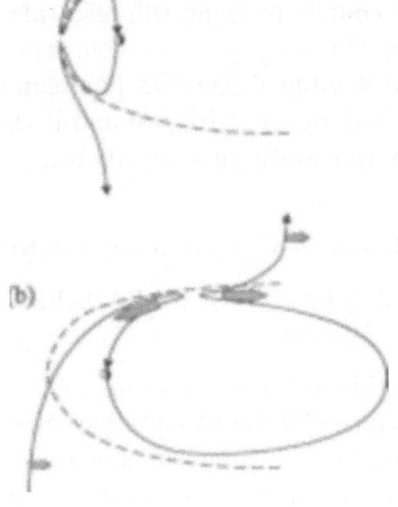

Figure 2.   Asymmetric reconnection for $B_z > 0$.

interplanetary field, reconnection takes place at the sub-solar point.
Points on the ground in each hemisphere map to field lines in the solar
wind which were previously connected. The effect on observed particle

spectra and subsequent convection can be expected to be very similar in both hemispheres because of the previous history of the plasma in the solar wind and within the magnetosphere.

- Dissimilar behaviour in each polar cap when the field lines are open. An idealised, and perhaps unrealistic, example is shown in Figure 2b. Here it is assumed that in the past the interplanetary field has been small enough for an extended period so that the magnetosphere is essentially closed, with the tail resulting largely from viscous drag. A strengthening of the northward magnetic field leads to enhanced reconnection in the lobes. As a result of asymmetry of the interplanetary field this is not the same in each hemisphere. The figure shows a field line just after it has merged with the interplanetary field. In the solar wind the field has a negative x component so that merging first takes place in the northern hemisphere. Clearly, after reconnection, the Maxwell curvature stresses on the field line connecting to the northern polar cap are very different from those connecting to the southern polar cap. The convection will be very different in the north and in the south, but it will still be strongly affected by the past history of the conjugacy of the field lines. Of course, merging will also take place in the southern hemisphere, but it will not be simultaneous; simultaneity of merging of points on the solar wind field line with points in northern and southern hemispheres only will occur if $B_x = 0$ and if there is exact symmetry between northern and southern hemispheres.

4. Radar Observations of Conjugate Phenomena

4.1. CONJUGATE MEASUREMENTS BY RADARS IN CONJUNCTION WITH OTHER INSTRUMENTS

While the emphasis in this review is on conjugate measurements in opposite polar caps, the importance of using satellite measurements conjugate to the HF radars in order to calibrate these instruments should not be neglected. It is important to be confident that the drift velocity observed by the HF radars is indeed the velocity of the $\mathbf{E} \times \mathbf{B}$ plasma drift. It is this that allows the electric field to be deduced. One such calibration was carried out by Baker *et al.* [2]. While observing the cusp region during a DMSP pass they were able to compare the transverse drift measured with DMSP and compare it with the drift observed by the radar. While the satellite showed large fluctuations on a small spatial scale, there was close agreement when the data were spatially averaged on a scale comparable to the resolution of the radar.

Another form of calibration is not of the parameters measured by the radar but their interpretation in order to identify magnetospheric regions.

Baker et al. [2, 3] showed that the nature of the spatial spectrum of the irregularities observed in the cusp and low latitude boundary layer was quite different from that inside the magnetosphere. They used particle data from overhead passes of DMSP, which clearly identified the different magnetospheric regions. The radar data at the magnetically conjugate footpoints at Goose Bay and Halley showed that the spectral width within the cusp was very much broader than elsewhere, and this defined a clear boundary of the cusp. This has been used in later work in order to delineate the position of the cusp in the radar field of view.

Rodger et al. [4] used the Halley radar in the southern hemisphere and the incoherent scatter EISCAT radar to study the conjugate behaviour of polar patches. The PACE radar in the south gave information about the motion of the patches over a wide field of view. The EISCAT data in the north did not give the spatial information provided by the PACE radar but provided information about the electron density and electron temperature which are not measured by HF radars. Both radars provided information about the motion of the equatorward edge of the patches. It was shown that the patches occur in both hemispheres at corresponding times. Although a one-to-one correspondence could not be established, about the same number of patches are formed in each hemisphere. The data set provides further support for hypothesis that the patches are formed as a result of transient processes on the day side.

4.2. THE PACE EXPERIMENT

The Polar Anglo-American Conjugate Experiment (PACE) represented the first extension of the current SuperDARN network beyond the original Goose Bay instrument[2]. Under the PACE agreement, which was the prototype for subsequent agreements, the construction of the Halley radar was undertaken, with a field of view in Antarctica conjugate (so far as the Earth's main field was concerned) to Goose Bay. Protocols for unified operations, data exchange, and scientific co-operation were set up in the agreement. Strictly, PACE describes the combined operations of the Goose Bay and Halley radars. Nevertheless, in common usage, the Halley radar has become known as the PACE radar, and this is a convenient usage.

One of the first efforts to exploit this conjugate capability [5] examined the effect on convection of the interplanetary magnetic field component B_y. They used the assumptions of Ruohoniemi *et al.* [6] to estimate the vector velocity from the line-of-sight velocity for each radar. The primary technique is to assume a quasi-steady flow along lines of geomagnetic latitude,

---

[2]An earlier HF radar at Schefferville, with a field of view overlapping that of Goose Bay, is no longer in existence.

and deduce the actual velocity from the component along the radar beam. They were able to relate their cusp convection patterns to those deduced theoretically for different B_y conditions. They studied reconfiguration of these patterns as B_y changed.

4.3. SHARE AND CONJUGATE MEASUREMENTS

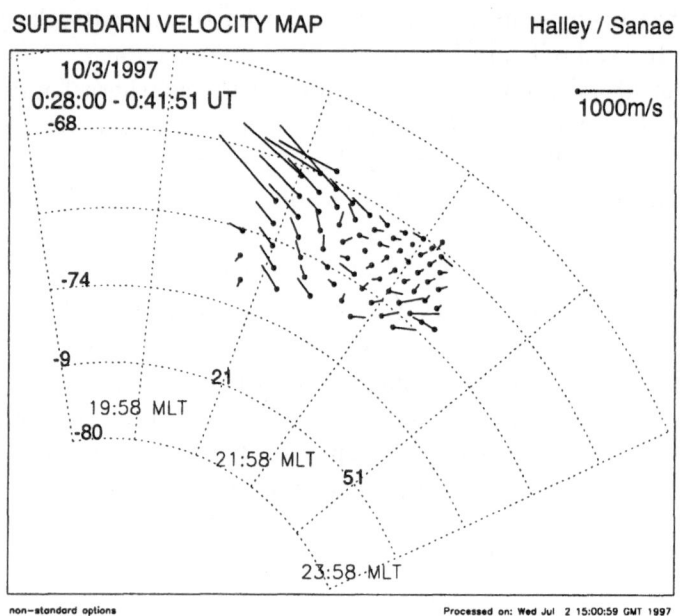

Figure 3. Fourteen minute average of the convection velocity in the field of view of the radar.

The Southern Hemisphere Auroral Radar Experiment (SHARE) represents an agreement, of the same nature as the PACE agreement, between the PACE experimenters and two South African Universities, to construct and operate a radar at Sanae 4, the South African base on the nunatak Vesleskarvet, in Antarctica. This represents the first opportunity for conjugate measurements which yield full vector information about the convection velocity. In the same way as the Halley radar has become known as PACE, the Sanae radar has become known as SHARE. A subsequent agreement incorporated the operations of the Syowa South radar, although successful combined observations with Halley and Sanae have not yet been achieved. Since then the SuperDARN agreement has integrated the operations of all the radars, and the Stokkseyri radar forms part of the conjugate set, Goose Bay, Stokkseyri, PACE, and SHARE.

An example of the capability of this system is work by Walker *et al.* [7] who have studied an event on March 9/10 1997 during which the inter-

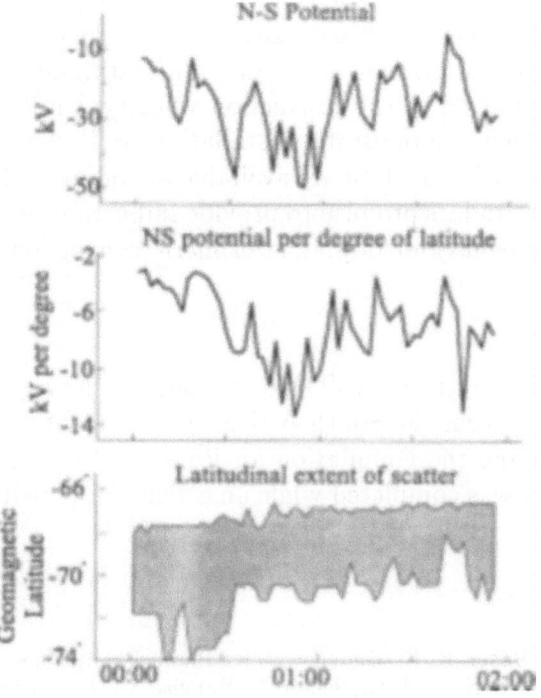

Figure 4. Potential drop across region of observation.

planetary conditions were extremely quiet. From 1800 UT on March 8 until
1200 UT on March 10 the total B never exceeded 3 nT. For much of that
time it was about 2 nT or less. For the whole period the B_z component
was northward except for a brief period on March 9 between 1330 UT and
1530 UT when it had a small southward component which averaged approximately 0.2 nT. The solar wind velocity was small and steady, between 300
and 340 km s^{-1} during this period. During this time the convection pattern was observed in both hemispheres. An example of the data from the
Halley–Sanae pair is shown in Figure 3. This represents a 14 minute average
of the radar data. The dominant feature is the very strong flow westward
and equatorward. This arises from a series of flow bursts, occurring quasi-
periodically on a time scale of about 8 min. The convection velocity during
these flow bursts peaks at more than 2 km s^{-1}, corresponding to electric
fields exceeding 100 mV m^{-1} in the ionosphere. The total potential drop
across the region of observation is shown in Figure 4.

5. Discussion and Conclusions

HF radars in the SuperDARN chain have already proved to be important devices for studying magnetospheric convection. Initially they were limited to monostatic installations which only gave one component of the convection velocity. Bistatic installations are now proliferating allowing determination of both components of the convection velocity. Until 1997 the only conjugate pair involved a monostatic installation at Halley. The SHARE radar now means that bistatic data are available at conjugate locations. Initial results from these radars promise to provide important new information about the conjugate behaviour of magnetospheric convection.

Acknowledgements

I am grateful for support from NATO to attend the Advanced Study Institute and for grants from the South African Department of Environment Affairs and Tourism and the Foundation for Research Development. The final draft of the paper was completed while on a visit to the British Antarctic Survey.

6. References

1. Greenwald, R. A., Baker, K. B., Dudeney, J. R., Pinnock, M., Jones, T. B., Thomas, E. C., Villain, J.-P., Cerisier, J.-C., Senior, C, Hanuise, C., Hunsucker, R. D., Sofko, G., Koehler, J., Nielsen, E., Pellinen, R., Walker, A. D. M., Sato, N., and Yamagishi, H., (1995) DARN/SuperDARN: A global view of high latitude convection, *Space Sci. Rev.*, **71**, 761-796.
2. Baker, K. B., Greenwald, R. A., Ruohoniemi, J. M., Dudeney, J. R., Pinnock, M., Newell P. T., Greenspan, M. E., and Meng, C.-I., (1990) Simultaneous HF-radar and DMSP observations of the cusp, *Geophys. Res. Lett.*, **17**, 1869–1872.
3. Baker, K. B., Dudeney, J. R., Greenwald, R. A., Pinnock, M., Newell, P. T., Rodger, A. S., Mattin, N., and Meng, C.-I., (1995) HF radar signatures of the cusp and low latitude boundary layer, *J. Geophys. Res.*, **100**, 7671–7695.
4. Rodger, A. S., Pinnock, M., Dudeney, J. R., Waterman, J., de la Beaujardière, O., and Baker, K. B., (1994) Simultaneous two hemisphere measurements of the presence of polar patches in the nightside ionosphere, *Ann. Geophysicae*, **12**, 642–648.
5. Greenwald, R. A., Baker, K. B., Ruohoniemi, J. M., Dudeney, J. R., Pinnock, M., Mattin, N., Leonard, J. M., and Lepping, R. P., (1995) Simultaneous conjugate observations of dynamic variations in high-latitude dayside convection due to changes in IMF B_y, *J. Geophys. Res.*, **95**, 8057–8072.
6. Ruohoniemi, J. M., Greenwald, R. A., Baker, K. B., Villain, J.-P., Hanuise, C., and Kelly, J., (1989) Mapping high-latitude convection with coherent HF radars, *J. Geophys., Res.*, **94**, 13463–13477.
7. Walker, A. D. M., Pinnock, M., Baker, K. B., Dudeney, J. R., and Rash, J. P. S., Strong flow bursts in the nightside ionosphere during extremely quiet conditions, submitted to *Geophys. Res. Lett.*, 1997.

CONJUGATE GROUND OBSERVATIONS AND POSSIBLE SOURCE REGIONS OF TWO TYPES OF PC 1-2 PULSATIONS AT VERY HIGH LATITUDES

L. P. DYRUD, M. J. ENGEBRETSON, J. L. POSCH
Augsburg College
2211 Riverside Ave.
Minneapolis, Minnesota 55454, USA

W. J. HUGHES
Boston University
Boston, Massachusetts 02215, USA

H. FUKUNISHI
Tohoku University
Sendai, Japan

R. L. ARNOLDY
University of New Hampshire
Durham, New Hampshire 03824, USA

P. T. NEWELL
The Johns Hopkins University Applied Physics Laboratory
Laurel, Maryland 20723, USA

Abstract This paper presents results from a study of one year's data from the recently installed Magnetometer Array for Cusp and Cleft Studies (MACCS) in Arctic Canada and from two stations of the Automated Geophysical Observatories (AGOs) in Antarctica. The magnetometer data from these conjugate arrays were used to study ULF waves in the Pc 1-2 (100-600 mHz) frequency band at cusp and polar cap latitudes (74° - 80° invariant latitude). In this paper we focus on the differences between observations in different hemispheres, spectral properties and latitudinal - local time distributions of Pc 1-2 events observed during 1994. The latitudinal trends and case studies are used to infer the source locations of the two major wave types we have observed. Broadband waves, with diffuse spectral character, were more prevalent at the higher latitudes. Waves with narrower bandwidth were much more common in our data set, and were the statistically dominant wave type at all the MACCS stations. These multistation observations, combined with data from DMSP satellite overpasses, suggest the possibility that these two wave types originate in quite

J. Moen et al. (eds.), Polar Cap Boundary Phenomena, 311–326.
© 1998 *Kluwer Academic Publishers.*

different regions near the magnetospheric boundary; the more narrowband waves in the subsolar and postnoon equatorial region, and the more broadband waves in the high latitude plasma mantle (and possibly at the poleward edge of the cusp). The case studies also reveal a connection between periods of IMF Bz northward and cutouts in broadband Pc 1-2 wave power, and hemispherical differences in Pc 1-2 observations are interpreted in terms of dipole tilt angle effects.

1. Introduction

Magnetic pulsations in the Pc 1-2 (100-600 mHz) frequency range are thought to be generated by electromagnetic ion cyclotron (EMIC) waves in the Earth's near magnetosphere [1]. The most common mechanism for the generation of EMIC waves is a temperature anisotropy within the distribution of hot protons [2]. EMIC waves are a non-negligible mechanism for the transport of energy within the magnetosphere. These waves can also heat thermal electrons via Landau damping, which in turn are a contributive source to electron heat flux into the ionosphere, and thus stable red arcs [3].

Pc 1-2 pulsations are observed on the ground over a broad range of L shells (~4 - 14). Using several different methods [4], there have been many studies attempting to classify Pc 1-2 by magnetospheric source regions [e.g., 5, 6, 7, 8, 9]. At lower latitudes (~L=6) they have been classified into subtypes by spectral character (with each subtype correlated with a specific source region in the magnetosphere) [5]. However, there is more debate surrounding the studies of Pc 1-2 observed at higher latitudes, i.e., in the region of the cusp and cleft, and their relationship to processes occurring in these boundary regions are still not well understood.

This paper presents results from a statistical study of Pc 1-2 pulsations observed at six cusp/cleft and polar cap stations (~74° -80° MLAT) for the year of 1994. Menk et al. [8] used data recorded over twelve summer days from six stations ranging in geomagnetic latitude from 62.1° to 80.8°. They reported that wave type varied with latitude, and they separated Pc 1-2 activity into six types of Pc 1-2 emissions, based on spectral appearance. Four of these wave types were most often observed, and appeared to originate, well equatorward of the cusp. The two highest latitude emission types, which are more relevant to this paper, are (i) narrowband Pc 1-2, which they observed mostly under and a few degrees equatorward of the plasma sheet boundary layer in the noon sector, and (ii) unstructured 0.15 - 0.4 Hz emissions, which they observed mostly within ~2° of the poleward edge of the cusp, with a duration correlating with the azimuthal width of the cusp.

Similar to Menk et al. [8] the results of the full-year study presented here reveal a latitudinal variation in Pc 1-2 wave type. Our observations show that higher latitude stations (~80° MLAT) more commonly observe wideband diffuse Pc 1-2 waves, and lower latitude stations (~74° MLAT) most commonly observe narrowband Pc 1-2 waves. Like Menk et al. [8], our study also used DMSP (Defense Meteorological Satellite Project) satellite overpasses to determine the magnetospheric regions mapping down to various stations during wave events. In agreement with previous

313

MACCS MAGNETOMETER NETWORK

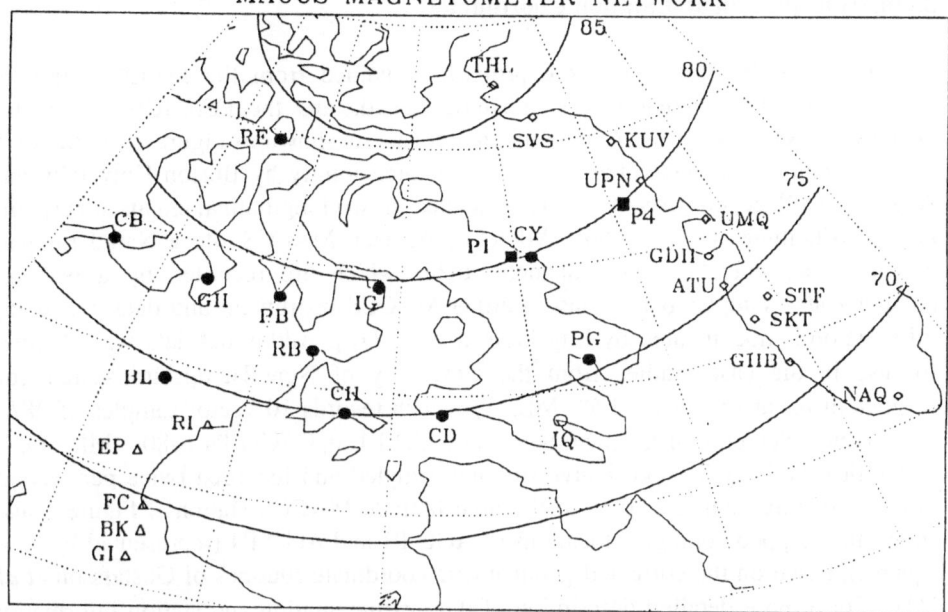

Figure 1. Map of Arctic Canada showing the MACCS stations (represented by solid circles). AGO P1 and P4 have been mapped using corrected geomagnetic coordinates [11] and superimposed on their conjugate Arctic sites. Magnetic latitude is shown every 5° by solid lines. Also shown are stations in the Greenland coastal chain (diamonds) and the Canopus array (triangles).

studies, our results also point to the subsolar, postnoon equatorial region as the source of more narrowband Pc 1-2 waves. However, we suggest the high latitude plasma mantle rather than the cusp proper as the dominant source of the more broadband waves.

This study incorporates a perspective rarely included in previous work, that is the long term comparison between magnetometer data in opposite hemispheres, normally termed as conjugate studies. Strict conjugacy usually refers to two stations at either footprint of the same closed magnetic field line [Walker, this edition]. However, the "conjugate" stations used in this study are nominally under the polar cap, so they can be expected to observe magnetic field lines connected to the magnetospheric tail lobe and plasma mantle rather than observing activity at both ends of a closed, dayside magnetic field line. These stations may still observe very related phenomena because of the earlier conjugacy of convecting field lines.

2. Instrumentation and Station Coordinates

The data used in this study were primarily obtained from the MACCS array of sensitive fluxgate magnetometers in Arctic Canada and the PENGUIn Automated Geophysical Observatories (AGOs) in Antarctica. As shown in Figure 1 the MACCS array, completed in September 1993 and operated jointly by Boston University and Augsburg College, consists of stations extended in two longitudinal chains at cusp and polar cap latitudes (~74° - 80° MLAT). At each MACCS site a Narod fluxgate magnetometer detects vector magnetic fields which are recorded by a personal computer at 2 samples per second. Further MACCS instrument and data processing information is documented by Engebretson et al. [10]. Additional data was obtained for use in the case studies from the University of New Hampshire search coil magnetometer at Iqaluit (~73.5° MLAT), which records 10 vector samples of dB/dt per second. Search coil magnetometers at AGO P1 and AGO P4 (~80° MLAT), in Antarctica, built by Tohoku University and sampled and recorded twice per second, provide roughly conjugate data to the high latitude MACCS stations. Figure 1 also shows the mapped conjugate locations of AGO P1 and AGO P4 (represented by solid squares), based on the corrected geomagnetic coordinate routines of Gustafsson et al. [11]. For a more detailed description of the instrumentation and station information see Dyrud et al. [12].

3. Statistical Study

As part of our routine processing, 0-1 Hz, 24 hour, color spectrograms were created using a 256 point Fast Fourier Transform for each available day for the six stations Clyde River (CR), Pelly Bay (PB), Gjoa Haven (GH), Cape Dorset (CD), AGO P1 (P1), and AGO P4 (P4). Continuous data were available from AGO P1 and AGO P4 after each station was deployed in January 1994, while MACCS stations suffered some periods of data loss. Computer hardware problems at Gjoa Haven resulted in significant data loss at that station. The daily spectrograms were then visually inspected by two researchers, and spectrograms from the six stations were viewed simultaneously for spectral comparison between stations. Pc 1-2 events were selected if their upper and lower frequency limits were between 0.1 and 0.6 Hz, if wave power was at least 1 decade higher than the background noise, and at least ~15 minutes in duration. Simultaneous Pc 1-2 events at distinctly different frequencies were recorded as separate events, and events with wave power cut-outs less than ~45 minutes (see Figure 3 for example) were recorded as one event.

3.1. LOCAL TIME AND OCCURRENCE DISTRIBUTIONS OF PC 1-2

Figure 2 displays all available data during 1994 for each station studied, in the form of occurrence plots of Pc 1-2 bandwidth vs. UT. The horizontal axis for all stations ranges from 00:00 to 24:00 UT, and each plot summarizes all of the Pc 1-2 events observed at these stations. The number of Pc 1-2 waves that fall into any given 10 mHz vs. 10 min. bin is displayed by grayscale level as indicated by the bar to the right of each panel. A vertical line in each panel indicates that station's local magnetic noon, and a horizontal line at 200 mHz bandwidth is used to separate events with somewhat different occurrence patterns, as will be discussed below. Because of several gaps in data coverage shown in Figure 2, it is conceivable that some of the differences between the occurrence patterns at different stations shown in Figure 2 could be due to seasonal variations. However, plots produced with data from only simultaneous data coverage show no significant differences in the distributions of events in either frequency or UT, and the same longitudinal and latitudinal trends are evident. The reader is referred to Dyrud *et al.* [12] for further discussion of the data handling. Figure 2 shows that Pc 1-2 waves at all six stations are centered around local noon (most within 3 hours or less). However, Pc 1-2 waves with bandwidth greater than 200 mHz are even more tightly centered around local noon. This result, that wider bandwidth Pc 1-2 waves are centered near local magnetic noon, is consistent with most earlier studies [e.g., 8].

Pc 1-2 waves appeared at the six stations on about half of the days studied or even less. The probability of occurrence of any type of Pc 1-2 wave activity at the six stations between 11:00 and 13:00 magnetic local time was 23% at AGO P4, 23% at AGO P1, 56% at Clyde River, 38% at Cape Dorset, 45% at Pelly Bay, and 33% at Gjoa Haven. The result that AGO P1 and P4, the two stations in the southern hemisphere, have a Pc 1-2 occurrence frequency that is approximately half that of their 80° MLAT counterpart Clyde River in the northern hemisphere is particularly unusual, and will be addressed in the following section.

3.2. VARIATION OF BANDWIDTH WITH LATITUDE

One of the chief findings of this statistical study is that Pc 1-2 bandwidth varies with latitude. We have found that high latitude stations (~80° MLAT) tend to observe more diffuse wideband waves and lower latitude stations tend to observe more discrete narrowband waves. The occurrence plot for Cape Dorset (MLAT = 74.6°) shows that there are very few wideband waves at this station at any time, and the fraction of waves increases from Gjoa Haven (MLAT = 78.3°) to Pelly Bay (MLAT = 78.7°) and Clyde River (MLAT = 79.7°). The percentage values of events wider than 200 mHz, as calculated by number of event hours. are as follows: of the 408.93 total event hours at AGO P4 31% are wider than 200 mHz, 38% of the 293.4 event hours are wider at AGO P1, 13% of 689.3 hours at Clyde River, 7% of 602.7 at Pelly Bay, 8% of 298.3 at Gjoa Haven, and 3% of 498.3 at Cape Dorset). Our observed percentages of Pc 1-2 events wider than 200 mHz for the 74.6° MLAT Cape Dorset station are reasonably consistent with the latitudinal trend reported by Popecki *et al.* [9] in a study using search coil magnetometers. They observed 6% of Pc 1-2 waves with bandwidth wider than 200 mHz at their 74.2° and -74.2° MLAT stations, and only 2% at the lower latitude station (-61° MLAT).

It is also evident in Figure 2 that there are a large number of Pc 1-2 events with bandwidth less than 100 mHz at Cape Dorset, and that there are fewer such waves at

316

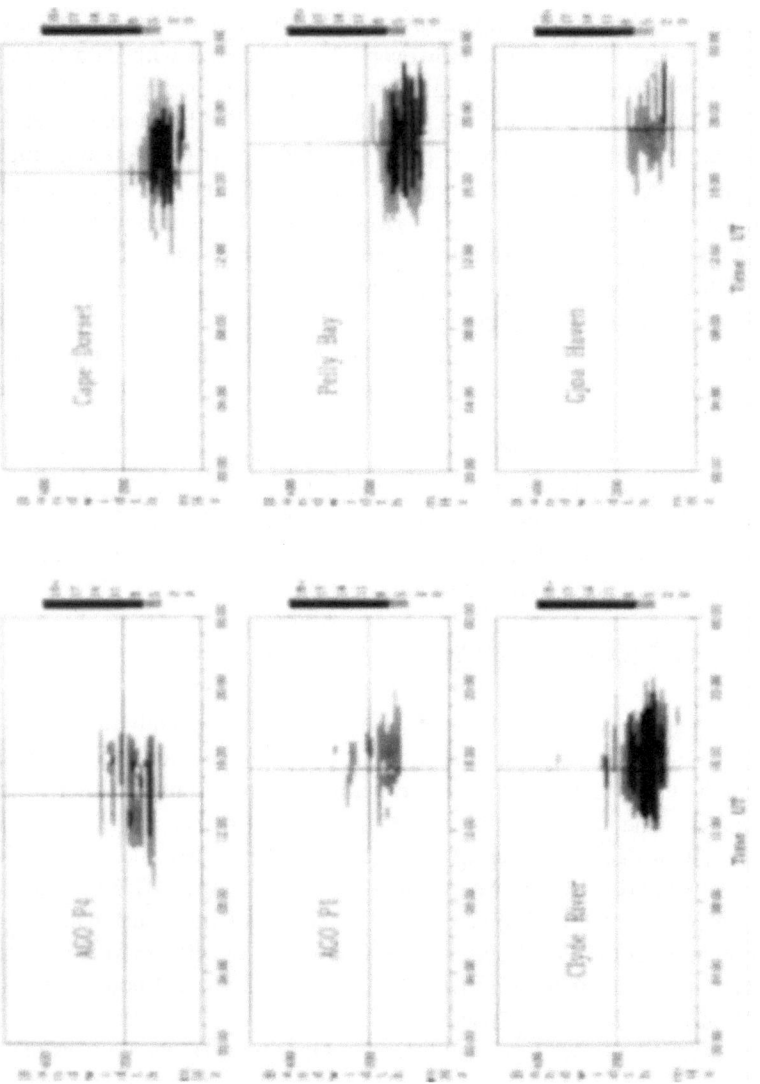

Figure 2. Occurrence plots of Pc 1-2 bandwidth vs. Universal Time using all available 1994 data from 6 stations. The vertical axis shows bandwidth in mHz from 0-500 and is broken into 10 mHz bins. The horizontal axis shows time in hours from 00:00 - 24:00 UT and is broken up into 10 minute bins. The number of Pc 1-2 waves that fall into any given 10 mHz vs. 10 min. bin is displayed by grayscale level as indicated by the bar on the right of each graph.

the higher latitude stations. Several studies have placed the source of narrowband Pc 1-2 somewhere in the closed field line regions of the magnetosphere. For example, Menk *et al.* [8] reported observing narrowband Pc 2 waves most commonly at their 70.3° MLAT station. They also reported observations of this wave type at 74.5° MLAT and 80.8° MLAT during magnetic activity levels of Kp = 3-4 and Kp = 1, respectively. Because of this statistical trend, they placed the source region of these waves under and a few degrees equatorward of the plasma sheet boundary layer in the noon sector. Popecki *et al.* [9] proposed a source region between the auroral zone and the plasmapause for Pc 1-2 observed at stations ranging in MLAT from ~62° - 74°. Hansen *et al.* [13] placed similar narrowband emissions in an equatorial plasmatrough source region. If the narrowband Pc 1-2 waves are generated in the closed field line region, it is then not surprising that Cape Dorset would observe more narrowband waves than the higher latitude stations Pelly Bay and Clyde River.

By the same reasoning, we suggest that the higher occurrence of wideband Pc 1-2 at the higher latitude stations (78° to 80° MLAT) is caused by a source region originating near such latitudes. In the following section we provide additional data from a case study that helps characterize this higher latitude source region. Although several earlier studies [8, 7, 14] proposed the cusp as the source region for wideband diffuse waves observed at high latitudes, the combination of ground and satellite data presented in the following section lead us to conclude that these waves originate in the plasma mantle (or poleward edge of the cusp) and not in the cusp itself.

3.3. HEMISPHERICAL DIFFERENCES

We attribute the significantly higher number of wideband Pc 1-2 events at the Antarctic AGOs P1 and P4 in comparison to Clyde River, despite their nearly identical invariant latitude, to the same mechanism responsible for the disparity of occurrence between these three stations. We believe that dipole tilt angle factors cause the overhead magnetospheric regions of the conjugate stations to be asymmetric. The variation of cusp latitude with respect to season is well known [15]. Figure 2 of Newell *et al.* [15], shows that the cusp in the winter hemisphere is on average ~2° lower than the summer hemisphere, but that at equinox the average cusp location should be at the same latitude, and therefore observations averaged over an entire year should show no difference between hemispheres. However, PACE and Goose Bay radar observations (A. S. Rodger, personal communication) show that there is an additional disparity in cusp latitude between the southern and northern hemisphere for the longitude sector where the MACCS and AGO stations are located. This disparity is that at equinox the average location of the northern hemisphere cusp is ~2-3° higher in latitude than its location in the southern hemisphere, and this latitudinal shift is fairly consistent throughout the year. This means that, even though Clyde River and AGO P1 and P4 are located at the same MLAT, Clyde River is most commonly located under the cusp and plasma mantle regions whereas the AGO stations are normally several degrees poleward of the cusp. Thus, AGO P1 and P4 are normally several hundred kilometers poleward from the ground location of the overhead cusp. This makes the probability of observing waves generated on closed field lines very unlikely, and since narrowband Pc 1-2 waves are commonly believed to be generated in such closed line field regions, it makes their observation much less likely. These arguments offer an explanation for AGO P1 and P4's lack of narrowband waves, and since this wave type is absent from the AGOs data set it also explains the disparity in wave occurrence between AGO P1, P4 and Clyde River. It is unlikely that these differences are due to the different instruments used at these stations, (search coil magnetometers at AGO P1 and P4, and fluxgate at Clyde River) because the search coil instrument sensitivity is better than that of the fluxgate's so one would expect any difference in observation to be in the direction of more waves observed by the search coils.

4. Case Study: November 18, 1994 (Day 94322)

A variation of Pc 1-2 bandwidth with latitude can be seen not only in the statistical results, but in individual cases as well. In this section we present an example of multistation observations of Pc 1-2 micropulsation events. Several days of magnetometer data and satellite data were examined on a case study basis, and the example displayed here is representative of both the case studies and the trends found in the statistical study. Figure 3 displays magnetometer data from eight stations, ranging from 73.5° to 80.0° invariant magnetic latitude, for November 18, 1994, and the Pc 1-2 waves observed display a definite variation with latitude. This day is an example of Pc 1-2 observed under average to quiet magnetospheric conditions; the 3 hour Kp values for this day are 1-, 1, 1-, 2+, 2, 2+, 2-, and 2-. The top and second panels of Figure 3 are search coil data from AGO P4 and P1 showing broadband Pc 1-2 activity from ~10:00-16:00 and ~11:00-19:00 UT, respectively. The third panel is fluxgate data from Clyde River which also shows similar broadband activity, but from ~13:30-19:30. The fourth and fifth panels of Figure 3 show fluxgate data from Pelly Bay and Gjoa Haven, respectively. Both panels show broadband Pc 1-2 waves similar to those appearing at P1, P4, and Clyde River, but at a later UT. The sixth panel of Figure 3 shows fluxgate data from Pangnirtung, which exhibits weak narrowband Pc 1-2 waves beginning near 13:30 UT and extending at least to 19:00 UT, when local interference obscures the natural signals. The seventh and eighth panels of Figure 3 are fluxgate data from Cape Dorset and search coil data from Iqaluit that show similar Pc 1-2 waves falling in frequency from ~13:30-17:00 UT, and with constant or rising frequency from 18:00-21:00, which are similar to those appearing at Pangnirtung at the same time. Pc 1 waves are also evident at Iqaluit from ~ 11:00 to 13:00 UT, possibly because of the higher sensitivity of search coil instruments to waves of higher frequency; no clear signal of such waves is evident at Cape Dorset or Pangnirtung. To summarize, the higher latitude stations P4, P1, Clyde River, Pelly Bay, and Gjoa Haven all show semi-structured broadband Pc 1-2, while the lower latitude stations Pangnirtung, Cape Dorset, and Iqaluit observe narrowband Pc 1-2 waves even during the broadband event at the high latitude stations. Thus, there are two distinctly

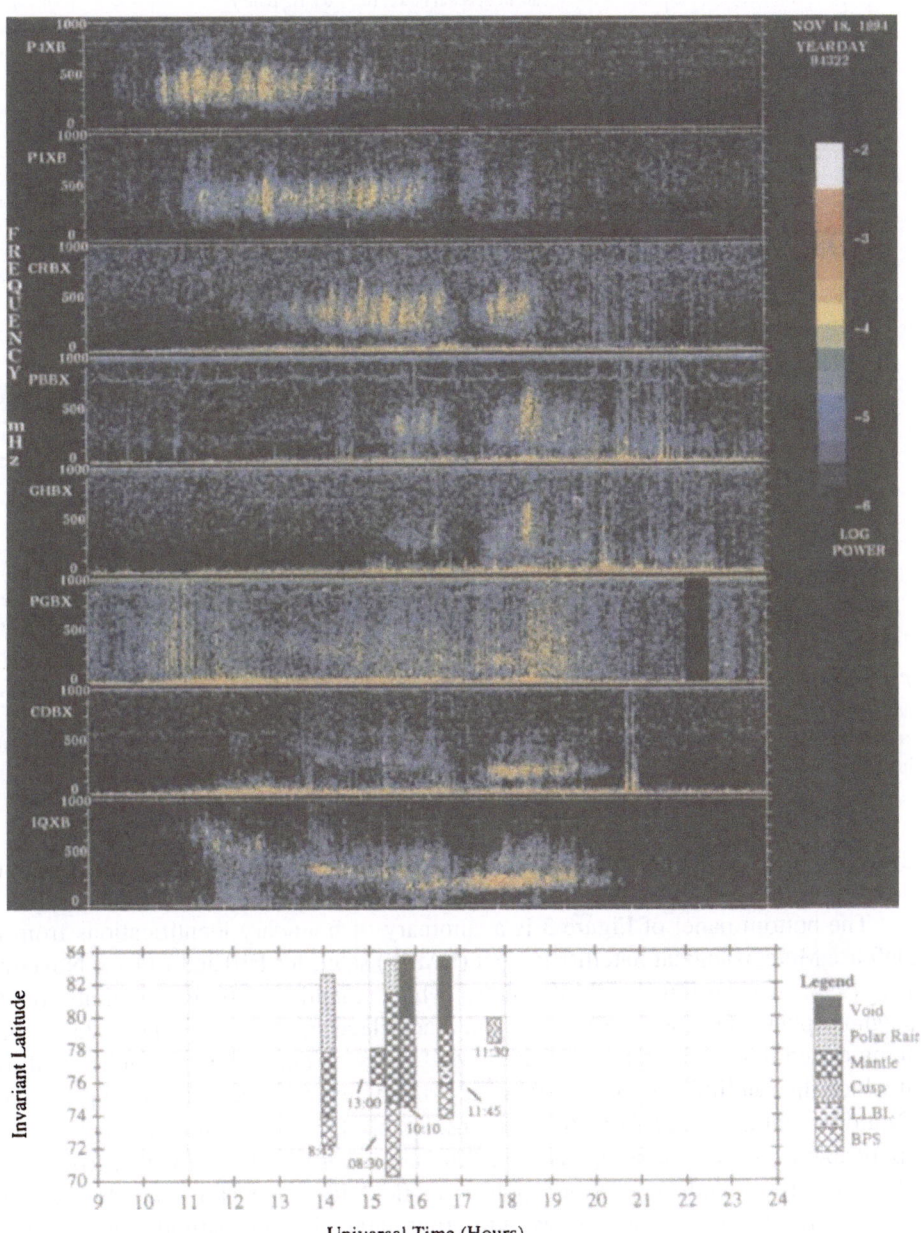

Figure 3. Ground and satellite data from 09:00 to 24:00 UT November 18, 1994 (Yearday 94322). All parts except the bottom panel are Fourier spectrograms of the Bx component from magnetometers at the stations indicated. The vertical axis of each spectrogram panel is in mHz, the horizontal axis is UT, and wave power is represented by color, according to the bar on the right. From top to bottom the spectrogram panels are 1) search coil magnetometer data from AGO P4 (P4) and 2) AGO P1 (P1); 3) fluxgate magnetometer data from Clyde River (CY), 4) Pelly Bay (PB), 5) Gjoa Haven (GH), Repulse Bay (RB), 6) Pangnirtung (PG), and 6) Cape Dorset (CD); and 8) search coil data from Iqaluit (IQ). The bottom panel summarizes DMSP observations of the invariant magnetic latitudes of magnetospheric regions as a function of the universal time of the satellite pass. The satellite's local time is printed next to each pass summary.

320

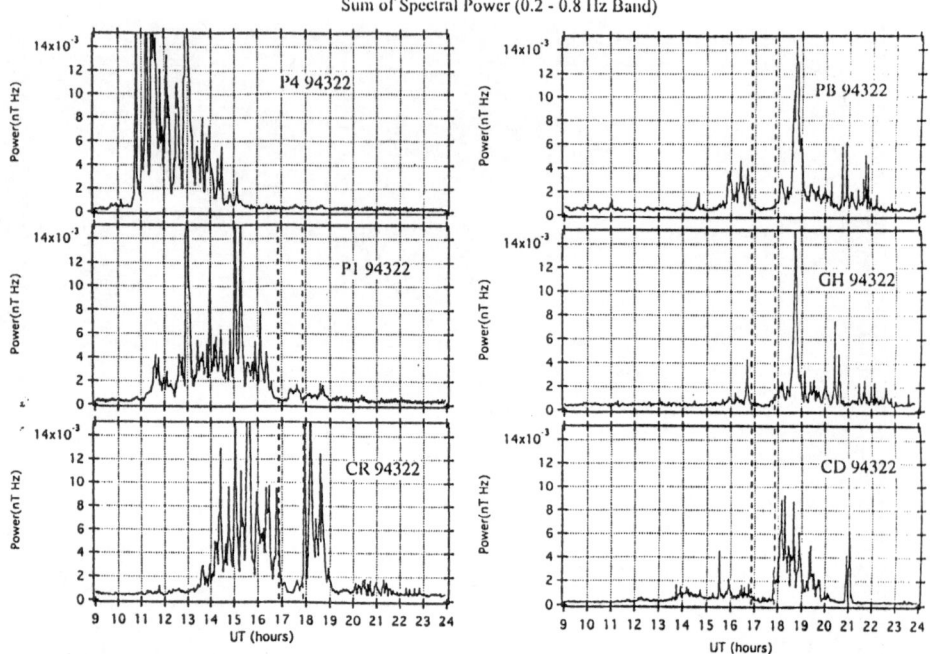

Figure 4. Plots of the sum of spectral power in the 0.2 - 0.8 Hz band as a function of time for Bx component data from AGO P4, AGO P1, Clyde River, Pelly Bay, Gjoa Haven, and Cape Dorset, on day 94322. The vertical dotted lines in each panel show the time inferred from GEOTAIL observations of the IMF when an interval of northward Bz impinges on the dayside magnetopause.

different types of Pc 1-2 waves appearing simultaneously on the ground, and the distinctions are clearly associated with latitude.

The bottom panel of Figure 3 is a summary of boundary identifications from the Defense Meteorological Satellite Project (DMSP) satellites F10 and F11 for November 18, 1994. The original DMSP plots, which identify different regions of the magnetosphere by their respective ion and electron signatures [15], have been summarized in the bar graphs of latitude vs. Universal Time. The magnetic local time at which the satellite was at 77° MLAT is printed next to each represented pass, and it should be realized that passes that are considerably far from local noon may not be as applicable as those near noon. The satellite passes for this day show that from 14:00-16:00 UT the mantle region and polar cap field lines are above nearly the entire MACCS region of latitude. As we will argue later, the high latitude plasma mantle may be a source region for wideband Pc 1-2 at very high latitudes. This may account for the associated broadband events at the high latitude stations. The pass at ~16:30 UT shows closed field lines moving up to ~76° and the final pass at 17:30 UT shows the cusp to be at very high latitudes (~80°). It is also at this time that the narrowband wave is seen at Pangnirtung and Cape Dorset (75.2°-74.6° MLAT). According to the DMSP satellite passes, open field line regions are over the MACCS and AGO stations during the observation of wideband Pc 1-2 waves, and closed field line regions are over stations during the observation of narrowband waves.

An additional point of interest is the ~40 minute cut-out in spectral power near 17:00 UT that appears at all stations. It appears to be relatively simultaneous across the stations of the MACCS array, affecting the broadband waves more than the narrowband ones, but the cut-out appears at a clearly different time at AGO P1 in the southern hemisphere. Figure 4 displays line plots of the sum of spectral power in the 0.2 - 0.8 Hz band vs. UT, for six stations from day 94322. The vertical dotted lines in each panel show the time inferred from GEOTAIL observations of the IMF when an interval of northward Bz impinges on the Earth (plot not shown). The second panel on the left side of Figure 3 shows the cut-out at AGO P1 from 16:40 to 17:18 UT, somewhat before the cutout at Clyde River and Cape Dorset, which is from 16:56 to 17:50 UT. The cutout at all four MACCS (northern hemisphere) stations appears to coincide, on the order of minutes, with the times of northward Bz inferred from the GEOTAIL data. One would expect changes of wave power stimulated by an IMF orientation change to be simultaneous in both hemispheres, if the wave source is located near the equatorial magnetopause. Since there is a time lag between the two hemispheres, the higher latitude waves may be generated on open field lines, and may further be associated with the non-zero value of IMF By during the interval (~ -5 nT).

Cutouts in Pc 1-2 wave power associated with northward turnings of the IMF were also observed on other days studied [12]. One day which was examined as closely as this day displayed a cutout in power that was simultaneous in both hemispheres. However, on that day there was no significant IMF By component.

5. Discussion

Data from November 18, 1994 displayed Pc 1-2 waves that are simultaneously wider in bandwidth at higher latitudes than at lower latitudes, and this latitudinal variation in bandwidth is consistent within the data set. There were 64 simultaneous Pc 1-2 events at Cape Dorset (~74° MLAT) and either Clyde River (~80° MLAT) or Pelly Bay (~79° MLAT). Seventeen of the 62 events were distinctly wider banded (+ 40 mHz or more) at the high latitude stations, than at Cape Dorset. There was not one simultaneous observation of a Pc 1-2 wave that was wider banded (+ 40 mHz or more) at Cape Dorset than at Pelly Bay or Clyde River. The remaining 47 simultaneous events were the same bandwidth (to within 40 mHz) at low and high latitudes. In addition to the evident latitudinal variation in Pc 1-2 bandwidth in the case studies and the larger statistical study, DMSP satellite data show that the plasma mantle and/or polar cap regions are over the MACCS and AGO stations during observations of wideband Pc 1-2 waves.

Direct comparison of Pc 1-2 wave type with DSMP satellite identifications of overhead magnetospheric regions is likely the most precise method of inferring the source region of Pc 1-2 wave types. However, the results of our statistical study also can be used to infer the source regions of the two types of Pc 1-2 activity we have observed. Table 1 from Newell *et al.* [15] show the average MLAT locations of the cusp equatorward and poleward boundaries at noon are 75.9° and 76.8° respectively. If the cusp proper is the generation source for wideband Pc 1-2 waves we might expect nearly equal distributions of wideband Pc 1-2 at both 74.5° and at 78.5°, and Clyde River at 79.7° should observe even fewer wideband waves than the stations at ~78° MLAT. However, such is not the case. The statistical evidence from this data set shows that wideband Pc 1-2 are most commonly observed at Clyde River (79.7° MLAT), slightly less often at Pelly Bay (78.7° MLAT) and hardly at all at Cape Dorset (74.6° MLAT).

Menk *et. al.* [8], specified a generation source for unstructured broadband Pc 1-2 to be within ~2° of the poleward edge of the cusp. Our observations do not contradict that statement, but qualify it: our statistical study suggests, and our case studies indicate, that the generation source for these waves is within ~2° poleward of the poleward edge of the cusp. It is important to note that we do not exclude the cusp as a possible source region for ground observations of Pc 1-2, but our observations indicate that if it exists, it is not as dominant as the plasma mantle source. Several previous studies have suggested the cusp to be a source of the Pc 1-2 waves observed in ground magnetometer data at high latitudes. However, there has been little discussion on the physical properties of such a cusp source and a paucity of relevant satellite data with which to test generation mechanisms. In what follows we outline some aspects of the physics of the high latitude cusp/mantle region that appear to us to constrain any such sources.

Although satellite passes through the low - and mid-altitude cusp have observed intense wave activity over a broad range of frequencies, including the Pc 1-2 band [16, 17], these investigations were not able to determine the source of the turbulence observed, its direction of propagation, or even whether the turbulence was propagating. Andre *et al.* [18] noted observations of low frequency waves, perpendicular to the magnetic field, in the cusp near the O^+ ion gyrofrequency. They suggested these waves were a source of heating for O^+ ions, but unfortunately their data did not really confirm that the waves were electromagnetic ion cyclotron (EMIC) waves due to the spin period of the satellite and other data acquisition problems.

It is well established that wave activity in the Pc 1-2 frequency range is a persistent feature of the cusp in satellite data. However, we have not been able to find a correlation between an overhead cusp location and ground observations of Pc 1-2 waves. We have examined ~ 30 cases in which the Freja satellite detected cusp signatures in the region overhead the MACCS lower latitude stations, but the MACCS stations detected no wave power in the Pc 1-2 range during any of these overflights. Thus, although it is common for satellites to observe wave activity in the cusp region, our data lead us to the conclusion that these waves are either electrostatic and/or not propagating earthward.

There is also an aspect of the ion cyclotron resonance instability [19], which is believed to be the source of Pc 1-2 waves, that appears to argue against the effectiveness of a cusp source. The cusp region contains large fluxes of

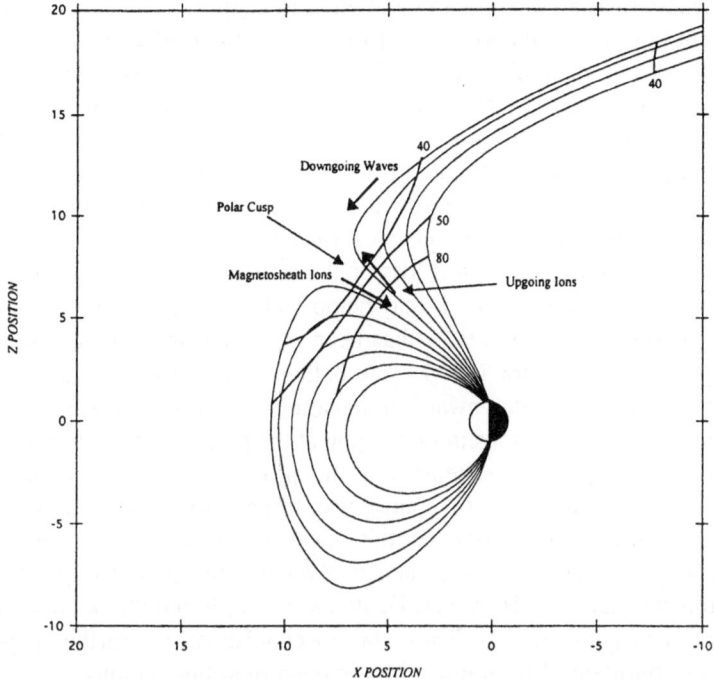

Figure 5. Noon-midnight cut through the Tsyganenko [21] magnetic field model, using the Kp=2 value appropriate for the event shown in Figure 3 at ~15:00 UT November 18, 1994 (94322). Magnetic field magnitudes are represented by the heavy contour lines, with values in nT. The figure indicates that field magnitudes in the high latitude mantle are comparable to those of the outermost closed field lines. The frequencies of waves observed near ~80° MLAT are consistent with such magnitudes. Schematic representations of a probable source mechanism for high latitude Pc 1-2 waves have also been superimposed on this figure.

magnetosheath plasma with energies of 100 eV for electrons and 1 keV for H^+ moving down the field line. If the H^+ is anisotropic, $T_\perp > T_z$, then wave generation below the H^+ gyrofrequency is possible by cyclotron resonance. More specifically however, this instability occurs with particles which are moving in a direction opposite to the waves, such that the wave frequency can be Doppler shifted up to the gyrofrequency of hot protons which provide the source of free energy for wave growth. The resonance condition leading to large growth rate is $w_r + k_{||} u_{||} = \Omega_H$ [2]. Because fluxes in the central cusp are dominated by downward-flowing particles, any ion cyclotron waves generated by such fluxes should be directed upwards. Thus any waves generated in this most commonly invoked version of the ion cyclotron instability would be propagating in the upward direction, and would not get to the ground without some reflection mechanism, such as reflection at the bi-ion resonance frequency, which is unlikely.

The same theoretical argument can be applied to suggest why the plasma mantle (and the adjacent poleward edge of the cusp) may be a more plausible source region for the Pc 1-2 waves we have observed at high latitudes. Many downgoing cusp ions mirror at low to mid altitudes [18], and as they spiral upward they are convected tailward into the mantle. Burch *et al.* [20], using data from the DE-1 satellite (with apogee of 4.9 RE) at the poleward edge of the cusp, also found only upward moving ions which had mirrored below the spacecraft. As these ions continue their upward motion to higher altitudes, they will convect further poleward as well. This upgoing reflected H$^+$ distribution will contain a loss cone (regions of positive gradient in the perpendicular direction, df/dv$_\perp$ > 0). It may be possible for such an upgoing loss cone to generate downgoing EMIC waves, as has been proposed by Denton *et al.*, [2] for the dayside equatorial region for L > 9. We believe it is reasonable, and is clearly consistent with our data, that downward propagating ion cyclotron waves should be generated by upgoing reflected ions in this region at the poleward edge of the cusp and in the adjacent mantle region. A schematic of this process is shown in Figure 5. Our observations that the "mantle wave" source was turned off when IMF Bz became positive are also clearly supportive of this mechanism, because northward IMF reverses the convection electric field, hence reversing the poleward flow of ions necessary for this mechanism. However, Denton *et al.* [2] found that a huge loss cone was needed for such generation. Thus, this mechanism needs further exploration before we can be confident of its applicability in open field line regions.

We can use the Tsyganenko [21] empirical magnetospheric field model to delimit possible regions of the mantle in which Pc 1-2 waves might be generated. In addition to the wave generation schematic, Figure 5 also shows some of the field lines generated by this model for conditions corresponding to our first example (shown in Figure 3). The figure shows a noon-midnight cut at 15:00 UT on November 18, 1994 using the Kp = 2 parameter appropriate for that day. Superposed on the field lines shown are heavy lines denoting contours of constant field magnitude. The figure indicates that field magnitudes in the high latitude mantle are comparable to those on the outermost closed dayside field lines, which is where lower latitude Pc 1-2 are commonly believed to be generated. Thus, if ion cyclotron waves are generated in the high altitude mantle they could be in the Pc 1-2 frequency range. In particular, we note regions in the high altitude mantle where the field magnitude is below 80 nT, as in the outer magnetosphere, as well as field minima, which for this day are below 40 nT, in field lines near the cusp in both the mantle and the dayside closed field region.

6. Summary

The multistation study presented here shows the capability of the conjugate arrays of MACCS and Antarctic statio.. for determining important characteristics of Pc 1-2 waves. Our statistical study of the available MACCS and AGO data for 1994 has revealed that there are two spectrally distinct types of Pc 1-2 waves in the MACCS and AGO data set - narrowband waves (bandwidth less than 150 mHz) and wideband waves (wider than 200 mHz). The occurrence variation of these two types with latitude and comparisons to DMSP satellite overpasses lead us to believe that the plasma mantle and not the cusp, as proposed in previous studies, is the source region for wideband Pc 1-2 observed at high latitudes. As the example in this paper showed, there is initial evidence linking sharp changes in IMF Bz from southward to northward with cutouts in Pc 1-2 wave power. We also found that an asymmetry in cusp latitude between the northern and southern hemispheres, caused AGO stations in Antarctica to be nominally several degrees poleward of the cusp. This caused these stations to observe far fewer narrowband waves than the their latitudinal counterpart in the northern hemisphere. This result suggests the conclusion that if a single station magnetometer observes very few narrowband Pc 1-2 waves, that station is probably several degrees poleward of the cusp. It also serves as a reminder of the importance of factoring dipole tilt angle effects into future magnetospheric models and high-latitude observational studies. Lastly, this study also shows the potential capabilities of large arrays of conjugate magnetometers to diagnose outer magnetospheric processes, and that fluxgate magnetometers may be used for studies involving pulsations in the Pc 1-2 range.

7. Acknowledgments

We thank Eftyhia Zesta of Boston University, John Samson and Gordon Rostoker of the University of Alberta, Jeff Cameron of Augsburg College, and the many volunteers at the Canadian MACCS sites for their help in developing and maintaining the MACCS array, and the personnel of Lockheed Martin and Antarctic Service Associates for their help in deploying and maintaining the U. S. AGOs in Antarctica. We also thank Mats Andre of the Swedish Institute of Space Physics, Brian Anderson and Robert Erlandson of JHU/APL, Brian Fraser of the University of Newcastle, Richard Horne of British Antarctic Survey, and Richard Denton of Dartmouth College for helpful discussions of high latitude wave sources, and we acknowledge S. Kokubun as Principal Investigator for the GEOTAIL Magnetometer and R. P. Lepping as Principal Investigator for the IMP 8 Magnetometer. The MACCS project is supported by National Science Foundation grant ATM 9401524 to Augsburg College and NSF grant ATM 9401733 to Boston University. The PENGUIn AGO project is supported by NSF grants OPP-8918689 and OPP-9529177 to the University of Maryland. Work at the University of New Hampshire is supported by NSF grant OPP-9217024, and work at the Johns Hopkins University Applied Physics Laboratory is supported by AFOSR grant F49620-96-1-0009.

8. References

1. Horne, R. B. and Thorne, R. M. (1993) On the preferred source region for the convective amplification of ion cyclotron waves, *J. Geophys. Res.*, **98**, 9233-9247.
2. Denton, R. E. , Hudson, M. K., and Roth, I. (1992) Loss-cone-driven ion cyclotron waves in the magnetosphere, *J. Geophys. Res.*, **97**, 12093-12103.
3. Cornwall, J. M., Coroniti, F. V., and Thorne, R. M. (1971) Unified theory of SAR arc formation at the plasmapause, *J. Geophys., Res.*, **94**, 4428- 4445.
4. Fukunishi, H. (1984) Pc 1-2 pulsations and related phenomena, in J. G. Roederer (Eds.), *ESA Achievements*

of the Internal Magnetospheric Study, ESA, Graz, Austria, pp. 437-447.

5. Fukunishi, H. Takeshi, K, Masayuki, K., and Kawamura, M. (1981) Classification of hydromagnetic emissions based on frequency-time spectra, *J. Geophys. Res.*, **86**, 9029-9039.

6. Bolshakova, O. B., Troitskaya, V. A., and Ivanov, K. G. (1980) High latitude Pc 1-2 geomagnetic pulsations and their connection with location of the dayside polar cusp, *Planet. Space Sci.*, **28**, 1-7.

7. Morris, R. J. and Cole, K. D. (1991) High latitude day-time Pc 1-2 continuous magnetic pulsations: a ground signature of the polar cusp and cleft projection, *Planet Space Sci.*, **39**, 1473-1491.

8. Menk, F. W., Fraser, B. J., Hansen, H. J., Newell, P. T., Meng, C.-I., and Morris, R. J. (1992) Identification of the magnetospheric cusp and cleft using Pc 1-2 ULF pulsations, *J. Atmos. Terr. Phys.*, **54**, 1021-1042.

9. Popecki, M., Arnoldy, R. L., Engebretson, M. J., and Cahill, L. J., Jr. (1993) High latitude ground observations of Pc 1-2 micropulsations, *J. Geophys. Res.*, **98**, 21481-21491.

10. Engebretson, M. J., Hughes, W. J., Alford, J. L., Zesta, E., Cahill, L. J. Jr., Arnoldy, R. L., and Reeves, G. D. (1995) Magnetometer array for cusp and cleft studies observations of the spatial extent of broadband ULF magnetic pulsations at cusp/cleft latitudes, *J. Geophys. Res.*, **100**, 19371-19286.

11. Gustafsson, G., Papitashvili, N. E., and Papitashvili, V. O. (1992) A revised corrected geomagnetic coordinate system for epochs 1985 and 1990, *J. Atmos. Terr. Phys.*, **54**, 1609- 1631.

12. Dyrud, L. P., Engebretson, M. J., Posch, J. L., Hughes, W. J., Fukunishi, H., Arnoldy, R. L., Newell, P. T., and Horne, R. B. (1997) Ground observations and possible source regions of two types of Pc 1-2 micropulsations at very high latitudes, *J. Geophys. Res.*, Paper No. 97.0041.

13. Hansen, H. J., Fraser, B. J., Menk, F. W., Hu, Y. D., Newell, P. T., and Meng, C. -I. (1991) High latitude unstructured Pc 1 emissions generated in the vicinity of the dayside auroral oval, *Planet. Space Sci.*, **39**, 709-719.

14. Kato, Y. and Tonegawa, Y. (1986) Pc 1 pulsations observed at Cambridge Bay in the cusp region and Fort Smith in the auroral region, *Mem. Nat. Inst. Pol. Res.*, **42**, 52-57.

15. Newell, P. T. and Meng, C-I. (1989) Dipole tilt angle effects on the latitude of the cusp and cleft/low latitude boundary layer, *J. Geophys. Res.*, **94**, 6949-6953.

16. Erlandson, R. E., Zanetti, L. J., and Potemra, T. A. (1988) Observations of electromagnetic ion cyclotron waves and hot plasma in the polar cusp, *Geophys. Res. Lett.*, **15**, 421-424.

17. Heppner, J. P., Liebrecht, M. C., Maynard, N. C., and Pfaff, R. F. (1993) High-latitude distributions of plasma waves and spatial irregularities from DE 2 alternating current electric field observations, *J. Geophys. Res.*, **98**, 1629-1652.

18. Andre, M., Crew, G. B., Peterson, W. J., Person, A. M., Pollock, C. J., and. Engebretson, M. J. (1990) Ion heating by broadband low-frequency waves in the cusp/cleft, *J. Geophys. Res.*, **95**, 20809-20823.

19. Kennel, C. F., and Petschek, H. E. (1966) Limit on stable trapped particle fluxes, *J. Geophys. Res.*, **71**, 1-28.

20. Burch, J. L., Reiff, P. H., Heelis, R. A., Winningham, J. D., Hanson, W. B., Gurgiolo, C., Menietti, J. D., Hoffman, R. A., and Barfield, J. N. (1982) Plasma injection and transport in the mid-altitude polar cusp, *Geophys Res. Lett.*, **9**, 921-924.

21. Tsyganenko, N. A. (1991) Methods for quantitative modeling of the magnetic field from Birkeland Currents, *Planet. Space Sci.*, **39**, 641-654.

STUDIES OF GEOMAGNETIC CONJUGACY AT VERY HIGH LATITUDES

C. G. MACLENNAN, L. J. LANZEROTTI and D. J. THOMSON
Bell Laboratories, Lucent Technologies
Murray Hill, NJ 07974 USA

Abstract. Studies of geomagnetic variations in conjugate regions have played an important role in understanding the Earth's space environment, especially since these measurements can often be used to distinguish between space and time variations on global scales. Largely because of logistics difficulties, conjugate studies at very high latitudes, including the magnetospheric cusp and polar cap regions, have not been extensively pursued over the years. This paper summarizes some current research on the nature of magnetic conjugacy at cusp and polar cap latitudes. We show that the daily Sq variations at conjugate cusp latitudes are appreciable, and that the ionosphere over South Pole station appears to have a larger ionization in austral summer than does the ionosphere over its conjugate region (Iqaluit, Canada) in the northern summer. We also show that the largest coherence between conjugate cusp areas (South Pole and Iqaluit) occurs in the period band ~200 to ~600 seconds. Using this coherence as an indicator during local morning and local afternoon hours in quiet geomagnetic conditions, the boundary between open and closed geomagnetic field lines can be deduced to lie between the South Pole/Iqaluit latitudes (~74° geomagnetic) and polar cap conjugate stations at ~80° geomagnetic.

1. Introduction

Upper atmosphere geophysical studies at very high geomagnetic latitudes are of considerable importance for understanding the structure and dynamics of the Earth's magnetosphere. Of special significance in this regard are studies that involve observations that are made in conjugate areas. Such studies can delineate the global extent of geomagnetic activity better than measurements obtained from largely unrelated sites in separate hemispheres. Indeed,

J. Moen et al. (eds.), Polar Cap Boundary Phenomena, 327–342.

conjugate studies from equatorial to auroral latitudes have been critical in the understanding of upper atmosphere phenomena ranging from optical aurora (e.g. DeWitt [1]; Wescott and Mather [2]; Belon et al. [3]; Makita et al. [4]) to particle aurora (e.g. Barcus et al. [5]; Hargreaves and Chivers [6]) to VLF phenomena (e.g. Sato and Kokubun [7]; Sato et al. [8]) to hydromagnetic waves (e.g. Sugiura and Wilson [9]; Nagata et al. [10]; Tonegawa and Fukunishi [11]; Lanzerotti [12]). An excellent summary of conjugate studies and their contributions to space research is contained in the volume edited by Sato [13].

While geomagnetic flux tubes can reasonably be mapped between hemispheres at low- to mid-latitudes, the concept of a conjugate 'point' becomes much less definable at higher latitudes. The conjugate 'point' is well known to vary with local time and with season, both through calculations with realistic geomagnetic field models (e.g. Ono [14]; Stassinopoulos et al. [15]) as well as through numerous observational investigations over the years (e.g. Wescott and Mather [2,16]; Bond [17]; Stenbaek-Nielsen et al. [18]; Belon et al. [19]; Rosenberg et al. [20]; Burns et al. [21]). These variations exist at mid to auroral latitudes, but are largest at high latitudes.

Conjugate studies at geomagnetic latitudes substantially above the auroral zone have been infrequent, largely because of the difficulty of maintaining equipment in often remote and harsh environments, and because the northern polar regions consist mainly of frozen sea, unlike the Antarctic (which is itself a difficult and expensive region to operate in). Nagata [22] provides an extensive review of conjugate phenomena up to the date of his paper about ten years ago.

Early measurements were made during the IGY at the station pairs Scott Base/Resolute Bay (invariant latitudes $-79.7°$ and $83.9°$, respectively) and Davis/Ny Ålesund (invariant latitudes $-74.6°$ and $75.7°$, respectively), although only little research from these sites has been reported in the literature. In the 1950s and 60s, the basic structure of the magnetosphere, including the dayside cusp regions and the connections of the polar cap to the geomagnetic tail, was poorly known. Thus any data obtained was difficult to place in context. Recently, Olson and Fraser [23] have begun conjugate studies of wave phenomena at the near-cusp stations Davis/Longyearbyen.

In recent years, the establishment of a permanent geophysical site with a variety of instrumentation at Iqaluit, Canada (Table 1), has provided the opportunity to carry out, on an almost continuous basis, conjugate studies with South Pole Station of many geophysical parameters (e.g. Rosenberg, [24]). More recently, the deployment of Automatic Geophysical Observatories (AGOs) in the Antarctic at geomagnetic latitudes higher than South Pole, and the set of magnetometers in the MACCS Array (Engebretson et al. [25]; Hughes and Engebretson [26]) at high latitudes in northern Canada, have provided researchers with a much improved set of instrumentation and locations to study high latitude geophysical phenomena.

This contribution to the Proceedings of the NATO Advanced Study Institute on Polar Cap Boundary Phenomena describes some recent studies of geomagnetic conjugacy and its variation at latitudes at or above the magnetospheric dayside cusp region. Data from two pairs of conjugate stations are used in this paper; the station locations are given in Table 1.

Table 1. Station Locations						
	Station	Geographic		Corr.Geomagnetic		UT of mid-
		Lat.	Long.	Lat.	Long.	night MLT
CY	Clyde River	70.5	291.4	79.5	18.5	3:51
IQ	Iqaluit	63.8	291.4	73.3	14.3	4:07
SP	South Pole	-90.0	0.0	-74.0	18.4	3:30
P1	AGO P1	-83.9	129.6	-80.1	16.8	3:44

2. Data

For this paper we have principally used data from three-axis fluxgate magnetometers operated by Bell Laboratories at Iqaluit, NWT, Canada, and at South Pole Station, Antarctica. Three field components are recorded: south-north (H), west-east (D), and vertical (Z), although only the H component data are used in this paper. In Section 5, data from the 80° geomagnetic latitude AGO P1 and the MACCS station at Clyde River are also incorporated into the analyses. Ten second data samples are used throughout.

Figure 1 is a map of the Antarctic continent, showing the locations of a number of relevant stations. Geographic latitudes of $-70°$ and $-80°$ are indicated with dotted lines, as are geographic longitudes in 60° intervals clockwise from 0° at the top. Geomagnetic coordinates (60° and 80° latitude contours, and longitude every 30°) are shown as solid lines. Locations of the US AGO stations are shown as open triangles. The nominal conjugate points of the northern stations at Iqaluit and Clyde River are shown as open circles, labeled IQ and CY. Permanent Antarctic stations from several nations are indicated with filled circles.

3. Sq at High Geomagnetic Latitudes

The study of the daily variations of the geomagnetic field as driven by solar illumination (the so-called Sq variations) is one of the oldest subjects in geomagnetism (e.g. Chapman and Bartels [27]) and has principally concentrated on measurements and modeling at equatorial to moderately high latitudes. Sq-type variations at polar latitudes (generally denoted Sq^P) in conjugate regions have not been investigated very extensively, largely because data have not been

330

available. One of the earliest studies was of data from the Second Polar Year [28]. Although Sq models such as those of Matsushita [29] imply information about the daily geomagnetic variations at very high latitudes, they were not intended to be reliable predictors of these variations; in fact, the variations at latitudes near 80° are found to be considerably larger than the models predict [30].

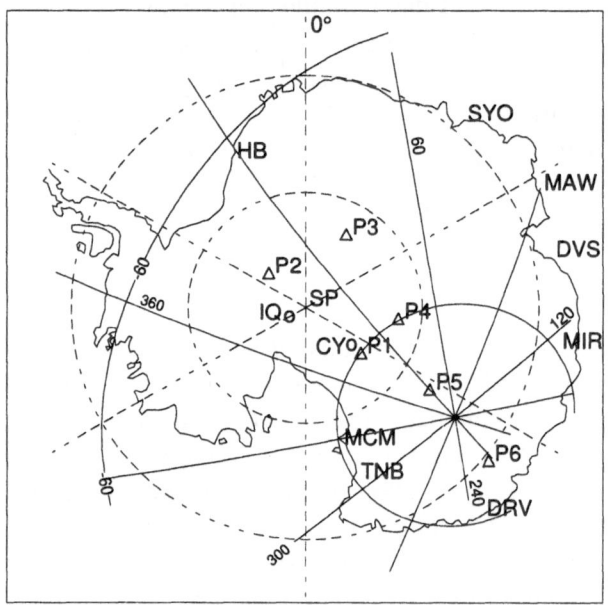

Figure 1.   Antarctic stations: Data from SP and P1 and from
the conjugate area stations IQ and CY are used in this report.

At polar latitudes, the Sq^P current system is composed of the atmospheric dynamo and disturbance fields driven by the interplanetary medium. Disturbances can occur in even very quiet times at polar latitudes (see Figure 6, below). The relative importance of the various contributions to Sq^P has a long, and at times controversial, history (e.g. Nagata and Kokubun [31]; Rostoker *et al.* [32]) and could use a revisit with current conjugate data.

Data from the South Pole and Iqaluit conjugate areas have been examined to obtain a perspective of the Sq profiles at high geomagnetic latitudes. Plotted in the left-hand panels of Figure 2 are daily variations of 30-min averages of the H-component of the magnetic field for South Pole (dashed line) and Iqaluit (solid line) averaged over days of 'quiet' geomagnetic activity in the months (top to bottom) February-March-April 1995 (FMA), May 1995-June 1994 (MJ), August-September-October 1995 (ASO), and November-December-January

1995 (NDJ). July was omitted because of instrument problems. The right-hand panels show the difference of the IQ minus SP variations as a function of UT. Local noon and local midnight are indicated in this and subsequent figures as open and filled triangles, respectively.

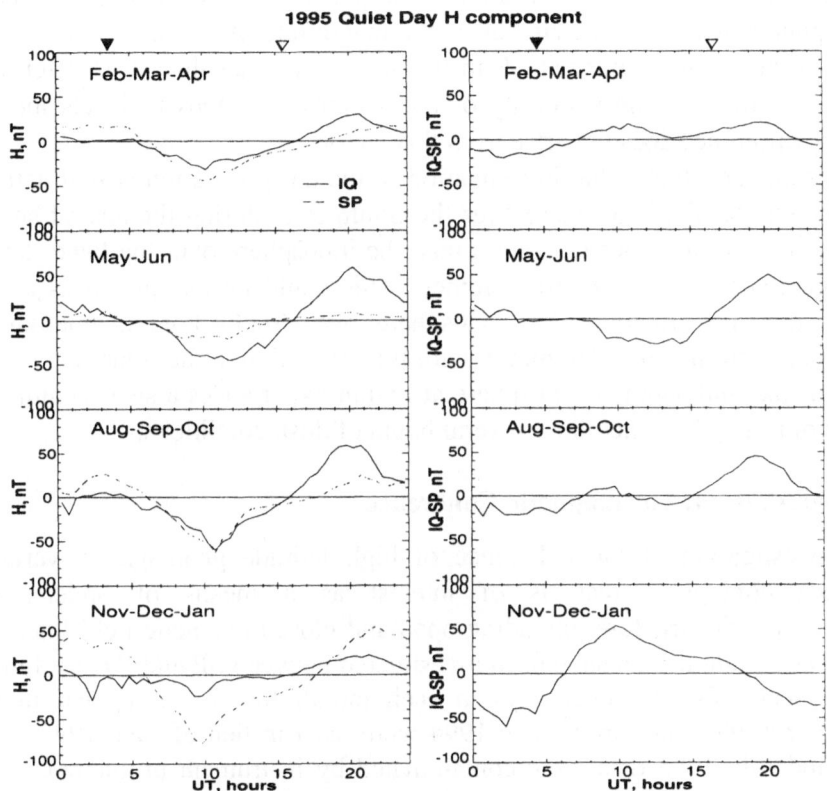

Figure 2. (Left panels) H-component daily variation in conjugate areas near 74° geomagnetic. *(Right panels)* Differences between conjugate area variations.

The data plotted for the two stations in Figure 2 are mean values obtained from the quietest days in the specified months. Eight days of data were used from FMA that had an average $\Sigma Kp = 4+$; six days from MJ with $\Sigma Kp = 7-$; 9 days from ASO with $\Sigma Kp = 8-$; 7 days from NDJ with $\Sigma Kp = 4+$. An ionospheric current of the order of 5×10^4 to 10×10^4 A is implied by the magnitude of the changes measured in the magnetic field variations. This magnitude is somewhat larger than that derived at these latitudes for a 'quiet sun' year (Campbell and Matsushita [33]).

In the equinox months, FMA and ASO, the daily mean profiles are very similar in the conjugate areas, as the IQ-SP differences show. This is somewhat unexpected since, while the two stations are at similar geomagnetic latitudes

(~74°), they differ by ~25° in geographic latitude. Thus, the amount of solar illumination and its incidence angle on the ionosphere are quite different for the two locations, even at the equinoxes.

The effects of solar illumination are evident in the MJ and NDJ months, as seen in the plots of Figure 2. For these months, in the austral winter and summer intervals, the Sq variation is larger in the northern and southern hemispheres, respectively. However, the magnitude of the daily variation is a factor of nearly two larger at SP in the austral summer than it is at IQ in the northern summer. The similarity of the geomagnetic activity levels indicates that the difference does not arise from this factor.

In summary, this initial investigation of the daily variation at high latitudes has shown that the ionosphere over the South Pole during the austral summer appears to have a higher ionization than the ionosphere over the lower latitude conjugate site in the northern summer. This could be interpreted as a solar-driven dynamo effect, as the ionosphere over South Pole is continuously illuminated during austral summer, with little variation in the solar zenith angle, whereas the mid-summer ionosphere at Iqaluit experiences a solar zenith angle swing of nearly 50°, including several hours of dusk conditions.

4. High Latitude Geomagnetic Coherence

The investigation of the coherence of high latitude geomagnetic variations between conjugate areas is of interest as a means of understanding magnetospheric structure, including open and closed magnetic field lines. The coherence relationships have been investigated between SP and IQ as a function of frequency for the quiet days in each month for an entire year in solar minimum (1995; data from June 1994 were used in lieu of June 1995 data as June and July 1995 data were contaminated by instrument problems). Cross-spectra, coherence, and phase were computed between the H-components at these stations for the defined quiet days.

A multi-taper spectral method was employed (Thomson [34-36]), using 10 Slepian sequences (prolate spheroidal wave functions) and a time bandwidth product of 6. Averages of cross-spectra for hour-long data segments were calculated for the defined quiet days, and the coherence and phase as a function of frequency and time were determined from that. The complex coherency, CC, is given as

$$CC(f) = \frac{S_{xy}(f)}{\left[S_{xx}(f)\,S_{yy}(f)\right]^{1/2}} \tag{1}$$

where $S_{xy}(f)$, $S_{xx}(f)$, and $S_{yy}(f)$ are the spectral amplitudes. The cross spectrum is the magnitude of $S_{xy}(f)$; for coherence we use magnitude-squared coherence

333

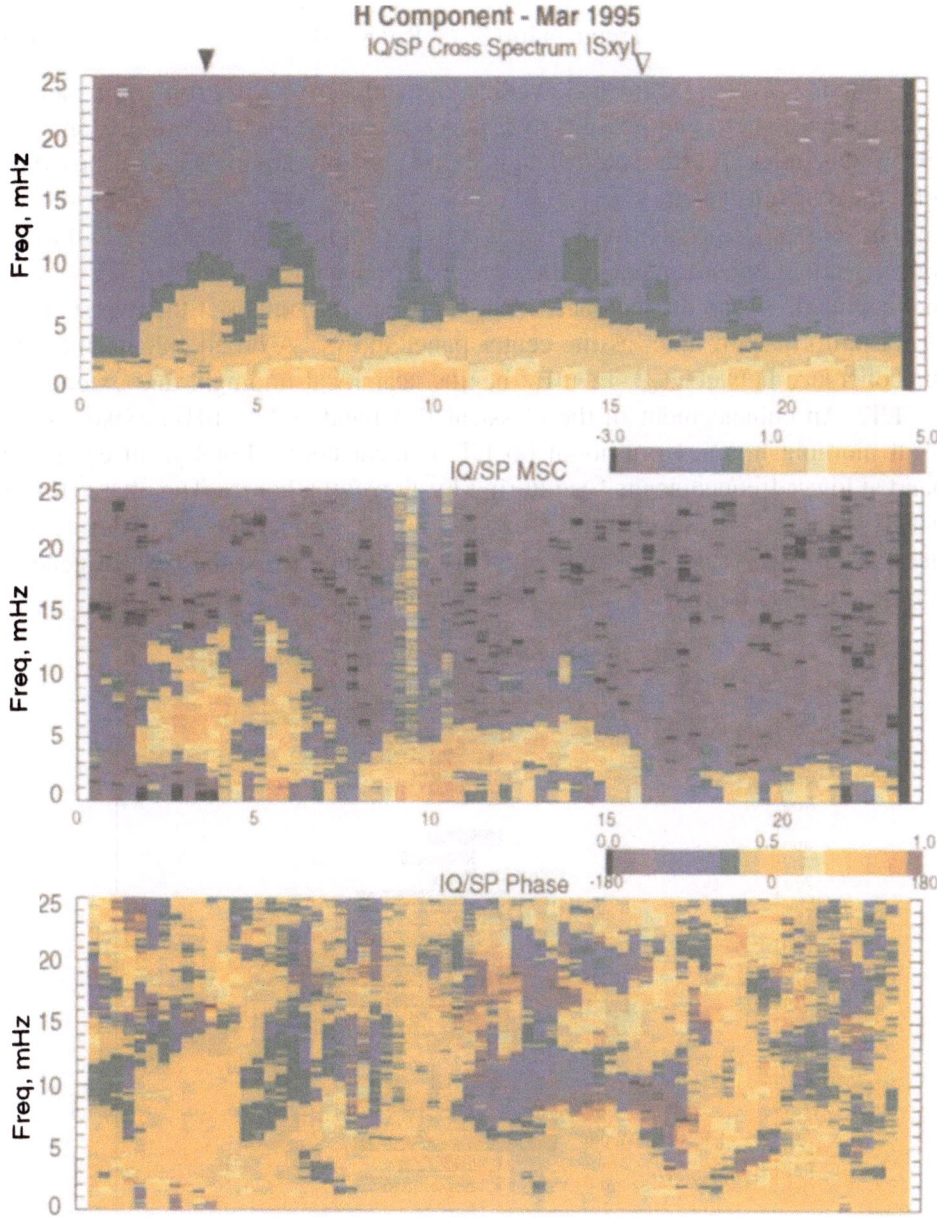

Figure 3. Cross-spectrum, magnitude-squared coherence and phase, averaged for three quiet days.

334

(MSC) where $MSC(f) = |CC(f)|^2$. The phase of the signals between the two locations is given by

$$arctan2(Im(S_{xy}), Re(S_{xy}))\qquad\qquad (2)$$

Plotted in Figure 3 are the average dynamic cross spectrum (top panel), MSC (center panel), and phase (lower panel) averaged for March 8, 21, and 22, 1995. One hour spectra were calculated and a twenty minute slide was used to form the dynamic spectra.

The cross spectral plot in the upper panel of Figure 3 shows, across all local times, a broad enhancement in the power at frequencies ≤10 mHz. However, the magnitude-squared coherence (MSC) between the signals at the two stations is not uniform with time, as the center panel shows. A broad enhancement in the coherence between ~3–12 mHz begins near local midnight and extends to ~5 LT. An enhancement in the classical Pc5 band (~2–8 mHz) exists in the local morning hours, from about 06 LT to local noon. Finally, an enhanced band at lower frequencies is seen during local evening hours. The phase (lower panel in Figure 3) is very close to zero at the frequencies, and during the intervals, of the highest coherences. This was found to be the case in general for the summary coherence results presented in Figure 4.

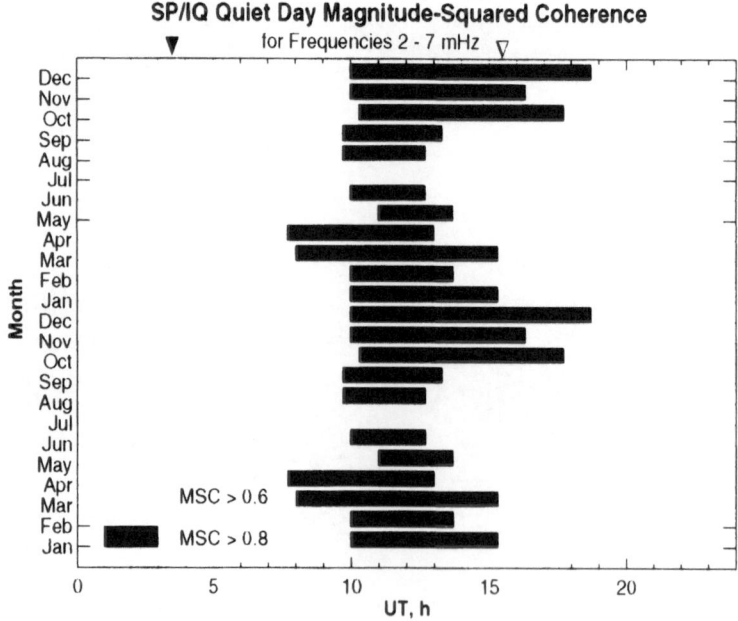

Figure 4. Seasonal dependence of Pc5 MSC in conjugate areas at two different levels.

Monthly average coherence results such as those shown in Figure 3 were examined for all of 1995 (with the exception of July and the substitution of June 1994, as noted above) in order to understand the seasonal dependence of the conjugate relationships. Plotted in Figure 4 are the intervals of coherence (two levels: >0.6 and >0.8) in the H-component in the frequency band 2–7 mHz as a function of month. The results are repeated in order to illustrate the northern winter season more clearly. It is significant to note that the local time of the onset of the high coherence (~5 LT) varies only slightly with season. This is in contrast to the extent of the high coherence interval during the remainder of the day. The interval of high coherence is very limited during northern summer months, ending well before local noon, but is much more extended during the months of northern winter, extending well into the local afternoon.

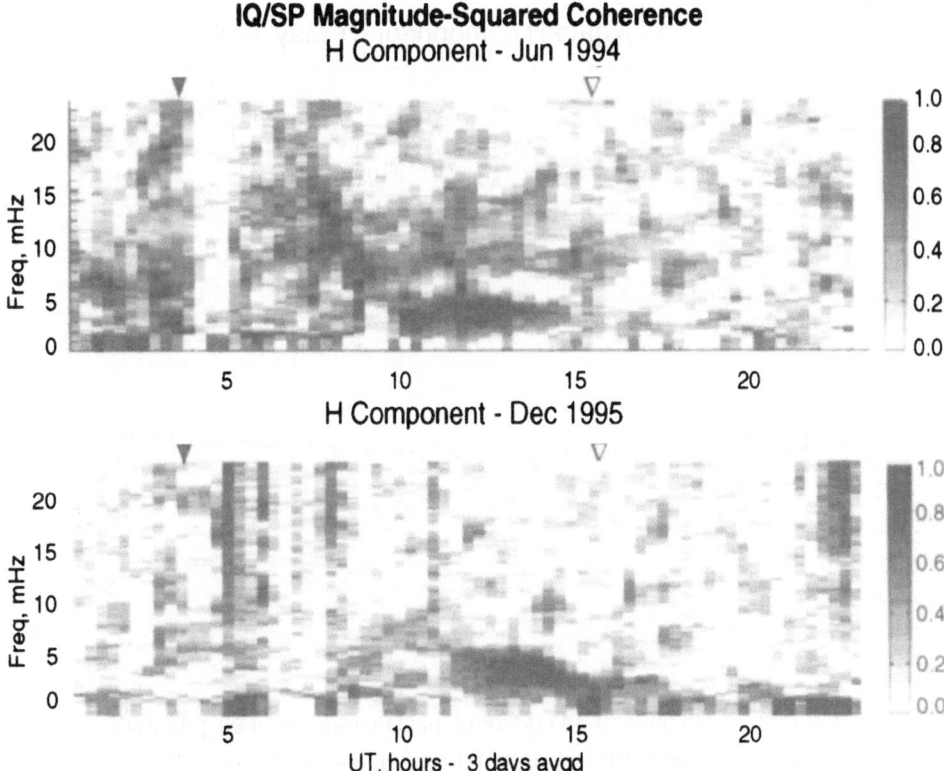

Figure 5. Seasonal comparison of quiet day coherence at near-cusp latitudes.

The strong seasonal effect on the conjugate coherency is further illustrated in Figure 5 for an average of three quiet days each in the months of June and December. In both months shown, high coherence in the Pc 5 band begins at 0900-1000 UT (0530-0630 LT). In the month of June, the high coherence decreases and then ends near local noon (1530 UT), while in December the high coherence continues, albeit with decreasing frequency, until several hours after local noon. The post-midnight broadband (10–15 mHz) coherence in June is absent in December.

5. Quiet Geomagnetic Day

The conjugate and single hemisphere spatial extent of the geomagnetic coherence was examined in detail for one quiet day (May 1, 1995; ΣKp = 5) for the SP/IQ conjugate pair and for a higher latitude pair at about the same geomagnetic longitude (see Figure 1), the AGO station P1 and its near conjugate Clyde River (CY). The geographical distance between SP and P1 is ~680 km and between IQ and CY is ~743 km.

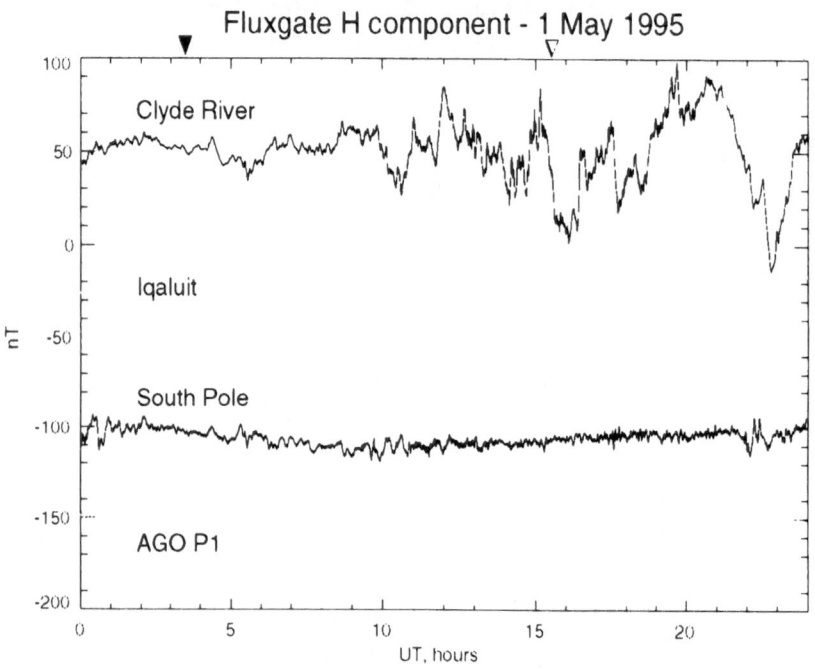

Figure 6. Magnetic field (H Component) variations at cusp latitude and polar cap conjugate areas on a geomagnetically quiet day.

The H component of the magnetic field data for this day for each of the four stations is shown in Figure 6. The vertical scale shows arbitrary zero levels for the traces. IQ shows some evidence of a slow variation over the day. A similar variation in the H-component magnitude is also evident at P1, while the SP trace is essentially flat. Thus, on this day, Sq^p is larger in the north than in the south. The most striking feature of these data, however, is the order of one hour and more disturbance field variations at CY, only 6° geomagnetically from IQ, which are much more pronounced than at the other locations. Similar disturbance fields are not seen in the conjugate area at P1, showing that the polar magnetic disturbance variations can be very non-conjugate, in the nominal conjugate area, even on a quiet day. The possibility that CY might be conjugate to a higher latitude southern location will be able to be investigated in the future using data from higher latitude AGO stations.

The coherences between the same hemisphere and conjugate hemisphere locations are shown in Figure 7. The left hand column of panels shows the coherences between stations in the same hemisphere (Antarctic stations in the upper left panel; northern hemisphere stations in the lower left panel) while the right hand set of panels shows the coherences for the conjugate pairs. The blank strip near 21 UT in the lower panels is due to a data gap at CY. There are several features evident in the panels:

Same hemisphere:
a. More, and broader-band, coherence is seen pre-midnight in the southern hemisphere than in the north.
b. The variations around local midnight at frequencies $\lesssim 10$ mHz are coherent in the individual hemispheres (SP/P1 and IQ/CY).
c. A broad band of high coherence is observed between IQ/CY at ~09 LT; in the south, coherence at SP/P1 at this time exists only in a narrow band between 2–5 mHz.

Conjugate hemispheres:
a. Except for a small interval just after local midnight, the conjugate pair CY/P1 shows almost no coherence throughout the day.
b. There is a varying, but persistent, band of coherence (~2 – 7mHz) between the conjugate pair SP/IQ throughout the day, except for a few hours pre-midnight.
c. An interval of higher coherence also exists between SP/IQ at f~16 – 18 mHz at a local time of ~17 – 18 UT.

The average coherences in the band 2–7 mHz for the four station pairs shown in Figure 7 were computed and are plotted as a function of time in Figure 8. The top and third from the top traces correspond to the same-hemisphere coherences, and the second and fourth from the top correspond to the conjugate coherences. Arrows indicate the right or left vertical scale for each station pair. Intervals when the coherence is above 0.3 in each trace are indicated by the shading. Several features are apparent:
a. Following local midnight, there is considerable coherence between same hemisphere stations as well as between the highest latitude conjugate pair, CY/P1.

338

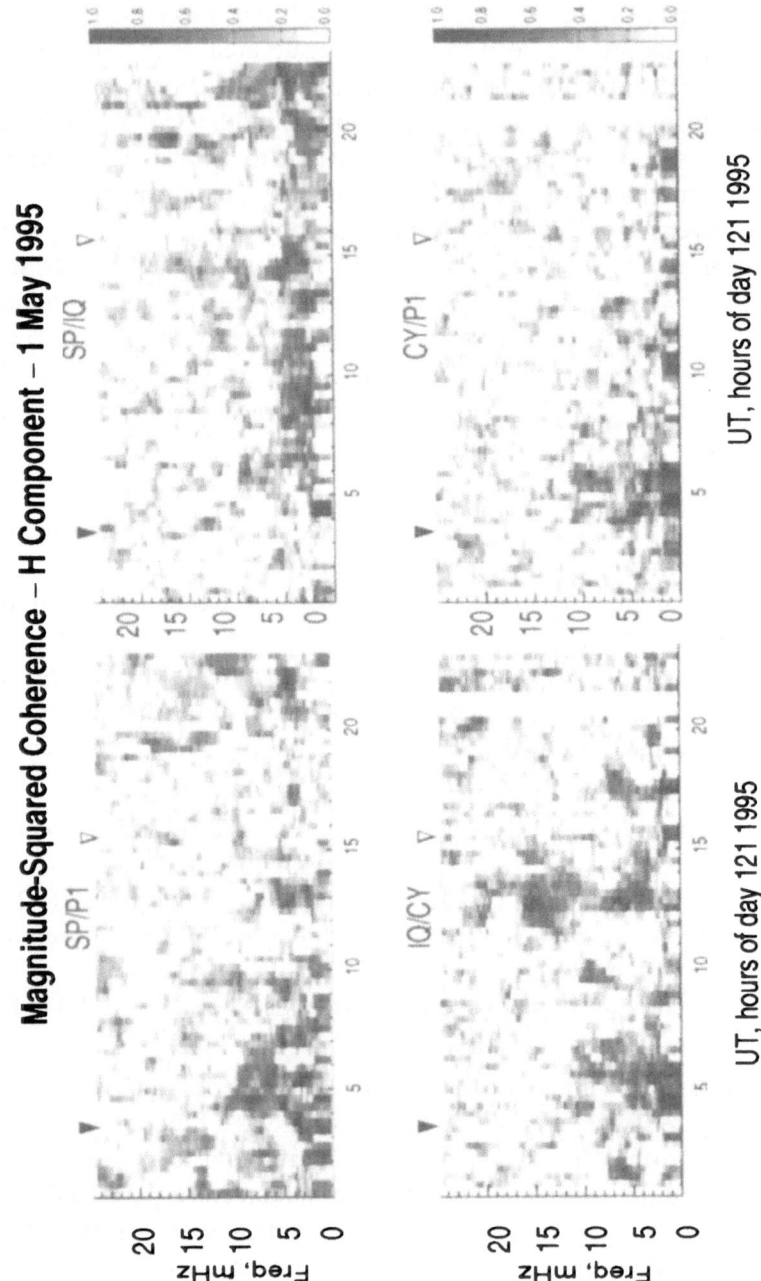

Figure 7. Same hemisphere and conjugate area MSCs of magnetic field variations on a geomagnetically quiet day.

b. There is coherence between three of the four station pairs (the exception being the high latitude conjugate CY/P1) prior to local noon.
c. The SP/IQ conjugate pair has the longest intervals of coherence of any of the sets of stations. These two stations are especially conjugate during the interval following local midnight to local noon, and during local afternoon hours from about 17–20 LT.

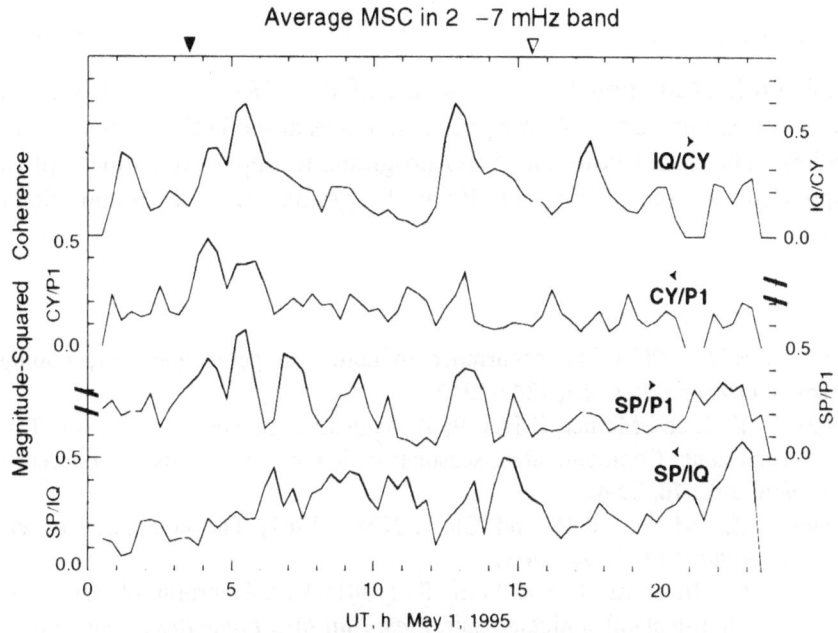

Average MSC in 2 –7 mHz band

Figure 8. Same hemisphere and conjugate area average MSCs. The shaded regions correspond to average MSC values ≥ 0.3.

From the above, it is interesting to make the following speculations:
a. The high coherence near local midnight in the same hemispheres, as well as between the two highest latitude conjugate areas (CY/P1), is related to geomagnetic activity produced in the magnetotail, perhaps corresponding to very high latitude 'substorms'. The SP/IQ pair may be on closed magnetosphere field lines at his time, but the high latitude geomagnetic activity, produced in the magnetotail, does not appear to be conjugately coherent at these latitudes (~74°). Alternatively, the lack of coherence between SP and IQ from about 2100 LT to about 0600 LT may reflect evidence of one of the conclusions of Wu *et al.* [37] that the westward electrojet 'flows at higher latitudes in the winter hemisphere than in the summer after about 2000 LT' (perhaps by as much as 4°).
b. The coherence between SP/IQ and the lack of coherence between CY/P1 during the local morning hours as well as in the late local afternoon suggest that the closed magnetic field boundary lies between 74° and 80°

latitude during these lengthy intervals. In addition, for conditions such as this with quiet-time solar wind flow, the two cusp-latitude stations are good indicators of high latitude conjugacy.

c. The coherence among most combinations of station pairs pre-local noon is likely to be a manifestation of the excitation of the nose region of the magnetosphere by the impinging solar wind.

6. Acknowledgments

We thank Prof. Mark Engebretson for use of the MACCS data from Clyde River which is sponsored by Atmospheric Sciences at the U.S. National Science Foundation. The South Pole and AGO programs in upper atmosphere physics are supported by the Office of Polar Programs, U.S. National Science Foundation.

7. References

1. DeWitt, R.N. (1962) The occurrence of aurora in geomagnetically conjugate areas, *J. Geophys. Res.*, **67**, 1347-1352.
2. Wescott, E.M. and Mather, K.B. (1965) Magnetic conjugacy from L=6 to L=1.4: 1. Auroral zone: Conjugate area, seasonal variations, and magnetic coherence, *J. Geophys. Res.*, **70**, 29-42.
3. Belon, A.E., Mather, K.B. and Glass, N.W. (1967) The conjugacy of visual aurorae, *Antarct.J.U.S.*, **2**, 124-127.
4. Makita, K., Hirasawa, T. and Fujii, R. (1981) Visual auroras observed at the Syowa Station-Iceland conjugate pair, *Mem Natl Inst. Polar Res., Spec. Issue*, **18**, 212-225.
5. Barcus, J.R., Brown, R.R., Karas, R.H., Bronstad, K., Trefall, H., Kodama, M. and Rosenberg, T.J. (1973) Balloon observations of auroral-zone X-rays in conjugate regions, *J. Atmos. Terr. Phys.*, **35**, 497-511.
6. Hargreaves, J.K., and Chivers, H.J.A. (1965) A study of auroral absorption events at the South Pole, 2. Conjugate properties, *J.Geophys.Res.*, **70**, 1093-1102.
7. Sato, N. and Kokubun, S. (1981) Interaction between ELF-VLF emissions and magnetic pulsations: Regular period ELF-VLF pulsations and their geomagnetic conjugacy, *J. Geophys. Res.*, **86**, 9-18.
8. Sato, N., Ayukawa, M. and Fukunishi, H. (1980) Conjugacy of ELF-VLF emissions near L=6, *J. Atmos. Terr. Phys.*, **42**, 911-928.
9. Sugiura, M. and Wilson, C.R. (1964) Oscillations of the geomagnetic field lines and associated magnetic perturbations at conjugate points, *J. Geophys. Res.*, **69**, 1211-1216.
10. Nagata, T., Kokubun, S. and Iijima, T. (1963) Geomagnetically conjugate relationships of giant pulsations at Syowa Base, Antarctica and Reykjavik, Iceland, *J. Geophys. Res.*, **68**, 4621-4625.

11. Tonegawa, Y. and Fukunishi, H. (1984) Harmonic structure of Pc3–5 magnetic pulsations observed at the Syowa-Husafell conjugate pair, *J. Geophys. Res.,* **89,** 6737-6748.

12. Lanzerotti, L.J. (1987) Conjugate studies of hydromagnetic waves, in N. Sato, (ed.), *Mem. Natl Inst. Polar Res., Spec. Issue,* **48,** 121-133.

13. Sato, N. (ed.) (1987) *Proceedings of the Nagata Symposium on Geomagnetically Conjugate Studies and the Workshop on Antarctic Middle and Upper Atmosphere Physics, Mem. Natl Inst. Polar Res., Spec. Issue,* **48,** National Institute of Polar Research, Tokyo.

14. Ono, T. (1987) Temporal variation of the geomagnetic conjugacy in Syowa-Iceland pair, in N. Sato, (ed.), *Mem. Natl Inst. Polar Res., Spec. Issue,* **48,** 46-57.

15. Stassinopoulos, E.G., Lanzerotti, L.J. and Rosenberg, T.J. (1984) Temporal variations in the Siple Station conjugate area, *J. Geophys. Res.,* **89,** 5655-5659.

16. Wescott, E.M. and Mather, K.B. (1963) Diurnal effects in magnetic conjugacy at very high latitude, *Nature,* **197,** 1259-1261.

17. Bond, F.R. (1969) Auroral morphological similarities at two magnetically conjugate stations: Buckles Bay and Kotzebue, *Aust.J.Phys.,* **22,** 421-433.

18. Stenbaek-Nielsen, H.C., Davis, T.N. and Glass, N.W. (1972) Relative motion of auroral conjugate points during substorms, *J. Geophys. Res.,* **77,** 1844-1858.

19. Belon, A.E., Maggs, J.E., Davis, T.N., Mather, K.B., Glass, N.W. and Hughes, G.F. (1969) Conjugacy of visual auroras during magnetically quiet periods, *J.Geophys.Res,* **74,** 1-28.

20. Rosenberg, T.J., Helliwell, R.A. and Katsufrakis, J.P. (1971) Electron precipitation associated with discrete very-low-frequency emissions, *J.Geophys.Res.,* **76,** 8445-8452.

21. Burns, G.B., McEwen, D.J., Eather, R.A., Berkey, F.T. and Murphree, J.S. (1990) Optical auroral conjugacy: Viking UV imager - South Pole station ground data, *J.Geophys.Res,* **95,** 5781-5790.

22. Nagata, T. (1987) Research of geomagnetically conjugate phenomena in Antarctica since the IGY, in N. Sato, (ed.), *Mem. Natl Inst. Polar Res., Spec. Issue,* **48,** 1-45.

23. Olson, J.V. and Fraser, B.J. (1995) A search for conjugate cusp Pc3, GEM poster, http://maxwell.gi.alaska.edu/GEM

24. Rosenberg, T. (1987) Cosmic noise absorption at South Pole and Frobisher Bay: Initial results, in N.Sato, (ed.), *Mem.Natl.Inst.Polar Res., Spec.Issue,* **48,** 161-170.

25. Engebretson, M.J., Hughes, W.J., Alford, J.L., Zesta, E., Cahill, L.J.Jr., Arnoldy, R.L. and Reeves, G. D. (1995) Magnetometer array for cusp and cleft studies: Observations of the spatial extent of broadband ULF magnetic pulsations at cusp/cleft latitudes, *J.Geophys.Res.,* **100,** 19371-19386.

26. Hughes, W.J., and Engebretson, M.J. (1997) MACCS: Magnetometer array for cusp and cleft studies, in M. Lockwood, H. J. Opgenoorth, M. N. Wild, and R. Stamper, (eds) *Satellite-Ground Based Coordination Sourcebook,* Rutherford Appleton Laboratory, in press.

27. Chapman, S., and Bartels, J. (1940) *Geomagnetism,* Oxford University Press, London and New York.

342

28. Nagata, T. and Mizuno, H. (1955) Sq field in the polar region on absolutely quiet days, *J.Geomag.Geoelect.,* **7,** 69-74.
29. Matsushita, S. (1967) Solar quiet and lunar daily variation fields, in S.Matsushita and W.Campbell (eds.), *Physics of Geomagnetic Phenomena, 1,* Academic Press, New York, pp.301-424.
30. Meloni, A., private communication.
31. Nagata, T. and Kokubun, S. (1962) An additional geomagnetic daily variation field (Sq^P field) in the polar region on geomagnetically quiet day, *J.Ionosph.Space Res.Japan,* **16,** 256-274.
32. Rostoker, G., Chen, A.J., Yasuhara, F., Akasofu, S.-I. and Kawasaki, K. (1974) High latitude equivalent current systems during extremely quiet times, *Planet. Space Sci.,* **22,** 427-437.
33. Campbell, W.H., and Matsushita, S. (1982) Sq currents: A comparison of quiet and active year behavior, *J.Geophys.Res.,* **87,** 5305-5308.
34. Thomson, D.J. (1982) Spectrum estimation and harmonic analysis, *Proc. IEEE,* **70,** 1055-1096.
35. Thomson, D.J. (1990) Quadratic inverse spectrum estimates, *Phil. Trans. Roy. Soc. Lond.* **A 332,** 539-597.
36. Thomson, D.J. and Chave, A.D. (1991) Jackknifed error estimates for spectra, coherences, and transfer functions, in S.Haykin (ed.), *Advances in Spectrum Analysis and Array Processing,* Prentice Hall, Chapter 2, pp. 58-113.
37. Wu, Q., Rosenberg, T.J., Lanzerotti, L.J., Maclennan, C.G. and Wolfe, A. (1991) Seasonal and diurnal variations of the latitude of the westward auroral electrojet in the nightside polar cap, *J.Geophys.Res,* **96,** 1409-1419.

A MODEL FOR THE HIGH LATITUDE ISOTROPIC BOUNDARY

M.J. ALOTHMAN AND T.A. FRITZ

Center for Space Physics
Boston University
725 Commonwealth Avenue
Boston, MA 02215-1401
USA

Abstract. The region of the high latitude trapping boundary for energetic particles has been studied for many years beginning with McDiarmid and Burrows [1] and Fritz [2]. Just equatorward of the cut-off boundary of the radiation belt is a region of isotropic pitch angle distributions for electrons and ions which varies in position and width with magnetic latitude, magnetic local time, and geomagnetic activity. Surprisingly it is present in some form for almost all satellite passes within a few hours of local midnight regardless of the level of geomagnetic activity. The cause of these isotropic pitch angle distributions is thought to be pitch angle scattering due to a sharp curvature in the field line geometry, [3], [4], and [5], violating the first adiabatic invariant. This curvature can be quantified by the ratio R of the minimum field line radius of curvature to the maximum radius of gyration for the particle of given energy and mass being studied. We have used a model of the magnetic field configuration calculated using a realistic field model given by an internal field (IGRF) and an external field (Tsyganenko T96) to determine the scattering regions for different energies and particles, based on the assumption that there is a minimum value of the ratio R that permits particles to complete their bounce motion without being scattered as they cross the field reversal region. By assuming that the particles, after the scattering process, will drift conserving the first and second adiabatic invariants, the location of the boundary as a function of energy and local time is found. The model results and predictions will be compared to energetic particle data from the NOAA/Tiros and GGS/Polar satellites.

J. Moen et al. (eds.), Polar Cap Boundary Phenomena, 343–354.

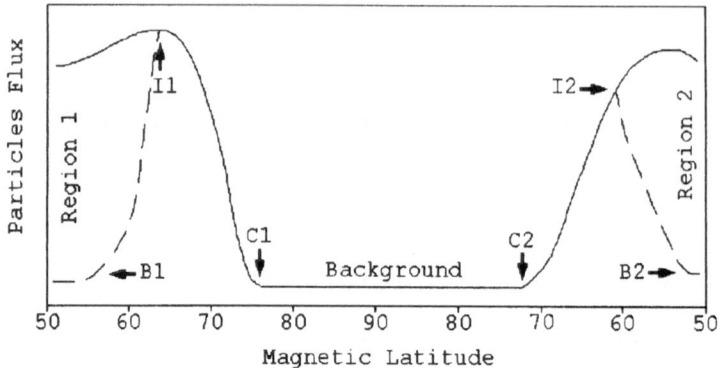

Figure 1. Morphology of the HLIB (see the text)

1. Introduction

The High Latitude Isotropic Boundary (HLIB) is defined as the equatorward edge of the region of isotropic pitch angle distribution equatorward of the cut-off boundary. The morphology of this region is illustrated in Figure 1. The solid line represents the flux of particles with 90° pitch angle (locally mirroring "trapped" particles) and the dashed line is that of particles with about 0° pitch angle (precipitating "loss cone" particles). At low and moderate magnetic latitudes, the particles exhibit an anisotropic distribution characteristic of a trapped population (regions 1 and 2 of Figure 1). As the latitude increases, the distribution breaks toward isotropy (points B1 and B2), reaching the HLIB a few degrees higher in latitude (I1 and I2). As the latitude increases, the flux drops until a cut-off boundary (points C1 and C2) is reached after which the background flux is observed. In some cases, however, the isotropic boundary is not observed, the cut-off boundary is reached before any isotropic distribution is observed. The HLIB exhibits a number of characteristics. The encounter frequency for the boundary for electrons varies from 100% for times around local midnight decreasing with increasing magnetic local time (MLT) reaching a minimum around dusk [2]. The latitude of the boundary was found to be about 7° lower in latitude around local midnight than local noon [6] and [2]. The boundary exhibits a steep energy gradient of around 10 to 20 MeV per unit L for electrons, with the boundary for higher energy occurring at a lower latitude (i.e. a negative gradient) [3] and [7].

A source of the isotropic distribution was suggested to be due to a sharp curvature in the geomagnetic field lines [3], [4], and [5]. If the radius of curvature becomes comparable to the radius of gyration of the particles, the first adiabatic invariant can be violated and the particles will be scattered in pitch angle producing an isotropic distribution. This condition is easily satisfied in the night sector in the region where the field changes from a dipole-like to a tail-like geometry and for this situation the ratio of the radius of curvature to the gyroradius for which an appreciable scattering will occur was found to be about 7 [5].

2. The Model

In this model the global geomagnetic field configuration is used to predict the latitude of the HLIB for different particles of different energies at different MLT. By tracing, the coordinate of each point along the field line $\mathbf{r}(x,y,z)$ and the magnetic field $\mathbf{B}(x,y,z)$ are calculated. At every point, the radius of curvature, R_c, can be calculated from the curvature, k,

$$\mathbf{k} = d^2\mathbf{r}(x, y, z)/ds^2 \tag{1}$$

and

$$R_c = 1/k \tag{2}$$

and the particle radius of gyration, r_g,

$$r_g = \sqrt{2mE} / qB(x, y, z) \tag{3}$$

Thus, at any given point along the field line there exists for particles of mass m and charge q, a minimum energy, E_{min}, such that the ratio of the radius of curvature to the gyroradius is equal to 7, violating the first adiabatic invariant, and is given by

$$E_{min} = (qBR_c)^2 / [2m(R_c/r_g)^2] \tag{4}$$

The minimum energy associated with a given field line is the minimum value of the minimum energy of all the points along this field line, which is a dependent on the minimum of the product BR_c. By tracing the field lines originating from a range of latitudes and MLTs (longitudes), the global magnetic field configuration is obtained, and hence, the scattering region, the source of the isotropic distribution, is found as a function of particle type, energy, latitude and MLT. By assuming that, after going through the scattering process, the particles will conserve their first and second adiabatic invariant, and by using the field line length as a representation of the second adiabatic invariant, their drift paths can be found. The HLIB latitude as a function of MLT, particle type, and energy is found as the contour of the shortest field line length intersecting the scattering region.

In this work, the global geomagnetic field configuration is obtained using IGRF for the internal component and Tsyganenko (T96) for the external one.

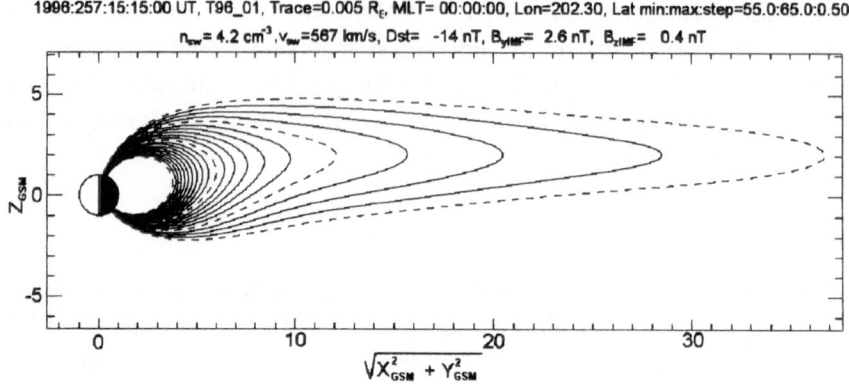

Figure 2. Geomagnetic field configuration according to IGRF and T96

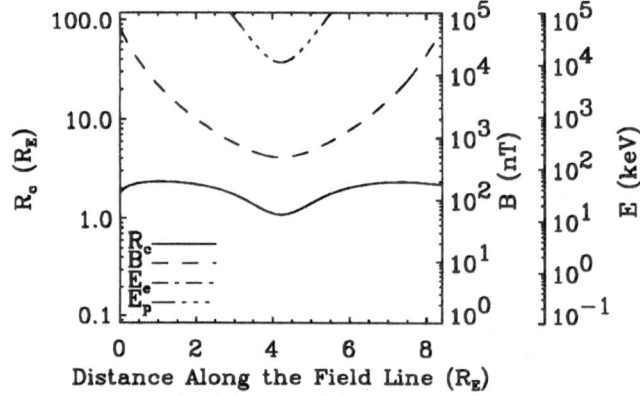

Figure 3. Profile of the field line originating from 0 MLT and 55° latitude

Figure 2 shows the configuration of the field lines originating from midnight MLT and magnetic latitudes from 55° to 65° in steps of 0.5° for September 13th 1996 (DoY 257) at 15:15:00 UT. The model was run for the actual solar wind conditions and the interplanetary magnetic field (IMF) values as measured by the Wind spacecraft and using the provisional D_{st} value obtained from World Data Center (C2) for Geomagnetism, Kyoto, Japan.

The profiles of four magnetic field lines originating from 55°, 60°, 63°, and 65° magnetic latitude, the four dashed lines in Figure 2, are given in Figures 3, 4, 5, and 6. The figures show the variation of radius of curvature, R_c, in units of earth radii (R_E), as a function of distance along the field line (solid line), the total magnetic field, B, in nano-Tesla (nT) (dashed line), as well as the minimum energy, E_{min}, for both protons and electrons in keV. The first field line given in Figure 3, resembles a dipole field with almost a constant curvature. However, as seen in Figure 4, a kink is starting to form at

347

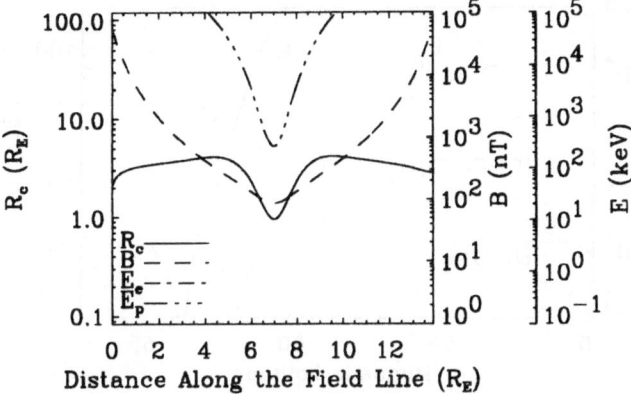

Figure 4. Profile of the field line originating from 0 MLT and 60° latitude

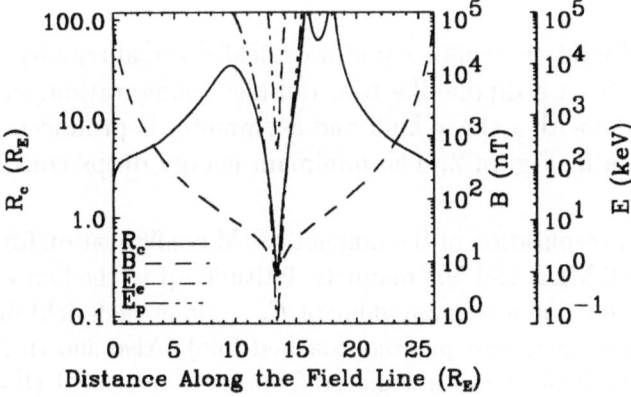

Figure 5. Profile of the field line originating from 0 MLT and 63° latitude

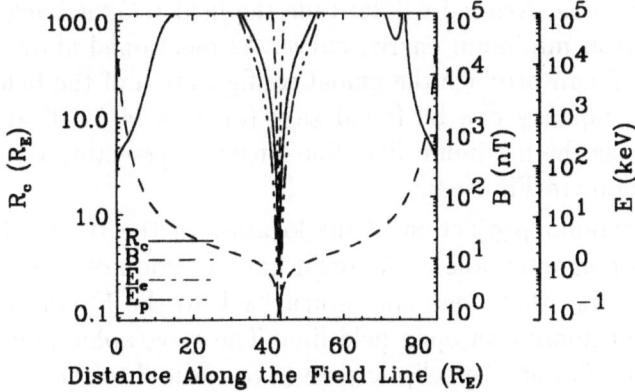

Figure 6. Profile of the field line originating from 0 MLT and 65° latitude

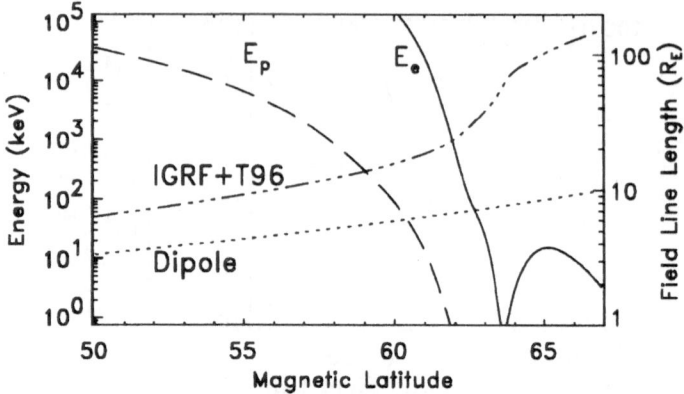

Figure 7. Profile of magnetic field lines originating from 0 MLT and magnetic latitude starting from 50° to the last closed field line in steps of 0.1°

the field line midpoint (magnetic equator or field reversal region) as the field starts to depart from a dipole-like to a tail-like configuration. In Figure 5, the field is tail-like with a sharp kink and asymmetry is produced due to the dipole tilt as seen in Figure 2. The minimum energy drops considerably in Figures 5 and 6.

Figure 7 is a presentation of the magnetic field configuration for field lines originating from 0 MLT and 50° magnetic latitude up to the last closed field line in steps of 0.1°. The minimum value of E_{min} along each field line is given for electrons (solid line) and protons (dashed line). Also shown is the field line length of each field line as calculated from the model field (IGRF+T96) and, for comparison, that of a dipole field (Dipole). As can be seen, at lower latitudes ($< 60°$) the model resembles a stretched dipole. However, between the latitudes of 60° and 64° a considerable departure occurs and the field becomes very tail like. Around 64° latitude the field relaxes back as can be seen in the electron minimum energy curve. As mentioned above, a grid in latitude and MLT can produce the global configuration of the field, and the latitude of the boundary can be found as a function of MLT and particle type and energy as the minimum field line length intersecting the scattering region. This is done in Figure 8.

Figure 8 is a polar projection of the location of the HLIB. The yellow area around the magnetic pole is the polar cap, a region of open field lines. Any traced field line that does not return back to the Earth in the other hemisphere is considered an open field line. The geographic pole is represented by a circled G and the ecliptic pole (axis normal to the ecliptic plane which is indicative of the subsolar point) is represented by a circled E. The two red curves enclose the region (above which and up to the polar cap) the field lines on which the particles will get scattered due to the curvature of

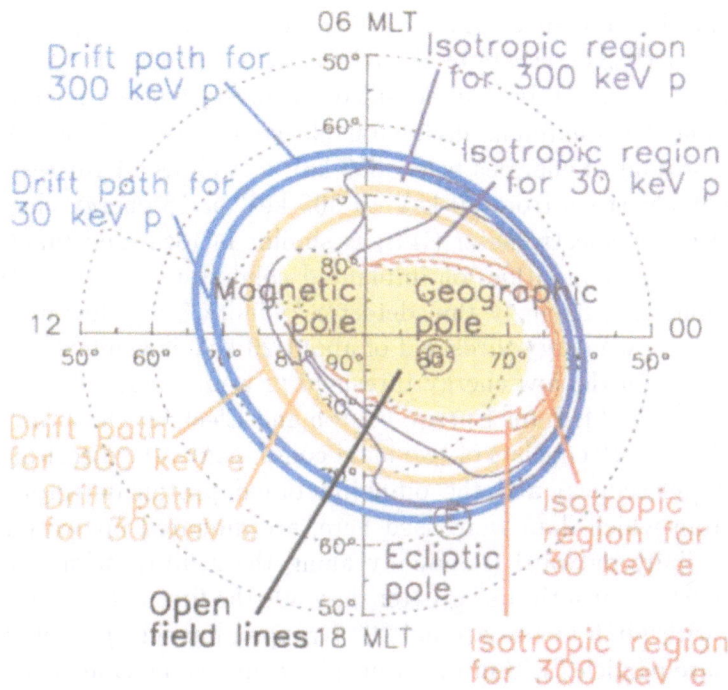

Figure 8. The location of the HLIB for electrons (orange) and protons (blue) as a function of MLT and energy (30 and 300 keV)

the field lines at the field reversal region. The lower latitude curve is for 300 keV electrons and the higher one is for 30 keV as indicated in the figure. The two orange curves are the contours of the minimum field line length (second adiabatic invariant) intersecting each scattering region, which sets a lower limit for the HLIB latitude. Similarly, the two dark blue curves enclose the scattering region for 300 and 30 keV protons, and the blue curves are the drift paths for these two energies. This picture satisfies the general conception of the HLIB configuration: the boundary for protons occurs at lower latitude than that for electrons, and for the same particle the lower the energy the higher the boundary.

3. Testing the Model

Measurement made by the Global Geospace Science (GGS) Polar and the National Oceanic and Atmospheric Administration NOAA-12 spacecrafts of the HLIB were compared to the model. Data from Polar's Comprehensive

Energetic Particle and Pitch Angle Distribution (CEPPAD, J.B. Blake, Principal Investigator) experiment's Imaging Electron Spectrometer (IES) and the Imaging Proton Spectrometer (IPS) were used. The Imaging Electron Spectrometer (IES) consists of three identical sensor heads with each head consisting of three ion-implanted silicon solid state strip detectors behind a pinslit and an aluminum mylar foil placed in front of each detector to eliminate protons of energies below 350 keV as well as a light response. The IES detects electrons from 30 keV to 500 keV in 15 energy channels. The Imaging Proton Spectrometer (IPS) is similar in form and function to the IES and uses a monolithic ion-implanted solid-state detector. As a result of employing an extremely thin detector "window" and low-noise support electronics, a low energy threshold of about 20 keV is achieved, and sixteen energy bins span the low energy threshold to a maximum of approximately 1.5 MeV [8]. The Medium Energy Proton and Electron Detector (MEPED) onboard NOAA-12 consists of two main components: the directional particle detectors (telescopes) and the omnidirectional proton detector. The telescopes, from which data were used here, are mounted in two pairs, one is looking radially outward (nominally along the zenith), while the other at 80° to the first. For latitudes greater than 30° the first detector will measure particles with pitch angles around 90° (trapped particles), while the second will measure particles within the atmospheric loss cone. One of each pair is a solid state electron detector with three integral energy channels, $> 30, > 100$, and > 300 keV. The other is a two-element solid-state proton detector with four differential energy channels, 30-80, 80-250, 250-800, and 800-2500 keV and one integral channel of > 2.5 MeV [9]. MEPED's three electron integral channels and the three lowest proton energy channels were used along with the nearest matching energy channels of the IES (40, 93, 285 keV) and the IPS (32, 76, 259 keV). Table 1 gives a comparison of Polar and NOAA-12 spacecrafts.

Two cases were compared when both NOAA-12 and Polar were crossing the HLIB were investigated, April 2nd and September 13th 1996, and the spacecrafts observations of the HLIB and the cut-off boundary were analyzed. The latter case is discussed below. NOAA-12 measurements of particles fluxes for 90° and 0° pitch angles are given in Figures 9 (poleward pass) and 10 (equatorward pass) with electron fluxes in the upper panel and proton fluxes in the lower one. The three arrows in each panel indicate the location of the modeled HLIB. Polar measurements are given in Figure 11 in a sector roll format, with electron fluxes in the upper three panels and the proton fluxes in the lower ones. As Polar spins around its axis, it measures fluxes at different pitch angles, and the roll modulation can bee seen between 14:30 and around 15:00 UT in the figure. Around 15:00 UT (slightly different for different panels) the roll modulation disappears as the distribu-

TABLE 1. Polar and NOAA-12 comparison

Spacecraft:	Polar	NOAA-12
Launch Date	February 24, 1996	May 14, 1991
Launch Vehicle	Delta II	ATLAS-E
Launch Site	Vandenberg AFB, CA	Vandenberg AFB, CA
Orbit	Polar, 1.8x9 R_E, apogee 20° from the North Pole	Sun Synchronous, LEO, circular 815 km
Inclination	85.9°	98.7°
Orbital Period	17.9 hours	101.35 minutes
Spin Period	6 seconds	Three-axis stabilized
Electron Detector	30 to 500 keV in 15 bins	> 30, > 100, > 300 keV
Proton Detector	20 keV to 1.5 MeV in 16 bins	30-2500 keV in 4 bins and > 2500 keV

tion becomes isotropic. The HLIB model was run around the universal time (15:15), for the observed solar wind IMF conditions for 30 keV electrons and protons (Figure 12), 100keV electrons and 80 keV protons (Figure 13), and 300 keV electrons and 250 keV protons (Figure 14). Magnetic field lines are traced from each spacecraft location down to the ionosphere and the foot of the field line (FOFL) is overlaid on the boundary. Along each spacecraft's FOFL three events are marked and color coded, the break toward isotropy is marked with a dotted line, reaching isotropy is marked with a thick solid line, and reaching the cut-off boundary is marked with a thin solid line. Events for electrons are marked in red and for protons are in blue.

The difference between the latitude where the isotropy is reached and the predicted one ($\Delta\Lambda_I = \Lambda_{measured} - \Lambda_{predicted}$) is given in Figure 15. In the figure the "+" is used for the electrons and "◇" are for protons. The red color for 30 keV particles, the blue for the 100keV electrons and 80 keV protons, and the green for 300 keV electrons and 250 keV protons. As seen, the model predicts the location of the boundary very reasonably for the electrons for both the inbound and the outbound passes. However, a noticable difference, up to 6°, is seen between the observed and the predicted latitude for the protons in the outbound pass in the morning sector. In order to predict the boundary with a higher precision, the evolution of the isotropic distribution originating for the midnight sector must be considered due to a number of interactions such as the wave-particle interactions specially in the dayside, the location of the magnetopause, the cross polar-cap electric field, the development of the loss cone as the particles bounce between their mirror points. Noting the eastward drift directions for the electrons and

352

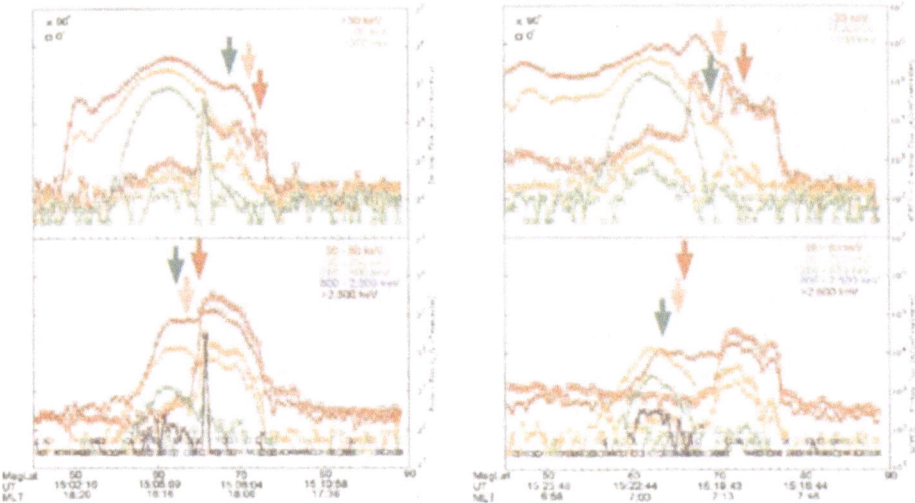

Figure 9. NOAA-12/MEPED measurement for Sep. 13th 1996 starting at 15:00 UT. Poleward pass of the boundary

Figure 10. NOAA-12/MEPED measurement for Sep. 13th 1996 starting at 15:14 UT. Equatorward pass of the boundary

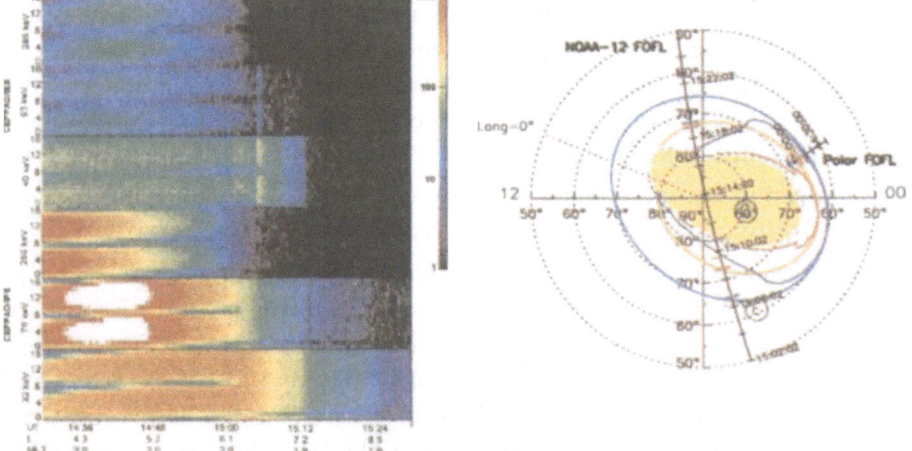

Figure 11. Polar/CEPPAD measurement for Sep 13th 1996 starting at 14:30 UT. Sector vs. time plot for the IES and IPS central sensor. Poleward pass of the boundary

Figure 12. HLIB measurement vs. modeling for 30 keV electrons and protons color coded as Figure 8.

the westward direction for protons, the greatest deviation of the observed HLIB location from the predicted one will occur in the afternoon sector for electrons, and in the morning sector for protons, as observed in the case discussed. Figure 16 shows the difference between the latitude where

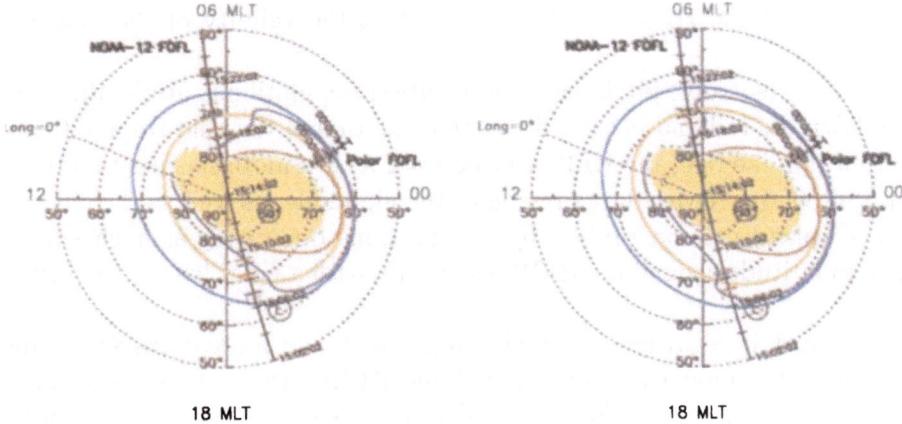

Figure 13. HLIB measurement vs. model-ing for 100 keV electrons and 80 keV pro-tons color coded as Figure 8.

Figure 14. HLIB measurement vs. mod-eling for 300 keV electrons and 250 keV protons color coded as Figure 8.

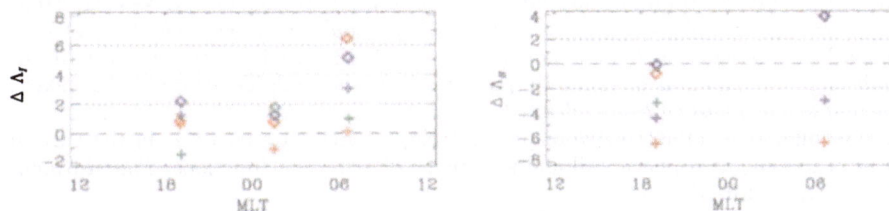

Figure 15. The difference between the ob-served and predicted HLIB latitude. The + are for electrons and ◇ are for protons.

Figure 16. The difference between the lat-itude where the break toward isotropy ob-served and predicted latitude.

the break toward isotropy occures and the predicted latitude of the HLIB $(\Delta\Lambda_B = \Lambda_{measured\ break} - \Lambda_{predicted})$ with the same color coding as the previous figure.

4. Conclusion

The observations of the HLIB agree reasonably well with the predictions of the model, particularly for the protons. In most of the cases of disagree-ment, the observed HLIB latitude was influenced by the drift of the particles through the dayside where the magnetopause and cusp can affect the particle drifts. Electrons may be governed by other factors since their agreement with the prediction of the model is generally poor. Since it is based on the mag-netic field model (IGRF and T96 in this case), this model has the limitation of the magnetic field model, the results may be improved or worsened by

354

the choice of the magnetic field model. In-other-words, this approach to the observation of the HLIB can be used to test the validity of the magnetic model.

Since the isotropic pitch angle distribution population is originating from deep down the tail and is caused by the transition of the field from a dipole-like to a tail-like one, the HLIB can be used as a remote sensing tool of the deep tail region. A very large region of the nightside, as well as the dayside, magnetosphere can be probed by observations done over a much smaller region in latitude around the HLIB, especially with a polar low earth orbiting satellite.

This work is regarded as a preliminary work and more cases are needed to evaluate the model and understand the HLIB. Also a better analysis is planned to utilize the highly resolved Polar/CEPPAD observations of the HLIB.

5. References

1. McDiarmid, I.B. and Burrows, J.R. (1964) High-latitude boundary of the outer radiation zone at 1000 km, *Canadian Journal of Physics* **42**, 616-626
2. Fritz, T.A. (1968) High-latitude outer-zone boundary region for (40-keV electrons during geomagnetically quiet periods, *Journal of Geophysical Research* **73**, 7245-7255
3. Imhof, W.L., Reagan, J.B., and Gaines, E.E. (1977) Fine-scale spatial structure in the pitch angle distribution of energetic particles near the midnight trapping boundary, *Journal of Geophysical Research* **82**, 5215-5221.
4. Tsyganenko, N.A. (1982) Pitch-angle scattering of energetic particles in the current sheet of the magnetospheric tail and stationary distribution functions, *Planetary and Space Science* **30**, 433-437
5. Sergeev, V.A. and Tsyganenko, N.A. (1982) Energetic particle losses and trapping boundaries as deduced from calculations with a realistic magnetic field model, *Planetary and Space Science* **30**, 999-1006.
6. O'Brien, B.J. (1963) A large diurnal variation of the geomagnetically trapped radiation, *Journal of Geophysical Research* **68**, 989-995.
7. Imhof, W.L., Chenette, D.L., and Gaines, E.E. (1997) Characteristics of electrons at the trapping boundary of the radiation belt, *Journal of Geophysical Research* **102**, 95-104.
8. Blake, J.B., et al. (1995) CEPPAD Comprehensive Energetic Particle and Pitch Angle Distribution Experiment on Polar, *Space Science Reviews* **71**, 531-562.
9. Raben, V.J., Evans, D.S., Sauer, H.H., Sahm, S.R., and Huynh, M. (1995) TIROS/NOAA satellite space environment monitor data archive documentation: 1995 update, NOAA Technical Memorandum ERL SEL-86.

PARTICLE BOUNDARIES DURING
A SOLAR ELECTRON EVENT

G. R. BIKKUZINA AND V. A. SERGEEV
Institute of Physics
St.Petersburg State University, St.Petersburg 198904, Russia

AND

T. BÖSINGER
Institute of Physical Sciences
University of Oulu, B.O.Box. 333, FIN-90571 Oulu, Finland

1. Introduction

From time to time, in consequence of solar flares, intense high energy particle flux intensities are observed at polar cap latitudes (these events are known as Solar Particle Events, or SPE). During electron rich SPEs, low altitude polar orbiting satellites detect a remarkably constant high energy particle flux which forms a so called solar plateau. It was shown that the electron flux in the plateau region is equal to the solar electron flux intensity in the solar wind [1]. A characteristics of this plateau is a sharp boundary at a latitude which is usually located well above the outer boundary of the radiation belt. It was long believed that this Solar electron Boundary (SB$_E$) of the plateau demarcates the boundary between open and closed magnetospheric field lines (OCB) [2]. This concept is still in use, for instance, in locating low energy precipitation structures observed at high latitudes [3], [4].

In spite of this practice, there exists nowadays a lot of evidence that solar electrons can penetrate deep into closed field line regions of the plasma sheet. As was first noticed by Evans and Stone [5] the particle fluxes observed at low altitudes in a region between the radiation belt and the SB$_E$ exhibit an energy spectrum similar to that of solar electrons. The authors Sergeev *et al* [6] compared particle fluxes in the equatorial plane as observed aboard ISEE-1 and by polar orbiting satellites in the polar cap and could show that the solar electron fluxes were nearly equal in the plasma sheet as compared to those in the polar cap. Moreover, the solar electrons

355

J. Moen et al. (eds.), Polar Cap Boundary Phenomena, 355–367.

were detected without any gap down to the radiation belt. Christon *et al* [7] showed that the high energy tail of the electron spectrum in the plasma sheet is the same as the one in the solar wind under quiet conditions. In addition, examples of the SB_E in disturbed conditions (with southward IMF), observed again much equatorward of the polar cap boundary can be found in a number of papers [8], [9]. All these observations provide evidence that solar electrons can penetrate deep into the plasma sheet maintaining thereby their flux intensities. This has a significant consequence: the solar electron plateau boundary (SB_E) should then reflect properties of the mechanism responsible for precipitating electrons from the outer part of the plasma sheet.

In most of the previous papers basically the precipitated SB_E fluxes were studied. When both, precipitated and trapped fluxes at low altitudes were available [8], [9], the trapped and precipitated fluxes were found to be equal in the plateau region. This proves that the precipitation is isotropic over the loss cone. This implies non-adiabatic particle scattering in the tail current sheet as the likely mechanism. It is capable to precipitate solar electrons on closed field lines thereby forming a sharp boundary at the edge of its operating realm. Previously, this mechanism was intensively investigated. It showed good agreement between observations and modeling. This investigation was, however, carried out in application to energetic protons [10], [11] rather than electrons.

Some relevant works have, fortunately, been also published with direct implication to energetic electrons. Imhof *et al* [12], Imhof [13], and Imhof *et al* [14] documented the isotropic precipitation of high energy electrons at the periphery of the outer radiation belt. They showed a distinct energy-dependent threshold for flux isotropy (i.e. isotropic boundary for electrons, IB_E) which is consistent with particle scattering in the tail current sheet. Most of Imhof's observations cover, unfortunately, only a narrow MLT sector (+/- 4h) from midnight. Moreover, their works do not include the analysis of SPEs.

In our analysis we will include SPEs and strive towards an united interpretation of the isotropic boundary of radiation belt electrons as well the plateau boundary of solar electrons. Practically speaking we will address the question what do NOAA's satellite measurements in conjunction with a specific SPE of September 1979 tell us about the spatial ordering of the PCB (Polar Cap Boundary), SB_E, IB_P, and IB_E (Isotropic Boundary of protons and electrons, respectively), what is the MLT dependency of these boundaries and how well do they follow magnetic field model predictions? As a result of this analysis we hope to be able to answer the following questions of principal importance:

– Do solar energetic electrons penetrate deeply into the plasma sheet?

- Does the SB_E correspond to the transition from loss cone to isotropic pitch-angle distribution?
- Does the MLT dependence of the SB_E correspond to model predictions of the IB_E?
- Does the current-sheet scattering mechanism apply to solar electrons to form the SB_E?

2. Data, Tools and Definitions

Low altitude polar orbiting satellite (TIROS-N and NOAA-6) data were used. The satellites made measurements at an altitude of 850 km and their orbital planes were aligned approximately along 03–15 MLT and 09–21 MLT (TIROS-N and NOAA-6, respectively). The MEPED instrument measures the energetic particle flux in two orthogonal directions covering the flux of the precipitating particles (pitch angles in the center of the loss cone) as well as the flux of locally trapped particles (pitch angles around 90 deg). In this study mainly one proton energy channel (30–80 keV, MP1) and three electron channels (> 30 keV, > 100 keV, > 300 keV; ME1, ME2, and ME3; respectively) were used. The TED instrument measures, among other things, the energy flux of precipitating auroral protons and electrons. The time resolution of the particle detectors is 2 seconds. For more information on the satellites and their instruments see [15].

Solar particle events (SPE), a consequence of solar flares, are rarely observed; in average only a few times per year. SPEs are characterized by high energy particle flux with constant intensity over the polar cap latitudes giving rise to a remarkable plateau with a sharp boundary at the outer boundary of the radiation belt.

In Figure 1 an example is shown of a solar electron event and related particle characteristics as seen in the data of TIROS-N during orbit 4772, on September 16th, 1979. The top panel shows UT variations in the total energy flux of auroral electrons and protons of 0.3–20 keV (in ergs cm^{-2} s^{-1}) which we used for estimation of the position and width of the auroral oval and the **Polar Cap Boundary** (PCB).

The panels below show integral particle fluxes from proton channel MP1 and the electron channels ME1, ME2, and ME3. The red traces denote locally trapped (90 deg pitch angle) and the green traces precipitating (radially inward directed) energetic particles. Note the typical plateau-type flux behaviour of the solar electrons (bottom panels) in the high latitude region of the satellite's trajectory. The "edge" of the solar electron plateau is called the **Solar electron Boundary** SB_E. Where the trapped and precipitating proton/electron flux intensities depart from being equal the **Isotropic Boundary for protons**, or **electrons** (IB_P or IB_E) is encountered. The

358

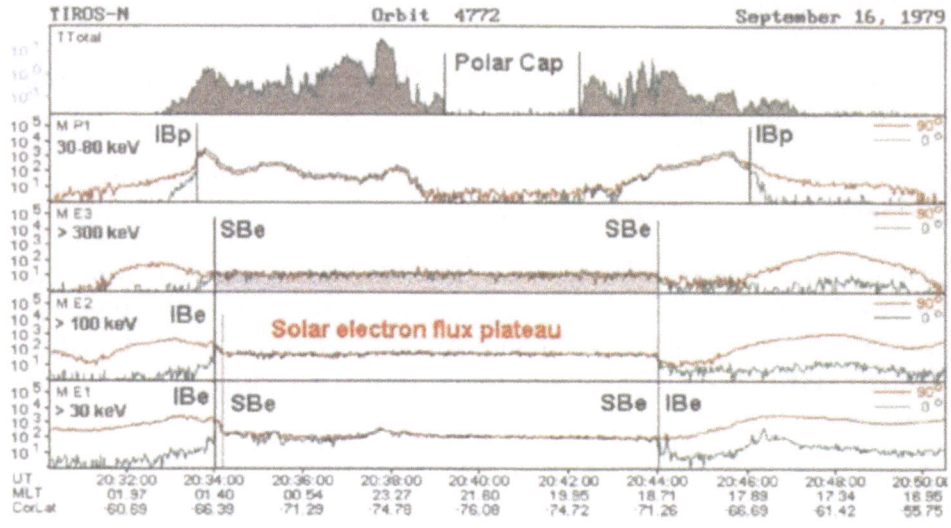

Figure 1. Particle data from the TIROS-N satellite pass of orbit 4772 on September 16th, 1979. For details see text.

vertical black lines in Figure 1 indicate in case of protons the (IB_P) and in case of electrons the (IB_E and/or SB_E). The latter was assigned to the latitude where the electron flux decreased (or increased) rapidly from its plateau level. By **regular scattering** scattering of particles due to the violation of their 1st adiabatic invariance is denoted.

3. Model Results

Using magnetospheric models (T89 and T96) we looked for the regions on closed field lines where the magnetic field is weak and the ratio of magnetic field curvature radius (R_c) to particle gyroradius (r_0) is $R_c/r_0 < 8$. Under this condition the particles will scatter isotropically over the loss cone and produce isotropic precipitation [10]. The Isotropic Boundaries discussed below correspond to the threshold condition $R_c/r_0 = 8$ for each species/energy. We traced these Isotropic Boundaries (IB) as well as the Last Closed Field Lines (LCFL) at all MLT meridians. All calculations were made for zero dipole angle.

The left side of Figure 2 shows the projection of the isotropic boundaries for electrons (IB_E, asterisks) and protons (IB_P, circles) into the ionosphere

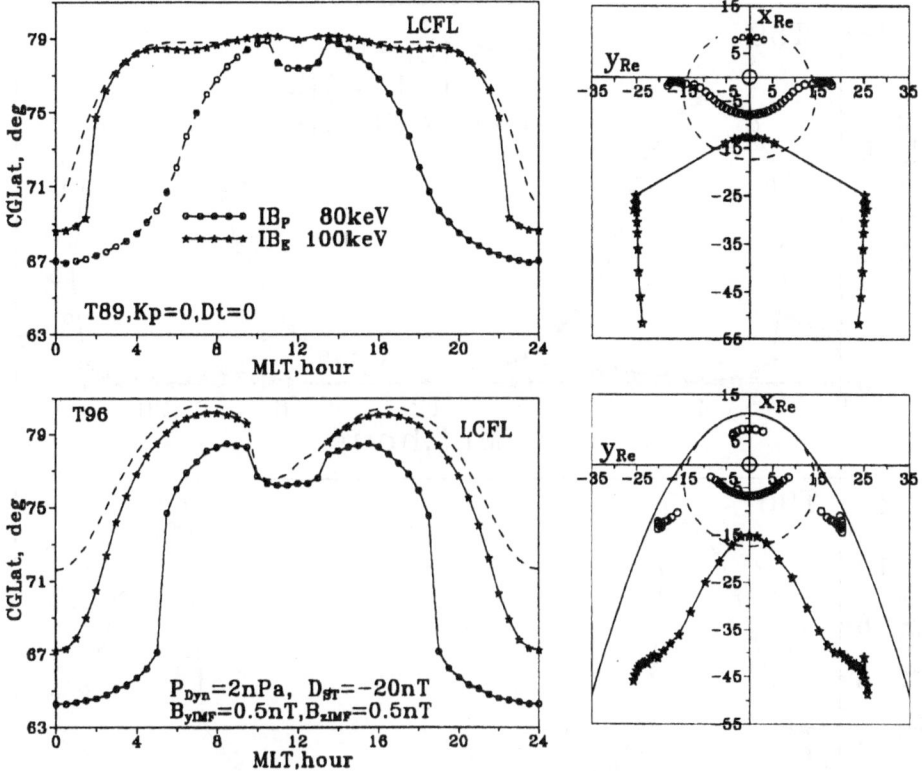

Figure 2. The isotropic boundary for electrons (full asterics) and protons (open circles) and the last closed field line LCFL (dashed line) as derived by the Tsyganenko magnetic field modes T89 and T96 with the input parameters as indicated. For details see text.

and the right side presents the same boundaries as mapped onto the equatorial plane. Figure 3 shows the same projection in a more appropriate (for the comparison with observation) format, i.e. the latitudinal differences between the isotropic boundaries of electrons and protons and the last closed field line (LCFL).

In spite of some differences between the results obtained by the two magnetic field models, the following common features can be noticed:

— (a) Both models predict an IB_P and IB_E in all MLT sectors (except for electrons near noon in T96, where the magnetic field strength in the high altitude cusp does not get small enough so that the 100 keV electrons would get non-adiabatic).

— (b) The MLT dependence of the IB_E is basically the same as for the IB_P, but the IB_E is predicted always at a higher latitude than IB_P, and always at a lower latitude than the LCFL.

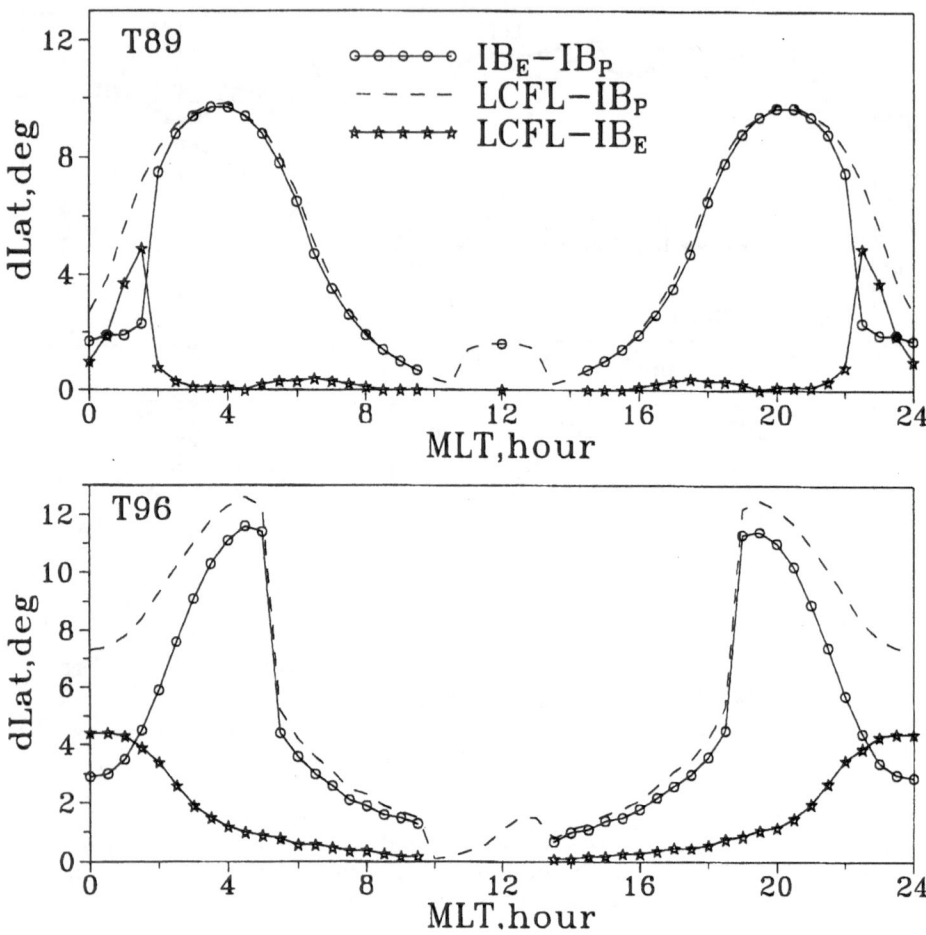

Figure 3. Results of model calculations as in Figure 2 but displayed as difference in latitude between the boundary pairs IB_E and IB_P, LCFL and IB_P, and LCFL and IB_E.

— (c) The latitudinal difference between LCFL-IB_E and IB_E-IB_P is smaller at midnight and noon and increases towards evening and morning MLT hours, with a maximum difference IB_E-IB_P at ~19 and ~05 MLT.

The magnetic field model calculations indicate where in the magnetosphere regular scattering could in principal occur but they do not tell us whether there are enough particles in these regions to produce a measurable isotropic precipitation. For instance, on the basis of T96 a region was found at the flanks of the distant magnetotail (cf. Figure 2) where the regular scattering of energetic electrons could occur, but the question is: are there enough energetic electrons? The dashed lines on the right-side figure boxes of Figure 2 indicate the last closed drift shell for zero pitch-angle particles (mirroring near the ionosphere). Having in mind the drift losses of particles, this line can serve as a proxy of the outer boundary of the radiation belt. Comparison of the last closed drift shell with the IB_E allows us to formulate two more model predictions.

- (d) Only in the narrow MLT sector, roughly between 21 and 03 hours MLT, the IB_E can intrude into the radiation belt, therefore, only in this MLT sector one could expect frequent observations of isotropic precipitation of energetic electrons. Outside of this sector the non-adiabatic scattering could hardly produce an observable isotropic precipitation of magnetospheric electrons (in the absence of solar electrons) because of the smallness of the energetic electron flux in these regions. Only during solar electron events, when solar electrons provide an intense enough particle source in the outer part of the magnetosphere, there exists an opportunity to observe the isotropic electron precipitation at any MLT. Only during such conditions we have a possibility to check our model predictions.
- (e) In the near-midnight sector, with solar electrons, when the isotropic boundary intrudes into the radiation belt, the IB_E flux level may well exceed the one in the plateau region. This means, correspondingly, that under such conditions, the IB_E may be encountered at slightly lower latitude than the SB_E.

4. Observations

The solar event of September 1979 was extraordinary in the sense that the observed solar proton flux was much smaller than the solar electron flux. This is quite unusual and has a very important consequence for our analysis. Since the electron detectors on the NOAA satellites suffer from contamination by high energy protons, the September 1979 event offered the unique possibility of having access to reliable electron data during an intense solar electron event.

Figure 1 illustrates - here for a midnight-afternoon pass of the satellite - typical patterns of particle fluxes. They are somewhat different near midnight and at dusk/dawn MLT sectors. A common feature for all MLT sectors is that the plateau boundary (SB_E) is definitely inside the auroral oval (equatorward of the polar cap boundary, PCB) but is at higher latitudes than the proton isotropic boundary (IB_P). In the dusk sector the sharp boundary of the electron flux plateau is - within the resolution - at the same latitude at all energies (ME1 to ME3) and it coincides with the transition from isotropic precipitation (in the plateau region) to the anisotropic one (loss cone pitch angle distribution), in other words, it coincides with the IBe. A quite different morphology is encountered in the midnight sector.

Here the IB_E (in ME1, ME2 channels) is observed at the periphery of the radiation belt and the SB_E is detected at a ~0.5 deg higher latitude than the IB_E (cf. Figure 1). Such patterns were observed in, roughly, half of all the near-midnight passes. The remaining half exhibits the patterns similar to the ones in the afternoon sector (but with smaller latitudinal separation between IB_P and SB_E).

An interesting detail is that the SB_E position of the most energetic electrons (channel ME3, 300 keV) is different from those in the two lower energy channels. It coincides, however, with the IB_E. In the statistical survey below we use the ME2 channel (> 100 keV) for the determination of the electron boundaries.

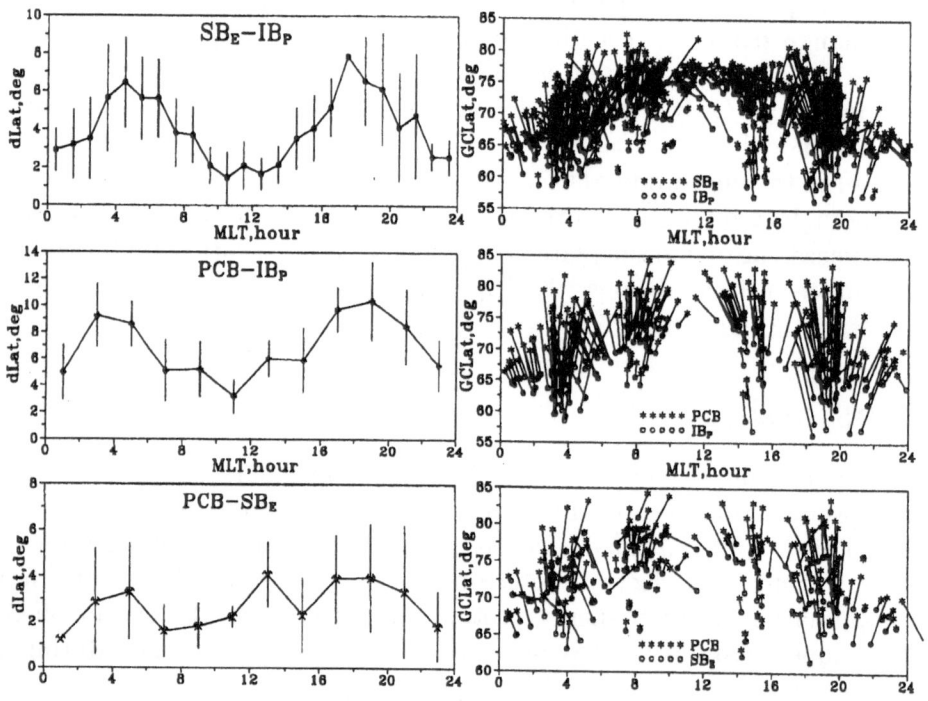

Figure 4. Left side: the three boundaries PCB, SB_E and IB_P in their latitudinal spacing to each other during the solar electron event. The error bars represent the standard deviations for each bin average (1 and 2 h, respectively). Right side: the orbit elements connection each boundary pair used.

The loci of the selected boundaries (IB_P, SB_E, PCB) and their latitudinal difference as a function of MLT are shown in Figure 4, that is with reference to IB_P and the latter, PCB, with reference to SB_E. In all panels (left-hand plot) the "butterfly-type" MLT-dependence - also to be noticed in the model calculations (cf. Figure 2) - can be realized: a minimum around local noon and two maxima, one at early morning and one at late evening. Moreover, the following features can be noticed:

First, the latitudinal ordering follows the pattern: first PCB, then IB_E and then IB_P (with decreasing latitude), which means in Figure 4, the latitudinal difference between PCB and IB_P is largest (central panel), the latitudinal difference between SB_E and IB_P is smaller but almost as large (upper panel) and the latitudinal difference between the PCB and the SB_E is smallest (lower panel). Second, there are indications of an asymmetry with respect to local noon with a shift of the symmetry line towards morning and with larger spacing (differences in latitude) at the evening side. Third, the statistics is best for IB_P and worst for the PCB. Forth, there are only few observations of the SB_E around noon and especially the pair of observation points PCB and SB_E along the satellites' pass is very scanty (cf. right side, bottom panel in Figure 4).

To characterize the MLT distribution of isotropic electron precipitation in the absence of solar electrons Figure 5 shows results obtained from seven months of data (from July 1979 to December 1980) from NOAA-6 and TIROS-N satellites. By normalizing the number of observed IB_E boundaries to the one of IB_P observations (which serves as an estimate of the total amount of crossings in each MLT bin), one immediately realizes that the chance to detect the electron isotropic precipitation are very low outside of the narrow near-midnight MLT sector (21-03 h MLT). In this near-midnight sector, in average, the difference IBe-IBp is about 2 deg in latitude, i.e. comparable to the one found for the SB_E-IB_P difference in the Figure 4.

5. Discussion

In the following we compare the LCFL in the models with "our" PCB (as determined by TED observations, see above) and the IB_E in the models with the SB_E (as determined by MEPED observations, see above). Doing so the confrontation of model predictions with observations can teach us the following:

5.1. AGREEMENTS BETWEEN MODELS AND OBSERVATIONS

The latitudes of the PCB, SB_E and IB_P as a function of MLT and their inter-latitudinal spacing follows basically the patterns predicted by the models if we take the observed PCB for the LCFL and the SB_E for the IB_E in the models. The "butterfly-type" MLT distribution with maximum spacing between SB_E and IB_P at $\sim$19 and $\sim$05 MLT is notable in both Figures 3 and 4. Therefore, the specific MLT-dependence of the SB_E follows the pattern predicted for the IB_E at the nightside, independently of activity conditions. As shown in Figure 1 the electron fluxes are isotropic

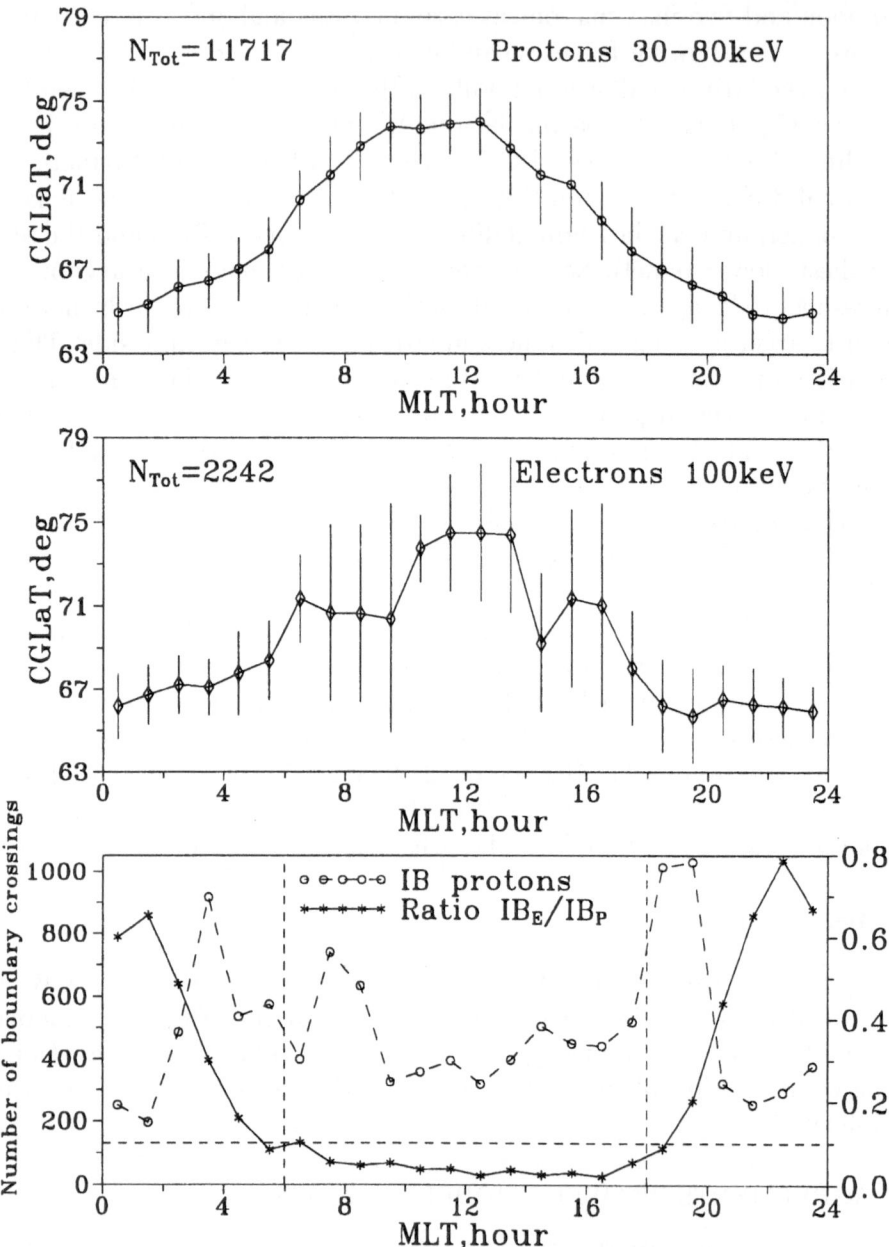

Figure 5. The IB$_P$ (top) and IB$_E$ (central) in absence of a solar electron event, derived from seven months of data, and the relative encounter frequency (bottom panel) of both boundaries, all quantities as a function of MLT. For details see text.

over the plateau region and the transition from isotropic to loss cone pitch-angle distribution (the isotropic boundary) always occurs at (or close to) the SB_E boundary. As the SB_E is obviously located inside of the auroral oval on closed field lines, it is a strong indication that the plateau boundary is a demarcation line between two domains with isotropic and empty loss cone distributions. In our solar electron event this was confirmed for both quiet and very disturbed conditions on a large statistics.

5.2. DISAGREEMENTS BETWEEN MODELS AND OBSERVATIONS

The basic disagreement between model predictions and observed properties is that the observed difference in the latitudes between PCB, SB_E, and IB_P are at all MLT and for each disturbance level ~30 to 50% smaller than the models do predict. We explain this being due to an inaccuracy of the models which implies that the models contain more magnetic flux in the outer parts of the plasma sheet than it is really the case. The physics of the SB_E at the dayside is a different story. As we have shown, the region of isotropic electron precipitation is limited to closed drift shells. The tricky interplay of a continuously waving, forth and back moving magnetopause with electron shadowing effects make the observations of the SB_E more scanty.

It can be anticipated that also the magnetopause current sheet is a region for particle scattering but this current sheet is a very thin structure and - taking into account the small gyroradius of electrons - the precipitation region must be very narrow. If it at all exists, it must be located near the sharp flux boundary constituting our PCB.

6. Conclusion

We investigated particle boundaries during the electron rich solar particle event of September 1979 and compared the observational results with model calculations. Based on these results the four questions formulated in the Introduction can now be answered as follows:

- We confirm earlier results and show that in all conditions the solar electrons penetrate deep into the plasma sheet so that their plateau boundary (SB_E) at all MLTs (with the exception of a narrow MLT sector near noon) is observed on closed field lines of the plasma sheet. The most clear confirmations are: (1) the SB_E is systematicly located equatorward of the PCB, and (2) near midnight, the solar electron plateau extended into the radiation belt in, roughly, half of all cases.
- We found that the solar electron plateau boundary always corresponds to the transition from isotropic to loss cone pitch-angle distribution

(with the exception of the night-side SB_E which merges the ou-ter boundary of the radiation belt.

- The MLT-dependence of the SB_E (and its latitudinal difference as compared to IB_P) has a specific shape which nicely correspond to the model predictions based on non-adiabatic scattering of energetic electrons in the equatorial current sheet.

- These facts all together allow us to conclude that the plateau boundary of solar electrons is formed by non-adiabatic scattering of solar electrons on closed field lines in the equatorial current sheet.

7. Acknowledgements

The work of G.R.B. was supported by a grant from the Center of International Mobility (CIMO) and support from the Dept. Phys. Sciences of the Univ. Oulu. The NOAA data were available from the WDC-A for STP in Boulder.

References

1. Lyons, L.R. and Williams, D.J. (1984) *Quantitative aspects of magnetospheric physics.* D. Reidel Publishing Company, Dordrecht, Holland.
2. Akasofu, S.A. (1977) *Physics of magnetospheric substorm.* D. Reidel Publishing Company, Dordrecht, Holland.
3. Gussenhoven, M.S., Hardy, D.A., Rich, F.J., Mullen, E.G. and Redus, R.H. (1990) Evidence that polar cap arcs occur on open field lines, *J. Geomagn. Geoelectr.*, **Vol. no. 42**, pp. 737–751.
4. Lyons, L.R., Lu, G., de la Beaujardiére, O. and Rich, F.J. (1996) Synoptic maps of polar caps for stable interplanetary magnetic field intervals during January 1992 geospace environment modeling campaign, *J. Geophys. Res.*, **Vol. no. 101**, pp. 27283–27298.
5. Evans, L.C. and Stone, E.C. (1972) Electron polar cap and the boundary of open geomagnetic field lines, *J. Geophys. Res.*, **Vol. no. 77**, pp. 5580–5584.
6. Sergeev, V.A., Kuznetsov, S.N. and Gotselyuk, Yu.V. (1987) The dynamics of the structure of the high-latitude magnetosphere according to data on solar electrons, *Geomagn Aeron.*, **Vol. no. 27**, pp. 380–385.
7. Christon, S.P., Mitchell, D.G., Williams, D.J., Frank, L.A., Huang, C.Y. and Eastman, T.E. (1988) Energy spectra of plasma sheet ions and electrons from ~50 eV/e to ~1 MeV during plasma temperature transitions, *J. Geophys. Res.*, **Vol. no. 93**, pp. 2562–2572.
8. Sergeev, V.A. and Bösinger, T. (1993) Particle dispersion at the night side boundary of the polar cap, *J. Geophys. Res.*, **Vol. no. 98**, pp. 233–241.
9. Yanin, A.G., Malkov, M.V., Sergeev, V.A., Pellinen, R.J., Aulamo, O., Vennerström, S., Friis-Christensen, E., Lassen, K, Danielsen, C., Craven, J.D., Deehr, C. and Frank, L.A. (1994) Features of steady magnetospheric convection, *J. Geophys. Res.*, **Vol. no. 99**, pp. 4039–4051.
10. Sergeev, V.A., Malkov, M. and Mursula, K. (1993) Testing the isotropic boundary algorithm method to evaluate the magnetic field configuration in the tail, *J. Geophys. Res.*, **Vol. no. 98**, pp. 7609–7620.

11. Sergeev, V.A., Bikkuzina, G.R. and Newell, P.T. (1997) Dayside isotropic precipitation of energetic protons, *Ann. Geophys.*, in press.
12. Imhof, W.L., Reagan, J.B. and Gaines, E.E. (1979) Studies of the sharply defined L-dependent energy threshold for isotropy at the midnight trapping boundary, *J. Geophys. Res.*, **Vol. no. 84**, pp. 6371–6384.
13. Imhof, W.L. (1988) Fine resolution measurements of the L-dependent energy threshold for isotropy at the trapping boundary, *J. Geophys. Res.*, **Vol. no. 93**, pp. 9743–9752.
14. Imhof, W.L., Chenette, D.L. and Gaines, E.E. (1997) Characteristics of electrons at the trapping boundary of the radiation belt, *J. Geophys.Res.*, **Vol. no. 102**, pp. 95–104.
15. Hill, V.J., Evans, D.S. and Sauer, H.H. (1985) TIROS/NOAA satellites space environment monitor. Archive tape documentation, *NOAA technical memorandum ERL*, **Vol. no. SEL-71**, pp. 1–77.

STRUCTURE ANALYSIS OF GEOSYNCHRONOUS SUBSTORM OSCILLATIONS

Ø. HOLTER
Department of Physics, University of Oslo
0316 Oslo, Norway

A. ROUX and S. PERRAUT
Centre d'étude des Environnements Terrestre et Planétaires
78140 Velizy, France

Abstract. Substorm events as observed by geostationary satellites are accompanied by large amplitude transient low frequency oscillations on the electric and magnetic field components. The "short" period oscillations ($\sim 40 - 100$ s) have been interpreted as standing waves confined to the thin current sheet which develops prior to the substorm onset. The signatures of these oscillations depend in particular on the position of the spacecraft relative to the current sheet, and on a number of additional physical parameters related both to the current sheet and the waves. Changes in these parameters with time give rise to both temporal and spatial effects. Synthetic signal analysis using wavelets can be employed to investigate signatures of oscillations under different physical conditions, and it can be a useful diagnostic tool in interpreting the nature of complicated real signals in a reasonably consistent manner. Wavelet analysis has been used to illustrate the signatures of oscillations recorded at different s/c positions, as would be the situation with multi-satellite systems.

1. Introduction

Associated with geosynchronous substorm events, large scale reconfigurations of the magnetic field from tail-like to dipole-like structures are observed. During this dipolarization process, large amplitude, low frequency transient oscillations are observed on the electric and magnetic field components. In a previous work by Holter *et al.* [1], the nature of these oscillations was investigated for a single substorm recorded on GEOS 2, January 25, 1979. It was suggested that these observed low frequency oscillations were standing waves of two types;

- short period (~ 40–100 s) standing waves confined inside a thin Current Sheet (CS),

- long period (~ 300–600 s) standing waves reflected at conjugate points in the ionosphere.

369

J. Moen et al. (eds.), Polar Cap Boundary Phenomena, 369–380.

A number of possible wave modes and harmonics may be excited during the substorm: The shear Alfvén wave, the slow magnetosonic wave, and the fast magnetosonic wave, with their higher harmonics.

The coupling of the shear Alfvén and slow magnetosonic waves via magnetic field curvature effects in a high β plasma has been discussed by Southwood and Saunders [2], Walker [3], and Ohtani et al. [4]. These waves are field aligned and thus obvious candidates for the confined "short–period" ($\sim 40 - 100$ s) transient oscillations observed during the most active period of the substorm. The appearance of these waves during the initial phase of the dipolarization, suggest that they may be important for the substorm development. They appear to be of the same nature as the ballooning modes confined on entire field lines between conjugate ionospheric reflection points [5,6,7]. The fast magnetosonic wave is not field aligned, and once generated, may propagate away in a direction normal to the magnetic field.

The appearance of transient oscillations within a wide frequency band, complicates the identification of the different modes with associated higher harmonics. The signature of standing waves in thin current sheets depends in particular on the location of the s/c relative to possible nodes/antinodes of the oscillating field quantities. Thus, depending on the s/c position, the signals recorded for different events may exhibit great variations, even if the nature of the events is basically similar.

In the present paper we shall study standing waves in environments as those expected in thin CS's during geosynchronous substorms. We shall investigate the signatures these oscillations may exhibit by applying wavelet analysis [8,9,10,11] to synthetically modelled magnetic and electric field oscillations confined between reflecting boundaries.

2. Method of Analysis

Since Fourier analysis does not provide any resolution in time, it is not well suited for studying the frequency-time behaviour of transient waves. With wavelet analysis, however, a presentation of signal amplitudes in the time-frequency domain is possible [8,9,10,11]. The use of wavelets gives an optimum time and frequency resolution for the signals.

The wavelet transform of a function $f(t)$ is defined by

$$W_s(a, \tau) = \int_{-\infty}^{\infty} f(t) h_{a\tau}^*(t) dt, \qquad (1)$$

where * denotes the complex conjugate, and

$$h_{a\tau}(t) = \frac{1}{\sqrt{a}} h(\frac{t - \tau}{a}), \qquad (2)$$

is the so called daugther wavelet constructed from a basic mother wavelet $h(t)$ by dilation and translation, i.e.

$$t \to \frac{t - \tau}{a},\tag{3}$$

where a is a scaling parameter. The translation of the analysing window in time is given by τ, its size being adjusted to the frequency analyzed by the scaling parameter a; short windows are used for high frequencies, long windows for low frequencies.

A great number of different (mother) wavelets are possible, provided they fulfill the following requirements:

- The admissibility condition (zero average);

- The energy must be finite;

- The Fourier transform must exist.

In terms of the Fourier transforms of $f(t)$ and $h(t)$, $F(\omega)$ and $H(\omega)$, the wavelet transform Eq. (1) can be written

$$W_s(a, \tau) = \frac{\sqrt{a}}{2\pi} \int_{-\infty}^{\infty} H^*(a\omega) F(\omega) e^{i\omega\tau} \, d\omega.\tag{4}$$

For the present analysis we have employed the Morlet (mother) wavelet [8] given by

$$h(t) = \frac{1}{\sqrt{2\pi}\sigma} e^{i\omega_0 t} e^{-\frac{t^2}{2\sigma^2}},\tag{5}$$

where ω_0 is the basic wavelet frequency and σ determines the width of the Gaussian envelope. The corresponding daughter wavelets are obtained by the transformation Eq. (3);

$$h_{a\tau}(t) = \frac{1}{\sqrt{2\pi a}\sigma} e^{i\omega_0(t-\tau)/a} e^{-\frac{(t-\tau)^2}{2a^2\sigma^2}}.\tag{6}$$

The Fourier transform of the mother wavelet is given by

$$H(\omega) = e^{-\sigma^2(\omega-\omega_0)^2/2}.\tag{7}$$

The complex wavelet transform coefficients $W_s(a, \tau)$, can be expressed in terms of the modulus $|W_s(a, \tau)|$ and the phase α,

$$W_s(a, \tau) = |W_s(a, \tau)| e^{i\alpha}.\tag{8}$$

In the wavelet diagrams we shall utilize in our analysis, time (or translation) and frequency (or scale) are given along the axes, while the modulus is given in colour. When interpreting these diagrams it should be kept in mind that they are subject to the uncertainty principle,

$$\Delta t \Delta \omega = constant,$$

where Δt and $\Delta \omega$ are the uncertainties in time and frequency, respectively. Thus, the resolution in time is better the higher the frequency is.

3. Substorm Events

A typical signature of the CS build up prior to the onset of a substorm is the gradual directional change of the magnetic field towards the magnetic tail. To identify this signature the satellite must be located somewhat away from the Magnetic Equatorial Surface (MES), which for GEOS 2 was normally the case. However, for certain time periods, GEOS 2 was close to the MES, and the magnetic field was mainly axial and decreasing as the CS was built up. The corresponding radial component of the magnetic field was very small.

Examples of non-equatorial and equatorial events are illustrated by the two geosynchronous substorm events of January 25 and July 17, 1979, presented in Figure 1. In these plots the axial and radial magnetic field components are shown together with the magnetic vector in the the the meridian plane. The coordinate system used is a satellite centered VDH-system with V radially outward from the Earth's center, H parallel to the Earth's spin axis, and D in the azimuthal east direction.

The principal differences between these two events, as most clearly seen from the vector plots, are that

- the event of January 25, 1979, as judged by the tilt of the magnetic field, was a non-equatorial event. The s/c was below and somewhat away from the MES, possibly close to the CS boundary,

- the event of July 17, 1979, was an equatorial event, with the s/c very close to the MES as indicated by the near axial direction of the magnetic field.

The start of the dipolarization process can be estimated by using the minimum value of the axial magnetic field component, B_H, as an indicator. Thus, we locate the onset of the January 25 event to $t \sim 420$ s (2017 UT) while the July 17 event onset is located to $t \sim 400$ s (2246:40 UT).

For the equatorial event in Figure 1, panel 4, the magnetic B_V-component is initially small and negative, then goes through zero at $t \sim 300$ s. B_V then remains small, but positive, until $t \sim 800$ s and then again becomes negative at $t \sim 900$ s. This variation of B_V indicates that the MES, defined by $B_V \approx 0$, is passing by the satellite from below at $t \sim 300$ s, and then again from above some 10 min later at $t \sim 900$ s. For the non-equatorial event in Figure 1, panel 3, the s/c is apparently some distance away from the MES. At the beginning of the dipolarization process, a reduction by a factor of two is recorded on the average B_V component in a time span of ~ 1 min. This sharp reduction would be expected if the s/c originally was below the MES close to the CS boundary, and the CS was displaced downwards. The weaker B_V component could thus be due to the s/c being closer to the MES as the dipolarization progressed. The apparent vertical displacement of the MES

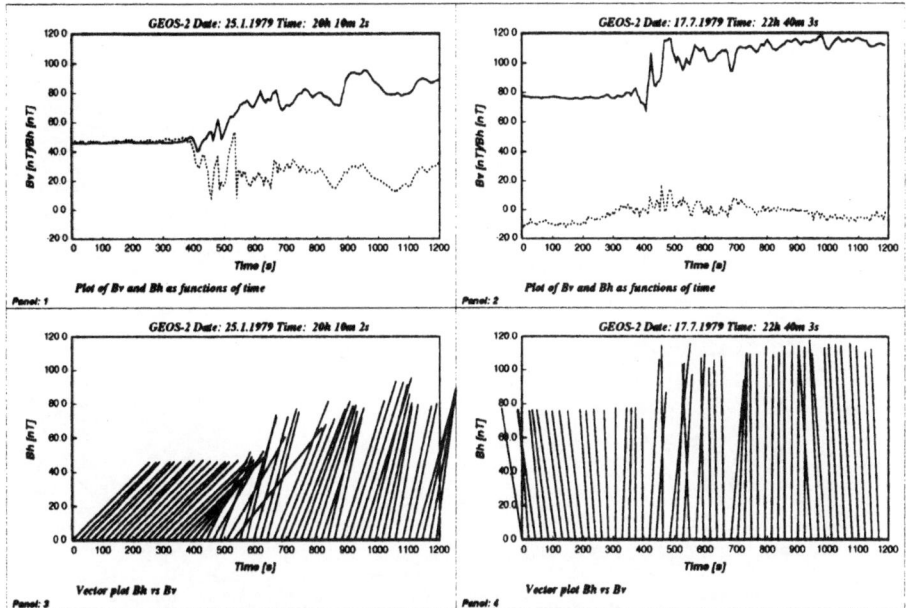

Figure 1: Time series for meridian magnetic field components for two geosynchronous substorm events recorded on GEOS 2. Non-equatorial event: 25.1.1979; equatorial event: 17.7.1979. Panel 1 and 2: Plots of the magnetic field components B_V (....) and B_H (—-) as functions of time. Panels 3 and 4: Magnetic field vectors in the meridian plane, B_H vs. B_V, as functions of time.

during the dipolarization may be significant for the interpretation of the observed oscillations.

The low frequency large amplitude oscillations observed on the magnetic field components during geosynchronous substorms are transient in nature and super-imposed on the overall reconfiguring magnetic field. To investigate the nature of these low frequency transient oscillations; i.e. their frequency, localization in time, modes and harmonics involved, wavelet transform analysis represents a valuable diagnostic technique.

As basis for the analysis of synthetic signals, we shall employ selected data from the GEOS 2 recorded event of January 25, 1979, which has been studied extensively by Roux et al.[12] and Holter et al.[1]. We shall only consider two of the field components; the radial magnetic field component, B_V, and the az-imuthal electric field component E_D. In the six panels of Figure 2 we present the detrended data time series, the Fourier spectra and the wavelet moduli for B_V and E_D, respectively. On the Fourier spectra (no time resolution) we observe for B_V (Figure 2 panel 4) a pronounced broad peak in the frequency range $\sim$ 18-25 mHz. In this frequency range there is almost no indication of an E_D-signal (Figure 2, panel 1). On the other hand, E_D exhibits a pronounced broad peak in

374

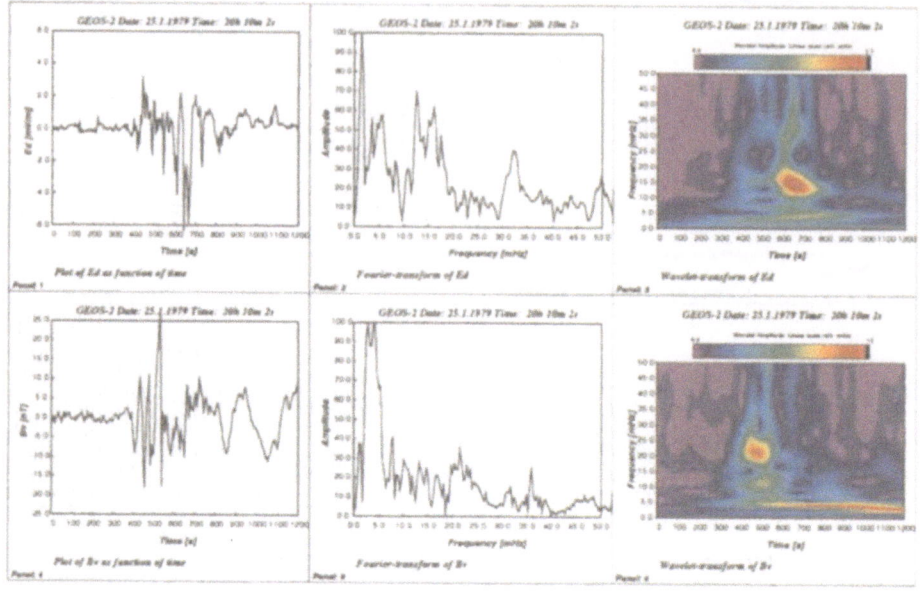

Figure 2: Plot showing the signals (panels 1 and 4) , the Fourier spectra (panels 2 and 5), and the wavelet diagrams (panels 3 and 6) for the detrended electric field, E_D, and magnetic field, B_V, respectively.

the frequency range ~ 10-18 mHz, where there is little indication of a B_V-signal. From the time series and the Fourier spectrum, we have little, if any, indication of the sequence of these signals (or lack of signals). However, the wavelet diagrams presented in panels 3 and 6 of Figure 2, make a separation in time possible.

On Figure 2, panel 6, which shows the modulus of the wavelet transform of B_V, we observe around the frequency ~ 22 mHz a short duration intensification from $t \sim 400$ s to $t \sim 540$ s, with an apparently decreasing frequency. The amplitude of B_V is small from $t \sim 600$ s for frequencies ~ 15 mHz. The modulus of the wavelet transform of E_D (Figure 2, panel 3) exhibits a minimum in E_D from $t \sim 400$ s to $t \sim 540$ s for frequencies ~ 22 mHz. This is followed by an intensification from $t \sim 540$ s to $t \sim 740$ s for frequencies ~ 15 mHz.

These results indicate a node structure and a decreasing frequency for the observed oscillations. The suggested interpretation of these signals by Holter *et al.* [1], was that they represent confined (standing) ballooning modes in the CS. As the CS is displaced and expands during the breakup, GEOS-2 may be located in a position where an antinode (node) and then a node (antinode) for the field components B_V (E_D) is moving past the s/c. The lowest frequency oscillations in the range ~2 – 5 mHz, which appeared after the dipolarization, and was interpreted as second harmonic standing waves along entire magnetic field lines, are not considered here.

In Figure 3 we have, in a qualitative manner, illustrated the amplitude variations expected for the different harmonics of standing waves confined inside the CS. Even if the oscillatory properties of the standing modes were known, their signatures could appear as widely different, depending on the location of the s/c relative to nodes/antinodes of the oscillating field quantities. The radial magnetic

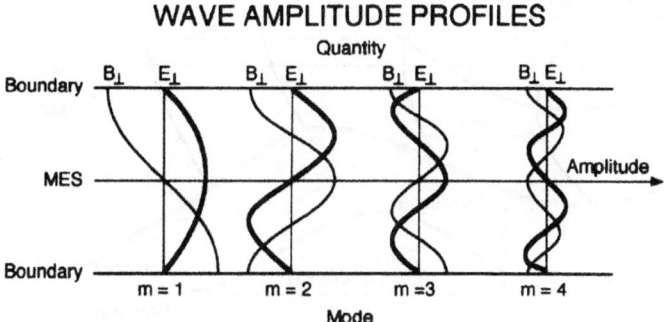

Figure 3: Qualitative illustration of electric and magnetic field amplitude profiles for four harmonic modes confined between reflecting boundaries.

netic field variations observed in fig 1, indicate a vertical displacement of the CS simultaneous with the dipolarization process and the associated low frequency oscillations. A vertical displacement of the CS implies that the s/c would record the oscillations at different positions relative to the MES, as a function of time. In Figure 4 this is qualitatively indicated for a multi satellite system consisting of four differently localized s/c's inside or close to a slow (oscillatory) vertically displaced and expanding CS. During substorms, which have a finite time duration as indicated in the lower part of Figure 4, the positions of s/c's relative to the MES and the CS boundaries will be continously changing. Thus, the signatures of substorm oscillations will be influenced by the s/c

- position relative to the MES, i.e. its position relative to nodes/antinodes;

- motion relative to the MES, i.e. temporary appearances of node/antinodes, and motion induced frequency shifts.

4. Synthetic Signal Analysis

Wave signals recorded during geosynchronous substorms depend on a number of physical parameters related both to the waves and the surrounding medium. Changes in these parameters result in a mixture of temporal and spatial effects. As a means of interpreting the nature of complicated real signals in a reasonably consistent manner, synthetic signal analysis can be useful. To establish signals which model real geosynchronous substorm signals, a number of input parameters

376

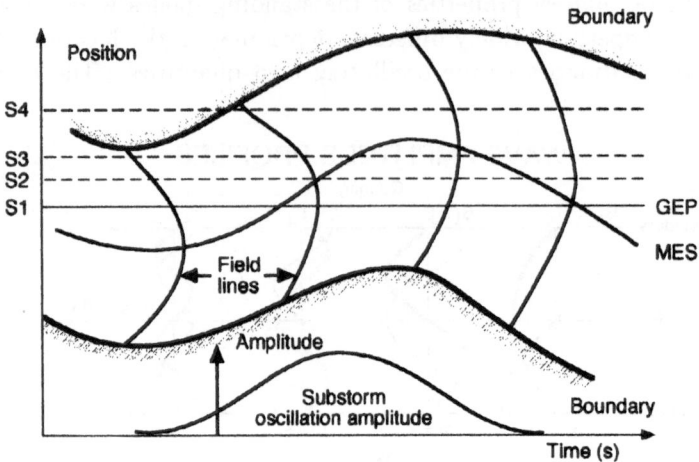

CURRENT SHEET
DISPLACEMENT and EXPANSION

Figure 4: Qualitative illustration of CS and MES long period oscillatory displacement and expansion relative to the GEP. Four s/c positions (S1.. S4) are indicated. The duration and amplitude of low frequency substorm oscillations are qualitatively indicated as function of time.

can be envisaged, i.e.
1. Satellite position parameters:
   - the satellite position relative to the Geographic Equatorial Plane (GEP),
   - the position of the Magnetic Equatorial Surface (MES) relative to the GEP, introduced as an oscillation (flapping) with specified
     - position amplitude as fraction of the (normalized) CS-thickness,
     - oscillation or flapping period.

2. Dipolarization process parameters:
   - the duration of the dipolarization;
   - changing propagation length L_B along magnetic field lines between reflecting boundaries caused by
     - shortening of field lines as they become more dipole like,
     - an increasing distance between reflecting boundaries as the CS expands.

3. Standing wave parameters:
   - the wave fundamental frequency,
   - the number of harmonics and the ratio of their amplitudes,
   - spatial wave amplitude profiles,
   - the wave amplitude envelopes; growth- and decay rates,
   - frequency shift ($f \sim v_A/L_B$) due to changing Alfvén velocity, v_A, caused by changes in density and magnetic field strength.

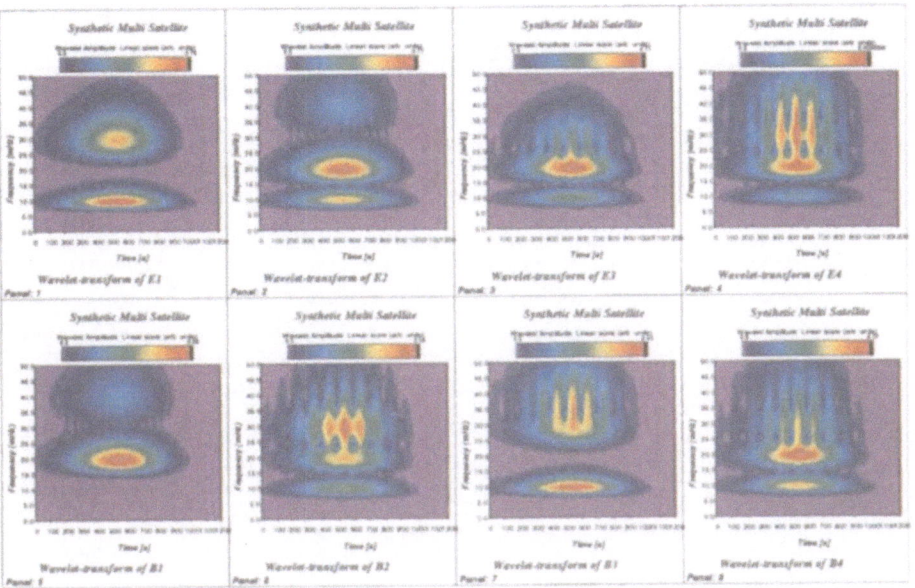

Figure 5: Wavelet diagrams for synthetic signals signals with four harmonics at four different positions inside a fixed CS-waveguide. The positions are at 1) the center, 2) the third harmonic node/antinode; 3) the second harmonic node/antinode, and 4) at the boundary.

As a basic synthetic signal we select an electromagnetic wave with four harmonics, confined as a standing wave between fixed reflecting boundaries. This situation is illustrated by the eight wavelet diagrams in Figure 5 for the oscillatory perpendicular electric and magnetic field components normal to the background magnetic field. In this case the first harmonic has a frequency 10 mHz, and the relative amplitudes of the four harmonics are arbitrary chosen as: 0.6/1.0/0.7/0.4. The four positions for which the wavelet diagrams have been presented are related to the MES and the CS boundary; half a wavelength of the first harmonic equals the distance between the reflecting boundaries along the magnetic field lines. In Figure 5 the diagrams in the panels 1 and 5 correspond to the center of the MES; panels 2 and 6 to the third harmonic node/antinode position; panels 3 and 7 to the second harmonic node/antinode position; and panels 4 and 8 to the reflecting boundary. The duration of the signal is ~ 600 s, beginning at $t \sim 400$ s. The wavelet diagrams presented in Figure 5 are clearly interpreted in the time frequency domain as harmonic oscillations of finite duration. The differences between the signals reflect the locations of the waves nodes/antinodes.

The basic signals as presented in Figure 5 can be modified in a way which, to a greater extent, models the situation encountered in connection with substorms. Although the model itself may not be fully adequate for reproducing real signals,

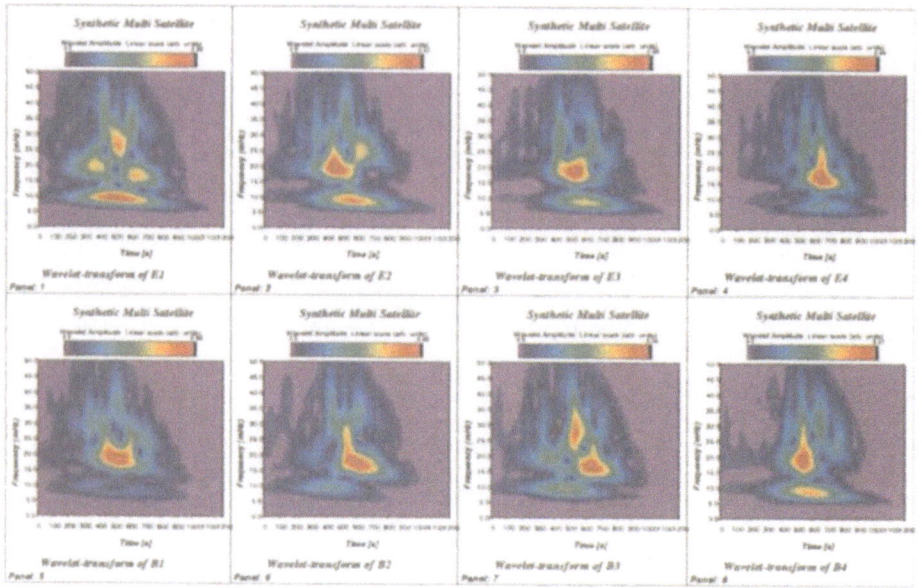

Figure 6: Wavelet diagrams for synthetic signals with four harmonics at four different positions inside a CS-waveguide with changing position and width. The original positions are at 1) the center, 2) the third harmonic node/antinode; 3) the second harmonic node/antinode, and 4) at the boundary.

synthetically produced wavelet diagrams may be used to clarify how changes in different physical parameter can alter a particular signal. The simple physical situation illustrated in Figure 5 is modified as follows: The MES is assumed to be displaced periodically relative to the GEP with period ~ 900 s and amplitude 0.2 times the original CS width. The CS width is assumed to double in the time interval $t \sim 400$ s to $t \sim 1000$ s. The phase of the CS displacement is specified such that the MES passes the GEP from below at $t \sim 510$ s, somewhat after the assumed onset of the oscillations at $t \sim 400$ s. With these parameter specifications the wavelet diagrams corresponding to those in Figure 6 are significantly modified as illustrated in Figure 6. The main changes are due to the fact that the nodes/antinodes of the oscillating quantities now move past the measuring positions 1) to 4).

In Figure 6, panel 1, we observe on E_1 at $t \sim 500$ s when the MES moves past the GEP (the assumed recording position), the first and third harmonic antinodes, and the second and fourth harmonic nodes. At the same instant we observe, in panel 5, the B_1 first and third harmonic nodes, and the second and fourth harmonic antinodes. Thus, the anticorrelation between the different harmonics of E_1 and B_1 is apparent. These wavelet diagrams could possibly reflect substorm situations encountered with equatorial events.

In Figure 6, the diagrams in panels 4 and 8, should be more relevant for non-equatorial events. In this case the CS boundary is passing position 4 from below at $t \sim 500$ s, and we see a delayed onset of the first harmonic E_4 (node) when compared to B_4 (antinode). For the second harmonic we have a B_4 antinode at $t \sim 520$ s and frequency ~ 20 mHz, and, somewhat later, an E_4 antinode at $t \sim 650$ s and frequency ~ 15 mHz. An antinode on one quantity clearly has a corresponding node on the other quantity. Although not optimized, the diagrams in Figure 6, panels 4 and 8, have qualitative similarities with the substorm wavelet diagrams in Figure 2. In that case the s/c was below the MES and the CS boundary was moving downwards, which, however, is similar to a situation where a s/c is above the MES with an upward moving CS. The possibility for reproducing synthetically the main elements of real substorm signals as presented in wavelet diagrams, may be useful for consistent interpretations of complex real signals. Since these signals are strongly position dependent, it could in particular be useful for events recorded by multi-satellite systems.

5. Conclusion

The transient low frequency oscillations observed in connection with substorms are interpreted as standing ballooning modes confined to the CS. This interpretation is based on the appearance of the node/antinode anticorrelation of the electric and magnetic fields, as seen on the wavelet diagrams for the non-equatorial event of January 25, 1979. The anticorrelation between the E_D- and B_V-oscillations is consistent with an interpretation of these oscillations as standing coupled slow magnetosonic – shear Alfvén waves. However, the variability of signals recorded during different substorms complicates the interpretation. The time series for the radial magnetic field suggests a vertical displacement of the magnetic equatorial surface with respect to the s/c during the substorm dipolarization process. This, in addition to a number of other parameters which change during this process, do not facilitate the interpretation of the oscillations in a consistent manner.

As a means of performing an adequate signal analysis, the combined use of wavelet transforms and synthetic signals can represent a valuable method. Thus, the importance of different parameters related to the position(s) of the s/c('s), the dipolarization process, and the wave properties can be analysed. In particular spatial and temporal effects observed on both single and multi satellite systems may be analysed in a more consistent manner with such a diagnostic technique.

6. References

1. Holter, Ø., Altman, C., Roux, A., Perraut, S., Pedersen, A., Pécseli, H., Lybekk, B., Trulsen, J., Korth, A., and Kremser, G. (1995) Characterization of low frequency oscillations at substorm breakup, *J. Geophys. Res.*, *100*, 19,109–19,119.

2. Southwood, D. J. and Saunders M.A. (1985) Curvature coupling of slow and Alfvén MHD waves in a magnetotail field configuration, *Planet. Space Sci.*, *33*, 127–134.

3. Walker, A.D.M. (1987) Theory of magnetospheric standing hydromagnetic waves with large azimuthal wave number, 1. Coupled magnetosonic and Alfvén waves, *J. Geophys. Res.*, *92*, 10,039–10,045.

4. Ohtani, S., Miura, A., and Tamao, T. (1989) Coupling between Alfvèn and slow magnetosonic waves in an inhomogeneous finite-β plasma – I. Coupled equations and physical mechanism, *Planet. Space Sci.*, *37*, 567–577.

5. Chan A.A., Xia, M., and Chen, L. (1994) Anisotropic Alfvén-ballooning modes in Earth's magnetosphere, *J. Geophys. Res.*, *99*, 17,351–17,366.

6. Cheng, C.Z., Chang, T.C., Lin, C.A., and Tsai, W. (1993) Magnetohydrodynamic theory of field line resonances in the magnetosphere, *J. Geophys. Res.* 98, 11,339–11,347.

7. Miura, A., Ohtani, S., and Tamao, T. (1989) Ballooning instability and structure of diamagnetic hydromagnetic waves in a model magnetosphere, *J. Geophys. Res.*, *94*, 15,231–15,242.

8. Grossmann, A., Kronland-Martinet, R., and Morlet, J. (1989) Reading and understanding continuous wavelet transforms, in *Wavelets, Time-Frequency Methods and Phase Space* (Combes, J.M. et al., eds), Springer Verlag.

9. Holter, Ø. (1995) Wavelet analysis of time series, in *Proc. of the Cluster Workshop on Data Analysis Tools*, Braunschweig, Germany 28-30 September 1994 (ESA SP-371) pp. 43-50.

10. Kaiser, G. (1994) *A Friendly Guide to Wavelets*, Birkhäuser Verlag.

11. Rioul, O. and Vetterli, M. (1991) Wavelets and Signal Processing, *IEEE SP MAGAZINE*, October 1991, 14-38.

12. Roux, A., Perraut, S., Robert, P., Morane, A., Pedersen, A., Korth, A., Kremser, G., Aparicio, D., Rodgers, D., and Pellinen, R. (1991) Plasma sheet instability related to the westward traveling surge, *J. Geophys. Res.*, *96*, 17,697–17,714.

SUBSTORMS AND THE INNER MAGNETOSPHERE: ONSET AND INITIAL EXPANSION

N. C. MAYNARD
Mission Research Corporation, Nashua, New Hampshire

G. M. ERICKSON
Boston University, Boston, Massachusetts

W. J. BURKE
Phillips Laboratory, Hanscom Air Force Base, Massachusetts

A. G. YAHNIN
Polar Geophysical Institute, Apatity, Russia

J. C. SAMSON
University of Alberta, Edmonton, Alberta, Canada

G. D. REEVES
Los Alamos National Laboratory, Los Alamos, New Mexico

M. NAKAMURA
University of Tokyo, Tokyo, Japan

V. V. KLIMENKO
Norilsk Observatory, Institute of Solar-Terrestrial Physics, Norilsk, Russia

1. Introduction

There are two competing views of substorm onset: the near-Earth neutral-line (NENL) model [1, 2] and near-Earth current disruption (NECD), the most extensive model for which is given by Lui [3]. While both models allow for reconnection at a near-Earth X-line (NEXL) during the substorm expansion phase, they differ as to the cause of substorm onset. (We use NEXL to refer to the location of reconnection and NENL to refer to the substorm model). In the NENL model substorm onset is caused by onset of fast reconnection near Earth. NEXLs have been experimentally observed to form near 35 R_E [4]. Some other mechanism, distinct from reconnection,

J. Moen et al. (eds.), Polar Cap Boundary Phenomena, 381–392.

operating even closer to Earth ($\sim$6–10 R_E) is assumed responsible for substorm onset in the NECD model. Various mechanisms have been proposed as responsible for substorm onset within the NECD model. These include a cross-field current instability [5], various roles for magnetosphere-ionosphere (M-I) coupling [6, 7], and ballooning [8, 9, 10, 11]. (For more information on the NENL-NECD debate, see Siscoe [12] and Erickson [13].) In defining the onset time, we follow the conventions established over the years from ground-based measurements: brightening and poleward expansion of the equatorward most arc and a sharp negative turning of the magnetic field H or X component, coinciding with strong Pi2 pulsation activity (e. g. [14]).

This paper summarizes results from two recent papers [7, 15], adds new results from the downward current region of the substorm current wedge (SCW), and ties these together into a possible onset and expansion scenario. Based on CRRES and ground-based data, Paper 1 [7] places onset at the inner edge of the plasma sheet, associated with the upward current leg of the SCW and strong M-I coupling. Paper 2 [15] focuses on the high-latitude boundary of the nightside auroral oval, using GEOTAIL and ground-based data, to provide insight as to when, where and how a NEXL influences substorm expansion. Section 2 provides a brief discussion of substorm onset from Paper 1. Section 3 expands our knowledge of the development of the SCW with observations of SCW downward currents at onset. The absence of the onset signatures observed near the downward currents, limits the M-I activity associated with onset to the upward current region. Section 4 synopsizes the results of Paper 2 regarding NEXL activation in the context of Paper 1. In the Discussion we suggest a sequence for substorm onset and expansion, including NEXL activation, that is consistent with NECD morphology and expands the onset scenario of Erickson et al. [16].

2. CRRES Observations Near Times of Substorm Onsets

The electrodynamics of the inner magnetosphere near times of substorm onsets have been investigated in Paper 1 using CRRES measurements of magnetic and electric fields, and energetic electron fluxes, in conjunction with ground-based observations. Six events were studied in detail, spanning the 2100 to 0000 MLT sector and L values from 5 to 7. All events occurred when CRRES was near the inner edges of the plasma sheet and the magnetospheric projection of the Harang discontinuity.

The dawn-to-dusk convection electric field was enhanced during the growth phase, extending into the expansion phase. Significant, low-frequency pulsations were observed, especially after onset, whose amplitudes were larger than the background electric fields. Electric field variations and field-aligned Poynting flux with periods in the Pi-2 range were consistent with bouncing Alfvén waves that provide electromagnetic communication between the ionosphere and magnetosphere. Magnetic signatures of field-aligned current filaments directed away from the ionosphere were seen in 3 events, from the upward leg of the SCW (orbits 540, 535 and 497). Observations in the other events occurred near the magnetic equator where magnetic perturbations from field-aligned current and Poynting flux parallel to **B** vanish.

These CRRES observations revealed several consistent features of onset and early expansion not previously seen: (1) reversal of the convection electric field from dawn-dusk to dusk-dawn provided the first signature related to substorm onset observed at CRRES, (2) energy transfer between the magnetosphere and ionosphere is observed, and the nature of the return of energy from the ionosphere appears to play a critical role in whether and how the substorm develops, (3) "explosive-growth-phase" (EGP) signatures are often observed after onset, and (4), electron injections at CRRES occur at least 5 minutes after onset determined by ground-based measurements.

Relative to the first two features, one or more dusk-to-dawn (eastward) electric field excursions were observed prior to local onset in nearly all the CRRES events near the upward leg of the SCW. Associated with a dusk-dawn excursion is a Poynting flux toward the ionosphere. Sometimes that flux was so small that nothing happened. During orbit 540, Poynting flux, presumably reflected from the ionosphere, returned in phase with electric field oscillations such that onset resulted. The measured Poynting flux into the ionosphere was 780 μW/m^2 at CRRES. Little or no Poynting flux returned following the dusk-dawn excursions of orbit 535, and only a local intensification or "pseudobreakup" occurred. Finally, during the activity of orbit 497, Poynting flux arrived from the ionosphere out of phase with magnetospheric electric field oscillations, and the westward electrojet turned off.

During event 540, ground onset occurred near Dixon at 1938 UT when the H component started to turn sharply negative. CRRES mapped close to Dixon. Dipolarization started at CRRES at 1939 UT. Passage of the upward leg of the SCW and associated EGP signature interrupted the dipolarization trend $\sim$3 minutes after local onset. This EGP signature resulted from stretching of the magnetic field owing to a westward polarization current at the head of an Alfvén wavefront propagating toward the ionosphere. Dipolarization resumed as the reflected wave returned to the vicinity of CRRES [16].

The injection of energetic (1–30 keV) electrons at CRRES always lagged substorm onset. The shortest time between onset and injection occurred during event 540, 5.5 minutes after the dusk-dawn electric field excursion associated with onset, 5 minutes after ground onset, about 4 minutes after

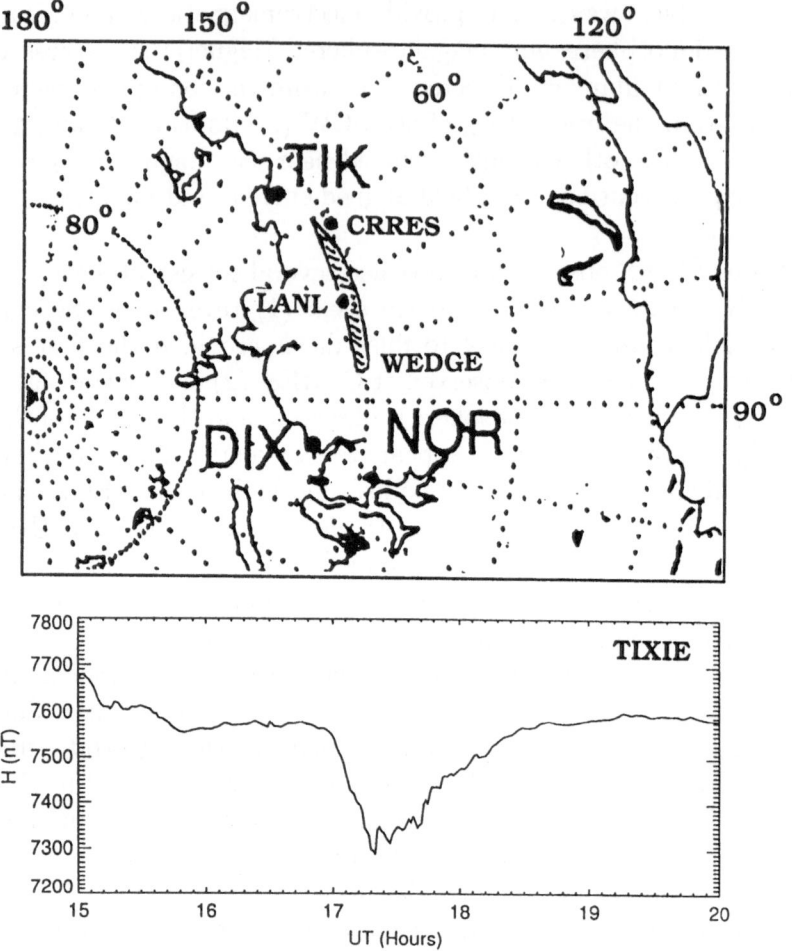

Figure 1. Ionospheric projections of the CRRES and LANL 095 orbital locations and the probable location of the SCW at the 1700 UT onset on 12 February 1991. The Tixie *H* component is shown in the lower panel.

the start of dipolarization, coincident with the arrival of the second wave of reflected Poynting flux, and at the end of an EGP signature.

3. Dynamics near the Downward Leg of the SCW

Having associated onset in Paper 1 with the inner-edge region of the plasma sheet, with the SCW upward current at the Harang discontinuity, and with strong M-I coupling, it is important to consider whether similar signatures occur near the downward wedge current and how the wedge expands. On 12 February 1991, during orbit 491, CRRES was inbound between 0030 and 0100 MLT at 15° magnetic latitude where it could detect magnetic

385

perturbations from field-aligned currents. The CRRES footprint mapped to the south and west of Tixie. Figure 1 shows the H component from the Tixie magnetogram and a map with the locations of Russian magnetometer stations, the footprints of CRRES and LANL 095, and our estimate of the substorm current wedge location based on ground-based magnetometer and optical measurements. Onset occurred at 1700 UT with the first auroral brightening to the east and slightly to the north of Norilsk as seen by all-sky cameras and with Pi2 pulsations. The Tixie H component immediately began decreasing. The Norilsk H component initially remained positive before turning negative at 1707 UT, coincident with enhanced Pi2 amplitudes. This is consistent with the upward current and the Harang discontinuity being to the east of Norilsk at 1700 UT. Significant positive D and Z variations at Tixie (not shown) accompanied the H component change, indicating that the eastward end of the electrojet and the downward wedge current was located to the south and west (see [17]).

The position of the geosynchronous LANL-095 satellite mapped to the middle of the wedge, located ~0.8 hr in MLT to the west and 0.5 R_E in L tailward of CRRES. LANL 095 detected a dispersionless injection of 50-300 keV electrons at 1708:00 UT. An injection arrived at CRRES at 1711:30 UT (see Figure 3 of [18]). Assuming the two populations are the same, the difference in timing is consistent with a 38 km/s eastward drift.

Figure 2 presents CRRES electron spectrograms for trapped and field-aligned number fluxes, covering the energy range from 100 eV to 30 keV, along with electric and magnetic field data. The lower end of the high-energy injection at 1711:30 UT is evident and is marked below in the fields traces. A second low-energy injection of electrons below 1 keV occurred at 1728 UT (also marked in the fields traces). Over the whole interval field-aligned fluxes in the low energy range were detected coming from the northern ionosphere.

The electric and magnetic field data are presented in coordinates which are spacecraft oriented and close to the GSE system. X is along the spin axis which nominally points to within a few degrees to the Sun. The Y and Z axes constitute the spin plane and are nominally parallel to Y_{GSE} and Z_{GSE}. E_Y and E_Z are measured quantities. E_X is calculated using the triaxial magnetometer measurements and the other two electric field components with the assumption that $\mathbf{E} \cdot \mathbf{B} = 0$. We note several characteristics of the data. (1) B_Y begins to decrease just after 1700 UT to a minimum near 1720 UT. This negative B_Y change is interpreted as arising from field-aligned currents into the ionosphere located tailward of CRRES. The amplitude increases faster after the initial injection at 1711:30 UT. The constant amplitude for 6 min after 1716 may result either from steady conditions or from a downward current which moves earthward and envelopes

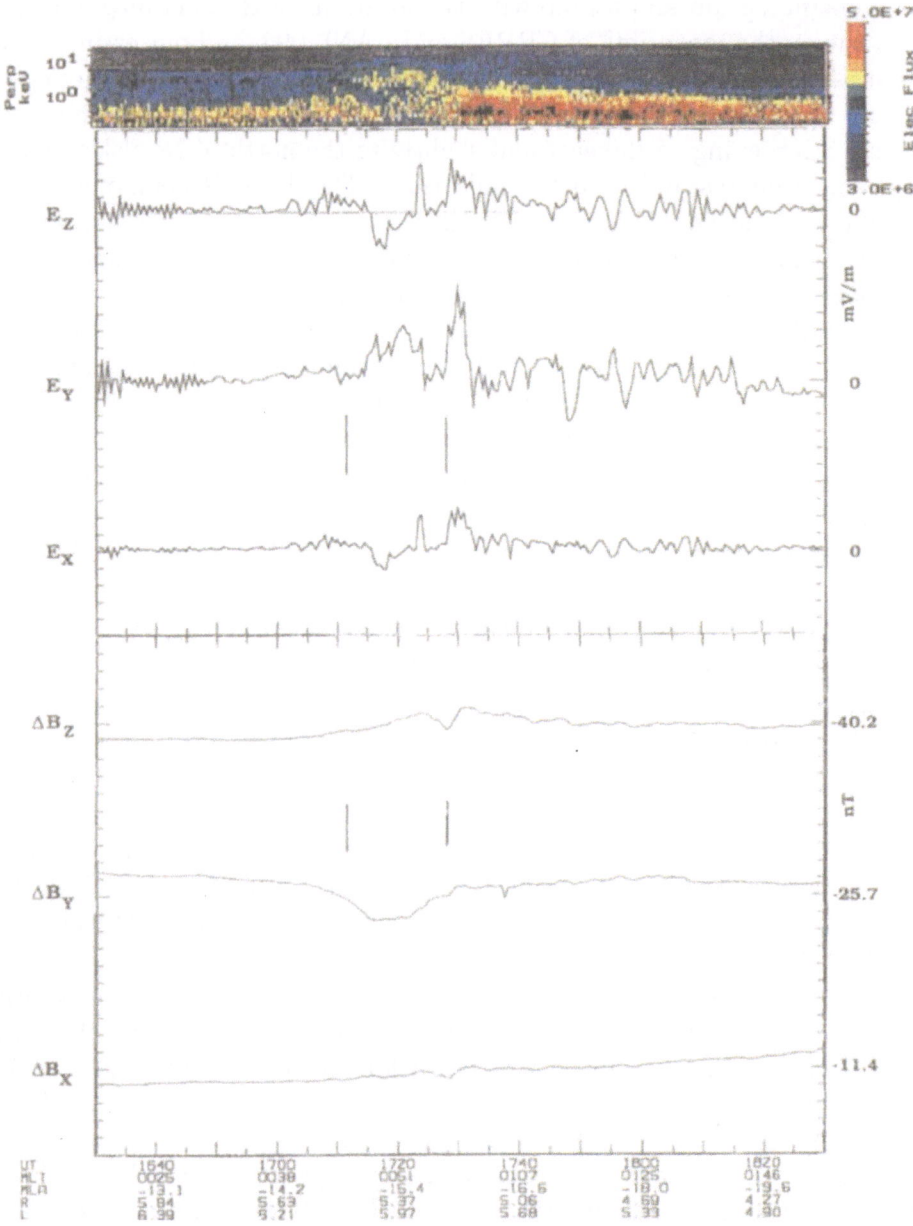

Figure 2. CRRES measurements of trapped electron fluxes, electric and magnetic field during orbit 491. The electric and magnetic field scales are 1 mv/m and 10 nT per division, respectively. The IGRF field has been subtracted from the measured magnetic field.

the spacecraft while still increasing in magnitude. (2) Dipolarization at CR-RES begins with a positive change in B_Z at 1708 UT (at the same time as a high energy injection at LANL 095), after the initial onset as determined from ground data. Note that dipolarization is interrupted starting at 1725 UT, resuming at 1728 UT with a second injection. Activity diminishes after 1731 UT. (3) Unlike observations in the upward current region [7], there is no clear electric field signature associated with onset and no significant field-aligned Poynting flux (not shown). Those features appear to be characteristic of the upward current segment of the wedge only. (4) Two positive enhancements in E_Y are seen at 1715 and 1728 UT. The first is at an increase in the rate of dipolarization and the second coincides with the resumption of dipolarization at the second injection. (5) E_Z is generally positive, with the exception of a wave-like structure between 1716 and 1720 UT, consistent with a projected equatorward ionospheric electric field at a post-midnight location, east of the Harang discontinuity.

In this event, activity expanded eastward and earthward to reach CR-RES. This is confirmed by the late start of dipolarization and a dispersion in the high-energy electrons injected to CRRES. While the downward field-aligned current is remotely sensed almost immediately, it takes 8 min for dipolarization effects to reach CRRES, indicating that it was initially outside of the wedge. The downward current region appears to be more diffuse than the upward wedge current. The source of the later, low-energy injection may be ionospheric electrons that were accelerated upward on field lines tailward of the spacecraft to supply the field-aligned current needed from the dark northern hemisphere ionosphere. Some of these would have to have been trapped near the equator and then convected to CRRES by the large increase in E_Y observed at 1728 UT. We note that the dip in the B_Z signature just before that injection has characteristics similar to the "explosive-growth-phase" signature of Ohtani et al. [19]; however it occurs almost at the end of the dipolarization. It is associated with the second, low-energy injection, which suggests that the EGP signature is more a result of the injection process than related to any onset mechanism.

4. NEXL Activity During Substorm Expansions

Expansion of activity poleward and the relation of this expansion to the projection of the plasma-sheet boundary layer, or the nightside open/closed magnetic field line boundary, provides clues to the relationship of NEXL activity to substorm expansion. In Paper 2 GEOTAIL plasma and field measurements at -90 R_E were compared with extensive ground-based, near-Earth, and geosynchronous measurements to study the relationships between auroral activity and magnetotail dynamics during the expansion

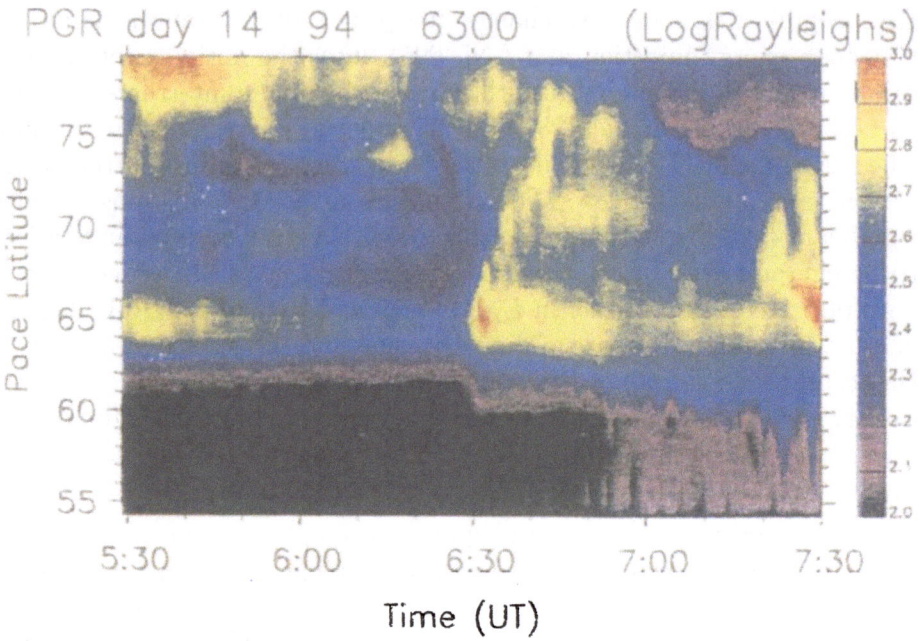

Figure 3. 630.0 nm emissions from CANOPUS on 14 January 1994 (from [15]).

phases of two substorms. The studied intervals are representative of intermittent, moderate activity. The observations are quite complex, and the
reader is referred to Paper 2 [15] for details. Five pertinent findings are
summarized as follows. (1) Reconnection of closed flux at a NEXL can occur intermittently for significant intervals of time (over 40 minutes in one
event) before reconnecting lobe flux. (2) Rapid poleward motions of the
poleward auroral boundary indicate high rates of reconnection of lobe flux,
most likely at a NEXL. (3) Equatorward motion of the poleward, auroral
boundary during periods of increased magnetic activity indicates residual
control of the lobe-flux reconnection by the distant X-line. NEXL activity
is on closed magnetic field lines at such times. (4) Multiple NEXL activations can result in multiple plasmoids without reconnecting lobe flux. (5)
Optical and GEOTAIL observations provide evidence that the onset of the
reconnection at the NEXL occurs after the beginning of activity detected
by ground magnetometers.

Optical measurements from the CANOPUS stations surrounding the
0629 UT onset on 14 January 1994 (Figure 3) help to illustrate these conclusions. Substorm onset occurred at 0629 UT just south of Gillam station
(invariant latitude 67.2°). This onset is clearly seen as an intensification

of 630.0 nm emissions near 65° MLAT (L-shell between 5 and 6) then gradually stepping poleward. The poleward boundary of auroral emissions delineates the last closed field line between the lobe and the plasma-sheet boundary layer. Note the two auroral enhancements at Churchill (69.5°), at 0636 UT and 0720 UT. Because the onset occurred at the low-latitude edge of precipitation while these later activations appeared poleward in a region where there were no previous emissions prior to 0636 UT, and because GEOTAIL observed the resulting plasmoids, Paper 2 interpreted these to have resulted from activations of NEXLs. During the expansion phase, the poleward auroral boundary steadily retreated equatorward while the active aurora expanded poleward. The poleward expanding auroral activity was totally on closed field lines. The poleward boundary retreated equatorward throughout the expansion phase, indicating that the distant X-line controlled lobe reconnection, but reconnection was at a slower rate than transport from the dayside. Reconnection at a NEXL progressed toward the lobe magnetic field lines, but did not engage the lobe until after 0720 UT when the poleward boundary started moving poleward.

5. Discussion

The observational results from Papers 1 and 2 and from Section 3 are consistent with the NECD morphology of substorms. Both CRRES observations and ground photometer measurements (e.g., Figure 3) indicate onset occurs at low L-shells, near the inner edges of the plasma sheet. When CRRES is in the onset MLT sector, we do not observe sharply enhanced inward flow at onset. Rather, a tailward displacement of the plasma initiates activity. Figure 4 provides a possible substorm flow chart which summarizes observations presented in Papers 1 and 2 and above.

Dusk-dawn excursions of the electric field prior to onset are consistent with the suggestion by Erickson and Heineman [10], based on global MHD stability, that onsets result from the tailward displacements of inner-edge flux tubes. Flux tube ballooning is often observed along the inner edges of the plasma sheet early in substorm expansions [20]. During CRRES orbit 540, onset appears to have been spawned from drift-Alfvén ballooning along the ion inner edges [16], as suggested by Roux et al. [8]. Events during orbit 540 clearly exhibit the presence of ballooning prior to onset. As the ballooning amplitude grew to levels where the net electric field reversed from westward to eastward, the equatorial magnetosphere changed from an electromagnetic load to a generator ($\mathbf{j} \cdot \mathbf{E} < 0$). At this time Poynting flux was observed flowing toward the ionosphere. Its reflection in or out of phase with the ballooning oscillations determines if onset or pseudobreakup results.

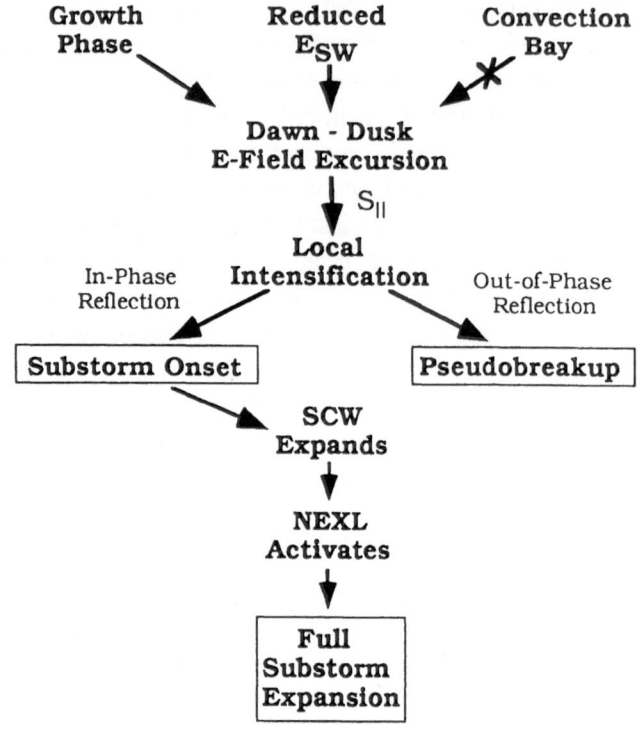

Figure 4. Substorm flow chart (adapted from Erickson *et al* [16]).

If we assume that information about near-geosynchronous onset propagates tailward at an average speed of 10 R_E/min, the signal to activate a NEXL reaches the vicinity of $35R_E$, a probable location for the NEXL (see e.g., Figure 4 of Nishida *et al.* [4]), in ~3 minutes. A similar interval is required for the information of NEXL activation to reach the ionosphere. A time delay of ~6 minutes is consistent with the time delay between the onset and the intensification at Churchill (Figure 3).

Possible consequences of the dusk-dawn electric field excursions being instrumental in substorm onset are the relationships to convection bays (steady magnetospheric convection events) [21], and solar-wind triggers [22] [23]. Outward displacement of inner-edge region flux tubes can result either from ballooning or from a reduction of the negative IMF B_z component or the $|B_y|$. The latter process, which is related to solar-wind triggers, can produce an overshielded configuration whereby the electric field reverses in the inner plasma sheet to move the inner edges tailward from where they started. During a convection bay, the westward convection electric field is enhanced several times over normal growth phase values, and presumably the eastward perturbation electric field resulting from ballooning cannot reverse the total electric field. This would be consistent with the lack of clear substorm phases with these events.

While the results presented here and in Papers 1 and 2 are supportive of NECD phenomenology for substorm onset, we should note that the sequence illustrated in Figure 4 may not be the only way substorms evolve. In situ case studies have been interpreted in support of the NENL model [24]. In other events [5] the cross-field current instability is cited as the primary means for NECD as opposed to our ballooning scenario with a critical role of M-I coupling.

References

1. Russell, C. T., and McPherron, R. L. (1973) The magnetotail and substorms, *Space Sci. Rev.* **15**, 205, 1973.
2. Hones, E. W., Jr. (1977) Substorm processes in the magnetotail: comments on 'On hot tenuous plasmas, fireballs, and boundary layers in Earth's magnetotail' by L. A. Frank, K. L. Ackerson, and R. P. Lepping, *J. Geophys. Res.* **82**, 5633.
3. Lui, A. T. Y. (1991) A synthesis of magnetospheric substorm models, *J. Geophys. Res.* **97**, 1849.
4. Nishida, A., Mukai, T., Yamamoto, T., Saito, Y., and Kokubun, S. (1996) Magnetotail convection in geomagnetically active times 1. Distance to the neutral line, *J. Geomag. Geoelectr.* **48**, 489.
5. Lui, A. T. Y., Mankofsky, A., Chang, C. L., Papadopoulos, K., and Wu, C. S. (1990) A current disruption mechanism in the neutral sheet: a possible trigger for substorm expansions, *Geophys. Res. Lett.* **17**, 745.
6. Kan, J. R. (1993) A global magnetosphere-ionosphere coupling model of substorms, *J. Geophys. Res.* **98**, 17,263.
7. Maynard, N. C., Burke, W. J., Basinska, E. M., Erickson, G. M., Hughes, W. J., Singer, H. J., Yahnin, A. G., Hardy, D. A., and Mozer, F. S. (1996) Dynamics of the inner magnetosphere near times of substorm onset, *J. Geophys. Res.* **96**, 7705.
8. Roux, A., *et al.* (1991) Plasma sheet instability related to the westward traveling surge, *J. Geophys. Res.* **96**, 17,697.
9. Samson, J. C., Wallis, D. D., Hughes, T. J., Creutzberg, F., Ruohoneimi, J. M., and Greenwald, R. A. (1992) Substorm intensifications and field-aligned resonances in the nightside magnetosphere, *J. Geophys. Res.* **97**, 8495.
10. Erickson, G. M, and Heinemann, M. (1992) A mechanism for magnetospheric substorms, in *Substorms 1*, ESA SP-335, ESTEC, Noordwijk, The Netherlands, pp. 587–592.
11. Voronkov, I., Rankin, R., Frycz, P., Tikhonchuk, V. T., and Samson, J. C. (1997) Coupling of shear flow and pressure gradient instabilities, *J. Geophys. Res.* **102**, 9639.
12. Siscoe, G. (1993) Recent activity in substorm research, *Adv. Space Res.* **13**, (4)165.
13. Erickson, G. M. (1995) Substorm theories: united they stand, divided they fall, *Rev. Geophys. Supplement* **33**, 685.
14. Rostoker, G., Akasofu, S.-I., Foster, J., Greenwald, R. A., Kamide, Y., Kawasaki, K., Lui, A. T. Y., McPherron, R. L., and Russell, C. T. (1980) Magnetospheric substorms - definition and signatures, *J. Geophys. Res.* **85**, 1663.
15. Maynard, N. C., *et al.* (1997) GEOTAIL measurements compared with the motions

of high-latitude auroral boundaries during two substorms, *J. Geophys. Res.* **102**, 9553.

16. Erickson, G. M., Burke, W. J., Heinemann, M., Samson, J. C., and Maynard, N. C. (1996) Towards a complete conceptual model of substorm onsets and expansions, in *Substorms 3*, ESA SP-339, ESTEC, Noordwijk, the Netherlands, pp. 423–428.

17. Orr, D., and Cramoysan, M. (1994) The location of substorms using mid-latitude magnetometer arrays, in Kan, J. R., Craven, J. D., and Akasofu, S.-I. (1994) *Substorms 2*, Geophysical Institute, Univ. of Alaska, Fairbanks, pp. 435- 438.

18. Reeves, G. D., Henderson, M. G., McLachlan, P. S., Belian, R. D., Friedel, R. H. W., and Korth, A. (1996) Radial propagation of substorm injections, in *Substorms 3*, ESA SP-339, ESTEC, Noorwijk, The Netherlands, pp. 579–584.

19. Ohtani, S., Takahashi, K., Zanetti, L. J., Potemra, T. A., McEntire, R. W., and Iijima, T. (1992) Initial signatures of magnetic field and energetic particle fluxes at tail reconfiguration: explosive growth phase, *J. Geophys. Res.* **97**, 19,311.

20. Pu Z., *et al.* (1997) MHD drift ballooning instability near the inner edge of the near-Earth plasma sheet and its application to substorm onset, *J. Geophys. Res.* **102**, 14,397.

21. Sergeev, V. A., Pulkkinen, T. I., Pellinen, J., and Tsyganenko, N. A. (1993) Hybrid state of the tail magnetic configuration during steady convection events, *J. Geophys. Res.* **99**, 23,571.

22. Rostoker, G. (1983) Triggering of expansion phase intensifications of magnetic substorms by northward turnings of the interplanetary magnetic field, *J. Geophys. Res.* **88**, 6981.

23. Lyons, L. R. (1995) A new theory for magnetospheric substorms, *J. Geophys. Res.* **100**, 19,069.

24. Baker, D. N., *et al.*. (1993) CDAW-9 analysis of magnetospheric events on May 3, 1986: Event C, *J. Geophys. Res.* **98**, 3815.

OBSERVATIONS OF TAILWARD STREAMING IONS IN THE NEAR-EARTH TAIL DURING A MAGNETOSPHERIC SUBSTORM

STEIN HÅLAND AND STEIN ULLALAND

Department of Physics,
University of Bergen, Allégt. 55, N-5007 Bergen, Norway

TADAYOSHI DOKE

Waseda University,
3-4-1 Okubo, Shinjuku-ku, Tokyo 169, Japan

GEOFF D. REEVES

Los Alamos National Laboratory,
NIS-2, Mail Stop D-436, Los Alamos, New Mexico, 87545, USA

BEREND WILKEN AND QUIGANG ZONG

Max-Planck-Institut für Aeronomie,
Postfach 20, D-37189 Katlenburg-Lindau, Germany

TATSUNDO YAMAMOTO

Institute of Space and Astronautical Science,
3-1-1, Yoshinodai, Sagamihara 229, Japan

Abstract. A tailward moving plasmoid (flux rope) was observed by the GEOTAIL spacecraft at [-22, 5, -3 R_E GSE] after a period with strong geomagnetic activity on November 27, 1995. The structure exhibited classic plasmoid like signatures; bipolar B_z, a strong core field and tailward streaming ions. Measurements from the solar wind and the ground show that the GEOTAIL-observations were correlated with a change in the IMF and a substorm onset. Together, these observations suggest that the flux rope was launched as a result of magnetic reconnection during substorm onset, and that the reconnection site was located earthward of -22 R_E. To our knowledge, observations of similar substorm associated flux rope formation and tailward streaming ions at such distances have not been reported earlier.

J. Moen et al. (eds.), Polar Cap Boundary Phenomena, 393–402.

1. Introduction

A still unresolved question in connection with magnetospheric substorms is the cause of the sudden onset and the location of the initiation region. One of the main paradigms, the Near-Earth Neutral Line (NENL) model [2, 3], suggests that the cross-tail current disruption and the sudden onset of the magnetospheric substorm are associated with magnetic field reconnection in the near-Earth tail (X_{GSM} in the range -10 to -25 R_E). During this reconnection process, it is believed that a part of the plasma sheet eventually disconnects and propagates tailward as a plasma bubble (plasmoid).

Such tailward moving plasmoid structures produce characteristic signatures in the plasma and the magnetic field. Tailward of the reconnection site, these features are often observed as a bipolar signature in the B_Z component of the magnetic field and fast tailward plasma flow [4]. The bipolar B_Z signature is also often accompanied by a strong core field - observed as a strong deflection in the B_Y component [5]. Such observations support the idea of helical shaped, three dimensional flux ropes rather than the conventional plasmoid picture [1].

Although the NENL model predicts magnetic reconnection and plasmoid formation in the near-Earth tail, most such observations come from the mid-tail and distant tail (see e.g. Moldwin and Hughes [6] and references therein). Sergeev *et al.* [7] reported magnetotail reconnection at X= -16 R_E prior to the onset of a small substorm, but observations from such distances are rare. Observations from the AMPTE/IRM satellite (apogee ~20 R_E) during a 8 month period concluded that magnetic neutral lines rarely or never form inside 20 R_E [8].

2. Instrumentation

The primary data basis for this study is the HEP-LD energetic particle spectrometer [9, 10] and the magnetic field experiment (MGF) [11] on board the GEOTAIL spacecraft. Data from the HEP instruments are available either as 3 second resolution (hereafter referred to as Editor A data) or as 24 second low resolution (Editor B). For the investigated event, high resolution Editor A data was available until 1849 UT.

The HEP-LD instrument measures energetic ions within the energy range 30 - 4000 keV. The spatial resolution is 16 azimuthal sectors and 3 polar look directions, covering the entire sphere during one spin period (3 seconds). In plasmoid mode, the HEP-LD energy range has been divided into 10 logarithmic spaced energy channels. The experiment does not distinguish between different ion species or different charge states when run in this mode. Energetic ion measurements are particularly well suited for bulk flow measurements and plasma sheet and boundary layer measurements, where the large gyroradius can be utilized for remote sensing of boundaries [12]. In this paper, we utilize the former feature to determine the bulk flow direction of energetic ions confined in the plasmoid.

In addition to data from the GEOTAIL spacecraft, data from WIND and three geosynchronous LANL spacecraft were available. IMP-8 suffered from data gaps during periods of the investigated event. An overview of the spacecraft locations is shown in Figure 1. Magnetic local time and GSE coordinates are given in brackets for each spacecraft.

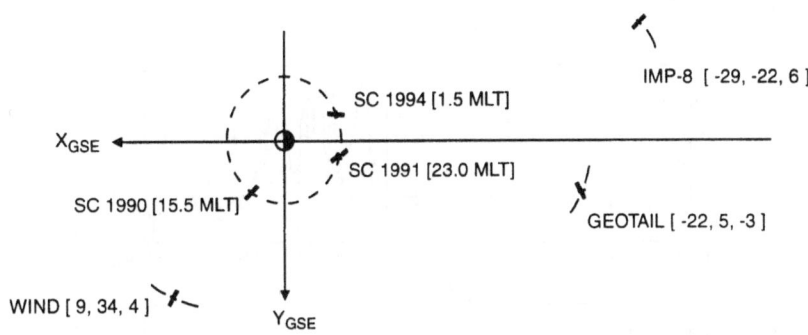

Figure 1. Equatorial projection of spacecraft locations on November, 27 at 1800 UT. GSE or MLT coordinates are given in brackets for each spacecraft.

3. Analysis

3.1. GEOTAIL OBSERVATIONS

On November 27, 1995, at 1800 UT, GEOTAIL was located at [-22, 5, -3 R_E GSE] in the geomagnetic tail. Figure 2 shows the magnetic field measured by GEOTAIL and Figure 3 shows key parameters from the HEP-LD experiment for the time period 1800 - 1830 UT this day. The magnetic field direction and the background flux level show that the spacecraft was inside the plasma sheet prior to the event. The low B_Z component combined with a B_X component near lobe strength prior to the event indicate a highly stretched and compressed field configuration.

The first flux rope signature is seen as a northward excursion of the B_Z component, followed almost immediately by enhanced flux level, best seen in the lower energy channels of the HEP-LD instrument. The fluctuations in B_X are most likely caused by a combination of internal currents in the flux rope and diamagnetic effects due to increased plasma density as the flux rope engulfs the spacecraft [13]. Similar fluctuations are seen in both the B_Y and B_Z components. This suggests a more complex structure than just simple helical loops of the magnetic field. Note that the B_Y perturbation has the same direction as the IMF. In Hughes and Sibeck [1], such correlations were interpreted as interplanetary magnetic field (IMF) penetration into the magnetosphere.

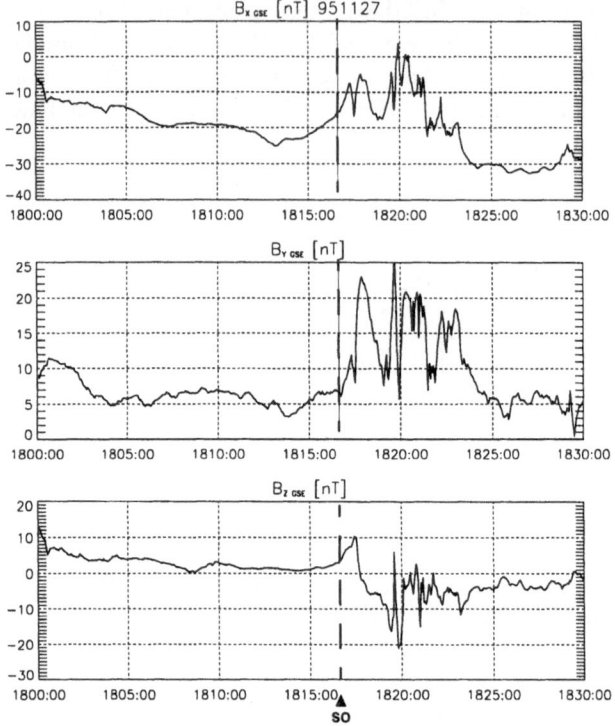

Figure 2. Geotail magnetic field components (3 second resolution). The upper panel (B_X) reveals a general decrease and large fluctuations. This can be attributed to combination diamagnetic effects and currents internal to the fluxrope. The B_Y component, shown in panel 2 demonstrates the 3D nature of the flux rope as suggested by Hughes and Sibeck [1]. The lower panel shows the typical plasmoid signature; a bipolar excursion of B_Z. Substorm onset, marked SO, was determined from ground magnetic pulsations (see Section 3.4).

Figure 3 presents energetic ion measurements from the HEP-LD instrument. The omnidirectional flux (panel 1) shows a peak starting around 1817 UT. The ions arrive simultaneously at all energies - indicating a nondispersive acceleration mechanism and no velocity filter effects. The azimuthal anisotropy, panel 3, shows that the flux enhancement is mainly caused by tailward streaming ions. A closer look at the azimuthal distribution (panel 4) shows that the increase is confined to a narrow azimuthal sector. The minor dawn/dusk asymmetry seen in panel 4 is due to the spacecraft position relative to the aberrant tail.

Together, these observations argue strongly in favour of a particle source initially located earthward of GEOTAIL. Furthermore, the lack of dispersion signatures shows that there were no velocity filters operating, or that the particle source was very close to the spacecraft.

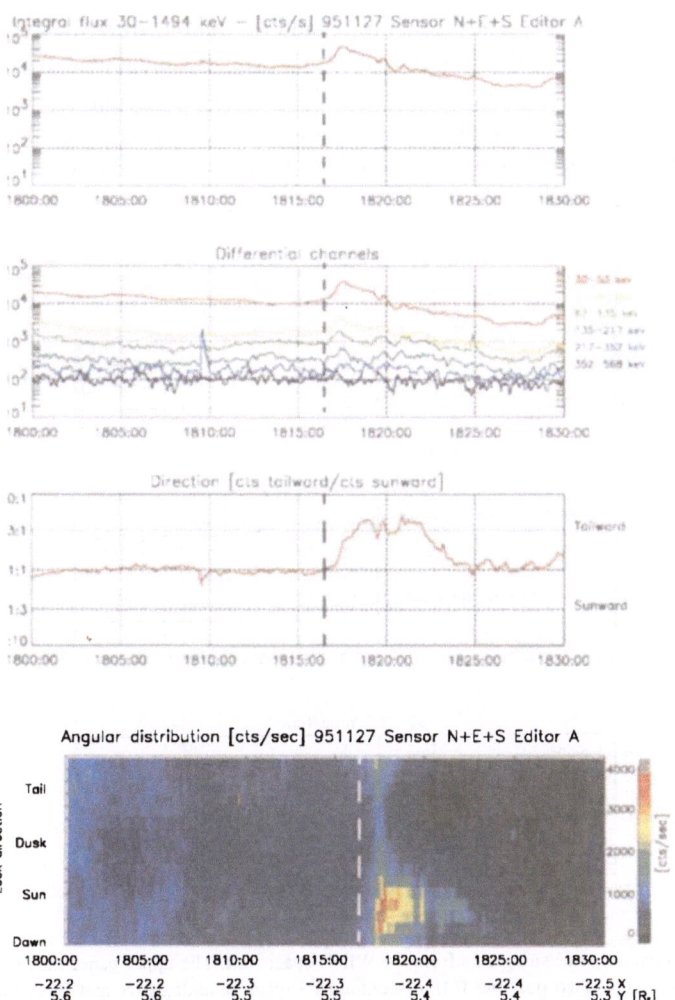

Figure 3. GEOTAIL HEP-LD key parameters (Editor A - 3 second resolution). Upper panel: Omnidirectional integral flux of ions within the energy range 30-1494 keV. Panel 2: Omnidirectional differential flux (6 lower energy channels only). Panel 3: Ratio between tailward streaming and sunward streaming ions. Lower panel: Azimuthal distribution spectrogram. Substorm onset is indicated SO.

3.2. OBSERVATIONS FROM THE INTERPLANETARY MEDIUM

It is well known that geomagnetic activity is closely correlated with solar activity and conditions in the interplanetary medium. In Lyons [14] it is claimed that *all* substorm

398

onsets are triggered by discontinuities in the interplanetary magnetic field, whereas others e.g. Henderson *et al.* [15] and Angelopoulos *et al.* [16] present events without any direct solar wind triggering.

Figure 4 shows interplanetary data (45 second resolution) from the WIND spacecraft, located at [9, 34, 4 R_E GSE]. Two possible IMF trigger candidates are indicated on the plot. The first candidate (A) is observed at 1753 UT. This discontinuity is characterized by a sharp northward turning of the IMF and an increase in the flow speed The second, although less pronounced candidate, marked B, is observed at 1802 UT.

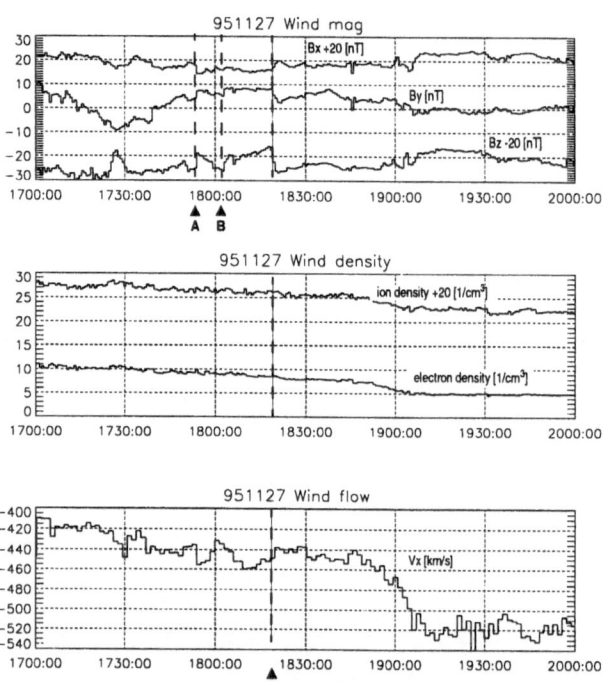

Figure 4. Interplanetary measurements from the WIND spacecraft. The upper panel shows the magnetic field components. Two possible IMF discontinuity trigger candidates (A and B) are indicated. Panel 2 shows the particle density for ions and electrons respectively, and the lower panel shows the X_{GSE} component of the flow speed. Note that artificial offsets have been used to separate the different curves. The time resolution is approximately 45 seconds. SO indicates substorm onset.

Sergeev *et al.* [17] found that the times of triggered substorm onsets depend on the orientation of the IMF discontinuity. On average, they found that substorm onsets occurred ~10 minutes after the first contact of the discontinuity with the bow shock. They found that the time delay was attributed to propagation effects in the magnetosphere.

Using the methods in Sergeev *et al.* [17], we found the total propagation time, including the propagation time between WIND and the bow shock to be 12-14 minutes for both trigger candidates. Substorm onset for the investigated event occurred at 1817 UT, i.e. ~15 minutes after trigger candidate B. Substorm onset may therefore be associated with an IMF discontinuity.

3.3. GEOSYNCHRONOUS OBSERVATIONS

Data from three LANL satellites were available during this event. Two of the satellites, SC 1991-080 and SC 1994-084 were located in the nightside while SC 1990-095 was located in the afternoon sector. Figure 5 shows electron fluxes for these spacecraft. Data from SC 1990-095, located in the afternoon sector shows dispersion signatures starting about 1830 UT. The signatures can be traced back to an injection in the midnight sector around onset.

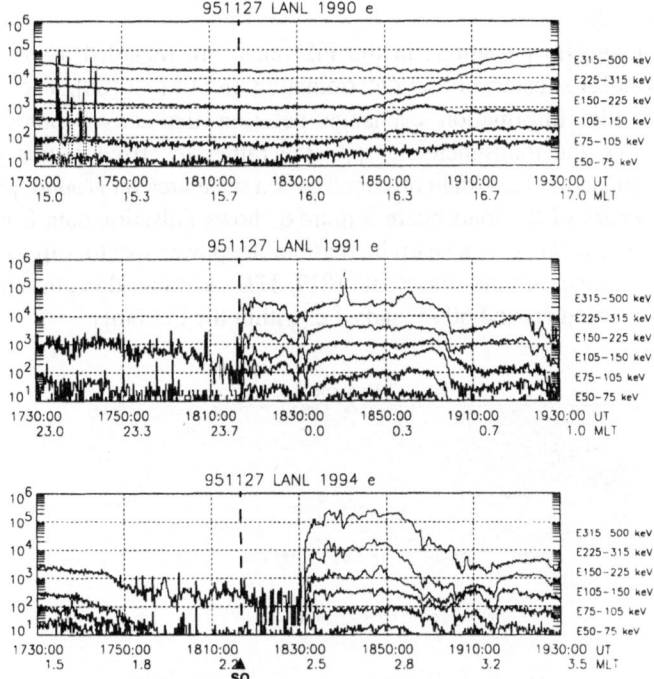

Figure 5. Electron data from the LANL geosynchronous spacecraft. The upper panel shows dispersion signatures from SC 1990-095 located in the afternoon sector. Panel 2 and 3 show data from SC 1991-080 and SC 1994-084 respectively. There are no dispersion in these latter signatures. The injection signature appears about 14 minutes later (~1831 UT) on SC 1994-084. The legends on the right hand side show the energy range for each channel. Substorm onset is indicated SO.

SC 1991-080 was located around 23.0 LT. A dispersionless injection-like signature is observed at 1817:00 UT, i.e. almost simultaneously with the ground based onset signatures. A similar, also dispersionless injection-like signature is observed at SC 1994-

084 about 1831 UT. The 14 minute delay between these spacecraft observations may either be attributed to the expansion of the current wedge, a new substorm intensification or local changes in the plasma sheet configuration.

3.4. GROUND BASED OBSERVATIONS

The field aligned currents associated with substorm expansion and cross tail current disruption are closed in the ionosphere (see e.g. Kaufmann [18]). On ground, these enhanced ionospheric currents will produce perturbations in the magnetic field at stations near the centre of the electrojet. However, the ionospheric currents are often very localized and time dependent [19]. For the investigated event, the observations from the magnetosphere (i.e. GEOTAIL and SC 1991-080) map to the northern part of Russia (~65 degrees geographic longitude) - a longitudinal sector with few available ground stations. The conjungate points on the southern hemisphere map to ~90 degrees geographic longitude.

On ground, magnetic pulsations in the Pi2 range are regarded as one of the most reliable indicators for substorm onset determination and timing [20]. The pulsations can be attributed to the information exchange between the active source area in the magnetosphere and the ionosphere. Despite the rather localized source region in the magnetosphere, pulsations are often observed over a wide area on ground, possibly due to waveguide properties of the ionosphere. Figure 6 shows pulsation data from Sodankyla, Finland, for this event. There is a sharp increase in the power spectra (upper panel) over a wide frequency range around onset at 1817 UT. A time domain analysis shows oscillations with period around 60 seconds, i.e. within the Pi2 range.

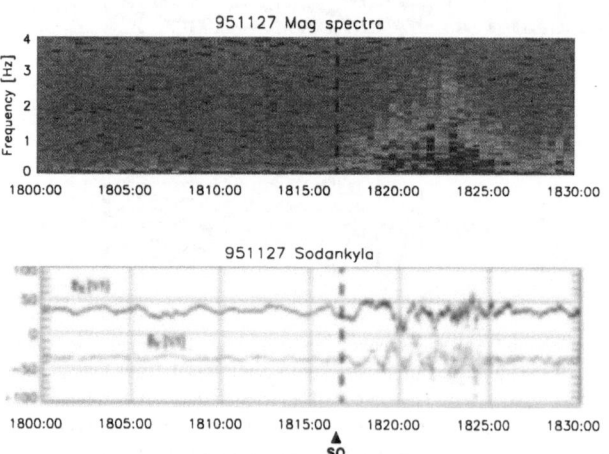

Figure 6. Magnetic pulsation from Sodankyla Geophysical Observatory for November 27, 1995. There is a sharp onset of pulsations over a wide frequency range at 1817 UT. A time domain analysis (lower panel) shows a burst of oscillations with approximately 60 second period, i.e. within the Pi2 range (40-150 second period time). Substorm onset is marked SO.

4. Summary and Discussion

Based on results from the ground based measurements and the GEOTAIL, WIND and LANL spacecraft, we conclude with following points:

* Tailward streaming energetic ions were observed at $X=-22\ R_E$.

* The magnetic field exhibited typical flux rope signatures; bipolar B_Z and a strong core field. The core field had the same direction as the IMF B_Y component.

* Injection-like signatures and drift echo signatures were observed at geosynchronous altitude.

* Magnetic pulsations over a wide frequency range, including the Pi2 range, were observed on ground.

Together, these observations show a substorm development associated with magnetic field reconnection and flux rope formation in the near-Earth magnetotail. Furthermore, observations from the interplanetary medium suggest that the substorm was triggered by a change in the interplanetary magnetic field. To our knowledge, such observations of flux rope formation and tailward streaming ions at distances earthward of -22 R_E have not been reported earlier. The results emphasize the importance of this region of the near-Earth tail for substorm dynamics.

5. Acknowledgements

This work was partially supported by the Norwegian Research Council (NFR) and the German Space Agency (DARA). We acknowledge the use of data from the WIND Solar Wind Experiment (PI: K. Ogilvie), WIND Magnetic Field Experiment, (PI: R. Lepping), WIND 3-D Plasma Analyzer (PI: R. Lin).

Magnetic pulsation data were provided by J. Kultima, Sodankyla Geophysical Observatory. The authors thank M. Jellestad and F. Both for data processing and R. Stadsnes for fruitful discussions and recommendations.

6. References

1. Hughes, W. J. and Sibeck, D. (1987) On the 3-dimensional structure of plasmoids. *Geophysical Research Letters*, 14, 636-639.

2. Hones, E. W. (1976) The magnetotail: its generation and dissipation. In D. J. Williams, editor, *Physics of Solar Planetary Environments*, AGU, Washington D.C. p 559–591.

3. Hones, E. W. (1979) Transient phenomena in the magnetotail and their relation to substorms. *Space Sci. Rev.*, 23, 393–410.

402

4. Richardson, I. G. and Cowley, S. W. H. (1985) Plasmoid associated energetic ion burst in the deep geomagnetic tail: properties of the boundary layer. *J. Geophys. Res.*, 90(A12), 12133–12158.

5. Slavin, J. A., Baker, D. N., Craven, J. D., Elphic, R. C., Fairfield, D. H., Frank, L. A., Galvin, A. B., Hughes, W. J., Manka, R. H., Mitchell, D. G., Richardson, I. G., Sanderson, T. R., Sibeck, D. J., Smith, E. J. and Zwickl, R. D. (1989) CDAW 8 observations of plasmoid signatures in the geomagnetic tail: an assessment. *J. Geophys. Res.*, 94(A11), 15153–15175.

6. Moldwin, M. B., and Hughes, W. J. (1992) On the formation and evolution of plasmoids: a survey of ISEE 3 geotail data. *J. Geophys. Res.*, 97(A12), 19259–19282.

7. Sergeev, V. A., Angelopoulos, V., Mitchell, D. G. and Russell, C. T. (1995) In situ observations of magnetotail reconnection prior to the onset of a small substorm. *J. Geophys. Res.*, 100(A10), 19121–19133.

8. Baumjohann, W., Paschmann, G. and Lühr, H. (1990) Characteristics of high-speed ion flows in the plasma sheet. *J. Geophys. Res.*, 95, 3801–3809.

9. Doke, T., Fujii, M., Fujimoto, M., Fujiki, K., Fukui, T., Gliem, F., Güttler, W., Hasebe, N., Hayashi, T., Ito, T., Itsumi, K., Kashiwagi, T., Kikuchi, J., Kohno, T., Kokubun, S., Livi, S., Maezawa, K., Moriya, H., Munakata, K., Murakami, H., Muraki, Y., Nagoshi, H., Nakamoto, A., Nagata, K., Nishida, A., Rathje, R., Shino, T., Sommer, H., Takashima, T,. Terasawa, T., Ullaland, S., Weiss, W., Wilken, B., Yamamoto,T., Yanagimachi, T., and Yanagita, S. (1994) The energetic particle spectrometer HEP onboard the GEOTAIL spacecraft. *J. Geomagn. Geoelectr.*, 46, 713–733.

10. Wilken, B., Güttler, W., Korth, A., Livi, S., Weiss, W., Gliem, F., Müller, A., Rathje, R., Fritz, T. A., Fennel, J. F., Borg, H., Grande, M., Hall, D., McKenna-Lawlor, S., Sarris, E. T., Tanskanen, P., Ullaland, S., Søraas, F., and Ersland, L. (1993) Rapid: The imaging energetic particle spectrometer on Cluster. In W. R. Burke, editor, *Cluster: mission, payload and supporting activities*, ESA Publication Division, ESTEC, Noordwijk, The Netherlands, p185-218.

11. Kokubun, S., Yamamoto, T., Acuna, M., Hayashi, K., Shiokawa, K., and Kawano, H. (1994) The GEOTAIL Magnetic Field Experiment. *J. Geomagn. Geoelectr.*, 46, 7–21.

12. Richardson, I. G., Cowley, S. W. H., Hones, E. W., and Bame, S. J. (1987) Plasmoid-associated energetic ion bursts in the deep geomagnetic tail: Properties of plasmoid and the postplasmoid plasma sheet. *J. Geophys. Res.*, 92(A12), 9997–10013.

13. Scholer, M., Baker, D. N., Bame, S. J., Baumjohann, W., Gloeckler, G., Ipavich, F. M., Smith, E. J. and Tsurutani, B. T. (1985) Correlated observations of substorm effects in the near-Earth region and the deep magnetotail. *J. Geophys. Res.*, 90, 4021-4026..

14. Lyons. L. R. (1995) A new theory of magnetospheric substorms. *J. Geophys. Res.*, 100, 19069-19081.

15. Henderson, M. G., Reeves, G. D., Belian, R. D., and Murphree, J. S. (1996) Observations of magnetospheric substorms occurring with no apparent solar wind/IMF trigger. *J. Geophys. Res.*, 101, 10773–10791.

16. Angelopoulos, V., Sergeev, V. S., Mozer, F. S., Tsuruda, K., Kokubun, S., Yamamoto, T., Lepping, R., Reeves, G. D., and Friis-Christensen, E (1996) Spontaneous substorm onset during a prolonged period of steady, southward interplanetary magnetic field. *J. Geophys. Res.*, 101(A11), 24583–24598.

17. Sergeev, V. A., Dmitrieva, N. P., and Bakova, E. S. (1986) Triggering of substorm expansion by the IMF directional discontinuities: Time delay analysis. *Planet. Space Sci.*, 34(11), 1109–1118.

18. Kaufmann, R. L. (1987) Substorm currents: Growth phase and onset. *J. Geophys. Res.*, 92, 7471–7486.

19. Rostoker, G., Akasofu, S.-I., Foster, J., Greenwald, R. A., Kamide, Y., Kawasaki, K., Lui, A. T. Y., McPherron, R. L., and Russell, C. T. (1980) Magnetospheric substorms - definitions and signatures. *J. Geophys. Res.*, 85, 1663–1668.

20. Saito, T. (1969) Geomagnetic pulsations. *Space Sci. Rev.*, 10, 319–412.

SPACE WEATHER - THE PRACTICE OF SPACE PHYSICS

N. C. MAYNARD
Mission Research Corporation, Nashua, NH

G. L. SISCOE
Boston University, Boston, MA

1. Introduction

Public awareness of space weather effects on space and terrestrial systems will grow rapidly in the coming years. Evidence of this can be seen in the media response to the two magnetic cloud driven storms in January and April of 1997. On January 11, in the latter stages of the magnetic storm associated with the magnetic cloud event, an AT&T Telstar geosynchronous communication satellite failed, the exact cause of which is still under investigation and may never be known. The casual association of the failure with enhancements in the energetic particle fluxes at geosynchronous orbit during the storm is easy and attracts attention. As a result articles appeared 12 days later in at least eight countries all over the world in newspapers, on radio and on television. The uniqueness of this event was that the SOHO spacecraft of the International Solar Terrestrial Physics (ISTP) Program, located in orbit about the libration point approximately 240 Earth radii in front of the Earth, was able to track the progress from the sun of the coronal mass ejection (CME) and resulting magnetic cloud responsible for the storm. SOHO scientists privately predicted the magnetic activity a day ahead at an ISTP scientific meeting at NASA. When SOHO observed another magnetic cloud approaching Earth in April, a press conference was held by NASA predicting a storm. Because the January event and the satellite failure was fresh in peoples minds, page 1 media articles were written, not about what was likely to happen, but emphasizing worst case effects. A significant storm happened as predicted on April 10, but the results of course did not match the worst case scenario. A subsequent article in a Boston Globe article on April 10, 1997, with the headline "Not much flare to this solar event experts say" presented a more realistic report. Most

J. Moen et al. (eds.), Polar Cap Boundary Phenomena, 403–414.

events will not be "killer" events, but all will produce space weather. The frequency of major events will increase as we approach the next solar cycle maximum.

Just as the medical doctor practices medicine by using the best knowledge of biology and medicine with limited observations and tests to diagnose patients problems and prescribe remedies or predict the course of a disease, some space physicists and technicians in the near future will apply the understanding that is now being developed of storms, substorms and the dynamics of the near-Earth environment to forecast space weather and its impacts on space systems and terrestrial systems. They will practice space physics by using limited observations coupled with numerical modeling to provide space weather forecasts to specific customers. Customers will pay for those forecasts and take actions based on them only when the forecasts are perceived to be accurate enough to derive economic benefit from the possible action. Missed forecasts then cause economic loss from either omission of a needed action or implementation of an unneeded action. For the practice of space physics it is important that the community develops capabilities for sound and accurate forecasts and does not oversell, or undersell, those capabilities.

The following sections provide a cursory overview of space weather effects, a synopsis of models currently existing or under development aimed at specification and forecasting space weather, and a general outlook for the future.

2. Impacts on the near-Earth environment

Technological systems in space and on the Earth's surface are subject to adverse effects from solar driven space weather effects. The increasing deployment of radiation-, current-, and field-sensitive technological systems over the last few decades, the increasing complexity of interlocking components such as those represented by the national electric power grid, and the increasing presence of systems in space combine to make society more vulnerable to solar-terrestrial disturbances [1]. The proceedings of the Solar-Terrestrial Predictions Workshops (e.g. *Solar- Terrestrial Predictions – IV*, [2]) provide source books of effects and attempts to mitigate those effects. A concise summary of effects has been provided by Maynard in a recent review article [3].

Large magnetic storms can occur throughout the 11 year solar cycle, although they are more prevalent near solar maximum. Figure 1 shows the sequence of events leading up to the major magnetic storm of March 24, 1991 [4]. The enhancement of the X-rays is followed by the solar proton enhancement and eventually the magnetic storm. The timing of some of the

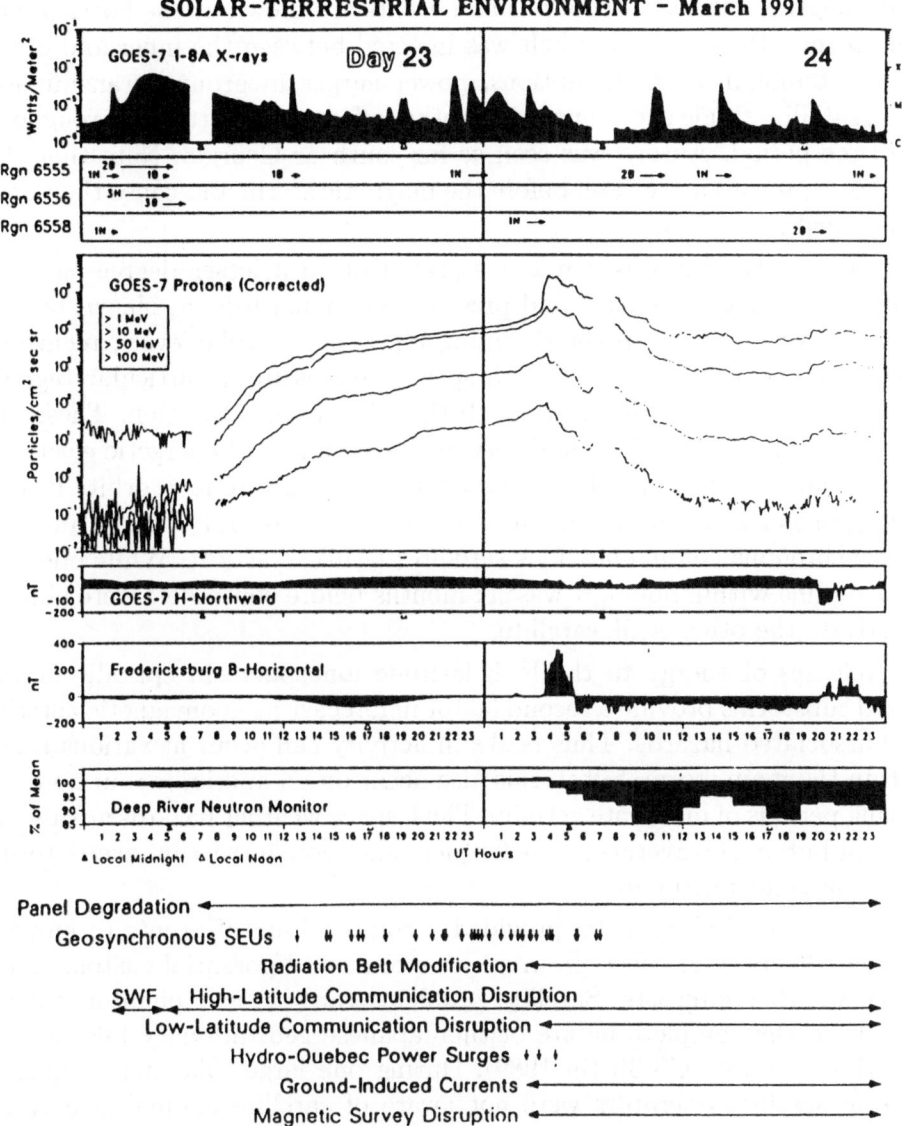

Figure 1. Illustration of the solar and terrestrial environment on 23-24 March 1991. Top: The 1–8 A soft solar X-ray flux as measured by the GOES-7 synchronous orbit spacecraft. Associated solar flares are indicated immediately below. Four selected integral proton flux energies from the GOES-7 spacecraft are shown in the center panel. The GOES-7 magnetometer is shown next followed by the Earth's magnetic field variations observed at Fredricksburg, Virginia. Below this is the cosmic radiation intensity, as measured by the neutron monitor at Deep river, Canada. A summary of spacecraft effects and other operational anomalies is listed at the bottom with the arrows indicating the times of degraded operation [4].

space weather related disruptions and events are shown at the bottom. On this storm a third radiation belt was injected between the inner and outer belts. Communication disruptions, power surges in ground transmission lines, satellite single event upsets and other effects are scattered throughout the time period. Aurora was seen as far south as Georgia. Note that the energetic protons arrive well before the magnetic storm and cause their own space weather effects.

Admittedly, this was a major storm. But to a lesser degree, smaller storms exhibit these effects and present similar hazards, as shown by the January, 1997, event. In the declining phase of the solar cycle, recurrent moderate storms occur when high speed streams from particular regions on the sun rotate by the Earth with the 27 day solar rotation. These are accompanied by several days of very enhanced fluxes of energetic electrons in the outer radiation belt as observed at geosynchronous orbit. It was during one of these events in January, 1994, that control of two Canadian geosynchronous communications satellites (Anik) was lost. While one was back on line within hours, it was six months before controllers were able to reactivate the other Anik satellite.

Releases of energy to the high latitude ionosphere in episodic events called substorms provide a second factor in forecasting geomagnetic activity and associated hazards. Thus peaks in activity can occur at various times within the main storm. Substorms also occur on an aperiodic routine basis during periods of moderate activity. Thus, space weather hazards are always present but are, on average, more frequent and more intense in severe storms and near solar maximum.

Allen and Wilkinson [1] provided a compendium of events to emphasize the diversity of effects from major storms and potential customers for space weather forecasts. Satellite problems range from "phantom" commands to the complete failure of the Japanese geostationary telecommunications satellite CS-3b (in 1989). During one large solar flare sequence certain satellite operators were not aware of satellite anomalies because their communication links to the satellites were inoperable due to the geomagnetic storm itself. On the ground, power grids can be overloaded from geomagnetically induced currents (GIC) caused by changing ionospheric currents . Transformer damage and service disruption result. GIC also enhances corrosion in long pipelines. Ionospheric heating from precipitating magnetospheric energetic particles and resistive closure of field aligned currents (Joule heating) changes the scale height of the neutral atmosphere and increases the drag on low orbiting satellites. This causes changes in normal satellite orbits which place satellite positions outside the expected "windows" of operation. Strong magnetic field gradients from field-aligned currents can also impact the orientation of low-altitude, high-inclination satellites who use the magnetic field as a part of their attitude control. Communication effects from ionospheric disturbances included the disruption of navigation systems, and HF (high frequency band) communication breakdown. Astronauts have experienced irritating "flashes" in their eyes during the Shuttle Atlantis mission to launch the NASA GALILEO satellite (to Jupiter), as energetic protons penetrated the optic nerves. These did not subside until the proton events ended. Radiation sensors on Concorde supersonic jets showed that passengers and crew received a radiation dose equivalent to a chest X-ray.

3. Space weather forecasting 101

The objectives of space weather model development range from post event analysis, through real time specification, through short term forecasts, to long term forecasts. Similarly, possible customer actions range from retrospective diagnostics, through planned or unplanned reaction to an event that is in progress, to planned mitigation in response to a forecast. Short term forecasts are of use to customers only if they are timely, accurate and reliable. Long term forecasts may be reliable and useful without having detailed accuracy.

A key to accurate, short–term (0 to 1 hr) forecasting of new disturbances is continuous, real-time solar wind data. Data taken at the libration point (240 Earth radii upstream), where the Earth's gravitational pull is balanced by that of the Sun, provide a 30 to 50 minute warning of when a shock or disturbance in the solar wind will encounter the magnetosphere. The precise time depends on the solar wind velocity. Solar wind velocity and density and the direction and magnitude of the interplanetary magnetic field (IMF) provide basic inputs to many models and as such, provide an equivalent lead time for the processes modeled. The WIND spacecraft currently provides solar wind and IMF measurements in front of the Earth at varying distances, some in real time, and is useful as a specification and forecast tool. The Advanced Composition Explorer (ACE) satellite will be launched in August, 1997, to become the first operational solar wind monitor at the libration point.

The development of space weather modeling has followed two tracks: (1) engineering a solution by combining empirical models and/or basic knowledge of high latitude electrodynamics with real time data and (2) physics-based approaches. Practicality sometimes mandates that the physics based models be adjusted ad hoc to produce timely results with limited resources. The following subsections provide brief synopses of several engineering and physics based models currently being developed. A look at some of these and other models in the light of predicting substorms can be found in the review of Siscoe and Maynard [5].

3.1. ENGINEERING A SOLUTION

The location of high latitude boundaries and boundary layers provide information relative to the state of energy storage and release in the magnetosphere. For instance the high latitude boundary of the aurora or the open/closed boundary on the nightside maps to the plasmasheet boundary layer. Its dynamics provide a measure of the balance of dayside merging with nightside reconnection and, combined with auroral observations, can provide clues as to whether the near-Earth or distant X-line is in control of reconnection of lobe flux [6] Flow across that boundary is a measure of the reconnection rate at whatever X-line is controlling lobe flux reconnection [7]. Rostocker and Nashi (private communication, 1997) recognized that the latitude of the poleward edge of the aurora provides a proxy relative to the amount of stored energy in the lobes. The equatorward edge provides a proxy of the location of the inner edge of the current sheet in the near tail. The Z component magnetometer variation from a meridian chain of ground magnetometers can be used to determine the poleward border and equatorward edge. They have developed a risk factor based on the difference between the measured lowest latitude and the average latitude of the poleward border for use by the Canadian power companies to assess the probability of a major GIC disturbance.

Another engineering approach for the prediction of GIC was used by Zannetti *et al.* [8] They noticed that increases in the irregularities in the magnetic field measured by the Freja satellite provided a monitor of the location of the electrojet currents. By comparing the lower border observed from real time passes over Scandanavia with the Iijima and Potemra [9] field-aligned current patterns, they would appropriately expand the patterns. Assuming persistence of the activity, they could predict with several hours lead when the current pattern would be expanded over northern US, thus making the northern tier power companies susceptible to experiencing GIC. During the life of Freja these measurements were made available over the internet to interested power companies. One notably successful prediction is given in the Zannetti et al. paper.

Accurate prediction of the auroral electrojet indices, AL, AU, and AE, from solar wind data is possible using an input-state phase space technique of *Vassiliadis* [10]. This technique constructs a phase space out of archival sequences which contain optimized coefficients for local-linear prediction filters. Taking an actual phase space condition, it looks for the nearest archival phase-space neighbors to determine the optimal prediction filter. Substorm dynamics as measured by the AE indices appear to be low dimensional

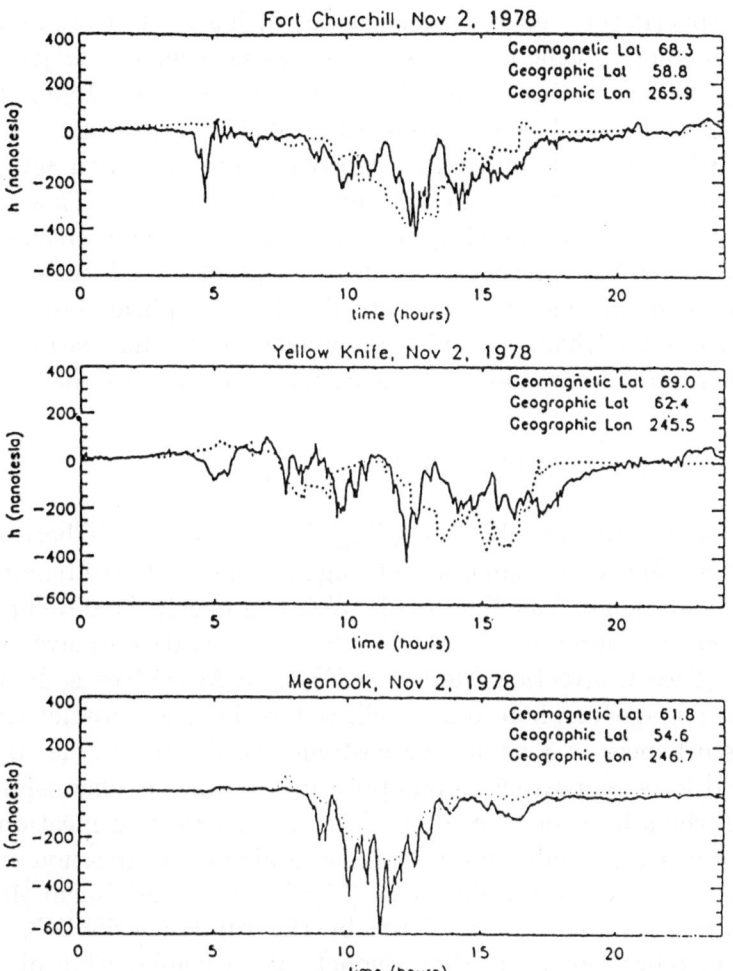

Figure 2. Comparison of predicted ground magnetometer H components with actual measured values for three Canadian stations at different latitudes and longitudes on November 2, 1978. This 45 min forecast was done with ISEE-3 data using the SWIFT algorithm [12].

and therefore amenable to prediction [11]. Maynard *et al.* [12] have developed an engineering hybrid forecast algorithm (Space Weather Ionospheric Forecast Technologies - SWIFT) to predict directly from solar wind data the global-scale high-latitude ionospheric current patterns by combining the Vassiliadis input-state space technique with neural network predictions of other ionospheric parameters from solar wind data drive and contort statistically-based empirical models of the electric field and conductivity into time varying patterns. It predicts the time-varying large-scale features of the ionospheric currents and electric fields with a one to two minute time step. The prediction time is dependent on the solar wind velocity and the distance of the monitoring satellite from the magnetopause. Figure 2 shows examples of end-to-end testing comparing SWIFT predictions of ground magnetic field variations under the ionospheric currents with actual magnetometer measurements at different auroral zone locations. This approach predicts most global scale and large mesoscale processes and forecasts major events with a low false alarm rate. It does not predict small scale structure and localized enhancements in currents. SWIFT is applicable to predicting GIC and associated hazards and communication and navigation hazards that result from increased ionospheric activity and absorption.

3.2. PHYSICS BASED APPROACHES

The magnetosphere-ionosphere coupling (MIC) approach is based on self consistent coupling of ionospheric and magnetospheric electrodynamics. Using the formulation of Vasyliunas [13], Wolf [14] began the development of a general purpose numerical research code which iteratively solves the MIC equations (Rice Convection Model - RCM). RCM evolves a disturbance in the inner magnetosphere quite well, but it does not predict the onset of the disturbance. In response to customer needs to be able to quickly specify and forecast particle populations in the inner magnetosphere, the Rice researchers have developed the Magnetospheric Specification Model (MSM) (specification only) and the Magnetospheric Specification and Forecast Model (MSFM) (see Bales et al. [15]; Freeman and Nagai [16]). The MSM is currently in operational use by the Air Force 55SWX. For the necessity of quick run time, these models use a combination of analytic and prespecified electric field patterns, a library of precomputed magnetic field patterns, and a few pragmatic assumptions to avoid iterating the calculations. Driven with K_P, DMSP data and solar wind data, they have reproduced magnetic storm particle fluxes at geosynchonous orbit to the degree that the average log of the error is 0.5 at energies below 100 kV. These particle fluxes are those responsible for surface charging and associated spacecraft anomalies.

Several global MHD codes have been under development over the past decade. They typically model the magnetosphere from 30 R_E in front to several hundred R_E down the tail. Usually the simulation is stopped at 3.5 R_E altitude and some parameterization representing the ionosphere is applied at that surface. Both Fedder et al.[17] and Raeder [18] have attempted to simulate substorms with some success. The dependability with which these codes capture real situations and the timing of onsets are at present not well understood or controlled (see Hesse [19]); Siscoe and

Maynard [5]). MHD does not encompass the drift physics of the RCM and MSFM; hence the two types of codes are complementary.

A next generation numerical model, the Integrated Space weather prediction Model (ISM), is currently under development at Mission Research Corporation. ISM architecture is based on an MHD core which uses fluid conservation equations for interpenetrating neutral and ion-electron plasma fluids which are coupled at the atomic level by collisional processes and chemistry. ISM solves the complete coupled system from the solar wind to the bottom of the ionosphere. As altitude increases, the neutral fluid effects continuously diminish and the equations evolve to the standard resistive MHD equations. The core MHD is augmented by several modules. A model encompassing RCM drift physics operates within the inner magnetosphere using the MHD calculated fields and returning to the MHD plasma fluid information. A Vlasov module determines field aligned potentials that must be added. High energy particles are treated in a separate module. Regulator modules provide for adjustments to properly simulate magnetopause properties and to switch into a substorm mode if needed. The code is the first to successfully meld MHD and drift physics and promises to be a powerful new tool, This model is targeted for operational use, and will provide comprehensive space weather predictions in all regions of near-Earth space [20] [21].

4. Trends for the future

Forecasting space weather is by no means new. The NOAA Space Environment Services Center (SESC) has been providing space weather information services since 1944. It currently monitors the Sun on a routine basis and issues alerts and warnings based on one to three day predictions of solar flares and on actual observations of the flares. On the basis of solar activity, it provides forecasts of future geomagnetic activity as reflected in the K_P and A_P (the linear version of K_P) indices. Geophysical data are also distributed from a variety of ground based and space based sources. A synopsis of observing and forecasting capabilities is given by Cliffswallow and Hirman [22]. Verification statistics show that geomagnetic forecasting is clearly a problem with only a 44% accuracy for a one day lead time and a 40% accuracy for a five day lead time. Despite accuracy problems SESC currently serves over 25 customer groups, the number of which has been steadily increasing over several decades as the awareness of space weather effects increases.

Increasing the accuracy and reliability of forecasts is essential. Based on the experience of meteorologists and the US National Weather Service, large scale numerical modeling is the key. Since the 1955 and the advent of computers, forecast accuracy has increased steadily from near 40% to better than 95% [23]. The increase has been fueled by increased computer capability which allowed more sophistication of the numerical models. As the accuracy and reliability have improved, more and more people have begun to make economic decisions based on forecasts. Meteorology has also developed a large industry which customizes the National Weather Service forecasts to the specific needs of individual customers.

Space weather forecasting is in its infancy with the first numerical models, including the MSM and MSFM, being implemented at the US Air Force Space Forecast Center. As these tools become operationally available, as advanced numerical models like ISM are developed for the future, and as computer capability continues to grow, we can expect increased accuracy of space weather forecasting to follow a similar curve to that experienced by the meteorologists. The implementation plan is in place for a National Space Weather Service, increasing the space weather forecasting capabilities with SESC being the operational link to civilian customers. It is anticipated that the numerical model forecasts of space weather will also need to be customized to individual users, with industry providing that function. More and more people will be "practicing space physics" for specific customers having special needs.

5. Summary

Space weather affects both space and terrestrial resources. In space the risk is increasing with the number of satellites growing rapidly, as well as the complexity of systems. On the ground the risk is also increasing. Power systems are more intertwined and economics drives toward maximum usage leaving no margin for added currents from GIC. In the next few years we will be approaching solar max with a large increase in storms. There is a definite economic and social need for accurate space weather forecasting.

Forecast tools are being developed. Space weather forecasting will not be taken seriously for large scale economic decision making until the forecasts have been shown to be reliable. Mitigation actions require a concept of operations, and those operational procedures will not be developed until a tool capability that can be relied upon is defined. Numerical modeling holds the long term key to improved accuracy and reliability. We are at the beginning of a new era of space weather "practice". As we "practice" space weather, it is important that we be both creative and realistic.

References

1. Allen, J. H., and Wilkinson, D. C. (1993) Solar-Terrestrial Activity Affecting Systems in Space and on Earth, in J. Hruska, M. A. Shea, D. F. Smart, and G. Heckman

(eds.), *Solar-Terrestrial Predictions - IV, Vol. 1*, NOAA, Boulder, pp. 75–107.

2. Hruska, J., Shea, M. A., Smart, D. F., and Hechman, G. (eds.) (1993) *Solar Terrestrial Predictions - IV*, (in 3 Volumes), NOAA, Boulder.

3. Maynard, N. C. (1995) Space weather prediction, *Rev. Geophys., Supplement* **33**, 547–557.

4. Shea, M. A., and Smart, D. F. (1993) Solar proton events: history, statistics and predictions, in J. Hruska, M. A. Shea, D. F. Smart, and G. Heckman, (eds.), *Solar-Terrestrial Predictions - IV, Vol. 2*, NOAA, Boulder pp. 48–70.

5. Siscoe, G. and Maynard, N. C. (1996) Predicting substorms, in it Proc. Third International Conference on Substorms (ICS-3), ESA SP-389, ESTEC, Noordwjick, pp. 633-638.

6. Maynard, N. C., Burke, W. J., Erickson, G. M., Nakamura, M., Mukai, T., Kokubun, S., Yamamoto, T., Jacobsen, B., Samson, J. C., Weimer, D. R., Reeves, G. D., and Lhr,, H. (1997) Geotail measurements compared with the motions of high-latitude auroral boundaries during two substorms, *J. Geophys. Res.* **102**, 9553-9572.

7. Blanchard, G. T., Lyons, L. R., de la Beaujardiere, O., Doe, R. A., and Mendillo, M. (1996) Measurement of the magnetotail reconnection rate, ıJ. Geophys Res. **101**, 15,265-15,276.

8. Zannetti, L., Potemra, T. A., Anderson, B. J., Erlandson, R. E., Holland, D. B., Acuna, M. H., Kappenman, J., Lesher, R., and Feero, B. (1994) Ionospheric currents correlated with geomagnetic induced currents: Freja magnetic field measurements and the sunburst monitor system, *Geophys. Res. Lett.* **21**, 1867–1870.

9. Iijima, T., and Potemra, T. A. (1978) Large scale characteristics of field-aligned currents associated with substorms, *J. Geophys. Res.* **83**, 599.

10. Vassiliadis, D. (1993) The input-state space approach to the prediction of auroral activity from solar wind variables, in J. Joselyn H. Lundstedt, and J. Trolinger (eds.), *Proceedings of the International Workshop on Artificial Intelligence in Solar Terrestrial Physics*, NOAA, Boulder, pp. 145–151.

11. Sharma, A. S. (1995) Assessing the magnetosphere's nonlinear behavior: its dimension is low, its predictability is high, *Rev. Geophys. Suppl.* **33**, 645–650.

12. Maynard, N. C., Baker, D. N., Freeman, J. W., Jr., Siscoe, G. L., and Vassiliadis, D. (1996) System and method for geomagnetic and ionospheric forecasting, US Patent Application Serial Number 08/583,428, January 5, 1996.

13. Vasyliunas, V. M. (1970) Mathematical models of magnetospheric convection and its coupling to the ionosphere, in B. M. McCormac (ed.), *Particles and Fields in the Magnetsphere*, D. Reidel Publishing Co., pp. 60–71.

14. Wolf, R. A. (1975) Ionosphere-magnetosphere coupling, *Space Sci. Rev.* **17**, 537–562.

15. Bales, B., et al. (1993) Status of the development of the Magnetospheric Specification and Forecast Model, in J. Hruska, M. A. Shea, D. F. Smart, and G. Heckman (eds.), *Solar-Terrestrial Predictions - IV, Vol. 2*, NOAA, Boulder, pp. 467–478.

16. Freeman, J., and Nagai, A. (1993) The Magnetospheric Specification and Forecast Model: Moving from real time to prediction, in J. Hruska, M. A. Shea, D. F. Smart, and G. Heckman (eds.), *Solar-Terrestrial Predictions - IV, Vol. 2*, NOAA, Boulder, pp. 524–539.

17. Fedder, J. A., Slinker, S. P., Lyon, J. G., and Elphinstone, R. D. (1995) Global numerical simulation of the growth phase and expansion phase onset for a substorm observed by Viking, *J. Geophys. Res.* **100**, 19,083–19, 093.

18. Raeder, J. (1994) Global MHD simulations of the dynamics of the magnetosphere: Weak and strong solar wind forcing, in J. R. Kan, J. D. Craven, and S.-I. Akasofu (eds.), *Substorms 2*, Geophysical Institute, Univ. Alaska, College, 561–568.

19. Hesse, M. (1995) The magnetotail's role in magnetospheric dynamics: Engine or exhaust pipe, *Rev. Geophys. Suppl.* **33**, 675–683.

20. White, W. W. (1997) Integrated space weather prediction model (ISM), paper given at the 1997 NSF GEM Workshop, Snowmass, CO, June, 1997.

21. Siscoe, G. (1997) ISM results: Global geometry of magnetopause reconnection for

414

IMF B_Y, paper given at the 1997 NSF GEM Workshop, Snowmass, CO, June 1997.

22. Cliffswallow, W. and Hirman, J. W. (1993) U. S. Space Weather Real-Time Observing and Forecasting Capabilities, in J. Hruska, M. A. Shea, D. F. Smart, and G. Heckman (eds.), *Solar-Terrestrial Predictions – IV, Vol. 1*, NOAA, Boulder, pp. 185–200.

23. McPherson, R. D. (1994) The National Centers for Environmental Prediction: Operational climate, ocean, and weather prediction for the 21st century, *Bull. Amer. Meteor. Soc.* **75**, 363–373.

A SUMMARY OF THE NATO ASI ON POLAR CAP BOUNDARY PHENOMENA

M. LOCKWOOD
Rutherford Appleton Laboratory, Chilton, Oxfordshire,
OX11 0QX, United Kingdom.

S. FUSELIER
Lockheed, Palo Alto Laboratories, USA

A.D.M. WALKER
Space Physics Research Institute, University of Natal,
Durban, South Africa

F. SØRAAS
Univ. of Bergen, Bergen, Norway

Abstract: The polar cap boundary is a subject of central importance to current magnetosphere-ionosphere research and its applications in "space weather" activities. The problems are that it has a number of definitions, and that the most physically meaningful definition (namely the open-closed field line boundary) is very difficult to identify in observations. New understanding of the importance of the structure and dynamics of the boundary region made the time right for a meeting reviewing our knowledge in this area. The Advanced Study Institute (ASI) on Svalbard in June 1997 discussed the boundary on both the dayside and the nightside, mapping magnetically to the dayside magnetopause and to tail plasma sheet/lobe interface, respectively. We held a "brainstorming" session, in which different ideas which arose from the presented papers were discussed and developed, and a summary session, in which session convenors gave a personal view of progress that has been made and problems which still need solving. Both were designed as ways of promoting further discussion. This paper attempts to distil some of the themes that emerged from these discussions.

1. Introduction

The ionospheric polar cap, and therefore its boundary, has a variety of definitions. Confusion and incorrect conclusions result if these definitions are used loosely and/or interchangeably. The term "polar cap boundary" has been used to mean: the open-closed field line boundary; the poleward edge of auroral luminosity (which depends on

J. Moen et al. (eds.), Polar Cap Boundary Phenomena, 415–432.

the wavelength and the threshold flux of the instrument) or particle precipitation (which depends on the species, energy, pitch angle of the particles and on the one-count level of the instrument); the convection reversal boundary (which may not be unique); the poleward boundary of isotropic particle trapping; or the poleward-most field-aligned current sheet. There is even some confusion over the term "open-closed boundary": it is here used to mean the magnetic topology separatrix. This may coincide with particle and wave signatures at all altitudes of the boundary surface, but only if there is no reconnection taking place on such field lines, producing boundary-normal flow and boundary-normal dispersion of features. Boundary-normal flow can be produced by either reconnection when, by definition, flux is transferred across that part of the boundary, or when the boundary is "adiaroic" (i.e. non-reconnecting) but in motion. However, in the case of moving adiaroic boundaries there is no mixing of the plasmas on the two sides of the boundary and all features are present at all altitudes and are convected together in the boundary-normal direction with the boundary itself. On the other hand, reconnection allows the mixing of the two plasma regimes along the newly-reconnected field lines and the particles and waves associated with this will be spatially dispersed in the direction normal to the open-closed boundary (the so-called "velocity filter effect"). In these cases, only hypothetical waves and particles with an infinite speed of field-parallel motion would be able to mark this boundary at all altitudes: real particles and waves (and any changes in their fluxes associated with the boundary) will be displaced downstream away from the boundary. These time-of-flight effects mean that the topological open-closed boundary has no observable signature in these cases and a boundary definition based, for example, on observed particles will depend on their energy, mass, and pitch angle. Here we allow the term "polar cap boundary" to be generic and to apply to all of the above definitions, whilst bearing in mind that there is much experimental uncertainty, temporal variability and theoretical significance in the relationship between the above, more precise definitions.

At the ASI on polar cap boundary phenomena, we discussed both the dayside open-closed boundary, which maps to the magnetopause, and the nightside open-closed boundary, which maps to the lobe/plasma sheet interface in the tail. In both cases, reconnection X-lines can appear in these surfaces and transfer magnetic flux across them in a steady or a time-varying manner and over a variety of spatial scales. Both have associated boundary layers: the low-latitude boundary layer (LLBL), cusp and mantle are associated with the magnetopause boundary, while the plasmasheet boundary layer (PSBL) and velocity-dispersed ion structures (VDIS) is associated with the cross-tail boundary. A number of general scientific themes ran throughout most of the ASI sessions. They were: the outer magnetospheric border - the bow shock, the magnetosheath and the magnetopause; how the interplanetary magnetic field (IMF) orientation and the solar wind control the structure and dynamics of the dayside cusp and cleft regions; magnetic topology and the occurrence and meaning of conjugate phenomena; the scale sizes involved in the magnetospheric boundary regions and their relation to energy transfer; how we observe and interpret boundary signatures, and nightside boundary structure and dynamics, particularly in relation to substorms.

In addition, there were recurrent themes which transcended these scientific areas, such as the importance of modelling, particularly with outputs designed to simulate observations by radars and satellites. In addition, all areas addressed future directions and the opportunities. In the near future these include the NASA Svalbard rocket campaign (described at the ASI by Pfaff), further rocket launches like CAPER from Andøya (a facility described by Bøen) and the launch of the Equator-S satellite. In the longer term are: the planned construction of the second antenna of the EISCAT Svalbard Radar (a facility reviewed by van Eyken) and the Polar Cap Radar, the extension of the SuperDARN network (reviewed by Walker and Villain), greater variety and coverage of other ground-based instruments and the Cluster 2 mission. A special workshop at the ASI, run by Farrugia, looked at the responses of various features to the magnetic cloud event of 6-12 January 1997.

As well as reports and reviews on recent research, the meeting contained some general tutorial lectures, such as that by Southwood on the development of our understanding of the polar cap boundary. Others were: by Meng, on polar cap boundaries seen from space; by Maynard [1], on "space-weather" applications of the research; by Potemra [2], on field-aligned current morphology and magnetospheric expansion and contraction due to solar wind variations, and by Cowley [3], on how radar observations in the vicinity of the polar cap boundary have led to a new, time-dependent view of how ionospheric convection is excited by the motion of the open-closed boundary. This view has been very important, not only in the interpretation of flow patterns, but also because it led to the prediction of cusp ion steps and hence their explanation in terms of pulsed magnetopause reconnection [4].

2. The bow shock, the magnetosheath and the magnetopause boundary layers

During the meeting, there were several discussions concerning modelling of dayside polar cap boundary processes. In particular, there was much debate concerning the magnetosheath plasma that enters the Earth's magnetosphere. Recent work on the Kelvin-Helmholtz (K-H) instability at the magnetopause was presented in which plasma depletion layer was included in the magnetosheath, adjacent to the magnetopause [5]. The net result was that the instability was enhanced because of the depletion layer. Later, Phan presented separate observations which directly related to this modelling effort. He showed a depletion layer well around the flanks of the magnetopause, near the dusk terminator, indicating the importance of this layer for northward IMF (also the very conditions that favour the K-H instability). An improved empirical model of the magnetopause shape was also presented which is applicable over a wider range of parameter space [6] and an analytic model of the response of the dayside magnetopause to solar wind pressure variations [7]. Another analytic model, of the magnetosheath structure, included the magnetic field in a self consistent way [8]: interestingly, the predictions for the magnetosheath were considerably different from those by the gas-dynamic model which is in common use, even though reconnection effects were not yet included.

There was considerable discussion of the source of ions for the cusp and dayside LLBL and how these sources relate to the topology of the field lines. The topology can define which mechanism or mechanisms are at work. It was almost universally agreed that the great predictive power of the open magnetosphere model shows that magnetic reconnection takes place in the magnetopause and cross-tail current sheets and that many observations are explicable with careful and proper application of this paradigm. The current sheet structure is also well reproduced by analytic considerations of time-dependent Petschek reconnection [9]. The main debates at the meeting were about how many of the observed phenomena can be explained by reconnection alone and where and how it takes place. In this respect, there are some important considerations that must be taken into account. For electrons, the two most important considerations are charge neutrality and "counter-streaming" distributions. Charge neutrality is particularly difficult because, while electrons are very mobile, large numbers cannot usually leave a source region without an equal number of ions to maintain charge neutrality (the only exceptions being when the voltage sources caused by reconnection call for a current towards the source). It is also important to note that electrons from multiple sources can be used to balance charge. Thus, the ionosphere can be a source of low energy electrons, which are observed by low altitude spacecraft in the cusp region (and several examples of this were shown in data from the FAST satellite). The second issue for electrons is the use of counter-streaming electrons as an indicator of closed magnetic topology. In fact, two ionospheric sources or mirror points (i.e., a closed magnetic field line) or one ionospheric mirror point with a continuous source of electrons, such as heating of sheath electrons at the magnetopause (i.e., an open magnetic field line) will both produce counter-streaming electron distributions. Several examples of counter-streaming electron distributions in the LLBL were shown at the ASI and were interpreted both as closed and open topologies. Lockwood [10] argued that not only the LLBL but the dayside BPS (boundary plasma sheet) precipitations were on open field lines, the energised ions in these regions being produced by reflection of magnetospheric CPS (central plasma sheet) ions off the magnetopause and/or the interior Alfvén wave launched by the reconnection site. He showed how the precipitations could be modelled and were also consistent with the arrival of the interior wave in the ionosphere and the convection reversal boundary. In this view, the longitudinal width of the cusp is set by the variation of density in the magnetosheath and not by the length of the reconnection X-line. This solves several anomalies, such as the fact that the full transpolar voltage is not seen across the cusp. Observations supporting this were presented by Moen who showed such particles seen by the NOAA-12 satellite, within transient dayside auroral events moving zonally and in a direct which was consistent with the IMF B_y polarity, indicating open field lines. Newell [11], on the other hand, whilst agreeing that there is an open LLBL equatorward of the cusp, argued that there was a distinct second type of LLBL on closed field lines on the dawn and dusk flank of the cusp. The key difference between the two types of LLBL was that only the former type exhibited a low-energy time-of-flight cut off, although both met the previous LLBL criteria.

For ions, an important new consideration for the cusp region is the distinction between magnetosheath and magnetospheric origins of plasma. For ions ~1 keV, this distinction is readily made by composition. Large amounts of alpha particles at energies between 100 V and 1 keV are a strong indicator of a magnetosheath ion population. Above 10 keV, this distinction is more difficult. Fuselier described how the Earth's quasi-parallel bow shock can be a source of ions in this energy range [12] as well as, or even instead of, the magnetosphere. The distinction may not always be made by composition because both sources will produce distributions with ~3-4% alpha concentrations at these energies; however, in strong storms we can also expect considerable fluxes of O^+ ions in the magnetospheric population. The observation of energetic ions in the cusp region was argued by Fritz to show that the turbulent cusp region, with its highly structured and variable diamagnetic effect, was an accelerator, possibly even in a closed magnetic field cusp topology. However, the open model predicts that such ions will be produced by reflection of CPS ions on newly-opened field lines by the magnetopause Alfvén waves and, even above several hundred keV, they could still come from the magnetosheath if an additional acceleration mechanism is present to elevate them to these energies. An important factor which was noted several times at the meeting (e.g. by Bösinger and by Fritz) was the evaluation of ion drift orbits in field models to allow for gradient-B and curvature drifts which may become especially distorted in cusp-like field configurations and confuse the use of particle observations to define the field -line topology. For example, magnetospheric like ions seen overlapping magnetosheath plasma on the dusk side, or a similar overlap on the dawnside with magnetosphere electrons, could result from non-MHD drifts onto open field lines and not signify a closed LLBL topology as is often assumed.

3. IMF and solar wind control of the cusp/cleft

At the ASI, there were several reviews of dayside cusp-cleft auroral morphology. A new feature of the IMF control of the aurora was presented by Sandholt, namely the existence of three distinct auroral types [13]. For southward IMF (clock angles in the Z-Y plane of 90-180°), a single peak in the luminosity-latitude profile of 630 nm emission was defined at low latitudes; for strongly northward IMF (clock angles of roughly 0-55°) there is a single peak at high latitudes and for intermediate cases (roughly 55-90°) both peaks were seen. This was interpreted as low-latitude reconnection, lobe reconnection and both taking place simultaneously, respectively. The topology of magnetic reconnection in the cusp for evolving IMF orientation during a northward turning was also studied by Lockwood, who pointed out that lobe reconnection occurred preferentially in the summer hemisphere, where the dipole axis is tilted towards the sun. In addition, the X-component also had a role in determining the likelihood of lobe reconnection taking place in a given hemisphere. These ideas were shown to be consistent with the observed evolution of the aurora and suggest that field line re-closure by reconnection at both lobe sites can take sometimes place: one lobe site generates "over-draped lobe" field lines from "old" open field lines (opened by a prior period of southward IMF) and the other later removes them by turning them into closed field lines. These concepts stress the need to measure cusp phenomena both in winter and in summer, because they may differ significantly. Flow channels which are co-incident with dayside auroral transients have been known for some time [14] but have mainly been reported for southward IMF: Nielsen presented some observations of a channel at very high latitudes seen by Sondrestrom and the Freja satellite which could be a high-latitude equivalent during northward IMF. Maynard [15] also discussed convection during northward IMF seen by the Polar satellite and concluded that the reconnection was patchy and sporadic giving time-varying convection. It should be noted that under these circumstances, the difference between two distorted cells and four cells in the NBz convection pattern becomes somewhat nominal because the patterns change in less than the period it would take a field line to flow all the way round any one instantaneous closed flow loop: the coalescing of two flow cells, even across field line topology boundaries, can occur because the boundaries move [2].

At the previous ASI in this series, there had been much debate about whether or not poleward-moving auroral transient events represented the ionospheric signatures of pulsed magnetopause reconnection. However, in the four years since then, the evidence that they were indeed caused this way has steadily accumulated. Some of this evidence was reviewed at the meeting, including the predominance of these events during southward IMF and the control of the zonal motion of the events by IMF B_y, in a manner consistent with the curvature force on newly-opened field lines. This association was further strengthened at this meeting by an analysis of the motion and field-aligned current structure of the events [16]. The strongest possible evidence that the optical events are signatures of reconnection pulses was provided by Farrugia and co-workers [17], who showed the first examples of cusp ion steps seen in close association with the poleward moving auroral transients. Cusp ion steps were a prediction of the Cowley-Lockwood model of ionospheric flow excitation [3], which had been the subject of considerable and heated debate at the previous meeting [18]. The model had been used to show how pulsed reconnection should generate steps in the cusp ion dispersion on the boundaries of poleward-moving events in the ionosphere [4]. This association had been seen before with poleward-moving regions of heated ionospheric electron gas seen by the EISCAT radar [19]. However, these observations were at equinox and so observations of the auroral transients themselves (expected to be caused by just such electron gas heating) had not been possible. Examples of cusp ion steps, seen by the Hydra instrument on Polar in down-going ions, were presented by Lockwood and co-workers who had fitted them with a model based on pulsed reconnection: as a test, they showed that this also fitted very well the stepped appearance of ions seen simultaneously moving upward, having been injected earlier and mirrored beneath the satellite [20]. Results like these mean that the cusp ion steps and poleward moving events are now generally considered to be much more direct evidence for reconnection pulses than magnetopause flux transfer event signatures (FTEs). There is still some debate about the interpretation of such FTE signatures; however, in the near future the Equator-S satellite should allow a proper analysis of the association of these ionospheric and magnetopause signatures.

A new and important feature of the flow excitation was discussed by Rodger, namely the possibility that super-Alfvénic electrons drive flow before the arrival of the interior Alfvén wave [21]. This allows the flow to ramp up linearly and be seen well equatorward of where the cusp precipitation first arrives. Examples were shown in a case exhibiting poleward-moving events (seen in HF radar data) and cusp ion steps (seen in a longitudinal satellite pass, giving a saw-tooth dispersion, as has modelled for pulsed reconnection over an extended X-line [22]). Recent studies reported by Lester

[23] show that the poleward-moving optical transients are indeed associated with these flow transients seen by the HF radar. The concept that there is a large delay between the opening of the field line and the arrival of the cusp electron precipitation was consistent with the cusp ion modelling presented by Lockwood (with maintenance of quasi-neutrality) for a low-latitude reconnection site and with examples of poleward-moving 630 nm transients presented by Smith. These events were seen poleward of a region of 557.7 nm emission (caused by an enhanced ring current following substorm activity), the poleward edge of which eroded equatorward prior to the onset of each 630 nm transient. The delay between opening the field line (first seen as a loss of energetic magnetospheric electrons and the steady 557.7 nm aurora they caused in this case) and the arrival of cusp precipitation could be seen as a latitudinal gap between the 557.7 nm and 630 nm emissions. Given that the flight time of the energetic magnetospheric electrons giving the 557.7 nm light is short (so their cross-field convective drift distance is very small), these observations are unique in directly monitoring the dayside polar cap boundary: the periodic erosion of that boundary was also consistent with the pulsed reconnection theory.

Poleward-moving events with a more continuous ion dispersion characteristics would be a signature of a different class of event, called "poleward-progressing convection disturbances", examples of which were discussed by Stauning [24]. These are explained in terms of steady reconnection with changes in the IMF B_y component and consequent changes in the curvature force on newly-opened field lines. However, reconnection pulses cannot be eliminated in many such cases: if the pulses are closer together than the time constants for flow excitation and decay then the poleward flow and progression of events would still be steady [3, 4], thus it is possible that some of these events are also associated with pulsed reconnection, possibly driven by the associated IMF B_y changes. A third class of events seen in the vicinity of the cusp/cleft is Travelling Convection Vortices (TCVs). Ernstrøm showed that the resolution of procedures like AMIE, in both time and space, was now adequate to resolve the larger examples of these events. From such an analysis, Moretto showed that these events may not actually travel as had previously been thought, but the variations seen could arise from transient enhancements of static current systems in response to solar wind pressure changes. Engebretson also pointed out that there were at least two other types of response to sudden changes in the solar wind.

Observations of polar cap patches were also widely discussed at the meeting, their formation being intimately linked with dayside polar cap boundary phenomena and IMF orientation, as reviewed by Carlson [25] and McEwen [26]. Anderson described the theory and model simulations whereby such patches are carved off the region of photo-ionised plasma on the dayside, by time-varying convection. In these simulations, the patches sometimes persisted and returned towards the dayside in the auroral zones. The way that the electric field pattern changes have been imposed in such simulations has been argued to be incorrect [27] because they do not make any consideration of the topology of the magnetic field lines and the effects of moving polar cap boundaries [3]. However the consequences that this has for patch production have not yet been investigated quantitatively. On the other hand, Rodger [28] invoked additional effects due to production (by soft precipitation) and loss (by enhanced electric fields) in the cusp itself. He pointed out that there was a terminology problem in this area, in that some workers referred to a distorted but continuous "tongue" of ionisation as patches, whereas others did not.

4. Conjugacy

Because of its implications for field-line topology, there was great interest in the conjugacy of phenomena at high latitudes. Yamagishi reviewed conjugate observations of auroras using TV cameras and imaging riometers [29]. The scanning photometer observations of substorms showed that break-up occurred at the conjugate latitude. The stepwise expansion in north and south were different and the small scale features were independent. There was a large longitude displacement for conjugate aurorae, whose sense depended on the sense of IMF B_y. While there was some agreement with predictions of the Tsyganenko model., it was fairly weak: given that there is lobe flux but no genuine and realistic open flux in that model (i.e. which threads the magnetopause), agreement at the higher latitudes would not be expected. A statistical study of Pc1-2 (100-600 mHz) ULF waves using one year's data from the MACCS (Magnetometer Array for Cusp and Cleft Studies) and two nominally conjugate Antarctic Automated Geophysical Observatories (AGOs) indicated that the plasma mantle may be a source of wide band waves observed on the ground [30]. When the IMF turns northward, this wave source is switched off. Seasonal and diurnal variations of dipole tilt angle cause inter-hemispherical asymmetries that may be responsible for statistical disparities between north and south.

Walker [31] reviewed the current status of the SuperDARN network of HF radars. The operational radars now cover a significant fraction of the polar cap, cusp, and auroral regions in the northern and southern hemispheres. The radars now allow conjugate observations of vector convection velocities over a large area. He presented preliminary results of a study, in which the IMF conditions were extremely quiet for an extended period. A closed convection cell was observed at 21-03 MLT, centred on magnetic latitude -70°. Equatorward of this cell strong westward flow bursts were observed with electric fields exceeding 100 mV m^{-1} in a region 200 km wide. Similar behaviour was observed in the northern polar cap.

Correlations of geomagnetic observations at South Pole and Iqaluit (formerly Frobisher Bay) were also reviewed a linear correlation study of the H-component in 1986 and 1995 [32]. A coherence analysis of the data showed that during quiet times same-hemisphere coherences were larger than conjugate coherences. A Pc5 band at cusp latitudes indicated closed field lines during local day. The interval of dayside conjugate coherence in the Pc5 band was more extended in local time during northern winter than during southern winter.

A most important insight was not so much what has been achieved, but rather the potential now available for conjugate studies. Instrumentation such as magnetometer arrays, imaging riometers, AGOs, and the SuperDARN radar chain are now widely deployed to observe both southern and northern polar cap regions. The capacity exists to distinguish space and time over large region and the opportunity exists to clarify the physical importance of a number of phenomena in which the observations in each hemisphere are affected by conjugate effects. Conjugate studies are important only if they add to our understanding of the physics. Examples are the inherent asymmetries in convection, and the polarisation of pulsations. Boundary mapping in space and time needs to be understood on a global scale. Many of these topics will be understood by exploiting the full range of instruments available, both in space and on the ground. Achieving this co-operation is as much a human as a technical problem.

5. Scale sizes and energy transfer

The scale sizes involved in dayside arcs were considered by Smith, who looked at factors such as viewing geometry and excited state lifetimes in the presence of neutral thermospheric winds. He concluded that 630 nm transients were not as large as they appeared in auroral images, unless the dimensions were measured near zenith. On the other hand, they also often extended beyond one or both edge of the field of view of a camera, making determination of their full size difficult. Optical and EISCAT radar observations were presented by Lanchester, showing that the majority of energy transfer in nightside arcs was in very narrow filaments (of order 100 m wide) embedded in broader arc regions [33]. The mechanism invoked was a shear in the flow and magnetic field giving a localised acceleration in a field-aligned current and was reproduced by a numerical model. The extreme spatial structure often present was stressed by the FAST data shown by Pfaff, which showed that narrow spike-like electric fields, wave bursts and alternate regions of upward and downward ion and electron motions. This structure had great implications of previous lower-resolution observations that found such features to be co-incident, rather than adjacent. Andersson and Kjus presented observations of, respectively, the meso-scale electron and ion structures and lower-hybrid waves seen by the Freja satellite.

Yamauchi [34] looked at field-aligned currents on a variety of scale sizes. He noted that the large-scale region 1-cusp currents are larger than the meso-scale currents seen in the cusp and argued that this eliminates the reconnection pulse concept of cusp structure. However, this argument assumes that the reconnection pulse model requires the large-scale currents to be the aggregated sum of all the mesoscale currents on the edge of each event. This is not the interpretation of those who use the model, who consider the large-scale currents to form as a response to the total amount of newly-opened flux, whether that flux be opened by steady reconnection at a lower rate or by short pulses of a higher rate. Thus apart from the weaker currents on the boundaries of the patches inside the cusp (associated with differences in their motion because they are at different stages of their evolution into the tail) the bulk of the current will be on the edges of the cusp (i.e. the region 1-cusp current pair) and almost the same as for steady-state.

It was agreed that assimilating the range of phenomena over the range of scale sizes now being observed in the magnetosphere (from the 100-m wide bunched electrons in arcs to the several R_E of reconnection X-lines and MHD phenomena) was a major challenge for future years.

6. Observing and interpreting boundaries

The conference discussed the inter-calibration of signatures near the polar cap boundary from satellites and from a variety of ground observations. One important new way to identify the cusp has been the use of the width of the backscatter spectrum seen by HF radars: this was reviewed by Lester [23] and Villain. These were compared with simultaneous optical and satellite data to show that there was indeed an association with cusp precipitation and aurorae but it was not always a straightforward one. However, there was, as discussed above in section 2, general agreement that the LLBL was open at least near noon, if not nearer the flanks as well. Thus the noon open-closed boundary is equatorward of both the cusp and the LLBL, and is not at the boundary of these two regions of HF backsctatter, as has often been assumed. The clearest identification of a dayside boundary was the periodically-eroding poleward-edge of enhanced 557.7 nm emission reported by Smith (discussed above). This was only possible in a period when the plasma sheet was exceptionally dense because of prior substorm activity. The importance of defining this boundary, its motion and orientation is that without such knowledge the ionospheric flow across it (i.e. the reconnection rate) cannot be properly measured and inferring the wrong topology of the field lines will lead to incorrect conclusions about the mechanisms involved.

As discussed in section 3, the consistency of the various poleward-moving features seen just poleward of this dayside open-closed boundary, strongly supports the pulsed reconnection concept. These features: include HF radar backscatter intensity patches and flow channels (of sense depending on IMF B_y), IS radar observations of flows and electron and ion temperature enhancements, 630 nm and 557.7 nm transients, magnetometer observations and steps in the cusp ion precipitation. In addition, Moen reported observations of pulses in the proton aurora in association with these features. Walker [35] presented a new method which can detect the effects of the cusp precipitation, and its structuring by reconnection pulses, namely ionospheric tomography using a network of total electron content monitors. Berry showed that the technique can also detect auroral arcs away from noon. Woch compared the precipitation signatures seen at middle and low altitudes. In particular, he drew attention to dispersed energetic ion signatures equatorward of the cusp. These have been discussed in terms of spatial structures, but Hall [36] demonstrated that some of them were associated with nightside injection events at geosynchronous orbit. To confirm this, examples where the dispersion had the opposite sense when the satellite moved in the opposite direction were presented.

At the nightside polar cap boundary, a situation highly analogous to the dayside is found with: dispersed ions (the VDIS - velocity dispersed ion structures); equatorward-drifting arcs; structured field-aligned currents; and an open-closed boundary that lies upstream (here poleward) of all these signatures. The major difference is that the reconnection site can lie considerably further away from the Earth than on the dayside, making the spatial separations of the velocity filter effect much greater. A uniquely clear identification of the nightside open-closed boundary was presented by Maezawa

[37] using data from the Akebono satellite. In this case, the IMF B_z and B_x were as required to give solar wind strahl electrons throughout the polar cap. The dispersed disappearance of these strahl electrons (starting with the highest energies) shows the field line closure and this dispersion is extended in velocity space to the VDIS, confirming them also to be on relatively recently-closed field lines, as has recently been predicted for a model of the PSBL [38]. The inferred reconnection site is of order 100 R_E down tail. What is startling about this example is how far poleward (about 2°) of the VDIS, or any major plasma sheet electron precipitation structure, this open closed boundary was found. This is very significant, both in terms of our interpretation of observations during substorms and for the size of the open field line region. Such a gap corresponds to a very large amount of magnetic flux. It implies that the open closed boundary can be well poleward of the poleward boundary of aurora or particle precipitation and that taking them to be the open-closed boundary seriously overestimates the true amount of open flux and the reconnection voltage differences associated with polar cap expansion and contraction. For example, the poleward expansions of substorms as far as the centre of the polar cap, as in the cases presented by McEwen [26], may not signify quite such massive field line closure as would have otherwise been inferred.

7. Nightside boundaries in time and space

Stadsnes described new observations from X-ray and ultraviolet imagers on the Polar satellite, showing precipitation patterns for various substorm conditions. The X-ray images were made by the PIXIE camera, the first two-dimensional X-ray camera flown on a satellite, which measured the electron-produced X-rays greater than 2 keV. The simultaneous ultraviolet images by the UVI camera on Polar measured electron precipitation regions at even lower energies. Observations were shown from three types of magnetospheric conditions: a westward travelling surge; a substorm expansion, and during the passage of a coronal mass ejection. It is expected that analysis of the multiwavelength observations will provide an important understanding of the electron precipitation regions as X-ray + UV images can separately show the precipitation regions for low-energy and high-energy electrons. The interpretation of global images at various wavelengths was reviewed by Brittachner who also made comparisons with in-situ particle data.

Søraas showed measurements of the precipitation of ions from the ring current during magnetic storms. The proton precipitation is very narrow in latitude extent and occurs at the plasmapause or at regions with a large gradient in the cold plasma density. A subauroral red (SAR) arc is often visible in the same region. In the storm main phase, the ion enhancement is accompanied by relativistic electron precipitation. The particle precipitation is thought to be caused by wave particle interaction, the ring current protons are unstable and the electrons are practically scattered by the ion cyclotron waves. The processes responsible for scattering the ions, and their implications for the decay of the ring current, are not yet understood. Fritz [39] reviewed measurements of the electron isotropic flux boundary, which has been used as an indicator of the boundary between open and closed field lines. He described a model developed by Seergeev and Tsyganenko that the electrons would not be simply trapped if the field line radius of curvature is greatly smaller than the electron gyroradius. Fritz tested this model with new data from POLAR and NOAA-12 and found good agreement. Bösinger also presented a study of these boundaries [40] using solar electrons during an SPE.

Statistical patterns of ion precipitation along the polar cap boundary during various IMF orientations, using observations from Akebono observations during 1989-1993, were presented by Asai. Soft ion precipitation is observed on the dawn side or dusk side along the polar cap boundary, where the convection flows are directed sunward. Such a feature was explained by Lockwood as being between the open-closed boundary and the point where the internal Alfvén wave arrives. Maezawa presented a statistical survey of particle and field observations magnetospheric tail by the Geotail satellite. the lobes and plasma sheet were shown to twist to up to a full 90° from the ecliptic plane in response to the IMF B_y component, consistent with the tension force on open field lines. In addition, fluxes of solar wind strahl electrons within the lobe were seen precipitating on the lobe boundary but precipitating and returning (after mirroring) in the central lobe, again consistent with a long tail of open flux. Saito presented wave observations from Geotail in the near-Earth tail and was able to associate them with Pc5 pulsations seen on the ground. This implies that such pulsations can be global cavity modes on open field lines and do not signify a closed topology.

There was much discussion about the location of the open closed boundary during substorms. This debate is central to understanding how substorm onset is initiated. Samson reviewed observations by meridian scanning photometers, which place substorm onset on closed field lines at the equatorward-most arc and at the inner edge of the cross-tail current. He invoked a near-Earth current sheet instability (such as ballooning) as a cause of onset. On the other hand, observations in the tail of plasmoids and travelling compression regions call for the existence of reconnection at a near-Earth neutral line (NENL) at some point in a substorm expansion phase and thus a key question is: does the NENL cause the destabilisation of the near-Earth current sheet, or is it the other way round ? [41]. Mukai showed a survey of 24 events in which Geotail had detected the onset of reconnection-driven flows in the centre of the mid-tail and very close to the time of substorm onset (within one or two minutes), implying that onset is caused by NENL formation. In many cases, the satellite would be in the wrong location to see the onset initially (and so there is a delay as the reconnection has to first expand so that the outflow region expands over the satellite). In other cases, the satellite may never see the reconnection-driven flows although they existed elsewhere. Given this, the number of cases showing such short delays was a surprise. There was discussion about the role of travel times, for example even Pi2's (frequently used as an indicator of the time of onset) take 1-2 min. to travel from the onset region and be detected at the Earth's surface. The Geotail observations certainly open up the possibility that X-lines in the mid-tail (15-25 R_E) do form at onset and could be responsible for onset, rather than forming in the later stages of an instability which forms close to the Earth and spreads tailward, turning into a NENL. Geostationary signatures of substorms and those in the dayside plasma sheet, equatorward of the cusp, were also surveyed [42; 36].

8. "Brainstorming"

The organised sessions, summarised briefly above, showed that the ionospheric polar cap boundary and its extension into the magnetosphere are areas of research in which there has been much recent activity and progress. Some of the key points made were new, whereas others were points that have been recognised for some time, but had taken on even greater importance in the light of new work. The brainstorming session contained a number of fascinating discussions, some of which are summarised here.

8.1 There was a discussion about how the "specularly-reflected" and "diffuse" ions in the magnetosheath would appear in the cusp precipitation. These are accelerated by interaction with the bow shock, and are in addition to the heated thermal "core" ion distribution in the sheath [12]. In particular, this mechanism could rival the internal Alfvén wave [10] as a way of producing this energised population which defines the dayside boundary plasma sheet (BPS) and the low-latitude boundary layer (LLBL). This idea generates some specific predictions. Firstly, the BPS and LLBL would be different for the quasi-parallel and quasi-perpendicular states of the shock and thus depend in an entirely unique way on IMF orientation. Secondly, the more-energetic ions of the LLBL and dayside BPS would have a solar-wind like composition, as opposed to being like the magnetospheric central plasma sheet (CPS) in the internal Alfvén wave.

8.2 Connections between the cusp and the nightside phenomena were discussed. Evidence for such connections (beyond mere coincidence of unrelated events) was generally considered not to be conclusive as yet, but is growing. This area was addressed by Pudovkin [43].

8.3 The existence of gaps between 630 nm auroras arose in two contexts. Firstly, there was much evidence presented in the meeting that there is an "intermediate" state between the northward-IMF and southward-IMF dayside red auroras [13], which showed a clear double peak in the intensity-latitude profile. These two peaks were generally attributed to reconnection taking place simultaneously at low-latitude and lobe sites (i.e. generating new open flux and reconfiguring existing open flux, respectively) when the IMF has clock angles that make it only weakly northward. Given the magnetic topology of the cusp, it is somewhat surprising that a minimum can be discerned between two peaks. The lobe reconnection site may be at quite large negative X (GSE coordinate), particularly for the winter solstice conditions required for the optical observations. However, the structure of the magnetosheath (with densities decreasing continuously with decreasing X) cannot explain the two-peaked latitudinal profiles in the aurora and clearly there are other factors influencing the electron precipitation. A similar question relates to the minima in 630 nm intensity between the poleward-moving transients. A simple model, in which electrons simply stream in from the magnetosheath along open field lines, would not predict this structure as each patch of newly-opened field lines is appended immediately equatorward of the prior patch. Evidence for this comes directly from the cusp ion steps. The bunching of the electron flux into narrower regions in dayside transients may be a weak version of that reported on the nightside [33]. The explanation of the bunching may also be linked to the periodic re-brightening of poleward-moving forms which is sometimes seen.

8.4 The phenomenon of overlapping injections at mid-altitudes [44] remains a mystery and may have a number of causes. Some scientists view these events as multiple, non-reconnection injections onto a closed field lines, by a mechanism which is as yet not defined. To understand the full implications of this, it must be noted that the ions are dispersed according to their field-aligned velocity (giving V-shaped energy-pitch angle dispersion and ramps of energy-position dispersion). This means that those injected in the second event (giving the higher energy trace) must pass through the magnetopause downstream of the entry point of those seen in the lower energy trace (as they are carried less distance by the convection electric field in the flight time to the satellite). Thus we require a source which moves over the magnetopause and turns on and off as it does so. This could occur with some kind of magnetopause wave, but is certainly not expected for any high-momentum patches in the magnetosheath. In addition, such patches are seen in solar wind but would be spread over the entire dayside magnetopause by the effect of the bow shock. The impulsive penetration picture of magnetosheath particle entry was discussed by Heikkila, Lemaire and Huba at the ASI, but these, and other difficulties were not answered. Some of the events seen in data from the Viking satellite can be seen to overlap in large pitch angles but do not overlap at all in field-parallel ions. The only explanation of such cases which does not involve unacceptable coincidences, is that they are really "steps" due to reconnection pulses, but with large gradient-B and curvature drifts giving overlap of ions of larger pitch angle ions [45]. This also explains the almost total lack of these type of overlapping injection in data from low-altitude satellites (other types, for example those caused by different ion species, are sometimes seen). In addition, several events only overlap at low flux levels. In these cases, a brief drop in magnetosheath densities can cause the fluxes observed at the satellite to fall below the one count level and this "hole" would be dispersed, in just the same way that the particles are, making a single dispersion ramp appear as two overlapping and separate dispersions. Such a cause would explain their occurrence near dawn and dusk, where magnetosheath densities are lower and fluxes closer to the one-count level. However, some events appear to have neither of these causes [46]. Other explanations which have been offered in terms of reconnection-driven particle entry include the possibility that the newly-opened field lines move slowly initially and then undergo considerable acceleration [47]: this may indeed produce two apparent populations at one time, but reproducing the full data sequences seen by Freja and Viking leads to a somewhat contrived scenario.

8.5 Counter-streaming electrons seen in association with magnetosheath plasma are often cited as evidence for a closed low-latitude boundary layer (LLBL). Such arguments assume the electrons to be ionospheric in origin. However, the balanced nature of these flows, even at the solstice, suggests that they are on open field lines and the source of the electrons is the magnetosheath: these electrons then mirror in the converging field lines and so are form balanced counter-streaming. In some cases, careful inspection shows that electrons appear first moving down and then moving back up the field line [48]. Such counter-streaming electrons are frequently seen in a layer on the magnetopause and coating "flux transfer events". An FTE has recently been identified as a partial transition into an open LLBL [49] and this analysis places the counter-streaming magnetopause electrons on open field lines. Ion distributions peaking near 90° can also be obtained from sheath plasma entering through a $\underline{J} \cdot \underline{E} < 0$ magnetopause, and do not signify closed field lines.

8.6 There was considerable discussion about the 8-minute average repetition time of magnetopause FTEs and the ionospheric auroral events and flow channels in the cusp/cleft region that are thought to be associated with them. It was noted with excitement that the forthcoming Equator-S satellite will provide the first opportunity to test this association for a decade. The only previous test was applied to Svalbard optical data and EISCAT radar data when ISEE-2 was at the dayside magnetopause, close to the end of that mission [50]: a limited association was deduced but far too few joint observations were made for this to be conclusive. The mean repeat time of magnetopause FTEs is the average of a skewed distribution (mode of about 3 min) [51] and there was discussion as to whether either of these periods represented a magnetospheric time constant, and if so how. Examples of reconnection pulses as rapid as every 3 min were presented, these having been derived from an analysis of cusp ion steps seen by Polar.

8.7 The locations of the reconnection X-lines on the magnetopause were also discussed. For northward IMF such sites are formed poleward of the magnetic cusp and result in already-opened field lines of the tail lobe being reconfigured (so that they again thread the dayside magnetopause, allowing high density sheath plasma to enter for a second time onto each field line). The topological validity of such configurations was discussed and eventually agreed. For southward IMF, the reconnection is somewhere between the two magnetic cusps where closed field lines are exposed to the boundary. There was discussion as to whether this reconnection was near the equatorial plane or at higher latitudes, near (but below) the magnetic cusp. At least some of the X-lines must form at low latitudes to give the observed accelerated flow and boundary-normal field directions on the dayside magnetopause. It was also noted that reconnection in the near cusp for one hemisphere would be in the far cusp for the other. Thus features explained by moving the reconnection site to the magnetic cusp in the near hemisphere must have observable consequences for the other hemisphere.

Acknowledgements: The authors are grateful to J. Samson, T. Mukai, A.S. Rodger, C.J. Farrugia, J. Woch, N. Maynard, P. Stauning, H.C. Carlson, A. Pedersen, and V.

Vasyliunas in helping to formulate this summary. They also thank A. Egeland and J. Moen for their excellent organisation of this ASI.

9. References

1. Maynard, N.C. and G.L Siscoe (1997) Space Weather - the practice of space physics, *this volume.*
2. Potemra, T.A. (1997) The dynamic magnetopause, *this volume.*
3. Cowley, S.W.H. (1997) Excitation of flow in the Earth's magnetosphere-ionosphere system: observations by incoherent scatter radar, *this volume.*
4. Lockwood, M. (1994) Ionospheric signatures of pulsed magnetopause reconnection, in *"Physical signatures of magnetopause boundary layer Processes", ed. J.A. Holtet and A. Egeland, NATO ASI Series C, Vol. 425,* Kluwer, 229-243.
5. Farrugia, C.J., F.T. Gratton, L. Bender, J.M. Quinn, R.B. Torbert, N.V. Erkaev and H.K. Biernat (1997) recent work on the Kelvin Helmholtz instability at the dayside magnetopause and boundary layer , *this volume.*
6. Kuznetsov, S.N. and A.V. Suvorova (1997) An empirical model of the magnetopause for broad ranges of solar wind pressure and IMF B_z , *this volume.*
7. Freeman, M.P. and C.J. Farrugia, Magnetopause motions in Newton-Busemann approach , *this volume.*
8. Erkaev, N.V., C.J. Farrugia, and H.K. Biernat (1997) Comparison of gasdynamics and MHD predictions for magnetosheath flow, *this volume.*
9. Biernat, H., V.S. Semenov, O.A. Drobysh, C.J. Farrugia, and N.V. Erkaev (1997) Time varying reconnection, *this volume.*
10. Lockwood, M. (1997) The relationship of dayside auroral precipitations to the open-closed separatrix and the pattern of convective flow, *J. Geophys. Res., 102,* 17475-17487.
11. Newell, P.T., and C.-I. Meng (1997) Open and closed low-latitude boundary layer, *this volume.*
12. Fuselier, S. (1997) Solar wind He^+ and H^+ distributions in the cusp for southward IMF, *this volume.*
13. Sandholt, P.E., C.J. Farrugia, J. Moen and B. Lybekk (1997) The dayside aurora and its regulation by the interplanetary magnetic field, *this volume.*
14. Lockwood, M., P.E. Sandholt, and S.W.H. Cowley (1989) Dayside auroral activity and momentum transfer from the solar wind, *Geophys. Res. Lett., 16,* 33-36.
15. Maynard, N.C., W.J. Burke, D.R. Wiemer, F.S. Mozer, J.D. Scudder, W.K. Peterson, R.P. Lepping and C.T. Russell (1997) Polar observations of cusp electrodynamics: evolution from 2- to 4- cell convection patterns, *this volume.*
16. Øieroset, M. and P.E. Sandholt (1997) Auroral and geomagnetic signatures of flux transfer events and associated current systems for positive and negative IMF B_z , *this volume.*
17. Farrugia, C.J., P.E. Sandholt, W.F. Denig, and R.B. Torbert (1997) Observation of a correspondence between poleward-moving auroral forms and stepped cusp ion precipitation, *J. Geophys. Res.,* in press.

18. Newell. P.T. and D.G. Sibeck (1994) Magnetosheath fluctuations, ionospheric convection and dayside ionospheric transients, *in "Physical signatures of magnetopause boundary layer Processes", ed. J.A. Holtet and A. Egeland, NATO ASI Series C, Vol. 425*, Kluwer, 245-261.

19. Lockwood, M., W.F. Denig, A.D. Farmer, V.N. Davda, S.W.H. Cowley & H. Lühr (1993) Ionospheric signatures of pulsed magnetic reconnection at the Earth's magnetopause, *Nature, 361 (6411)*, 424-428.

20. Lockwood, M., C.J. Davis, T.G. Onsager, and J.A. Scudder (1997) Modelling signatures of pulsed magnetopause reconnection in cusp ion dispersion signatures seen at middle altitudes, *Geophys. Res. Lett.*, in press.

21. Rodger, A.S. (1997) Ionospheric signatures of magnetospheric processes, *this volume.*

22. Lockwood, M., and C.J. Davis (1996) On the longitudinal extent of magnetopause reconnection bursts, *Annales Geophys., 14,* 865-878.

23. Lester, M. (1997) Coherent-scatter radar observations of the dayside cusp, *this volume.*

24. Stauning, P., Ionospheric radiowave absorption processes in the dayside polar cap boundary regions, *this volume.*

25. Carlson, H.C., Jr. (1997) Response of the polar cap ionosphere to changes in the solar wind and IMF, *this volume.*

26. McEwen, D. (1997) Polar cap phenomena and their relation to boundary layers and the IMF, *this volume.*

27. Lockwood, M. (1993) Modelling the high-latitude ionosphere for time-varying plasma convection, *Proc. Inst. Elec. Eng. - H, 140,* 91-100.

28. Rodger, A.S. (1997) Polar patches - outstanding issues, *this volume.*

29. Yamagishi, H., Y. Fujita, N. Sato, P. Stauning, M. Nishino and K. Makita (1997) Conjugate features of auroras observed by TV cameras and imaging riometers at auroral zone and polar cap conjugate pair stations, *this volume.*

30. Dyrud, L.P., M.J. Engebretson, J.L. Posh, W.H Hughes, H. Fukinishi, R.L. Arnoldy and P.T. Newell (1997) Conjugate ground observations and possible source regions of two types of Pc1-2 pulsations at very high latitudes, *this volume.*

31. Walker, A.D.M. (1997) HF radars as a tool for conjugate studies of magnetospheric phenomena, *this volume.*

32. Maclennan, C.G., L.J. Lanzerotti, and D.J. Thomsen (1997) Studies of geomagnetic conjugacy at very high latitudes, *this volume.*

33. Lanchester, B.L., M.H. Rees, D. Lummerzheim, A. Otto, H. Frey, and L. Kaila (1997) Large fluxes of auroral electron in filaments of 100 m width, *J. Geophys. Res., 102,* 9741-9748.

34. Yamauchi, M., R. Lundin, L. Eliasson, S. Ohtani, and J.H. Clemmons (1997) relationship between large-, meso- and small-scale field-aligned currents and their current carriers, *this volume.*

35. Walker, I.K., J. Moen, C.N. Mitchell, L. Kersley, and P.E. Sandholt (1997) magnetopause reconnection observed using ionospheric tomography. *Geophys. Res. Lett,* in press.

36. Hall, A.M., M. Lockwood, C.H. Perry, M. Grande, B. Kellet, M. Lester, G. Reeves, H.E. Spence, J. Woch and J. Fennell (1997) Dayside Polar observations

432

of dispersed, substorm-associated particle injection features in the vicinity of the cusp, *Annales Geophys.*, in press.

37. Shirai, H., K. Maezawa, M. Fujimoto, T. Mukai, T. Yamamoto, Y. Saito, S. Kokubun and N. Kaya (1997) Drop-off in the polar rain near the plasma sheet boundary, *J. Geophys. Res., 102*, 2271-2278.

38. Onsager, T.G., and T. Mukai (1996) The structure of the plasma sheet and its boundary layers, *J. Geomag. Geoelect., 48*, 687-698.

39. Alothman, M.J. and T.A. Fritz (1997) A model for the high-latitude isotropic boundary, *this volume.*

40. Bikkuzina, G.R., V.S. Sergeev and T. Bösinger (1997) Particle boundaries during a solar electron event, *this volume.*

41. Maynard, N.C., G.M. Erickson, W.J. Burke, A.G. Yahnin, J.C. Samson, J.D. Reeves and V.V. Klimenko (1997) Substorms and the inner magnetosphere: onset and initial expansion , *this volume.*

42. Holter, Ø., A. Roux and S. Perrault (1997) Structural analysis of geosychronous substorm oscillations, *this volume.*

43. Pudovkin, M.I. and A. Egeland (1997) Large-scale electric fields in the dayside magnetosphere, *this volume.*

44. Woch, J., and R. Lundin (1991) Temporal magnetosheath plasma injection observed with Viking: a case study, *Ann. Geophys., 9,* 133-142.

45. Lockwood, M., and M.F. Smith (1994) Low- and mid-altitude cusp particle signatures for general magnetopause reconnection rate variations: I - Theory, *J. Geophys. Res., 99*, 8531-8555.

46. Norberg, O., M. Yamauchi, L. Eliasson and R. Lundin, Freja observations of multiple injection events in cusp (1994) *Geophys. Res. Lett.* 21, 1919-1922.

47. Lockwood, M. (1995) Overlapping cusp ion injections: an explanation invoking magnetopause reconnection, *Geophys. Res. Lett, 22*, 1141-1144.

48. Takahashi, K., D.G. Sibeck, P.T. Newell and H.E. Spence (1991) ULF waves in the low-latitude boundary layer and their relationship with magnetospheric pulsations: a multi-satellite observation, *J. Geophys. Res., 96*, 9503-9519.

49. Lockwood, M., and M.A. Hapgood (1997) How the Magnetopause Transition Parameter Works, *Geophys. Res. Lett., 24*, 373-376.

50. Elphic, R.C., M. Lockwood, S.W.H. Cowley, and P.E. Sandholt (1990) Flux transfer events at the magnetopause and in the ionosphere, *Geophys. Res. Lett., 17*, 2241-2244.

51. Lockwood, M., and M.N. Wild (1993) On the quasi-periodic nature of magnetopause flux transfer events, *J. geophys. Res.*, 98, 5935-5940.